Lecture Notes in Computer Science   7226

Commenced Publication in 1973
Founding and Former Series Editors:
Gerhard Goos, Juris Hartmanis, and Jan van Leeuwen

Alwyn E. Goodloe Suzette Person (Eds.)

NASA Formal Methods

4th International Symposium, NFM 2012
Norfolk, VA, USA, April 3-5, 2012
Proceedings

 Springer

Volume Editors

Alwyn E. Goodloe
Suzette Person
NASA Langley Research Center
MS 130, Hampton VA 23681, USA
E-mail:{a.goodloe, suzette.person}@nasa.gov

ISSN 0302-9743   e-ISSN 1611-3349
ISBN 978-3-642-28890-6   e-ISBN 978-3-642-28891-3
DOI 10.1007/978-3-642-28891-3
Springer Heidelberg Dordrecht London New York

Library of Congress Control Number: 2012933117

CR Subject Classification (1998): D.2.4, D.2, D.3, F.3, D.1

LNCS Sublibrary: SL 2 – Programming and Software Engineering

Typesetting: Camera-ready by author, data conversion by Scientific Publishing Services, Chennai, India

Printed on acid-free paper

Springer is part of Springer Science+Business Media (www.springer.com)

Preface

This publication contains the proceedings of the 4th NASA Formal Methods Symposium (NFM 2012), held April 3–5, 2012, in Norfolk, VA, USA. The NASA Formal Method Symposium is a forum for theoreticians and practitioners from academia, industry, and government, with the goal of identifying challenges and providing solutions to achieving assurance in mission- and safety-critical systems. Within NASA, for example, such systems include autonomous robots, separation assurance algorithms for aircraft, Next Generation Air Transportation (NextGen), and autonomous rendezvous and docking for spacecraft. Rapidly increasing code size and emerging paradigms, such as automated code generation and safety cases, bring new challenges and opportunities for significant improvement. Also gaining increasing importance in NASA applications is the use of more rigorous software test methods and code analysis techniques, founded in theory.

The focus of the symposium is understandably on formal methods, their foundation, current capabilities, as well as their current limitations. The NASA Formal Methods Symposium is an annual event that was created to highlight the state of the art in formal methods, both in theory and practice. The series was originally started as the Langley Formal Methods Workshop, and was held under that name in 1990, 1992, 1995, 1997, 2000, and 2008. In 2009, the first NASA Formal Methods Symposium was organized by NASA Ames Research Center, and took place at Moffett Field, CA. This year, the symposium was organized by NASA Langley Research Center, and held in Norfolk, VA.

The topics covered by NFM 2012 included but were not limited to: theorem proving, symbolic execution, model-based engineering, real-time and stochastic systems, model checking, abstraction and abstraction refinement, compositional verification techniques, static and dynamic analysis techniques, fault protection, cyber security, specification formalisms, requirements analysis, and applications of formal techniques.

Two types of papers were considered: regular papers describing fully developed work and complete results or case studies, and short papers describing tools, experience reports, and work in progress or preliminary results. The symposium received 93 submissions (66 regular papers and 27 short papers), of which the committee selected 36 papers (26 regular papers and 10 short papers). All submissions went through a rigorous review process.

In addition to the refereed papers, the symposium featured three invited talks and a panel session. The invited talks were presented by Andrew Appel from Princeton University, on "Verified Software Toolchain," Patrick Cousot from École Normale Supérieure, Paris, and New York University, on "Formal Verification by Abstract Interpretation," and Cesare Tinelli from the University of Iowa, on "SMT-Based Model Checking." The panel, composed of Mike

Lowry (NASA Ames), Klaus Havelund (NASA/JPL), and Ricky Butler (NASA Langley), discussed the history and current application of formal methods at NASA.

The organizers are grateful to the authors for submitting their work to NFM 2012 and to the invited speakers for sharing their insights. NFM 2012 would not have been possible without the collaboration of the Steering Committee, Program Committee, and external reviewers, and the general support of the NASA Formal Methods community. Special thanks go to Raymond Meyer for the graphical design of NFM 2012 visual material and the NFM 2012 website, which can be found at http://shemesh.larc.nasa.gov/nfm2012/index.html.

January 2012   Alwyn Goodloe
                                                               Suzette Person

Organization

Program Committee

Nikolaj Bjorner	Microsoft Research, USA
Jonathan P. Bowen	Museophile Limited, UK
Julia Braman	NASA- Johnson Space Center, USA
Ricky Butler	NASA Langley Research Center, USA
Rance Cleaveland	University of Maryland, USA
Darren Cofer	Rockwell Collins, USA
Ewen Denney	SGT/NASA Ames, USA
Dino Distefano	Queen Mary, University of London, UK
Jin Song Dong	National University of Singapore, Singapore
Jean-Christophe Filliatre	CNRS, France
Dimitra Giannakopoulou	NASA Ames, USA
Alwyn Goodloe	NASA Langley Research Center, USA
Eric Goubault	CEA/Saclay, France
George Hagen	NASA Langley Research Center, USA
John Hatcliff	Kansas State University, USA
Klaus Havelund	Jet Propulsion Laboratory, California Institute of Technology, USA
Mats Heimdahl	University of Minnesota, USA
Gerard Holzmann	JPL, USA
Joe Hurd	Galois, Inc., USA
Bart Jacobs	Katholieke Universiteit Leuven, Belgium
Ken Mcmillan	Cadence Berkeley Labs, USA
Eric Mercer	Brigham Young University, USA
Cesar Munoz	National Aeronautics and Space Administration, USA
Anthony Narkawicz	NASA Langley, USA
Natasha Neogi	National Institute of Aerospace, USA
Corina Pasareanu	CMU/NASA Ames Research Center, USA
Charles Pecheur	UC Louvain, Belgium
Suzette Person	NASA Langley Research Center, USA
Kristin Yvonne Rozier	NASA Ames Research Center, USA
Natarajan Shankar	SRI International, USA
Oleg Sokolsky	University of Pennsylvania, USA
Sofiene Tahar	Concordia University, USA
Oksana Tkachuk	Fujitsu Laboratories of America
Willem Visser	Stellenbosch University, South Africa
Michael Whalen	University of Minnesota, USA
Virginie Wiels	ONERA / DTIM, France
Jim Woodcock	University of York, UK

Additional Reviewers

Ancona, Davide
Aridhi, Henda
Ayoub, Anaheed
Belt, Jason
Beringer, Lennart
Bouissou, Olivier
Brotherston, James
Busard, Simon
Combefis, Sebastien
Cruanes, Simon
Cuoq, Pascal
Denman, William
Di Vito, Ben
Dubreil, Jeremy
Florian, Mihai
Gawanmeh, Amjad
Gui, Lin
Haucourt, Emmanuel
Holloway, C. Michael
Jourdan, Jacques-Henri
Khan-Afshar, Sanaz
King, Andrew
Le Gall, Tristan
Lemay, Michael
Leslie, Rebekah

Liu, Liya
Mahboubi, Assia
Mhamdi, Tarek
Miller, Sheena
Miner, Paul
Mullier, Olivier
Namjoshi, Kedar
Owre, Sam
Pai, Ganesh
Paskevich, Andrei
Rocha, Camilo
Rozier, Eric
Rungta, Neha
Sanán, David
Schrammel, Peter
Shi, Ling
Siminiceanu, Radu
Smans, Jan
Song, Songzheng
Spitters, Bas
Tan, Tian Huat
Vanoverberghe, Dries
Wang, Shaohui
Zhang, Shaojie
Zheng, Manchun

Table of Contents

SMT-Based Model Checking

Cesare Tinelli*

Department of Computer Science
The University of Iowa
cesare-tinelli@uiowa.edu

It is widely recognized that the field of model checking owes much of its great success and impact to the use of symbolic techniques to reason efficiently about the reachable states of a hardware or software system. Traditionally, these techniques have relied on propositional encodings of transition systems and on propositional reasoning engines such as BDDs and SAT solvers. More recently, a number of these techniques have been adapted, and new ones have been devised, based instead on first-order encodings and reasoners for Satisfiability Modulo Theories (SMT).

SMT is an area of automated deduction that studies methods for checking the satisfiability of first-order formulas with respect to some logical theory T of interest. For being theory-specific and restricting their language to certain classes of formulas (such as quantifier-free formulas), these specialized methods can be implemented in solvers that are in practice more powerful than SAT solvers and more efficient than general-purpose theorem provers. The most sophisticated SMT solvers combine together and integrate in a fast propositional engine several *theory solvers*, decision procedures each focused on checking the satisfiability of conjunctions of literals in a particular theory—such as, for instance, linear integer or rational arithmetic, the theory of equality over uninterpreted function symbols, of bit-vectors, of arrays, and so on.

SMT encodings of model checking problems provide several advantages over propositional encodings. For instance, they are more natural and close to the level of abstraction of the original system; they allow one to model finite-state systems compactly; and they can be used to model infinite-state systems directly, without resorting to finite state abstractions. At the same time, they largely fall within logical fragments that are efficiently decidable.

This talk will highlight a few model checking approaches and techniques based on SMT encodings and relying on SMT solvers as their main reasoning engine. We will see that SMT-based model checking methods blur the line between traditional (propositional) model checking and traditional (first or higher order) deductive verification. More crucially, they combine the best features of both by offering the scalability and scope of deductive verification while maintaining the high level of automation of propositional model checking.

---

* The author's research on this subject is largely the result of past and on-going collaborations with C. Barrett, M. Deters, G. Hagen, P.-L. Garoche, Y. Ge, T. Kahsai, S. Miller, and M. Whalen, and was made possible in part by the support of grant #FA9550-09-1-0517 from the Air Force Office of Scientific Research and grants #1049674 and #0551646 from the National Science Foundation.

A. Goodloe and S. Person (Eds.): NFM 2012, LNCS 7226, p. 1, 2012.
© Springer-Verlag Berlin Heidelberg 2012

Verified Software Toolchain

Andrew W. Appel

Princeton University

Abstract. The software toolchain includes static analyzers to check assertions about programs; optimizing compilers to translate programs to machine language; operating systems and libraries to supply context for programs. Our *Verified Software Toolchain* verifies with machine-checked proofs that the assertions claimed at the top of the toolchain really hold in the machine-language program, running in the operating-system context, on a weakly-consistent-shared-memory machine.

Our verification approach is modular, in that proofs about operating systems or concurrency libraries are oblivious of the programming language or machine language, proofs about compilers are oblivious of the program logic used to verify static analyzers, and so on. The approach is scalable, in that each component is verified in the semantic idiom most natural for that component.

Finally, the verification is *foundational:* the trusted base for proofs of observable properties of the machine-language program includes only the operational semantics of the machine language, not the source language, the compiler, the program logic, or any other part of the toolchain—even when these proofs are carried out by source-level static analyzers.

In this paper I explain the construction of a a verified toolchain, using the Coq proof assistant. I will illustrate with shape analysis for C programs based on separation logic.

A. Goodloe and S. Person (Eds.): NFM 2012, LNCS 7226, p. 2, 2012.

Formal Verification by Abstract Interpretation

Patrick Cousot

CIMS NYU, New York, USA
CNRS–ENS–INRIA, Paris, France

Abstract. We provide a rapid overview of the theoretical foundations and main applications of abstract interpretation and show that it currently provides scaling solutions to achieving assurance in mission- and safety-critical systems through verification by fully automatic, semantically sound and precise static program analysis.

Keywords: Abstract interpretation, Abstraction, Aerospace, Certification, Cyber-physical system, Formal Method, Mission-critical system, Runtime error, Safety-critical system, Scalability, Soundness, Static Analysis, Validation, Verification.

1 Abstract Interpretation

Abstract interpretation [9,10,11,12,13] is a theory of abstraction and constructive approximation of the mathematical structures used in the formal description of programming languages and the inference or verification of undecidable program properties.

The design of an inference or verification method by abstract interpretation starts with the formal definition of the semantics of a programming language (formally describing all possible program behaviors in all possible execution environments), continues with the formalization of program properties, and the expression of the strongest program property of interest in fixed point form.

The theory provides property and fixed point abstraction methods than can be constructively applied to obtain formally verified abstract semantics of the programming languages where, ideally, only properties relevant to the considered inference or verification problem are preserved while all others are abstracted away.

Formal proof methods for verification are derived by checking fixed point by induction. For property inference in static analyzers, iterative fixed point approximation methods with convergence acceleration using widening/narrowing provide effective algorithms to automatically infer abstract program properties (such as invariance or definite termination) which can then be used for program verification by fixed point checking.

Because program verification problems are undecidable for infinite systems, any fully automatic formal method will fail on infinitely many programs and, fortunately, also succeed on infinitely many programs. An abstraction over-approximates the set of possible concrete executions and so may include executions not existing in the concrete. This is not a problem when such fake

A. Goodloe and S. Person (Eds.): NFM 2012, LNCS 7226, pp. 3–7, 2012.

executions do not affect the property to be verified (e.g. for invariance the execution time is irrelevant). Otherwise this may cause a false alarm in that the property is violated by an inexistent execution. In this case, the abstraction must be refined to better distinguish between actual and fake program executions.

To maximize success for specific applications of the theory, it is necessary to adapt the abstractions/approximations so as to eliminate false alarms (when the analysis is too imprecise to provide a definite answer to the verification problem) at a reasonable cost. The choice of an abstraction which is precise enough to check for specified properties and imprecise enough to be scalable to very large programs is difficult. This can be done by refining or coarsening general-purpose abstractions.

A convenient way to adjust the precision/cost ratio of a static analyser consists in organizing the effective abstract fixed point computation in an abstract interpreter (mainly dealing with control) parameterized by abstract domains (mainly dealing with data). These abstract domains algebraically describe classes of properties and the associated logical operations, extrapolation operators (widening and narrowing needed to over-approximate fixed points) and primitive transformers corresponding to basic operations of the programming language (such as assignment, test, call, etc).

To achieve the desired precision, the various abstract domains can combined by the abstract interpreter, e.g. with a reduced product [28], so as to eliminate false alarms at a reasonable cost.

Several surveys of abstract interpretation [1,7,19,21] describe this general methodology in more details.

2   A Few Applications of Abstract Interpretation

Abstract interpretation has applications in the syntax [22], semantics [14], and proof [20] of programming languages where abstractions are sound (no possible case is ever omitted in the abstraction) and complete (the abstraction is precise enough to express/verify concrete program properties in the abstract without any false alarm) but in general incomputable (but with severe additional hypotheses such as finiteness). Full automation of the verification task requires further sound but incomplete abstractions as applied to static analysis [9,30], contract inference [27], type inference [6], termination inference [23] model-checking [8,15,16], abstraction refinement [29], program transformation [17] (including watermarking [18]), combination of decision procedures [28], etc.

3   Applications to Assurance in Mission- and Safety-Critical Systems

Abstract interpretation has been successful this last decade in program verification for mission- and safety-critical systems. Significant applications of abstract interpretation to aerospace systems include e.g. airplane control-command [31,34,35] and autonomous rendezvous and docking for spacecraft [5].

An example is Astrée [1,2,3,4,24,25,26] (www.astree.ens.fr) which is a static analyzer to verify the absence of runtime errors in structured, very large C programs with complex memory usages, and involving complex boolean as well as floating-point computations (which are handled precisely and safely by taking all possible rounding errors into account), but without recursion or dynamic memory allocation. Astrée targets embedded applications as found in earth transportation, nuclear energy, medical instrumentation, aeronautics and space flight, in particular synchronous control/command such as electric flight control.

Astrée reports any division by zero, out-of-bounds array indexing, erroneous pointer manipulation and dereferencing (null, uninitialized and dangling pointers), integer and floating-point arithmetic overflow, violation of optional user-defined assertions to prove additional run-time properties (similar to assert diagnostics), code it can prove to be unreachable under any circumstances (note that this is not necessarily all unreachable code due to over-approximations), read access to uninitialized variables. Astrée offers powerful annotation mechanisms, which enable the user to make external knowledge available to Astrée, or to selectively influence the analysis precision for individual loops or data structures. Detailed messages and an intuitive GUI help the user understand alarms about potential errors. Then, true runtime errors can be fixed, or, in case of a false alarm, the analyzer can be tuned to avoid them. These mechanisms allow to perform analyses with very few or even zero false alarms. Astrée is industrialised by AbsInt (www.absint.com/astree).

AstréeA [32,33] is built upon Astrée to prove the absence of runtime errors and data races in parallel programs. Asynchrony introduces additional difficulties due to the semantics of parallelism (such as the abstraction of process interleaving, explicit process scheduling, shared memory model, etc).

4   Conclusion

Abstract interpretation has a broad spectrum of applications from theory to practice. Abstract interpretation-based static analysis is automatic, sound, scalable to industrial size software, precise, and commercially supported for proving the absence of runtime errors. It is a premium formal method to complement dynamic testing as recommended by DO-178C/ED-12C (http://www.rtca.org/doclist.asp).

References

1. Bertrane, J., Cousot, P., Cousot, R., Feret, J., Mauborgne, L., Miné, A., Rival, X.: Static analysis and verification of aerospace software by abstract interpretation. In: AIAA Infotech@Aerospace 2010, Atlanta, Georgia, April 20-22. American Institute of Aeronautics and Astronautics (2010)
2. Bertrane, J., Cousot, P., Cousot, R., Feret, J., Mauborgne, L., Miné, A., Rival, X.: Static analysis by abstract interpretation of embedded critical software. ACM SIGSOFT Software Engineering Notes 36(1), 1–8 (2011)

3. Blanchet, B., Cousot, P., Cousot, R., Feret, J., Mauborgne, L., Miné, A., Monni-aux, D., Rival, X.: Design and Implementation of a Special-Purpose Static Program Analyzer for Safety-Critical Real-Time Embedded Software. In: Mogensen, T.Æ., Schmidt, D.A., Sudborough, I.H. (eds.) The Essence of Computation. LNCS, vol. 2566, pp. 85–108. Springer, Heidelberg (2002)

4. Blanchet, B., Cousot, P., Cousot, R., Feret, J., Mauborgne, L., Miné, A., Monniaux, D., Rival, X.: A static analyzer for large safety-critical software. In: Proceedings of the ACM SIGPLAN 2003 Conference on Programming Language Design and Implementation 2003, San Diego, California, USA, June 9-11, pp. 196–207. ACM (2003)

5. Bouissou, O., Conquet, E., Cousot, P., Cousot, R., Feret, J., Ghorbal, K., Goubault, E., Lesens, D., Mauborgne, L., Miné, A., Putot, S., Rival, X., Turin, M.: Space software validation using abstract interpretation. In: Proc. of the Int. Space System Engineering Conf., Data Systems in Aerospace (DASIA 2009), Istambul, Turkey, vol. SP-669, pp. 1–7. ESA (May 2009)

6. Cousot, P.: Types as abstract interpretations. In: POPL, pp. 316–331 (1997)

7. Cousot, P.: The calculational design of a generic abstract interpreter. In: Broy, M., Steinbrüggen, R. (eds.) Calculational System Design. NATO ASI Series F. IOS Press, Amsterdam (1999)

8. Cousot, P.: Partial Completeness of Abstract Fixpoint Checking. In: Choueiry, B.Y., Walsh, T. (eds.) SARA 2000. LNCS (LNAI), vol. 1864, pp. 1–25. Springer, Heidelberg (2000)

9. Cousot, P., Cousot, R.: Abstract interpretation: A unified lattice model for static analysis of programs by construction or approximation of fixpoints. In: Proceedings of the 4th ACM SIGPLAN-SIGACT Symposium on Principles of Programming Languages, POPL 1977, Los Angeles, California, USA, January 17-19, pp. 238–252 (1977)

10. Cousot, P., Cousot, R.: Systematic design of program analysis frameworks. In: Proceedings of the 6th ACM SIGPLAN-SIGACT Symposium on Principles of Programming Languages, POPL 1979, San Antonio, Texas, USA, January 17-19, pp. 269–282 (1979)

11. Cousot, P., Cousot, R.: Abstract interpretation and application to logic programs. J. Log. Program. 13(2&3), 103–179 (1992)

12. Cousot, P., Cousot, R.: Abstract interpretation frameworks. J. Log. Comput. 2(4), 511–547 (1992)

13. Cousot, P., Cousot, R.: Comparing the Galois Connection and Widening/Narrowing Approaches to Abstract Interpretation. In: Bruynooghe, M., Wirsing, M. (eds.) PLILP 1992. LNCS, vol. 631, pp. 269–295. Springer, Heidelberg (1992)

14. Cousot, P., Cousot, R.: Inductive definitions, semantics and abstract interpretation. In: POPL, pp. 83–94 (1992)

15. Cousot, P., Cousot, R.: Refining model checking by abstract interpretation. Autom. Softw. Eng. 6(1), 69–95 (1999)

16. Cousot, P., Cousot, R.: Temporal abstract interpretation. In: Proceedings of the 4th ACM SIGPLAN-SIGACT Symposium on Principles of Programming Languages, POPL 2000, Boston, Massachusetts, USA, January 19-21, pp. 12–25 (2000)

17. Cousot, P., Cousot, R.: Systematic design of program transformation frameworks by abstract interpretation. In: POPL, pp. 178–190 (2002)

18. Cousot, P., Cousot, R.: An abstract interpretation-based framework for software watermarking. In: Jones, N.D., Leroy, X. (eds.) Proceedings of the 31st ACM SIGPLAN-SIGACT Symposium on Principles of Programming Languages, POPL 2004, Venice, Italy, January 14-16, pp. 173–185. ACM (2004)

19. Cousot, P., Cousot, R.: Basic concepts of abstract interpretation. In: Jacquard, R. (ed.) Building the Information Society, pp. 359–366. Kluwer Academic Publishers (2004)
20. Cousot, P., Cousot, R.: Bi-inductive structural semantics. Inf. Comput. 207(2), 258–283 (2009)
21. Cousot, P., Cousot, R.: A gentle introduction to formal verification of computer systems by abstract interpretation. In: Esparza, J., Grumberg, O., Broy, M. (eds.) Logics and Languages for Reliability and Security. NATO Science Series III: Computer and Systems Sciences, pp. 1–29. IOS Press (2010)
22. Cousot, P., Cousot, R.: Grammar semantics, analysis and parsing by abstract interpretation. Theor. Comput. Sci. 412(44), 6135–6192 (2011)
23. Cousot, P., Cousot, R.: An abstract interpretation framework for termination. In: Field, J., Hicks, M. (eds.) Proceedings of the 39th ACM SIGPLAN-SIGACT Symposium on Principles of Programming Languages, POPL 2012, Philadelphia, Pennsylvania, USA, January 22-28, pp. 245–258. ACM (2012)
24. Cousot, P., Cousot, R., Feret, J., Mauborgne, L., Miné, A., Monniaux, D., Rival, X.: The ASTREÉ Analyzer. In: Sagiv, M. (ed.) ESOP 2005. LNCS, vol. 3444, pp. 21–30. Springer, Heidelberg (2005)
25. Cousot, P., Cousot, R., Feret, J., Mauborgne, L., Miné, A., Rival, X.: Why does Astrée scale up? Formal Methods in System Design 35(3), 229–264 (2009)
26. Cousot, P., Cousot, R., Feret, J., Miné, A., Mauborgne, L., Monniaux, D., Rival, X.: Varieties of static analyzers: A comparison with Astrée. In: First Joint IEEE/IFIP Symposium on Theoretical Aspects of Software Engineering, TASE 2007, Shanghai, China, June 5-8, pp. 3–20. IEEE Computer Society (2007)
27. Cousot, P., Cousot, R., Logozzo, F.: Precondition Inference from Intermittent Assertions and Application to Contracts on Collections. In: Jhala, R., Schmidt, D. (eds.) VMCAI 2011. LNCS, vol. 6538, pp. 150–168. Springer, Heidelberg (2011)
28. Cousot, P., Cousot, R., Mauborgne, L.: The Reduced Product of Abstract Domains and the Combination of Decision Procedures. In: Hofmann, M. (ed.) FOSSACS 2011. LNCS, vol. 6604, pp. 456–472. Springer, Heidelberg (2011)
29. Cousot, P., Ganty, P., Raskin, J.-F.: Fixpoint-Guided Abstraction Refinements. In: Riis Nielson, H., Filé, G. (eds.) SAS 2007. LNCS, vol. 4634, pp. 333–348. Springer, Heidelberg (2007)
30. Cousot, P., Halbwachs, N.: Automatic discovery of linear restraints among variables of a program. In: Proceedings of the 5th ACM SIGPLAN-SIGACT Symposium on Principles of Programming Languages, POPL 1978, Tucson, Arizona, USA, January 23-25, pp. 84–96 (1978)
31. Delmas, D., Souyris, J.: Astrée: From Research to Industry. In: Riis Nielson, H., Filé, G. (eds.) SAS 2007. LNCS, vol. 4634, pp. 437–451. Springer, Heidelberg (2007)
32. Miné, A.: Field-sensitive value analysis of embedded C programs with union types and pointer arithmetics. In: ACM SIGPLAN/SIGBED Conf. on Languages, Compilers, and Tools for Embedded Systems (LCTES 2006), pp. 54–63. ACM Press (June 2006)
33. Miné, A.: Static Analysis of Run-Time Errors in Embedded Critical Parallel C Programs. In: Barthe, G. (ed.) ESOP 2011. LNCS, vol. 6602, pp. 398–418. Springer, Heidelberg (2011)
34. Souyris, J.: Industrial experience of abstract interpretation-based static analyzers. In: Jacquart, R. (ed.) IFIP 18th World Computer Congress, Topical Sessions, Building the Information Society, Toulouse, France, August 22-27, pp. 393–400. Kluwer (2004)
35. Souyris, J., Delmas, D.: Experimental Assessment of Astrée on Safety-Critical Avionics Software. In: Saglietti, F., Oster, N. (eds.) SAFECOMP 2007. LNCS, vol. 4680, pp. 479–490. Springer, Heidelberg (2007)

Quantitative Timed Analysis
of Interactive Markov Chains

Dennis Guck[1], Tingting Han[2],
Joost-Pieter Katoen[1], and Martin R. Neuhäußer[3]

[1] RWTH Aachen University, Germany
[2] University of Oxford, UK
[3] Saarland University, Germany

Abstract. This paper presents new algorithms and accompanying tool support for analyzing interactive Markov chains (IMCs), a stochastic timed $1\frac{1}{2}$-player game in which delays are exponentially distributed. IMCs are compositional and act as semantic model for engineering formalisms such as AADL and dynamic fault trees. We provide algorithms for determining the extremal expected time of reaching a set of states, and the long-run average of time spent in a set of states. The prototypical tool IMCA supports these algorithms as well as the synthesis of ε-optimal piecewise constant timed policies for timed reachability objectives. Two case studies show the feasibility and scalability of the algorithms.

1 Introduction

Continuous-time Markov chains (CTMCs) are perhaps the most well-studied stochastic model in performance evaluation and naturally reflect the random real-time behavior of stoichiometric equations in systems biology. LTSs (labeled transition systems) are one of the main operational models for concurrency and are equipped with a plethora of behavioral equivalences like bisimulation and trace equivalences. A natural mixture of CTMCs and LTSs yields so-called *interactive Markov chains* (IMCs), originally proposed as a semantic model of stochastic process algebras [18,19]. As a state may have several outgoing action-transitions, IMCs are in fact stochastic real-time $1\frac{1}{2}$-player games, also called continuous-time probabilistic automata by Knast in the 1960's [21].

IMC usage. The simplicity of IMCs and their compositional nature —they are closed under CSP-like parallel composition and restriction— make them attractive to act as a semantic backbone of several formalisms. IMCs were developed for stochastic process algebras [18]. Dynamic fault trees are used in reliability engineering for safety analysis purposes and specify the causal relationship between failure occurrences. If failures occur according to an exponential distribution, which is quite a common assumption in reliability analysis, dynamic fault trees are in fact IMCs [4]. The same holds for the standardized Architectural Analysis and Design Language (AADL) in which nominal system behavior is extended with probabilistic error models. IMCs turn out to be a natural semantic model

A. Goodloe and S. Person (Eds.): NFM 2012, LNCS 7226, pp. 8–23, 2012.

for AADL [5]; the use of this connection in the aerospace domain has recently been shown in [26]. In addition, IMCs are used for stochastic extensions of Statemate [3], and for modeling and analysing industrial GALS hardware designs [12].

IMC analysis. The main usage of IMCs so far has been the compositional generation and minimization of models. Its analysis has mainly been restricted to "fully probabilistic" IMCs which induce CTMCs and are therefore amenable to standard Markov chain analysis or, alternatively, model checking [1]. CTMCs can sometimes be obtained from IMCs by applying weak bisimulation minimization; however, if this does not suffice, semantic restrictions on the IMC level are imposed to ensure full probabilism. The CADP toolbox [11] supports the compositional generation, minimization, and standard CTMC analysis of IMCs. In this paper, we focus on the *quantitative timed analysis* of arbitrary IMCs, in particular of those, that are non-deterministic and can be seen as stochastic real-time $1\frac{1}{2}$-player games. We provide algorithms for the expected time analysis and long-run average fraction of time analysis of IMCs and show how both cases can be reduced to stochastic shortest path (SSP) problems [2,15]. This complements recent work on the approximate time-bounded reachability analysis of IMCs [27]. Our algorithms are presented in detail and proven correct. Prototypical tool support for these analyses is presented that includes an implementation of [27]. The feasibility and scalability of our algorithms are illustrated on two examples: A dependable workstation cluster [17] and a Google file system [10]. Our IMCA tool is a useful backend for the CADP toolbox, as well as for analysis tools for dynamic fault trees and AADL error models.

Related work. Untimed quantitative reachability analysis of IMCs has been handled in [11]; timed reachability in [27]. Other related work is on continuous-time Markov decision processes (CTMDPs). A numerical algorithm for time-bounded expected accumulated rewards in CTMDPs is given in [8] and used as building brick for a CSL model checker in [7]. Algorithms for timed reachability in CTMDPs can be found in, e.g. [6,24]. Long-run averages in stochastic decision processes using observer automata ("experiments") have been treated in [14], whereas the usage of SSP problems for verification originates from [15]. Finally, [25] considers discrete-time Markov decision processes (MDPs) with ratio cost functions; we exploit such objectives for long-run average analysis.

Organization of the paper. Section 2 introduces IMCs. Section 3 and 4 are devoted to the reduction of computing the optimal expected time reachability and long-run average objectives to stochastic shortest path problems. Our tool IMCA and the results of two case studies are presented in Section 5. Section 6 concludes the paper.

2 Interactive Markov Chains

Interactive Markov chains. IMCs are finite transition systems with action-labeled transitions and Markovian transitions which are labeled with a positive real number (ranged over by λ) identifying the rate of an exponential distribution.

Definition 1 (Interactive Markov chain). *An* interactive Markov chain *is a tuple* $\mathcal{I} = (S, Act, \rightarrow, \Longrightarrow, s_0)$ *where* S *is a nonempty, finite set of states with initial state* $s_0 \in S$, Act *is a finite set of actions, and*

- $\rightarrow \subseteq S \times Act \times S$ *is a set of* action *transitions and*
- $\Longrightarrow \subseteq S \times \mathbb{R}_{>0} \times S$ *is a set of* Markovian *transitions.*

We abbreviate $(s, \alpha, s') \in \rightarrow$ by $s \xrightarrow{\alpha} s'$ and $(s, \lambda, s') \in \Longrightarrow$ by $s \xRightarrow{\lambda} s'$. IMCs are closed under parallel composition [18] by synchronizing on action transitions in a TCSP-like manner. As our main interest is in the analysis of IMCs, we focus on so-called *closed* IMCs [20], i.e. IMCs that are not subject to any further synchronization. W.l.o.g. we assume that in closed IMCs all outgoing action transition of state s are uniquely labeled, thereby naming the state's nondeterministic choices. In the rest of this paper, we only consider closed IMCs. For simplicity, we assume that IMCs do not contain deadlock states, i.e. in any state either an action or a Markovian transition emanates.

Definition 2 (Maximal progress). *In any closed IMC, action transitions take precedence over Markovian transitions.*

The rationale behind the maximal progress assumption is that in closed IMCs, action transitions are not subject to interaction and thus can happen immediately, whereas the probability for a Markovian transition to happen immediately is zero. Accordingly, we assume that each state s has either only outgoing action transitions or only outgoing Markovian transitions. Such states are called interactive and Markovian, respectively; we use $IS \subseteq S$ and $MS \subseteq S$ to denote the sets of interactive and Markovian states. Let $Act(s) = \{\alpha \in Act \mid \exists s' \in S. s \xrightarrow{\alpha} s'\}$ be the set of enabled actions in s, if $s \in IS$ and $Act(s) = \{\bot\}$ if $s \in MS$. In Markovian states, we use the special symbol $\bot$ to denote purely stochastic behavior without any nondeterministic choices.

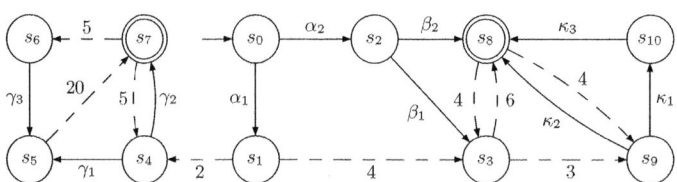

Fig. 1. An example IMC

Example 1. Fig. 1 depicts an IMC $\mathcal{I}$, where solid and dashed lines represent action and Markovian transitions, respectively. The set of Markovian states is $MS = \{s_1, s_3, s_5, s_7, s_8\}$; IS contains all other states. Nondeterminism between action transitions appears in states s_0, s_2, s_4, and s_9.

A sub-IMC of an IMC $\mathcal{I} = (S, Act, \rightarrow, \Longrightarrow, s_0)$, is a pair (S', K) where $S' \subseteq S$ and K is a function that assigns each $s \in S'$ a set $\varnothing \neq K(s) \subseteq Act(s)$ of actions such that for all $\alpha \in K(s)$, $s \xrightarrow{\alpha} s'$ or $s \xRightarrow{\lambda} s'$ imply $s' \in S'$. An *end component* is a sub-IMC whose underlying graph is strongly connected; it is *maximal* w.r.t. K if it is not contained in any other end component (S'', K).

Example 2. In Fig. 1, the sub-IMC (S', K) with state space $S' = \{s_4, s_5, s_6, s_7\}$ and $K(s) = Act(s)$ for all $s \in S'$ is a maximal end component.

IMC semantics. An IMC without action transitions is a CTMC; if $\Longrightarrow$ is empty, then it is an LTS. We briefly explain the semantics of Markovian transitions. Roughly speaking, the meaning of $s \overset{\lambda}{\Longrightarrow} s'$ is that the IMC can switch from state s to s' within d time units with probability $1 - e^{-\lambda d}$. The positive real value λ thus uniquely identifies a negative exponential distribution. For $s \in MS$, let $\mathbf{R}(s, s') = \sum\{\lambda \mid s \overset{\lambda}{\Longrightarrow} s'\}$ be the *rate* to move from state s to state s'. If $\mathbf{R}(s, s') > 0$ for more than one state s', a competition between the transitions of s exists, known as the race condition. The probability to move from such state s to a particular state s' within d time units, i.e. $s \Longrightarrow s'$ wins the race, is

$$\frac{\mathbf{R}(s, s')}{E(s)} \cdot \left(1 - e^{-E(s)d}\right), \tag{1}$$

where $E(s) = \sum_{s' \in S} \mathbf{R}(s, s')$ is the *exit rate* of state s. Intuitively, (1) states that after a delay of at most d time units (second term), the IMC moves probabilistically to a direct successor state s' with discrete branching probability $\mathbf{P}(s, s') = \frac{\mathbf{R}(s, s')}{E(s)}$.

Paths and schedulers. An infinite path π in an IMC is an infinite sequence:

$$\pi = s_0 \xrightarrow{\sigma_0, t_0} s_1 \xrightarrow{\sigma_1, t_1} s_2 \xrightarrow{\sigma_2, t_2} \cdots$$

with $s_i \in S$, $\sigma_i \in Act$ or $\sigma_i = \bot$, and $t_i \in \mathbb{R}_{\geq 0}$. The occurrence of action α in state s_i in π is denoted $s_i \xrightarrow{\alpha, 0} s_{i+1}$; the occurrence of a Markovian transition after t time units delay in s_i is denoted $s_i \xrightarrow{\bot, t} s_{i+1}$. For $t \in \mathbb{R}_{\geq 0}$, let $\pi @ t$ denote the set of states that π occupies at time t. Note that $\pi @ t$ is in general not a single state, but rather a set of states, as an IMC may exhibit immediate transitions and thus may occupy various states at the same time instant. Let *Paths* and *Paths** denote the sets of infinite and finite paths, respectively.

Nondeterminism appears when there is more than one action transition enabled in a state. The corresponding choice is resolved using *schedulers*. A scheduler (ranged over by D) is a measurable function which yields for each finite path ending in some state s a probability distribution over the set of enabled actions in s. For details, see [27]. A stationary deterministic scheduler is a mapping $D : IS \to Act$. The usual cylinder set construction yields a σ-algebra $\mathfrak{F}_{Paths}$ of subsets of *Paths*; given a scheduler D and an initial state s, $\mathfrak{F}_{Paths}$ can be equipped with a probability measure [27], denoted $\Pr_{s,D}$.

Zenoness. The time elapsed along an infinite path $\pi = s_0 \xrightarrow{\sigma_0, t_0} s_1 \xrightarrow{\sigma_1, t_1} \cdots$ up to state n is $\sum_{i=0}^{n-1} t_i$. Path π is non-Zeno whenever $\sum_{i=0}^{\infty} t_i$ diverges to infinity; accordingly, an IMC $\mathcal{I}$ with initial state s_0 is non-Zeno if for all schedulers D, $\Pr_{s_0,D}\{\pi \in Paths \mid \sum_{i=0}^{\infty} t_i = \infty\} = 1$. As the probability of a Zeno path in a finite CTMC —thus only containing Markovian transitions— is zero [1], IMC $\mathcal{I}$ is non-Zeno if and only if no strongly connected component with states $T \subseteq IS$ is reachable from s_0. In the rest of this paper, we assume IMCs to be non-Zeno.

Stochastic shortest path problems. The (non-negative) SSP problem considers the minimum expected cost for reaching a set of goal states in a discrete-time Markov decision process (MDP).

Definition 3 (MDP). $\mathcal{M} = (S, Act, \mathbf{P}, s_0)$ *is a* Markov decision process, *where* S, Act *and* s_0 *are as before and* $\mathbf{P} : S \times Act \times S \to [0, 1]$ *is a transition probability function such that for all* $s \in S$ *and* $\alpha \in Act$, $\sum_{s' \in S} \mathbf{P}(s, \alpha, s') \in \{0, 1\}$.

Definition 4 (SSP problem). *A non-negative stochastic shortest path problem (SSP problem) is a tuple* $\mathcal{P} = (S, Act, \mathbf{P}, s_0, G, c, g)$, *where* $(S, Act, \mathbf{P}, s_0)$ *is an MDP,* $G \subseteq S$ *is a set of goal states,* $c : S \setminus G \times Act \to \mathbb{R}_{\geq 0}$ *is a cost function and* $g : G \to \mathbb{R}_{\geq 0}$ *is a terminal cost function.*

The infinite sequence $\pi = s_0 \xrightarrow{\alpha_0} s_1 \xrightarrow{\alpha_1} s_2 \xrightarrow{\alpha_2} \ldots$ is a path in the MDP if $s_i \in S$ and $\mathbf{P}(s_i, \alpha_i, s_{i+1}) > 0$ for all $i \geqslant 0$. Let k be the smallest index such that $s_k \in G$. The accumulated cost along π of reaching G, denoted $C_G(\pi)$, is $\sum_{j=0}^{k-1} c(s_j, \alpha_j) + g(s_k)$. The minimum expected cost reachability of G starting from s in the SSP $\mathcal{P}$, denoted $cR^{\min}(s, \Diamond G)$, is defined as

$$cR^{\min}(s, \Diamond G) = \inf_D \mathbb{E}_{s,D}(C_G) = \inf_D \sum_{\pi \in Paths_{abs}} C_G(\pi) \cdot \mathrm{Pr}_{s,D}^{abs}(\pi),$$

where $Paths_{abs}$ denotes the set of (time-abstract) infinite paths in the MDP and $\mathrm{Pr}_{s,D}^{abs}$ the probability measure on sets of MDP paths that is induced by scheduler D and initial state s. The quantity $cR^{\min}(s, \Diamond G)$ can be obtained [2,13] by solving the following linear programming problem with variables $\{x_s\}_{s \in S \setminus G}$: maximize $\sum_{s \in S \setminus G} x_s$ subject to the following constraints for each $s \in S \setminus G$ and $\alpha \in Act$:

$$x_s \leqslant c(s, \alpha) + \sum_{s' \in S \setminus G} \mathbf{P}(s, \alpha, s') \cdot x_{s'} + \sum_{s' \in G} \mathbf{P}(s, \alpha, s') \cdot g(s').$$

3   Expected Time Analysis

Expected time objectives. Let $\mathcal{I}$ be an IMC with state space S and $G \subseteq S$ a set of goal states. Define the (extended) random variable $V_G : Paths \to \mathbb{R}_{\geq 0}^{\infty}$ as the elapsed time before first visiting some state in G, i.e. for infinite path $\pi = s_0 \xrightarrow{\sigma_0, t_0} s_1 \xrightarrow{\sigma_1, t_1} \cdots$, let $V_G(\pi) = \min \{t \in \mathbb{R}_{\geq 0} \mid G \cap \pi @ t \neq \varnothing\}$ where $\min(\varnothing) = +\infty$. The minimal expected time to reach G from $s \in S$ is given by

$$eT^{\min}(s, \Diamond G) = \inf_D \mathbb{E}_{s,D}(V_G) = \inf_D \int_{Paths} V_G(\pi) \, \mathrm{Pr}_{s,D}(d\pi).$$

Note that by definition of V_G, only the amount of time before entering the first G-state is relevant. Hence, we may turn all G-states into absorbing Markovian states without affecting the expected time reachability. Accordingly, we assume for the remainder of this section that for all $s \in G$ and some $\lambda > 0$, $s \xrightarrow{\lambda} s$ is the only outgoing transition of state s.

Theorem 1. *The function $eT^{\min}$ is a fixpoint of the Bellman operator*

$$[L(v)](s) = \begin{cases} \dfrac{1}{E(s)} + \displaystyle\sum_{s' \in S} \mathbf{P}(s,s') \cdot v(s') & \text{if } s \in MS \setminus G \\ \displaystyle\min_{s \xrightarrow{\alpha} s'} v(s') & \text{if } s \in IS \setminus G \\ 0 & \text{if } s \in G. \end{cases}$$

Intuitively, Thm. 1 justifies to add the expected sojourn times in all Markovian states before visiting a G-state. Any non-determinism in interactive states (which are, by definition, left instantaneously) is resolved by minimizing the expected reachability time from the reachable one-step successor states.

Computing expected time probabilities. The characterization of $eT^{\min}(s, \Diamond G)$ in Thm. 1 allows us to reduce the problem of computing the minimum expected time reachability in an IMC to a non-negative SSP problem [2,15].

Definition 5 (SSP for minimum expected time reachability). *The SSP of IMC $\mathcal{I} = (S, Act, \rightarrow, \Longrightarrow, s_0)$ for the expected time reachability of $G \subseteq S$ is $\mathcal{P}_{eT\min}(\mathcal{I}) = (S, Act \cup \{\bot\}, \mathbf{P}, s_0, G, c, g)$ where $g(s) = 0$ for all $s \in G$ and*

$$\mathbf{P}(s, \sigma, s') = \begin{cases} \dfrac{\mathbf{R}(s,s')}{E(s)} & \text{if } s \in MS \wedge \sigma = \bot \\ 1 & \text{if } s \in IS \wedge s \xrightarrow{\sigma} s' \\ 0 & \text{otherwise, and} \end{cases}$$

$$c(s, \sigma) = \begin{cases} \dfrac{1}{E(s)} & \text{if } s \in MS \setminus G \wedge \sigma = \bot \\ 0 & \text{otherwise.} \end{cases}$$

Intuitively, action transitions are assigned a Dirac distribution, whereas the probabilistic behavior of a Markovian state is as explained before. The reward of a Markovian state is its mean residence time. Terminal costs are set to zero.

Theorem 2 (Correctness of the reduction). *For IMC $\mathcal{I}$ and its induced SSP $\mathcal{P}_{eT\min}(\mathcal{I})$ it holds:*

$$eT^{\min}(s, \Diamond G) = cR^{\min}(s, \Diamond G)$$

where $cR^{\min}(s, \Diamond G)$ denotes the minimal cost reachability of G in SSP $\mathcal{P}_{eT\min}(\mathcal{I})$.

Proof. According to [2,15], $cR^{\min}(s, \Diamond G)$ is the unique fixpoint of the Bellman operator L' defined as:

$$[L'(v)](s) = \min_{\alpha \in Act(s)} c(s,\alpha) + \sum_{s' \in S \setminus G} \mathbf{P}(s,\alpha,s') \cdot v(s') + \sum_{s' \in G} \mathbf{P}(s,\alpha,s') \cdot g(s').$$

We prove that the Bellman operator L from Thm. 1 equals L' for SSP $\mathcal{P}_{eT\min}(\mathcal{I})$. By definition, it holds that $g(s) = 0$ for all $s \in S$. Thus

$$[L'(v)](s) = \min_{\alpha \in Act(s)} c(s,\alpha) + \sum_{s' \in S \setminus G} \mathbf{P}(s,\alpha,s') \cdot v(s').$$

For $s \in MS$, $Act(s) = \{\perp\}$; if $s \in G$, then $c(s, \perp) = 0$ and $\mathbf{P}(s, \perp, s) = 1$ imply $L'(v)(s) = 0$. For $s \in IS$ and $\alpha \in Act(s)$, there exists a unique $s' \in S$ such that $\mathbf{P}(s, \alpha, s') = 1$. Thus we can rewrite L' as follows:

$$
[L'(v)](s) = \begin{cases} c(s, \perp) + \displaystyle\sum_{s' \in S \setminus G} \mathbf{P}(s, \perp, s') \cdot v(s') & \text{if } s \in MS \setminus G \\[2mm] \displaystyle\min_{s \xrightarrow{\alpha} s'} c(s, \alpha) + v(s') & \text{if } s \in IS \setminus G \\[2mm] 0 & \text{if } s \in G. \end{cases} \tag{2}
$$

By observing that $c(s, \perp) = \frac{1}{E(s)}$ if $s \in MS \setminus G$ and $c(s, \sigma) = 0$, otherwise, we can rewrite L' in (2) to yield the Bellman operator L as defined in Thm. 1. $\square$

Observe from the fixpoint characterization of $eT^{\min}(s, \diamondsuit G)$ in Thm. 1 that in interactive states—and only those may exhibit nondeterminism—it suffices to choose the successor state that minimizes $v(s')$. In addition, by Thm. 2, the Bellman operator L from Thm. 1 yields the minimal cost reachability in SSP $\mathcal{P}_{eT^{\min}}(\mathcal{I})$. These two observations and the fact that stationary deterministic policies suffice to attain the minimum expected cost of an SSP [2,15] yields:

Corollary 1. *There is a stationary deterministic scheduler yielding $eT^{\min}(s, \diamondsuit G)$.*

The uniqueness of the minimum expected cost of an SSP [2,15] now yields:

Corollary 2. *$eT^{\min}(s, \diamondsuit G)$ is the unique fixpoint of L (see Thm. 1).*

The uniqueness result enables the usage of standard solution techniques such as value iteration and linear programming to compute $eT^{\min}(s, \diamondsuit G)$.

4   Long-Run Average Analysis

Long-run average objectives. Let $\mathcal{I}$ be an IMC with state space S and $G \subseteq S$ a set of goal states. We use $\mathbf{I}_G$ as an indicator with $\mathbf{I}_G(s) = 1$ if $s \in G$ and 0, otherwise. Following the ideas of [14,22], the fraction of time spent in G on an infinite path π in $\mathcal{I}$ up to time bound $t \in \mathbb{R}_{\geq 0}$ is given by the random variable (r. v.) $A_{G,t}(\pi) = \frac{1}{t} \int_0^t \mathbf{I}_G(\pi@u) \, du$. Taking the limit $t \to \infty$, we obtain the r. v.

$$
A_G(\pi) = \lim_{t \to \infty} A_{G,t}(\pi) = \lim_{t \to \infty} \frac{1}{t} \int_0^t \mathbf{I}_G(\pi@u) \, du.
$$

The expectation of A_G for scheduler D and initial state s yields the corresponding long-run average time spent in G:

$$
\text{LRA}^D(s, G) = \mathbb{E}_{s,D}(A_G) = \int_{Paths} A_G(\pi) \Pr{}_{s,D}(d\pi).
$$

The minimum long-run average time spent in G starting from state s is then:

$$
\text{LRA}^{\min}(s, G) = \inf_D \text{LRA}^D(s, G) = \inf_D \mathbb{E}_{s,D}(A_G).
$$

For the long-run average analysis, we may assume w.l.o.g. that $G \subseteq MS$, as the long-run average time spent in any interactive state is always 0. This claim follows directly from the fact that interactive states are instantaneous, i.e. their sojourn time is 0 by definition. Note that in contrast to the expected time analysis, G-states cannot be made absorbing in the long-run average analysis.

Theorem 3. *There is a stationary deterministic scheduler yielding* $\text{LRA}^{\min}(s, G)$.

In the remainder of this section, we discuss in detail how to compute the minimum long-run average fraction of time to be in G in an IMC $\mathcal{I}$ with initial state s_0. The general idea is the following three-step procedure:

1. Determine the maximal end components $\{\mathcal{I}_1, \ldots, \mathcal{I}_k\}$ of IMC $\mathcal{I}$.
2. Determine $\text{LRA}^{\min}(G)$ in maximal end component $\mathcal{I}_j$ for all $j \in \{1, \ldots, k\}$.
3. Reduce the computation of $\text{LRA}^{\min}(s_0, G)$ in IMC $\mathcal{I}$ to an SSP problem.

The first phase can be performed by a graph-based algorithm [13] which has recently been improved in [9], whereas the last two phases boil down to solving linear programming problems. In the next subsection, we show that determining the LRA in an end component of an IMC can be reduced to a long-run ratio objective in an MDP equipped with two cost functions. Then, we show the reduction of our original problem to an SSP problem.

4.1 Long-Run Averages in Unichain IMCs

In this subsection, we consider computing long-run averages in *unichain* IMCs, i.e. IMCs that under any stationary deterministic scheduler yield a strongly connected graph structure.

Long-run ratio objectives in MDPs. Let $\mathcal{M} = (S, Act, \mathbf{P}, s_0)$ be an MDP. Assume w.l.o.g. that for each state s there exists $\alpha \in Act$ such that $\mathbf{P}(s, \alpha, s') > 0$. Let $c_1, c_2 : S \times (Act \cup \{\bot\}) \to \mathbb{R}_{\geqslant 0}$ be cost functions. The operational interpretation is that a cost $c_1(s, \alpha)$ is incurred when selecting action α in state s, and similar for c_2. Our interest is the ratio between c_1 and c_2 along a path. The *long-run ratio* $\mathcal{R}$ between the accumulated costs c_1 and c_2 along the infinite path $\pi = s_0 \xrightarrow{\alpha_0} s_1 \xrightarrow{\alpha_1} \ldots$ in the MDP $\mathcal{M}$ is defined by[1]:

$$\mathcal{R}(\pi) = \lim_{n \to \infty} \frac{\sum_{i=0}^{n-1} c_1(s_i, \alpha_i)}{\sum_{j=0}^{n-1} c_2(s_j, \alpha_j)}.$$

The minimum long-run ratio objective for state s of MDP $\mathcal{M}$ is defined by:

$$R^{\min}(s) = \inf_D \mathbb{E}_{s,D}(\mathcal{R}) = \inf_D \sum_{\pi \in Paths_{abs}} \mathcal{R}(\pi) \cdot \Pr_{s,D}^{abs}(\pi).$$

---

[1] In our setting, $\mathcal{R}(\pi)$ is well-defined as the cost functions c_1 and c_2 are obtained from non-Zeno IMCs, as explained below. This entails that for any infinite path π, $c_2(s_j, \alpha_j) > 0$ for some index j.

From [13], it follows that $R^{\min}(s)$ can be obtained by solving the following linear programming problem with real variables k and x_s for each $s \in S$: Maximize k subject to the following constraints for each $s \in S$ and $\alpha \in Act$:

$$x_s \leqslant c_1(s, \alpha) - k \cdot c_2(s, \alpha) + \sum_{s' \in S} \mathbf{P}(s, \alpha, s') \cdot x_{s'}.$$

Reducing LRA objectives in unichain IMCs to long-run ratio objectives in MDPs. We consider the transformation of an IMC into an MDP with 2 cost functions.

Definition 6. *Let $\mathcal{I} = (S, Act, \rightarrow, \Rightarrow, s_0)$ be an IMC and $G \subseteq S$ a set of goal states. The induced MDP is $\mathcal{M}(\mathcal{I}) = (S, Act \cup \{\bot\}, \mathbf{P}, s_0)$ with cost functions c_1 and c_2, where*

$$\mathbf{P}(s, \sigma, s') = \begin{cases} \frac{\mathbf{R}(s, s')}{E(s)} & \text{if } s \in MS \wedge \sigma = \bot \\ 1 & \text{if } s \in IS \wedge s \xrightarrow{\sigma} s' \\ 0 & \text{otherwise,} \end{cases}$$

$$c_1(s, \sigma) = \begin{cases} \frac{1}{E(s)} & \text{if } s \in MS \cap G \wedge \sigma = \bot \\ 0 & \text{otherwise,} \end{cases} \qquad c_2(s, \sigma) = \begin{cases} \frac{1}{E(s)} & \text{if } s \in MS \wedge \sigma = \bot \\ 0 & \text{otherwise.} \end{cases}$$

Observe that cost function c_2 keeps track of the average residence time in state s whereas c_1 only does so for states in G. The following result shows that the long-run average fraction of time spent in G-states in the IMC $\mathcal{I}$ and the long-run ratio objective $R^{\min}$ in the induced MDP $\mathcal{M}(\mathcal{I})$ coincide.

Theorem 4. *For unichain IMC $\mathcal{I}$, $LRA^{\min}(s, G)$ equals $R^{\min}(s)$ in MDP $\mathcal{M}(\mathcal{I})$.*

Proof. Let $\mathcal{I}$ be a unichain IMC with state space S and $G \subseteq S$. Consider a stationary deterministic scheduler D on $\mathcal{I}$. As $\mathcal{I}$ is unichain, D induces an ergodic CTMC $(S, \mathbf{R}, s_0)$, where $\mathbf{R}(s, s') = \sum\{\lambda \mid s \xRightarrow{\lambda} s'\}$, and $\mathbf{R}(s, s') = \infty$ if $s \in IS$ and $s \xrightarrow{D(s)} s'$.[2] The proof now proceeds in three steps.

⟨1⟩ According to the ergodic theorem for CTMCs [23], almost surely:

$$\mathbb{E}_{s_i}\left(\lim_{t \to \infty} \frac{1}{t} \int_0^t \mathbf{I}_{\{s_i\}}(X_u)\, du \right) = \frac{1}{z_i \cdot E(s_i)}.$$

Here, random variable X_t denotes the state of the CTMC at time t and $z_i = \mathbb{E}_i(T_i)$ is the expected return time to state s_i where random variable T_i is the return time to s_i when starting from s_i. We assume $\frac{1}{\infty} = 0$. Thus, in the long run almost all paths will stay in s_i for $\frac{1}{z_i \cdot E(s_i)}$ fraction of time.

⟨2⟩ Let μ_i be the probability to stay in s_i in the long run in the embedded discrete-time Markov chain $(S, \mathbf{P}', s_0)$ of CTMC $(S, \mathbf{R}, s_0)$. Thus $\boldsymbol{\mu} \cdot \mathbf{P}' = \boldsymbol{\mu}$ where $\boldsymbol{\mu}$ is the vector containing μ_i for all states $s_i \in S$. Given the probability μ_i of staying in state s_i, the expected return time to s_i is

$$z_i = \frac{\sum_{s_j \in S} \mu_j \cdot E(s_j)^{-1}}{\mu_i}.$$

---

[2] Strictly speaking, ∞ is not characterizing a negative exponential distribution and is used here to model an instantaneous transition. The results applied to CTMCs in this proof are not affected by this slight extension of rates.

$\langle 3 \rangle$ Gathering the above results now yields:

$$\text{LRA}^D(s, G) = \mathbb{E}_{s,D}\Big(\lim_{t \to \infty} \frac{1}{t} \int_0^t \mathbf{I}_G(X_u)\, du \Big) = \mathbb{E}_{s,D}\Big(\lim_{t \to \infty} \frac{1}{t} \int_0^t \sum_{s_i \in G} \mathbf{I}_{\{s_i\}}(X_u)\, du \Big)$$

$$= \sum_{s_i \in G} \mathbb{E}_{s,D}\Big(\lim_{t \to \infty} \frac{1}{t} \int_0^t \mathbf{I}_{\{s_i\}}(X_u)\, du \Big) \overset{\langle 1 \rangle}{=} \sum_{s_i \in G} \frac{1}{z_i \cdot E(s_i)}$$

$$\overset{\langle 2 \rangle}{=} \sum_{s_i \in G} \frac{\mu_i}{\sum_{s_j \in S} \mu_j E(s_j)^{-1}} \cdot \frac{1}{E(s_i)} = \frac{\sum_{s_i \in G} \mu_i E(s_i)^{-1}}{\sum_{s_j \in S} \mu_j E(s_j)^{-1}}$$

$$= \frac{\sum_{s_i \in S} \mathbf{I}_G(s_i) \cdot \mu_i E(s_i)^{-1}}{\sum_{s_j \in S} \mu_j E(s_j)^{-1}} = \frac{\sum_{s_i \in S} \mu_i \cdot (\mathbf{I}_G(s_i) \cdot E(s_i)^{-1})}{\sum_{s_j \in S} \mu_j \cdot E(s_j)^{-1}}$$

$$\overset{(\star)}{=} \frac{\sum_{s_i \in S} \mu_i \cdot c_1(s_i, D(s_i))}{\sum_{s_j \in S} \mu_j \cdot c_2(s_j, D(s_j))} \overset{(\star\star)}{=} \mathbb{E}_{s,D}(\mathcal{R})$$

Step $(\star)$ is due to the definition of c_1, c_2. Step $(\star\star)$ has been proven in [13].

By definition, there is a one-to-one correspondence between the schedulers of $\mathcal{I}$ and its MDP $\mathcal{M}(\mathcal{I})$. Together with the above results, this yields that $\text{LRA}^{\min} = \inf_D \text{LRA}^D(s)$ in IMC $\mathcal{I}$ equals $R^{\min}(s) = \inf_D \mathbb{E}_{s,D}(\mathcal{R})$ in MDP $\mathcal{M}(\mathcal{I})$.   □

To summarize, computing the minimum long-run average fraction of time that is spent in some goal state in $G \subseteq S$ in unichain IMC $\mathcal{I}$ equals the minimum long-run ratio objective in an MDP with two cost functions. The latter can be obtained by solving an LP problem. Observe that for any two states s, s' in a unichain IMC, $\text{LRA}^{\min}(s, G)$ and $\text{LRA}^{\min}(s', G)$ coincide. In the sequel, we therefore omit the state and simply write $\text{LRA}^{\min}(G)$ when considering unichain IMCs. In the next subsection, we consider IMCs that are not unichains.

4.2   Reduction to a Stochastic Shortest Path Problem

Let $\mathcal{I}$ be an IMC with initial state s_0 and maximal end components $\{\mathcal{I}_1, \ldots, \mathcal{I}_k\}$ for $k > 0$ where IMC $\mathcal{I}_j$ has state space S_j. Note that being a maximal end component implies that each $\mathcal{I}_j$ is also a unichain IMC. Using this decomposition of $\mathcal{I}$ into maximal end components, we obtain the following result:

Lemma 1. *Let* $\mathcal{I} = (S, Act, \rightarrow, \Longrightarrow, s_0)$ *be an IMC,* $G \subseteq S$ *a set of goal states and* $\{\mathcal{I}_1, \ldots, \mathcal{I}_k\}$ *the set of maximal end components in* $\mathcal{I}$ *with state spaces* $S_1, \ldots, S_k \subseteq S$. *Then*

$$\text{LRA}^{\min}(s_0, G) = \inf_D \sum_{j=1}^k \text{LRA}_j^{\min}(G) \cdot \text{Pr}^D(s_0 \models \Diamond S_j),$$

where $\text{Pr}^D(s_0 \models \Diamond S_j)$ *is the probability to eventually reach some state in* S_j *from* s_0 *under scheduler* D *and* $\text{LRA}_j^{\min}(G)$ *is the long-run average fraction of time spent in* $G \cap S_j$ *in unichain IMC* $\mathcal{I}_j$.

We finally show that the problem of computing minimal LRA is reducible to a non-negative SSP problem [2,15]. This is done as follows. In IMC $\mathcal{I}$, each maximal end component $\mathcal{I}_j$ is replaced by a new state u_j. Formally, let $U = \{u_1, \ldots, u_k\}$ be a set of fresh states such that $U \cap S = \varnothing$.

Definition 7 (SSP for long run average). *Let $\mathcal{I}$, S, $G \subseteq S$, $\mathcal{I}_j$ and S_j be as before. The SSP induced by $\mathcal{I}$ for the long-run average fraction of time spent in G is the tuple $\mathcal{P}_{LRA^{\min}}(\mathcal{I}) = \left(S \setminus \bigcup_{i=1}^{k} S_i \cup U, Act \cup \{\bot\}, \mathbf{P}', s_0, U, c, g\right)$, where*

$$
\mathbf{P}'(s, \sigma, s') = \begin{cases} \mathbf{P}(s, \sigma, s'), & \text{if } s, s' \in S \setminus \bigcup_{i=1}^{k} S_i \\ \sum_{s' \in S_j} \mathbf{P}(s, \sigma, s') & \text{if } s \in S \setminus \bigcup_{i=1}^{k} S_i \wedge s' = u_j, u_j \in U \\ 1 & \text{if } s = s' = u_i \in U \wedge \sigma = \bot \\ 0 & \text{otherwise.} \end{cases}
$$

Here, $\mathbf{P}$ is defined as in Def. 6. Furthermore, $g(u_i) = \mathrm{LRA}_i^{\min}(G)$ for $u_i \in U$ and $c(s, \sigma) = 0$ for all s and $\sigma \in Act \cup \{\bot\}$.

The state space of the SSP consists of all states in the IMC $\mathcal{I}$ where each maximal end component $\mathcal{I}_j$ is replaced by a single state u_j which is equipped with a $\bot$-labeled self-loop. The terminal costs of the new states u_i are set to $\mathrm{LRA}_i^{\min}(G)$. The transition probabilities are defined as in the transformation of an IMC into an MDP, see Def. 6, except that for transitions to u_j the cumulative probability to move to one of the states in S_j is taken. Note that as interactive transitions are uniquely labeled (as we consider closed IMCs), $\mathbf{P}'$ is indeed a probability function. The following theorem states the correctness of the reduction.

Theorem 5 (Correctness of the reduction). *For IMC $\mathcal{I}$ and its induced SSP $\mathcal{P}_{LRA^{\min}}(\mathcal{I})$ it holds:*

$$
\mathrm{LRA}^{\min}(s, G) = cR^{\min}(s, \Diamond U)
$$

where $cR^{\min}(s, \Diamond U)$ is the minimal cost reachability of U in SSP $\mathcal{P}_{LRA^{\min}}(\mathcal{I})$.

Example 3. Consider the IMC $\mathcal{I}$ in Fig. 1 and its maximal end components $\mathcal{I}_1$ and $\mathcal{I}_2$ with state spaces $S_1 = \{s_4, s_5, s_6, s_7\}$ and $S_2 = \{s_3, s_8, s_9, s_{10}\}$, respectively. Let $G = \{s_7, s_8\}$ be the set of goal states. For the underlying MDP $\mathcal{M}(\mathcal{I})$, we have $\mathbf{P}(s_4, \gamma_1, s_5) = 1$, $c_1(s_4, \gamma_1) = c_2(s_4, \gamma_1) = 0$, $\mathbf{P}(s_7, \bot, s_4) = \frac{1}{2}$, $c_1(s_7, \bot) = c_2(s_7, \bot) = \frac{1}{10}$, and $\mathbf{P}(s_5, \bot, s_7) = 1$ with $c_1(s_5, \bot) = 0$ and $c_2(s_5, \bot) = \frac{1}{20}$. Solving the linear programming problems for each of the maximal end components $\mathcal{I}_1$ and $\mathcal{I}_2$, we obtain $\mathrm{LRA}_1^{\min}(G) = \frac{2}{3}$, $\mathrm{LRA}_1^{\max}(G) = \frac{4}{5}$, and $\mathrm{LRA}_2^{\max}(G) = \mathrm{LRA}_2^{\min}(G) = \frac{9}{13}$. The SSP $\mathcal{P}_{LRA^{\min}}(\mathcal{I})$ for the complete IMC $\mathcal{I}$ is obtained by replacing $\mathcal{I}_1$ and $\mathcal{I}_2$ with fresh states u_1 and u_2 where $g(u_1) = \frac{2}{3}$ and $g(u_2) = \frac{9}{13}$. We have $\mathbf{P}'(s_1, \bot, u_1) = \frac{1}{3}$, $\mathbf{P}'(s_2, \beta_2, u_2) = 1$, etc. Finally, by solving the linear programming problem for $\mathcal{P}_{LRA^{\min}}(\mathcal{I})$, we obtain $\mathrm{LRA}^{\min}(s_0, G) = \frac{80}{117}$ by choosing α_1 in state s_0 and γ_1 in state s_4. Dually, $\mathrm{LRA}^{\max}(s_0, G) = \frac{142}{195}$ is obtained by choosing α_1 in state s_0 and γ_2 in state s_4.

5   Case Studies

5.1   Tool Support

What is IMCA*?* IMCA (Interactive Markov Chain Analyzer) is a tool for the *quantitative* analysis of IMCs. In particular, it supports the verification of IMCs against (a) timed reachability objectives, (b) reachability objectives, (c) expected time objectives, (d) expected step objectives, and (e) long-run average objectives. In addition, it supports the minimization of IMCs with respect to strong bisimulation. IMCA synthesizes ε-optimal piecewise constant timed policies for (a) timed reachability objectives using the approach of [27], and optimal positional policies for the objectives (b)–(e). Measures (c) and (e) are determined using the approach explained in this paper. IMCA supports the plotting of piecewise constant policies (on a per state basis) and incorporates a plot functionality for timed reachability which allows to plot the timed reachability probabilities for a state over a given time interval.

Input format. IMCA has a simple input format that facilitates its usage as a back-end tool for other tools that generate IMCs from high-level model specifications such as AADL, DFTs, PRISM reactive modules, and so on. It supports the bcg-format, such that it accepts state spaces generated (and possibly minimized) using the CADP toolbox [11]; CADP supports a LOTOS-variant for the compositional modeling of IMCs and compositional minimization of IMCs.

Implementation Details. A schematic overview of the IMCA tool is given in Fig. 2. The tool is written in C++, consists of about 6,000 lines of code, and exploits the GNU Multiple Precision Arithmetic Library[3] and the Multiple Precision Floating-Point Reliable Library[4] so as to deal with the small probabilities that occur during discretization for (a). Other included libraries are QT 4.6 and LP-solve[5] 5.5. The latter supports several efficient algorithms to solve LP problems; by default it uses simplex on an LP problem and its dual.

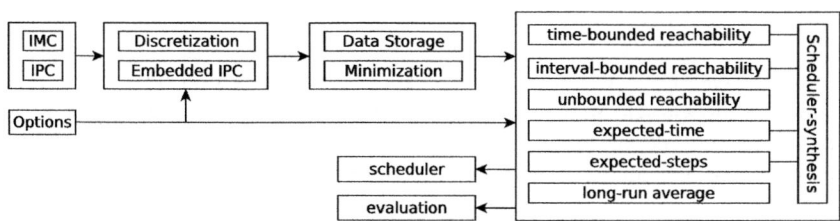

Fig. 2. Tool functionality of IMCA

---

[3] http://gmplib.org/

[4] http://www.mpfr.org/

[5] http://lpsolve.sourceforge.net/

Table 1. Computation times for the workstation cluster

N	# states	# transitions	$\lvert G \rvert$	$eT^{\max}(s, \Diamond G)$ time (s)	$\Pr^{\max}(s, \Diamond G)$ time (s)	$\mathrm{LRA}^{\max}(s, G)$ time (s)
1	111	320	74	0.0115	0.0068	0.0354
4	819	2996	347	0.6418	0.1524	0.3629
8	2771	10708	1019	3.1046	1.8222	11.492
16	8959	36736	3042	35.967	18.495	156.934
32	38147	155132	12307	755.73	467.0	3066.31
52	96511	396447	30474	5140.96	7801.56	OOM

5.2   Case Studies

We study the practical feasibility of IMCA's algorithms for expected time reachability and long-run averages on two case studies: A dependable workstation cluster [17] and a Google file system [10]. The experiments were conducted on a single core of a 2.8 GHz Intel Core i7 processor with 4GB RAM running Linux.

Workstation cluster. In this benchmark, two clusters of workstations are connected via a backbone network. In each cluster, the workstations are connected via a switch. All components can fail. Our model for the workstation cluster benchmark is basically as used in all of its studies so far, except that the inspection transitions in the GSPN (Generalized Stochastic Petri Net) model of [17] are immediate rather than —as in all current studies so far— stochastic transitions with a very high rate. Accordingly, whenever the repair unit is available and different components have failed, the choice which component to repair next is nondeterministic (rather than probabilistic). This yields an IMC with the same size as the Markov chain of [17]. Table 1 shows the computation times for the maximum expected reachability times where the set G of goal states depends on the number N of operational workstations. More precisely, G is the set of states in which none of the operational left (or right) workstations connected via an operational switch and backbone is available. For the sake of comparison, the next column indicates the computation times for unbounded reachability probabilities for the same goal set. The last column of Table 1 lists the results for the long-run average analysis; the model consists of a single end component.

Google file system. The model of [10] focuses on a replicated file system as used as part of the Google search engine. In the Google file system model, files are divided into chunks of equal size. Several copies of each chunk reside at several chunk servers. The location of the chunk copies is administered by a single master server. If a user of the file system wants to access a certain chunk of a file, it asks the master for the location. Data transfer then takes place directly between a chunk server and the user. The model features three parameters: The number M of chunk servers, the number S of chunks a chunk server may store, and the total number N of chunks. In our setting, $S = 5000$ and $N = 100000$, whereas M varies. The set G of goal states characterizes the set of states that offer at least service level one. We consider a variant of the GSPN model in [10] in which the

Table 2. Computation times for Google file system ($S = 5000$ and $N = 100000$)

M	# states	# transitions	$|G|$	$eT^{\min}(s, \Diamond G)$ time (s)	$\Pr^{\min}(s, \Diamond G)$ time (s)	$\text{LRA}^{\min}(s, G)$ time (s)
10	1796	6544	408	0.7333	0.9134	4.8531
20	7176	27586	1713	16.033	48.363	173.924
30	16156	63356	3918	246.498	271.583	2143.79
40	28736	113928	7023	486.735	1136.06	4596.14
60	64696	202106	15933	765.942	1913.66	OOM

probability of a hardware or a software failure in the chunk server is unknown. This aspect was not addressed in [10]. Table 2 summarizes the computation times for the analysis of the nondeterministic Google file system model.

6   Conclusions

We presented novel algorithms, prototypical tool support in IMCA, and two case studies for the analysis of expected time and long run average objectives of IMCs. We have shown that both objectives can be reduced to stochastic shortest path problems. As IMCs are the semantic backbone of engineering formalisms such as AADL error models [5], dynamic fault trees [4] and GALS hardware designs [12], our contribution enlarges the analysis capabilities for dependability and reliability. The support of the compressed bcg-format allows for the direct usage of our tool and algorithms as back-end to tools like CADP [11] and CORAL [4]. The tool and case studies are publicly available at http://moves.rwth-aachen.de/imca. Future work will focus on the generalization of the presented algorithms to Markov automata [16], and experimentation with symbolic data structures such as multi-terminal BDDs by, e.g. exploiting PRISM for the MDP analysis.

Acknowledgment. This research was supported by the EU FP7 MoVeS and MEALS projects, the ERC advanced grant VERIWARE, the DFG research center AVACS (SFB/TR 14) and the DFG/NWO ROCKS programme. We thank Silvio de Carolis for the bcg-interface and Ernst Moritz Hahn for his help on the Google file system.

References

1. Baier, C., Haverkort, B.R., Hermanns, H., Katoen, J.-P.: Model-checking algorithms for continuous-time Markov chains. IEEE TSE 29, 524–541 (2003)
2. Bertsekas, D.P., Tsitsiklis, J.N.: An analysis of stochastic shortest path problems. Mathematics of Operations Research 16, 580–595 (1991)
3. Böde, E., Herbstritt, M., Hermanns, H., Johr, S., Peikenkamp, T., Pulungan, R., Rakow, J., Wimmer, R., Becker, B.: Compositional dependability evaluation for STATEMATE. IEEE TSE 35, 274–292 (2009)

4. Boudali, H., Crouzen, P., Stoelinga, M.: A rigorous, compositional, and extensible framework for dynamic fault tree analysis. IEEE TSDC 7, 128–143 (2009)
5. Bozzano, M., Cimatti, A., Katoen, J.-P., Nguyen, V., Noll, T., Roveri, M.: Safety, dependability and performance analysis of extended AADL models. The Computer Journal 54, 754–775 (2011)
6. Brázdil, T., Forejt, V., Krcál, J., Kretínský, J., Kucera, A.: Continuous-time stochastic games with time-bounded reachability. In: FSTTCS. LIPIcs, vol. 4, pp. 61–72. Schloss Dagstuhl (2009)
7. Buchholz, P., Hahn, E.M., Hermanns, H., Zhang, L.: Model Checking Algorithms for CTMDPs. In: Gopalakrishnan, G., Qadeer, S. (eds.) CAV 2011. LNCS, vol. 6806, pp. 225–242. Springer, Heidelberg (2011)
8. Buchholz, P., Schulz, I.: Numerical analysis of continuous time Markov decision processes over finite horizons. Computers & OR 38, 651–659 (2011)
9. Chatterjee, K., Henzinger, M.: Faster and dynamic algorithms for maximal end-component decomposition and related graph problems in probabilistic verification. In: Symp. on Discrete Algorithms (SODA), pp. 1318–1336. SIAM (2011)
10. Cloth, L., Haverkort, B.R.: Model checking for survivability. In: QEST, pp. 145–154. IEEE Computer Society (2005)
11. Coste, N., Garavel, H., Hermanns, H., Lang, F., Mateescu, R., Serwe, W.: Ten Years of Performance Evaluation for Concurrent Systems Using CADP. In: Margaria, T., Steffen, B. (eds.) ISoLA 2010, Part II. LNCS, vol. 6416, pp. 128–142. Springer, Heidelberg (2010)
12. Coste, N., Hermanns, H., Lantreibecq, E., Serwe, W.: Towards Performance Prediction of Compositional Models in Industrial GALS Designs. In: Bouajjani, A., Maler, O. (eds.) CAV 2009. LNCS, vol. 5643, pp. 204–218. Springer, Heidelberg (2009)
13. de Alfaro, L.: Formal Verification of Probabilistic Systems. PhD thesis, Stanford University (1997)
14. de Alfaro, L.: How to specify and verify the long-run average behavior of probabilistic systems. In: LICS, pp. 454–465. IEEE CS Press (1998)
15. de Alfaro, L.: Computing Minimum and Maximum Reachability Times in Probabilistic Systems. In: Baeten, J.C.M., Mauw, S. (eds.) CONCUR 1999. LNCS, vol. 1664, pp. 66–81. Springer, Heidelberg (1999)
16. Eisentraut, C., Hermanns, H., Zhang, L.: On probabilistic automata in continuous time. In: LICS, pp. 342–351. IEEE Computer Society (2010)
17. Haverkort, B.R., Hermanns, H., Katoen, J.-P.: On the use of model checking techniques for dependability evaluation. In: SRDS, pp. 228–237. IEEE CS (2000)
18. Hermanns, H. (ed.): Interactive Markov Chains. LNCS, vol. 2428. Springer, Heidelberg (2002)
19. Hermanns, H., Katoen, J.-P.: The How and Why of Interactive Markov Chains. In: de Boer, F.S., Bonsangue, M.M., Hallerstede, S., Leuschel, M. (eds.) FMCO 2009. LNCS, vol. 6286, pp. 311–338. Springer, Heidelberg (2010)
20. Johr, S.: Model Checking Compositional Markov Systems. PhD thesis, Saarland University (2007)
21. Knast, R.: Continuous-time probabilistic automata. Information and Control 15, 335–352 (1969)
22. López, G.G.I., Hermanns, H., Katoen, J.-P.: Beyond Memoryless Distributions: Model Checking Semi-Markov Chains. In: de Luca, L., Gilmore, S. (eds.) PAPM-PROBMIV 2001. LNCS, vol. 2165, pp. 57–70. Springer, Heidelberg (2001)
23. Norris, J.: Markov Chains. Cambridge University Press (1997)

24. Rabe, M.N., Schewe, S.: Finite optimal control for time-bounded reachability in CTMDPs and continuous-time Markov games. Acta Inf. 48, 291–315 (2011)
25. von Essen, C., Jobstmann, B.: Synthesizing systems with optimal average-case behavior for ratio objectives. In: iWIGP. EPTCS, vol. 50, pp. 17–32 (2011)
26. Yushtein, Y., Bozzano, M., Cimatti, A., Katoen, J.-P., Nguyen, V.Y., Noll, T., Olive, X., Roveri, M.: System-software co-engineering: Dependability and safety perspective. In: SMC-IT, pp. 18–25. IEEE Computer Society (2011)
27. Zhang, L., Neuhäußer, M.R.: Model Checking Interactive Markov Chains. In: Esparza, J., Majumdar, R. (eds.) TACAS 2010. LNCS, vol. 6015, pp. 53–68. Springer, Heidelberg (2010)

Lessons Learnt from the Adoption
of Formal Model-Based Development

Alessio Ferrari[1], Alessandro Fantechi[1,2], and Stefania Gnesi[1]

[1] ISTI-CNR, Via G. Moruzzi 1, Pisa, Italy
{firstname.lastname}@isti.cnr.it
http://www.isti.cnr.it/
[2] DSI, Università degli Studi di Firenze, Via di S.Marta 3, Firenze, Italy
fantechi@dsi.unifi.it
http://www.dsi.unifi.it/~fantechi/

Abstract. This paper reviews the experience of introducing *formal mo-del-based design* and *code generation* by means of the Simulink/Stateflow platform in the development process of a railway signalling manufacturer. Such company operates in a standard-regulated framework, for which the adoption of commercial, non qualified tools as part of the development activities poses hurdles from the verification and certification point of view. At this regard, three incremental intermediate goals have been de-fined, namely (1) identification of a safe-subset of the modelling language, (2) evidence of the behavioural conformance between the generated code and the modelled specification, and (3) integration of the modelling and code generation technologies within the process that is recommended by the regulations.

These three issues have been addressed by progressively tuning the usage of the technologies across different projects. This paper summarizes the lesson learnt from this experience. In particular, it shows that formal modelling and code generation are actually powerful means to enhance product safety and cost effectiveness. Nevertheless, their adoption is not a straightforward step, and incremental adjustments and refinements are required in order to establish a *formal model-based process*.

Introduction

The adoption of formal and semi-formal modelling technologies into the different phases of development of software products is constantly growing within industry [4,29,27]. Designing model abstractions before getting into hand-crafted code helps highlighting concepts that can hardly be focused otherwise, enabling greater control over the system under development. This is particularly true in the case of embedded *safety-critical* applications such as aerospace, railway, and automotive ones. These applications, besides dealing with code having increasing size and therefore an even more crucial role for safety, can often be tested only on the target machine or on *ad-hoc* expensive simulators.

A. Goodloe and S. Person (Eds.): NFM 2012, LNCS 7226, pp. 24–38, 2012.

Within this context, recent years have seen the diffusion of graphical tools to facilitate the development of the software before its actual deployment. Technologies known as *model-based design* [32] and *code generation* started to be progressively adopted by several companies as part of their software process.

The development of safety-critical software shall conform to specific international standards (e.g., RTCA/DO-178B [30] for aerospace, IEC-61508 [22] for automotive and CENELEC/EN-50128 [7] for railway signalling in Europe). These are a set of norms and methods to be used while implementing a product having a determined safety-related nature. In order to certify a product according to these standards, companies are required to give evidence to the authorities that a development process has been followed that is coherent with the prescriptions of the norms.

Introducing model-based design tool-suites and the code generation technology within a standard-regulated process is not a straightforward step. The code used in safety-critical systems shall conform to specific quality standards, and normally the companies use *coding guidelines* in order to avoid usage of improper constructs that might be harmful from the safety point of view. When modelling is adopted, the generated code shall conform to the same standard asked to the hand-crafted code. Concerning the tools, the norms ask for a certified or *proven-in-use* translator: in absence of such a tool, a strategy has to be defined in order to assess the equivalence between the model and the generated code behaviour. The modelling and code generation technologies are then required to be integrated with the established process, that shall maintain its coherence even if changes are applied.

With the aim of establishing guidelines for a formal model-based development process, in this paper we review a series of relevant experiences done in collaboration with a railway signalling manufacturer operating in the field of Automatic Train Protection (ATP) systems. Inside a long-term effort of introducing formal methods to enforce product safety, indeed the company decided to adopt the Simulink/Stateflow tool-suite to exploit formal model-based development and code generation within its own development process [2,15]. The decision was followed by four years of incremental actions in using commercial tools to build a formal model-based process focused on code generation. Details of these actions, have been published elsewhere [15,17,18,16]. We are here interested instead to give a global view of the overall experience.

The paper is structured as follows. In Sect. 1 some background is given concerning *formal methods, model-based design* and the existing approaches integrating the two technologies. In Sect. 2 the research problem of introducing code generation from formal models in a safety-critical domain is expressed and discussed. In Sect. 3 the projects where formal model-based development has been employed are presented, together with the goals progressively achieved with respect to the main research problem. In Sect. 4 the advantages and the critical aspects of code generation are evaluated. Sect. 5 draws final conclusions and remarks.

1   Formal Model-Based Design

In 1995, Bowen and Hinchey published the Ten Commandments of Formal Methods [5], a list of guidelines for applying formal techniques, edited according to their experience in industrial projects [20]. Ten years later, the authors review their statements, and they witness that not so much have changed [6]: the industrial applications that demonstrate the feasibility and the effectiveness of formal methods are still limited, though famous projects exist, which show that the interest in these methods is not decreased. Among them, it is worth citing the Paris Metro onboard equipment [3], where the B method has been employed, and the Maeslant Kering storm surge barrier control system [33], where both the Z and the Promela notations have been used.

The comprehensive survey of Woodcock et al. [35] confirms that industries are currently performing studies on formal methods applications, but still perceive them as experimental technologies.

While formal methods have struggled for more than twenty years for a role in the development process of the companies, the *model-based design* [32] paradigm has gained ground much faster. The defining principle of this approach is that the whole development shall be based on graphical model abstractions, from which an implementation can be manually or automatically derived. Tools supporting this technology allow simulations and tests of the system models to be performed before the actual deployment. The objective is not different from the one of formal methods, which is detecting design defects before the actual implementation. However, while formal methods are perceived as rigid and difficult, model-based design is regarded as closer to the needs of the developers, which consider graphical simulation more intuitive than formal verification.

This trend has given increasing importance to tools such as the SCADE suite [11], a graphical modelling environment mostly used in aerospace and based on the Lustre synchronous language, Scicos [23], an open source platform for modelling and simulating control systems, and the two tools ASCET [14] and AutoFocus [21], both oriented to automotive systems and using block notations for the representation of distributed processes.

In this scenario, the safety-critical industry has progressively seen the clear establishment of the Simulink/Stateflow [25] platform as a *de-facto* standard for modelling and code generation. The Simulink language uses a block notation for the definition of continuous-time dynamic system. The Stateflow notation is based on Harel's Statecharts [19] and supports the modelling and animation of event-based discrete-time applications. The integration of the two languages allows a flexible representation of hybrid systems, while tools such as Simulink Coder [25] and TargetLink [12] support automatic source code generation from the models. These features, strengthened by the large amount of associated toolboxes to analyse the different aspects of an application, has enabled a cross-domain spread of the platform.

Nevertheless, since the languages and tool-suite are not formally based, their full employment for the development of safety-critical applications poses challenges from the verification and certification point of view: how to ensure that the generated code is compliant with the modelled application? How to integrate model-based practices with traditional certified processes? These are all questions that started pushing industries and researchers toward an integration between model-based design and formal techniques [1,29]. The main goal is to take profits from the flexibility of the first and the safety assurance of the latter, going toward the definition of *formal model-based design* methods.

Large size companies have been the first to employ formal model-based practices. Already in 2006, Honeywell started defining an approach for the translation of Simulink models into the input language of the SMV model checker [26]. Airbus has used the model checking capabilities of the SCADE suite for ten years in the development of the controllers for the A340-500/600 series [4]. The most complete, integrated methodology is probably the one currently practiced by Rockwell Collins [27]. The process implemented by this company of the avionic sector starts from Simulink/Stateflow models to derive a representation in the Lustre formal language. Then, formal verification is performed by means of different engines, such as NuSMV and Prover, followed by code generation in C and ADA.

The main contribution of the current paper with respect to the related work is the in-depth focus on the code generation aspect, together with the evaluation of the advantages given by the introduction of this technology in a medium-size company. Our objective is to give a clear picture of how the adoption of code generation affects the overall development process.

2   Problem Statement

The company considered in this paper operates in the development of safety-related railway signalling systems. Inside an effort of adopting formal methods within its own development process, the company decided to introduce system modelling by means of the Simulink/Stateflow tools [2], and in 2007 decided to move to code generation [15].

Formal modelling with automatic code generation were seen as breakthrough technologies for managing projects of increasing size, and for satisfying the requirements of a global market in terms of product flexibility.

In order to achieve this goal, the company contacted experts from academia, expected to give guidance and support along this paradigm-shift in the development process. This paper reviews a series of experiences done in this collaboration, inside a four year reasearch activity started at the end of 2007, with the aim to address the following:

Problem Statement

Define and implement a methodology for the adoption of the code generation technology from formal models by a railway signalling company

---

During the research activity, the problem statement has been decomposed into the following sub-goals.

Goal 1 - modelling language restriction. The code used in safety-critical systems shall conform to specific quality standards, and normally the companies use coding guidelines in order to avoid usage of improper constructs that might be harmful from the safety point of view. When modelling and auto-coding are adopted, the generated code shall conform to the same standard asked to the hand-crafted code. Hence, the identification of a safe subset of the adopted modelling language is required for the production of code compliant with the guidelines and that can be succesfully integrated with the existing one.

Goal 2 - generated code correctness. Safety-critical norms ask for a certified or *proven-in-use* translator. In absence of such a tool, like in the case of the available code generators for Simulink/Stateflow, a strategy has to be defined in order to ensure that the code behaviour is fully compliant to the model behaviour, and no additional improper functions are added during the code synthesis phase. The objective is to perform the verification activities at the level of the abstract model, minimizing or automating the operations on the code.

Goal 3 - process integration. Product development is performed by companies by means of processes, which define a framework made of tasks, artifacts and people. Introduction of new technologies in an established process requires adjustments to the process structure, which shall maintain its coherence even if changes are applied. This is particularly true in the case of safety-critical companies, whose products have to be validated according to normative prescriptions. Hence, a sound process shall be defined in order to integrate modelling and code generation within the existing process.

3   Projects and Achievements

Addressing the problem statement issued above started with the objective of introducing model-based design and code generation within the development process of the company. The specific projects, summarized in the following, were subsequently selected as test-benches for the incremental introduction of such technologies. Each goal expressed in Sect. 2 is evaluated according to its progressive refinement during the projects.

The first experiments have been performed during **Project 1**, involving the development of a simple Automatic Train Protection (ATP) system.

ATP systems are typically embedded platforms that enforce the rules of signaling systems by adding an on-board automatic control over the speed limit imposed to trains along the track. In case of dangerous behaviour acted by the driver (e.g., speed limit or signalling rules violation) the system is in charge of regulating the speed by enforcing the brakes until the train returns to a safe state (i.e., the train standing condition or a speed below the imposed limit).

During Project 1, an ATP system was developed from scratch with the support of the Simulink/Stateflow tool-suite. A Stateflow model was designed in collaboration with the customer in order to define and assess the system requirements [2]. This experience, completed with the successful deployment of the system, allowed the assessment of the potentials of modelling for prototype definition and requirements agreement.

The actual research on code generation started when the hand-crafted system was already deployed and operational. The Stateflow model, formerly employed for requirements agreement, has been used as a prototype platform for the definition of a first set modelling language restrictions in the form of *modelling guidelines* [15]. The model passed through a refactoring path according to the guidelines defined, and proper code synthesis of the single *model units* was achieved through the Stateflow Coder[1] tool.

At the end of 2007, the system evolved in a new version. The refactored model substituted the original one for the definition of the new system specifications. Still, code generation was not employed in the actual development process, and the product remained an hand-crafted system also in its new version, since a proper V&V process for its certification against regulations was not defined yet.

Project 2, involved an ATP system about ten times larger in terms of features compared with Project 1. With this project, the company put into practice its acquired experience with code generation. The set of internal guidelines edited during Project 1 has been integrated with the public MAAB recommendations[2] [24] for modelling with Simulink/Stateflow.

Furthermore, a preliminary process for code verification has been defined [18]. The defined process was structured as follows. First, an internally developed tool was used to check modelling standard adherence, a sort of static analysis performed at model level. Then, functional unit-level verification was performed by means of a two-phase task made of *model-based testing* [13] and *abstract interpretation* [9]. The first step checks for functional equivalence between model and code. The second step, supported by the Polyspace tool [10,25], is used to assess the absence of runtime errors. Due to the timing of the project, both guidelines verification and model-based testing have been only partially employed. The process had to be adjusted with *ad-hoc* solutions, mostly based on traditional code testing, in order to address the problem of a non-certified code generator.

Project 3, concerning an ATP system tailored for metro signalling, has been the first complete instance of a formal development process [17,16]. Besides the already adopted technologies, a hierarchical derivation approach has been employed. Simulink and Stateflow are proper tools to represent the low-level aspects of a system, while they offer poor support for reasoning at the software architecture level. Therefore, we resorted to adopt the UML notation to model the software architecture of the system. The approach starts from such an UML

---

[1] The tool is currently distributed as part of Simulink Coder.

[2] Set of guidelines developed in the automotive domain for modelling with Simulink/Stateflow. The current version of the MAAB recommendations is 2.2, issued in 2011. The project adopted the 2.0 version, issued in 2007.

representation, and requires deriving unit-level requirements and a formal representation of them in the form of Stateflow diagrams. During the project, the modelling guidelines have been updated with a set of restrictions particularly oriented to define a formal semantics for the Stateflow language (see Sect. 3.1). These restrictions have enforced the formal representation of the requirements.

A new code generator, Real Time Workshop Embedded Coder[3], has been introduced, which permitted to generate also the integration code between the different generated units. With the previously adopted Stateflow Coder, code integration was performed manually. The adoption of the new generator allowed a further automation and speed-up of the development.

Within the project, the goal of ensuring correctness of the generated code in absence of a certified translator has been addressed by combining a model-based testing approach known as *translation validation* [8] with abstract interpretation. Translation validation has been performed with an internally developed framework. This framework supports *back-to-back* [34] model-code execution of unit level tests, and the assessment of consistency between model and code coverage. Abstract interpretation with Polyspace has been performed with a strategy analogous to the one already applied for Project 2.

Project 3 has also marked the start of the first structured experiments with formal verification by means of Simulink Design Verifier [25]. The evaluation was particularly oriented to verify whether this technology, employed at the level of the model units, could actually replace model-based unit testing with a substantial cost reduction. The first results have been encouraging. The experiments have shown that about 95% of the requirements can be verified with the tool to achieve a cost reduction of 50% to 66% in terms of man/hours [16]. The remaining requirements, for which this cost gain cannot be achieved, can be verified through model-based testing. The company is currently devising strategies to systematically employ formal verification in the development process.

Many of the development, verification and certification issues related to formal model-based development appeared only during its actual deployment: the goals planned at the beginning of the research have been addressed after progressive tuning of the strategy across the different projects. Table 1 summarizes the technologies incrementally introduced during the projects. *Italics* indicate partial adoption of a technology or partial achievement of a goal.

3.1   Goal 1 - Modelling Language Restriction

The first goal was to identify a proper subset of the Simulink/Stateflow language: the idea was that C code generated from models in this subset would be compliant with the guidelines defined by the company in accordance with the quality standard required by the norms. With Project 1, this problem was addressed by first analysing the violations of the quality standard issued by the code generated from the original model. Then, sub-models have been defined, on which the evaluation could be performed more easily. The translation of

---

[3] Currently renamed Simulink Coder.

Table 1. Summary of the results achieved during the projects

Year	Project	Technologies (Full or *Partial* Adoption)	Goal
2007-2008	Project 1	Modelling guidelines (25) *Code generation (Stateflow Coder R2007b)*	1
2008-2010	Project 2	Modelling guidelines + MAAB (43) Code generation (Stateflow Coder R2007b) *Guidelines verification* *Model-based testing* Abstract interpretation (Polyspace 7.0)	1 *2* *3*
2009-2011	Project 3	Modelling guidelines + MAAB (43) Semantics restrictions UML + hierarchical derivation Code generation (RTW Embedded Coder R2010a) Translation Validation Abstract interpretation (Polyspace 8.0) *Formal Verification (Simulink Design Verifier R2010a)*	1 2 3

single graphical constructs, and of combination of them, have been evaluated and classified. Proper modelling guidelines have been defined in order to avoid the violations experienced. The activity led to the definition of a preliminary set of 25 guidelines for creating models targeted for code generation [15].

With Project 2, where code generation has been actually employed for the development of the whole application logic software, a more systematic study has been performed. The preliminary set of guidelines had in fact the limit of being derived from a specific model, and could lack of generality. A comparison with the experience of other safety-critical domains was needed. Actually, in the automotive sector a set of accepted modelling rules equivalent to the MISRA [28] ones for C code had emerged, that is, the MAAB guidelines [24], defined by OEMs[4] and suppliers of the automotive sector to facilitate model exchange and commissioning. The preliminary set was extended by adapting the MAAB guidelines to the railway domain. This new set, composed of 45 guidelines in total, prompted further restrictions. These restrictions were not only limited to enforce generation of quality code, but were also oriented to define well-structured models.

A further step was performed during Project 3: in order to ease a formal analysis and a formal representation of the requirements, it was decided to complete the modelling style guidelines by restricting the Stateflow language to a semantically unambiguous set. To this end, the studies of Scaife et al. [31], focused on translating a subset of Stateflow into the Lustre formal language, have been used. These studies brought to the definition of a formal semantics for Stateflow [17], which constrains the language to an unambiguous subset. A set of guidelines has been defined for enforcing the development of design models in accordance to this subset of the language. The models produced are semantically independent from the simulation engine, and a formal development process could actually take place.

---

[4] Original Equipment Manufacturers.

3.2   Goal 2 - Generated Code Correctness

The second goal was to address the problem of a non-certified, neither proven-in-use, translator. The objective was to ensure the code to be fully compliant with the model behaviour, and to guarantee that no additional improper functions are added during the code synthesis phase. The approach adopted, preliminarily defined during Project 2, but refined and fully applied only on Project 3, consisted in implementing a model-based testing approach known as *translation validation* [8], and completing it with static analysis by means of abstract interpretation [9].

Translation validation consists of two steps: (1) a model/code back-to-back execution of unit tests, where both the model and the corresponding code are exercised using the same scenarios as inputs and results are checked for equivalence; (2) a comparison of the structural coverage obtained at model and at code level. The first step ensures that the code behaviour is compliant with the model behaviour. The second one ensures that no additional function is introduced in the code: tests are performed until 100% of decision coverage is obtained on the models. If lower values are obtained for the code, any discrepancy must be assessed and justified.

Model-based testing with translation validation ensures equivalence between model and code, but cannot cover all the possible behaviours of the code in terms of control-flow and data-flow. In particular, it lacks in detecting all those runtime errors that might occur only with particular data sets, such as division by zero and buffer overflow. For this reason, translation validation has been completed with abstract interpretation by means of the Polyspace tool. The main feature of the tool is to detect runtime errors by performing static analysis of the code.

Since the correctness of the source is not decidable at the program level, the tools implementing abstract interpretation work on a conservative and sound approximation of the variable values in terms of intervals, and consider the state space of the program at this level of abstraction. Finding errors in this larger approximation domain does not imply that the bug also holds in the program. The presence of *false positives* after the analysis is actually the drawback of abstract interpretation that hampers the possibility of fully automating the process.

Already within Project 2, a two-steps procedure has been defined for the usage of the tool to address the problem of false positives: (1) a first analysis step is performed with a large over-approximation set, in order to discover systematic runtime errors and identify classes of possible false positives that can be used to restrict the approximation set; (2) a second analysis step is performed with a constrained abstract domain, derived from the first analysis, and the number of uncertain failure states to be manually reviewed is drastically reduced.

3.3   Goal 3 - Process Integration

The third goal was integrating the modelling and code generation technologies into a coherent development process. Also concerning this issue, a sound process was finally achieved only with Project 3, after incremental adjustments. The introduction of modelling and the need to ensure consistency between models and code, has prompted changes also to the verification and validation activities, which had to be tailored according to the new technology. On the other hand, it has allowed working on a higher level of abstraction, and different methods and tools have been combined to achieve a complete formal development.

The final process is an enhanced V-based development model, as depicted in Fig. 1. The process embeds two verification branches: one for the activities performed on the models, the other for the tasks concerning source code and system.

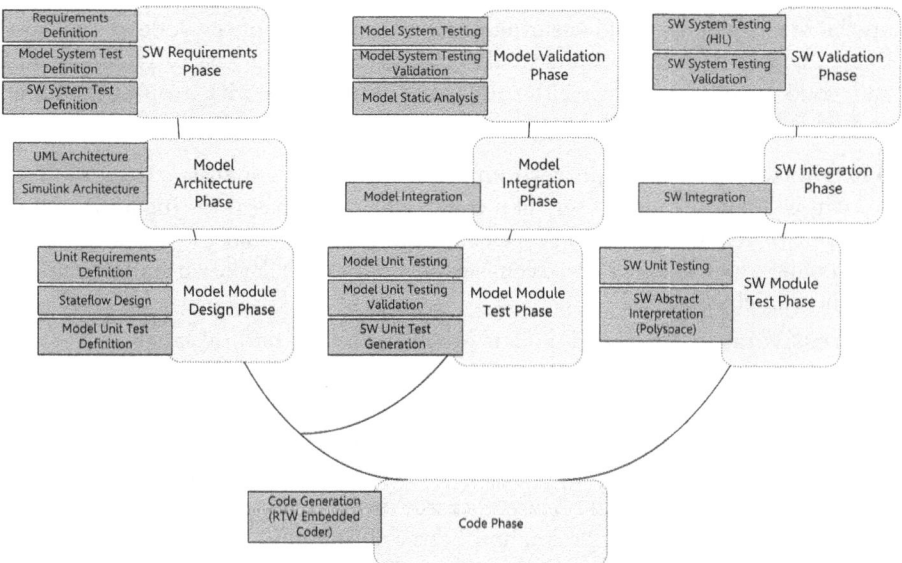

Fig. 1. Overview of the formal model-based development process adopted

From system-level software requirements, tests are defined to be perfomed both at model-integration level and at system-level (**SW Requirements Phase**). Then, a UML architecture is defined in the form of a component diagram. The diagram is then manually translated into a Simulink architecture (**Model Architecture Phase**). During the design phase, system requirements are decomposed into unit requirements apportioned to the single architectural components. Furthermore, the Stateflow models are defined according to these unit requirements, following the style-guidelines and the semantics restrictions (**Model Module Design Phase**).

Functional unit testing (**Model Module Test Phase**) and system testing (**Model Validation Phase**) are performed on the models by using the Simulink simulator before generating code through RTW Embedded Coder (**Code Phase**). After code generation, translation validation is performed, followed by static analysis by means of abstract interpretation (**SW Module Test Phase**). The application code is then integrated with operating system and drivers (**SW Integration**

Phase), and *hardware-in-the-loop* (HIL) is used to perform system tests according to the system requirements (**SW Validation Phase**). The whole process is supported by coherent documentation: this is auto-generated by means of Simulink Report Generator, a Simulink toolbox, using the comments edited by the developers on the models.

4   Lessons Learnt

Code generation was introduced following the intuition that defining a formal model of the specifications, and automatically producing code, allows speeding-up the development, while ensuring greater correctness of the code at the same time. The intuition has been actually confirmed by the practice. The modelling and code generation showed the following advantages with respect to hand-crafted code.

Abstraction. Models require working at a higher level of abstraction, and they can be manipulated better than code. The model-based testing approach, in the two versions put into practice during Project 2 and Project 3, gives the advantage of defining test scenarios at component level without disrupting the model structure.

Expressiveness. Graphical models are closer to the natural language requirements. At the same time, they are an unambiguous mean to exchange or pass artefacts among developers. This observation has been enlightened by the Project 3 experience, where the project passed from the hands of its first main developer to another developer within one month only

Cohesion & Decoupling. The generated software is composed by modules with higher internal cohesion and better decoupling. Interfaces among functionalities are based solely on data, and the control-flow is simplified since there is no cross-call among different modules. Decoupling and well-defined interfaces have helped in easing the model outsourcing, which is a relevant aspect when developing with time-to-market constraints.

Uniformity. The generated code has a repetitive structure, which facilitates the automation of the verification activities. When strict modelling guidelines are defined, one could look at the generated code as if it would be the software always written by the same programmer. Therefore, any code analysis task can be tailored on the artificial programmer's design habits. As a witness for this observation, consider that the full two-step Polyspace procedure (see Sect. 3.2) resulted profitable on the generated code only, since systematic analysis on hand-crafted code was made harder by its variable structure and programming style.

Traceability. Software modules are directly traceable with the corresponding blocks of the modelled specification. Traceability is a relevant issue in the development of safety-critical systems, since any error has to be traced back to the process task, or artefact defect, that produced it. The formal development approach introduced, with the support of RTW Embedded Coder, has allowed defining navigable links between the single code statements and the requirements.

Control. The structured development has given greater control over the components, producing in the end software with less bugs already before the verification activities, as witnessed by the bug reduction evaluation measured during Project 3 (from 10 to 3 bugs per module) [17].

Verification Cost. When passing from traditional code unit testing based on *structural* coverage objectives, to testing based on *functional* objectives aided with abstract interpretation, it was possible to reduce the verification cost of about 70% [18]. The recent experiments with formal verification have shown that this cost can be further reduced by 50-66% [16].

The main drawback encountered in introducing code generation has been the *size* and overall *complexity* of the resulting software. Though these aspects were not complicating the verification activities, they posed challenges from the performance point of view.

ATP systems do not have hard real-time constraints, however they are reactive systems that, might a failure occur, shall activate the brakes in a limited amount of time in order to reach the safe state. The reaction time is influenced by the main execution time, which resulted four times higher in the first experiments. In the discussed case, the hardware upgrade actually solved the problem. However, with the design of new, more complex systems, this issue has to be taken into account while defining the hardware architecture.

The hardware designer shall consider that the code is larger in size, and there is less flexibility in terms of optimizations at source level (we recall that optimizations at compiler level are not recommended for the development of safety-critical systems): when designing the platform, a larger amount of memory has to be planned if one wants to employ code generation.

Though consistent cost improvements have been achieved on the verification activities, manual test definition is still the bottleneck of the process, requiring about 60-70% of the whole unit-level verification cost.

Preliminary experiments with formal verification applied at unit-level have shown that this technology might considerably reduce the verification cost for the majority of the requirements. However, further analysis is required before introducing formal verification as part of the process.

Some lessons have been learned also from the *knowledge transfer* point of view. The research activity has been performed according to the following research management model.

On one side there is a research assistant who comes from the university and is fully focused on the technology to be introduced. On the other side there is an internal development team, which puts the research into practice on real projects when the exploratory studies are successful.

The results obtained across these four years would have not been possible through intermittent collaborations only. Moreover, they would have been hardly achieved if just an internal person would have been in charge of the research. In order to separate the research from the time-to-market issues, the independence of the research assistant from the development team has to be preserved. Large companies can profit from dedicated internal research teams, or even entire research divisions. Instead, medium-size companies often have to employ the same personnel for performing research explorations, which are always needed to stay on the market, and for takeing care of the day-by-day software development. We argue that the research management model adopted in the presented experience, based on an academic researcher independently operating within a company, can be adapted to other medium-size companies with comparable results.

5   Conclusion

The research activity reported in this paper started with the objective of introducing the formal design and code generation technologies within the development process of a railway signalling manufacturer. At the end of the experience, these techniques have radically changed the whole process in terms of design tasks and in terms of verification activities. In particular, formal model-based design has opened the door to model-based testing, has facilitated the adoption of abstract interpretation, and has allowed performing the first successful experiences with formal verification.

This methodology shift required four years and three projects to be defined and consolidated. Most of the implications of the introduction of code generation could not be foreseen at the beginning of the development, but had to be addressed incrementally. This tuning has been facilitated by the flexibility of the toolsuite adopted: given the many toolboxes of Matlab, there was no need to interface the tool with other platforms to perform the required software process tasks (e.g., test definition, tracing of the requirements, document generation)[5].

However, we believe that the success of the experience has been mainly driven by the research management model followed. The presence of an independent reseacher operating within the company has been paramount to ensure that research was performed without pressure, while research results were properly transferred to the engineering team. The experience showed that also for medium-size companies, such as the one considered in this paper, it is possible to perform research when a proper model is adopted.

Research is essential to address the new market requirements. Along with the experience reported here, the company started to enlarge its business, previously focused in Italy, towards foreign countries, such as Sweden, China, Kazakhstan and Brazil, and the introduction of formal model-based development had actually played a relevant role to support this evolution.

The considerations made in this paper have mainly concerned a formal model-based design process based on commercial tools, in a given application domain.

---

[5] The reader can note that most of the tool support referred in this paper comes from a single vendor. It is not at all the intention of the authors to advertise for such vendor. However, we have to note that interfacing with a single vendor is a preferential factor for industry, and in this case has influenced the choice of the tools.

Assuming different tools and different application domains, it is not so immediate that the same considerations still hold. As future work, we are launching the study of a similar development process based on UML-centered tools: in this case the flexibility will not be given by an integrated toolsuite, but by the Unified Modelling notation itself, even if open-source or free tools will be adopted. Different application domains will be addressed as well.

Acknowledgements. The authors thank Daniele Grasso and Gianluca Magnani from General Electric Transportation Systems for the effort put in several of the reported experiences. This work has been partially funded by General Electric Transportation Systems with a grant to University of Florence.

References

1. Adler, R., Schaefer, I., Schuele, T., Vecchié, E.: From Model-Based Design to Formal Verification of Adaptive Embedded Systems. In: Butler, M., Hinchey, M.G., Larrondo-Petrie, M.M. (eds.) ICFEM 2007. LNCS, vol. 4789, pp. 76–95. Springer, Heidelberg (2007)
2. Bacherini, S., Fantechi, A., Tempestini, M., Zingoni, N.: A Story About Formal Methods Adoption by a Railway Signaling Manufacturer. In: Misra, J., Nipkow, T., Karakostas, G. (eds.) FM 2006. LNCS, vol. 4085, pp. 179–189. Springer, Heidelberg (2006)
3. Behm, P., Benoit, P., Faivre, A., Meynadier, J.-M.: Météor: A Successful Application of B in a Large Project. In: Wing, J.M., Woodcock, J. (eds.) FM 1999. LNCS, vol. 1708, pp. 369–387. Springer, Heidelberg (1999)
4. Bochot, T., Virelizier, P., Waeselynck, H., Wiels, V.: Model checking flight control systems: The Airbus experience. In: ICSE Companion, pp. 18–27. IEEE (2009)
5. Bowen, J.P., Hinchey, M.G.: Ten commandments of formal methods. IEEE Computer 28(4), 56–63 (1995)
6. Bowen, J.P., Hinchey, M.G.: Ten commandments of formal methods...ten years later. IEEE Computer 39(1), 40–48 (2006)
7. CENELEC. EN 50128, Railway applications - Communications, signalling and processing systems - Software for railway control and protection systems (2011)
8. Conrad, M.: Testing-based translation validation of generated code in the context of IEC 61508. Formal Methods in System Design 35(3), 389–401 (2009)
9. Cousot, P., Cousot, R.: Abstract interpretation: A unified lattice model for static analysis of programs by construction or approximation of fixpoints. In: POPL, pp. 238–252 (1977)
10. Deutsch, A.: Static verification of dynamic properties. Polyspace Technology, white paper (2004)
11. Dormoy, F.X.: Scade 6: a model based solution for safety critical software development. In: ERTS 2008, pp. 1–9 (2008)
12. dSPACE. Targetlink (December 2011), http://www.dspaceinc.com
13. El-Far, I.K., Whittaker, J.A.: Model-based software testing. Encyclopedia of Software Engineering 1, 825–837 (2002)
14. ETAS. Ascet (December 2011), http://www.etas.com
15. Ferrari, A., Fantechi, A., Bacherini, S., Zingoni, N.: Modeling guidelines for code generation in the railway signaling context. In: NFM 2009, pp. 166–170 (2009)

16. Ferrari, A., Grasso, D., Magnani, G., Fantechi, A., Tempestini, M.: The Metrô Rio ATP Case Study. In: Kowalewski, S., Roveri, M. (eds.) FMICS 2010. LNCS, vol. 6371, pp. 1–16. Springer, Heidelberg (2010); journal special issue (to appear, 2012)

17. Ferrari, A., Grasso, D., Magnani, G., Fantechi, A., Tempestini, M.: The Metrô Rio ATP Case Study. In: Kowalewski, S., Roveri, M. (eds.) FMICS 2010. LNCS, vol. 6371, pp. 1–16. Springer, Heidelberg (2010)

18. Ferrari, A., Magnani, G., Grasso, D., Fantechi, A., Tempestini, M.: Adoption of model-based testing and abstract interpretation by a railway signalling manufacturer. IJERTCS 2(2), 42–61 (2011)

19. Harel, D.: Statecharts: A visual formalism for complex systems. Science of Computer Programming 8(3), 231–274 (1987)

20. Hinchey, M.G., Bowen, J.: Applications of formal methods. Prentice-Hall (1995)

21. Huber, F., Schätz, B., Schmidt, A., Spies, K.: Autofocus: A Tool for Distributed Systems Specification. In: Jonsson, B., Parrow, J. (eds.) FTRTFT 1996. LNCS, vol. 1135, pp. 467–470. Springer, Heidelberg (1996)

22. IEC. IEC-61508, Functional safety of electrical/electronic/programmable electronic safety-related systems (April 2010)

23. INRIA. Scicos: Block diagram modeler/simulator (December 2011), http://www.scicos.org/

24. MAAB. Control algorithm modeling guidelines using Matlab, Simulink and Stateflow, version 2.0 (2007)

25. MathWorks. MathWorks products and services (December 2011), http://www.mathworks.com/products/

26. Meenakshi, B., Bhatnagar, A., Roy, S.: Tool for Translating Simulink Models into Input Language of a Model Checker. In: Liu, Z., Kleinberg, R.D. (eds.) ICFEM 2006. LNCS, vol. 4260, pp. 606–620. Springer, Heidelberg (2006)

27. Miller, S.P., Whalen, M.W., Cofer, D.D.: Software model checking takes off. Commun. ACM 53(2), 58–64 (2010)

28. MISRA. Guidelines for the use of the C language in critical systems (October 2004)

29. Mohagheghi, P., Dehlen, V.: Where is the Proof? - A Review of Experiences from Applying MDE in Industry. In: Schieferdecker, I., Hartman, A. (eds.) ECMDA-FA 2008. LNCS, vol. 5095, pp. 432–443. Springer, Heidelberg (2008)

30. RTCA. DO-178B, Software considerations in airborne systems and equipment certification (December 1992)

31. Scaife, N., Sofronis, C., Caspi, P., Tripakis, S., Maraninchi, F.: Defining and translating a "safe" subset of Simulink/Stateflow into Lustre. In: EMSOFT, pp. 259–268. ACM (2004)

32. Selic, B.: The pragmatics of model-driven development. IEEE Software 20(5), 19–25 (2003)

33. Tretmans, J., Wijbrans, K., Chaudron, M.R.V.: Software engineering with formal methods: the development of a storm surge barrier control system revisiting seven myths of formal methods. Formal Methods in System Design 19(2), 195–215 (2001)

34. Vouk, M.A.: Back-to-back testing. Inf. Softw. Technol. 32, 34–45 (1990)

35. Woodcock, J., Larsen, P.G., Bicarregui, J., Fitzgerald, J.S.: Formal methods: Practice and experience. ACM Comput. Surv. 41(4) (2009)

Symbolic Execution of Communicating and Hierarchically Composed UML-RT State Machines

Karolina Zurowska and Juergen Dingel

School of Computing
Queen's University
Kingston, ON, Canada
{zurowska,dingel}@cs.queensu.ca

Abstract. The paper introduces a technique to symbolically execute hierarchically composed models based on communicating state machines. The technique is modular and starts with non-composite models, which are symbolically executed. The results of the execution, symbolic execution trees, are then composed according to the communication topology. The composite symbolic execution trees may be composed further reflecting hierarchical structure of the analyzed model. The technique supports reuse, meaning that already generated symbolic execution trees, composite or not, are used any time they are required in the composition. For illustration, the technique is applied to analyze UML-RT models and the paper shows several analyses options such as reachability checking or test case generation. The presentation of the technique is formal, but we also report on the implementation and we present some experimental results.

1   Introduction

This paper is concerned with the analysis of models of reactive systems in the context of Model-Driven Development (MDD). In MDD, development is centered around the creation and successive refinement of models until code can be generated from them automatically. MDD has been used in different domains, but has been most successful for the development of reactive systems. Several MDD tools exist including IBM Rational®Software Architect - Real Time Edition (IBM RSA RTE)[1] [2] and IBM Rational Rhapsody® [1] and Scade Suite [3] from Esterel Technologies. However, more research into MDD is needed, e.g., to determine how to best support MDD with suitable model analyses.

The use of symbolic execution for model analysis has already been suggested. For instance, the work in [18,8,23,5] considers state machines, one of the most important model types in the MDD of reactive systems. In this work we extend the use of symbolic execution from individual state machines [26] to collections of communicating and possibly hierarchically composed state machines. More

---

[1] IBM, Rational and Rhapsody are trademarks of International Business Machines Corporation, registered in many jurisdictions worldwide.

A. Goodloe and S. Person (Eds.): NFM 2012, LNCS 7226, pp. 39–53, 2012.

precisely, we are interested in models that are structured as shown in Figure 1. Every model has a state machine defining its (top-level) behavior; models may contain submodels (called *parts*) along with *ports* and *connectors* that define how models may communicate with each other.

Using a particular MDD technique (UML-RT [21] introduced in Section 2), this paper presents an approach for the symbolic execution of the composite models (in Section 3), shows how the resulting execution trees can be used for analysis, and briefly describes a prototype implementation and its use on three sample models (in Section 4). In our approach, symbolic execution proceeds recursively over the structure of the model shown in Figure 1. Symbolic execution trees for non-composite models are obtained as described in [26], while symbolic execution trees for composite models are obtained by composing the symbolic execution trees of their parts in such a way that their communication topology and thus their ability to asynchronously exchange messages (signals) is fully respected. The compositional nature of the execution not only avoids any structure-destroying "flattening" operation, but also allows leveraging the repeated occurrence of any parts of the model: analysis is sped up by storing and reusing execution trees of repeated parts. Although our formalization, implementation, and examples target UML-RT, the approach should also be applicable to other modeling languages as long as a tool for the symbolic execution of non-composite models is available.

While some work on the symbolic execution of models exists (also shown in Section 5), symbolic execution of programs and source code is much more thoroughly researched. For instance, recent work has produced many different versions of symbolic program execution and some even have proved successful in industrial contexts [19]. Modular approaches have also been suggested to deal with, e.g., calls to external modules [14,24], concurrency [9,12], or distribution [13]. Of these approaches, only the last is based on the combination of symbolic execution trees.

2   UML-RT Models and Their Symbolic Execution

2.1   Overview of the UML-RT Modeling Language

The UML-RT modeling language is used to model real-time and embedded systems [21] and is one of UML 2 [4] profiles. The development of UML-RT models is supported by IBM RSA RTE [2], which also allows automatic code generation.

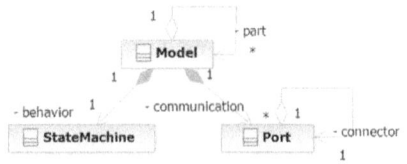

Fig. 1. Hierarchical organization of models

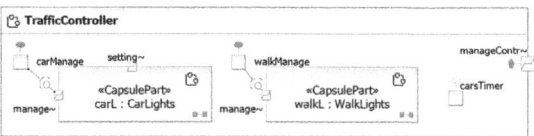

(a) UML-RT structure diagram

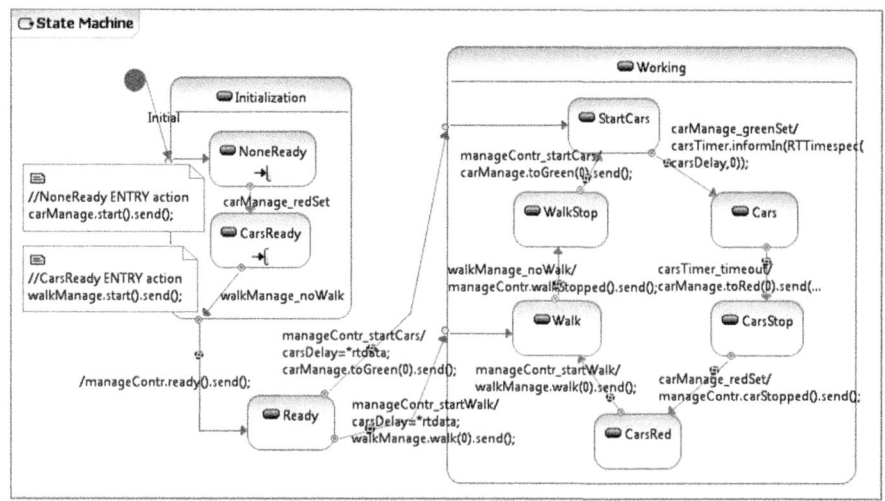

(b) UML-RT State Machine

Fig. 2. Capsule `TrafficController` with its structure (a) and state machine (b). Labels of transitions are of the forms: `port_signal` or `port_signal/action_code`

A UML-RT model consists of *capsules*. A capsule (referred as a model in Figure 1) is an entity which communicates with other capsules only by sending and receiving signals and only through its *ports*. Each port has a type specified with a *protocol*, which identifies signals sent or received by implementing ports. Ports may be connected with *connectors* and connected ports must implement the same protocol. Capsules are organized hierarchically and each capsule may contain a number of instances of other capsules, called *parts*.

The behavior of a capsule is specified with UML-RT State Machines [21]. The state machines in UML-RT are a special case of UML 2 State Machines [4] with some simplifications (e.g., no orthogonal states) and some additional refinements (e.g., to support executability). A UML-RT State Machine has hierarchical states and guarded transitions, which are triggered by signals received on ports; action code is used to send signals (`send()` method), to set timers (`informIn()` method) and to get values of input parameters of signals (using `*rtdata`).

Example 1. Figure 2 presents the structure and the behavior of the capsule `TrafficController`, which models a control system of traffic lights. Figure 2(a) shows the structure of the capsule, which consists of two parts: `carL` (instance of

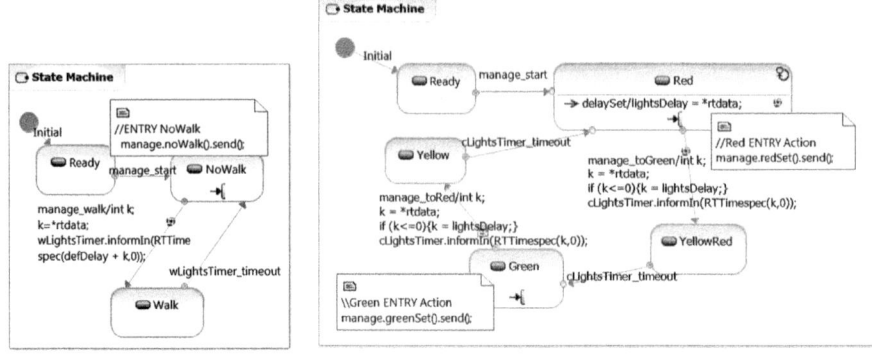

(a) `WalkLights` state machine

(b) `CarLights` state machine

Fig. 3. UML-RT State Machines for parts in capsule `TrafficController` in Figure 2 (labels of transitions as in Figure 2)

capsule `CarLights`) and `walkL` (instance of capsule `WalkLights`) that are responsible for car and pedestrian lights, respectively. Additionally, the capsule has protected ports (`carManage` and `walkManage` that connect it to the parts), an external port `manageContr` and a timer `carsTimer`. The behavior of `TrafficController` is shown in Figure 2(b). After the `default` transition, the `carL` part is initialized in the entry code of the `NoneReady` state by sending the `start()` signal over the `carManage` port. This signal is received through the connected port `manage` in the `CarLights` capsule (Figure 3(b)).

The behaviors of `CarLights` and `WalkLights` capsules are given in Figure 3. The internal structure of these capsules (not shown) is straightforward including only timers (`cLightsTimer` and `wLightsTimer`).

2.2   Symbolic Execution of UML-RT Capsules

Symbolic execution of UML-RT models, such as the one in Example 1, can be performed for a model as a whole or it can use the inherent modularity of the models. In this paper we use the latter approach, which goes beyond the analysis of non-composite state machines (as in [8,23,5]). Our approach is outlined in Algorithm 1. Initially, the UML-RT State Machine of a capsule is symbolically executed producing symbolic execution tree SET_C. If the capsule is non-composite, then SET_C (and the updated map *Trees*) is the result (line 4). However, if the capsule contains parts then the tree for each capsule C_i of the part must be provided. Such tree is either retrieved from the *Trees* map (line 8) or if it is the first occurrence of the capsule, the capsule is symbolically executed in a recursive call (line 9). When all trees are gathered then the communication topology in the capsule is determined (line 11) and trees are composed (line 12). The algorithm returns a tree *CSET* and an updated map of all trees

---

Algorithm 1. Symbolic execution of a UML-RT capsule

Require: a capsule C with a set *parts* of n parts and a state machine SM_C
Require: a map *Trees* : $Capsules \rightarrow SET$
Ensure: $CSET = \mathcal{SET}(SET_C, SET_1, ..., SET_n)$
Ensure: *Trees* is a compued map $Capsules \rightarrow SET$
    $SET_C \leftarrow$ perform symbolic execution of SM_C
    if n=0 **then**
3:      $Trees[C \leftarrow SET_C]$
      return SET_C, *Trees*
    else
6:      **for all** $part_i \in parts$ **do**
        $C_i \leftarrow$ capsule of $part_i$
        $SET_i \leftarrow Trees(C_i)$
9:        **if** SET_i is not computed **then**
          $SET_i \leftarrow$ perform symbolic execution of C_i using *Trees*
        $conn \leftarrow$ get connectors between ports from the structure of C
12:     $CSET \leftarrow compose_{conn}(SET_C, SET'_1, ..., SET'_N)$
      $Trees[C \leftarrow CSET]$
      return $CSET$, *Trees*

Trees. The composed tree *CSET* satisfies the requirements of the composition $\mathcal{SET}(SET_C, SET_1, ..., SET_n)$ as stated in Definition 6. In the algorithm nested capsules are dealt with by the composition operation, the communication topology is represented by the relation *conn* (both explained in Section 3) and reuse of symbolic execution trees is achieved via the *Trees* map.

The following section provides the details of composing trees, that is, line 12 in Algorithm 1. The details of the method to symbolically execute UML-RT State Machines (line 1) is presented elsewhere [26] and is omitted here. Moreover, the following sections do not aim to provide formal semantics of the UML-RT modeling language, since it is out of scope for the work presented here.

3   Symbolic Execution of Communicating State Machines and Their Hierarchies

In this section we define the symbolic execution tree for a state machine, then we introduce composite trees.

3.1   Symbolic Execution Tree (SET) for a State Machine

We assume here that a state machine has states (called *locations*) and each transition between states is labeled with an *input action* and a sequence of *output actions*. Both types of actions can have variables associated with them. Additionally, each transition may be guarded and may update values of *attributes*, that is, variables accessible during the entire execution of a state machine.

During symbolic execution, variables (inputs, outputs or attributes) are mapped to *symbolic values*, which are first-order logic terms [22] that involve the operators available for the type of variable. For a set of variables we define a symbolic valuation as in Definition 1.

Definition 1. For a set of variables $X = \{X_1, ..., X_N\}$ a *symbolic valuation* is a function $val^s : X \to \Phi$ that maps each variable X_i to a term Φ_i that has the type of X_i. The set of all possible valuations of X is denoted with $Val^s[X]$.

Beside the symbolic valuation of variables, we use first-order formulas as path constraints PC. These constraints need to be satisfied for a particular execution path in a state machine to be feasible.

Definition 2 (Symbolic Execution Tree SET). Let SM be a state machine with locations (states) L, input and output actions A_I and A_O with their variables AV_I and AV_O and with attributes A. For such a state machine its *symbolic execution tree* is a tuple $\mathcal{SET}(SM) = (S^s, InVars^s, Tr^s, s_0^s, \prec^s)$, where:

- S^s is a set of *symbolic states*, and each state is a tuple containing a location, a valuation of attributes A and a set of path constraints, that is, $S^s \subseteq (L \times Val^s[A] \times \mathbb{P}(PC))$ ($\mathbb{P}$ denotes the powerset operation). For a symbolic state $s = (l, val, pc)$ the following projection is defined: $pc(s) = pc$,
- $InVars^s$ is a set of *mappings* that assign a unique variable (different from all other variables) to each input variable of some input action $a_i \in A_I$. This variable is a symbolic value that represents the input in the symbolic execution. Each mapping is a special kind of symbolic valuation, i.e., $InVars^s \subseteq Val^s[AV_I']$ with $AV_I' \subseteq AV_I$. We will denote all symbolic values (variables) from $InVars^s$ with $sv(InVars^s) = \{v_i^s : \exists iv \in InVars^s. \exists v_i \in AV_I : iv(v_i) = v_i^s\}$. For a mapping $iv \in InVars^s$ let iv^{-1} be the inverse mapping, i.e., the one that assigns an input variable to a symbolic value (variable),
- Tr^s is a *transition relation* $Tr^s \subseteq (S^s \times A_I \times InVars^s \times Seq(A_O \times Val^s[AV_O']) \times S^s)$, where $AV_O' \subseteq AV_O$, Seq is a possibly empty sequence of pairs (a_o, val_o), with $a_o \in A_O$ an output action and val_o a symbolic valuation of the variables in a_o. Each transition $(s, a_i, val_i, seq, s') \in Tr^s$ is obtained from a transition in SM,
- $s_0^s \in S^s$ is the initial symbolic state $(l_0, v_0, \emptyset)$ with l_0 being an initial location and v_0 an initial valuation in SM,
- $\prec^s$ is a *subsumption relation* $\prec^s \subseteq (S^s \times S^s)$. For $s_1 = (l_1, v_1^s, pc_1)$ and $s_2 = (l_2, v_2^s, pc_2)$ we say that s_2 subsumes s_1, $(s_1, s_2) \in \prec^s$, if both have the same location $l_1 = l_2$ and the same symbolic valuation $v_1^s = v_2^s$ and all path constraints of s_2 are included in those of s_1, i.e., $pc_2 \subseteq pc_1$.

The details of generating a SET for a UML-RT State Machine are in [26]. In the technique presented there we used the following assumptions:

- action code in transitions and locations is represented in a compact way with functions, which result from symbolic execution of code,
- loops in the action code are executed to some predefined bounds, therefore for models with loops the resulting SET might not be exhaustive,
- a subsumption relation $\prec^s$ is used to make the SET finite, even in the presence of cyclic behavior in the UML-RT State Machines.

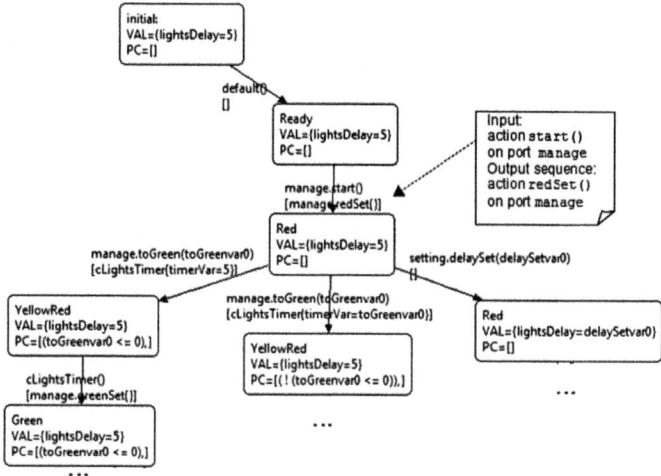

Fig. 4. Part of the symbolic execution tree ($SET_{CarLights}$) for a state machine in Figure 3(b)

Example 2. Figure 4 presents a part of a symbolic execution tree for the UML-RT State Machine given in Figure 3(b). After a default action there is a symbolic state for the location Ready and then, after receiving the start signal, for the location Red. From this state two actions are possible: delaySet, which changes the attribute lightsDelay, and toGreen. With the second action a symbolic input variable toGreenvar0 is received. If it is greater than zero, then it is used as a delay for yellow-red lights (setting the timer cLightsTimer).

3.2   Composing Symbolic Execution Trees

The goal of composing symbolic execution trees is to symbolically represent an execution of models structured as in Figure 1. Such models, beside the behavioral specification given with a state machine, contain a set of *parts P*, a function *p* that maps those parts to a model (which indicates the type of a part) and a connection relation (in Definition 3). Due to space limitations the formalization of this notion of models is omitted here, but we introduce the concepts necessary to show their symbolic execution.

In order to represent connections between parts of a model, a relation *conn* is defined. To avoid unnecessary details, mappings between variables of input and output actions that a connection gives rise to are not presented (it is assumed that such mappings are inferred from the relation between actions).

Definition 3 (Connection relation). Let $P = \{P_1, ..., P_n\}$ be a set of parts in some model M. Each part $P_i \in P$ is mapped to a model with a state machine SM_i, which has sets of actions A_{Ii}, A_{Oi} for $i = 0, 1, ..., n$ (with SM_0 being a state machine of M). A *connection* is a relation $conn \subseteq (\bigcup_i A_{Oi}) \times ((\bigcup_i A_{Ii}) \times P \cup \{\sigma\})$, where σ denotes „self". We say that an output and an input action are *connected* $(a_1, (a_2, p)) \in conn$ iff $a_1 \in A_{Ok} \Rightarrow (a_2 \in A_{Il} \wedge (p = P_l \vee p = \sigma) \wedge k \neq l)$ for some $k, l \in \{0, 1, ..., n\}$. *conn* is a one-to-one partial function, hence, an action can be connected to at most one other action. We will define a set *opn* (abbreviates "open") to include all input actions that are not targeted in the relation *conn*. That is, $opn = \{a \in \bigcup_i A_{Ii} : \forall (a_1, (a_2, p)) \in conn : a_2 \neq a\}$.

Queues are used to store actions waiting to be handled by the appropriate model. Hence, a queue is a sequence of tuples consisting of an input action, a symbolic valuation of its input variables, which represents symbolic values received with a connected output action, and a receiving tree.

Definition 4 (Queue). Let P^+ be a set of parts of some model M and recursively of parts of those parts and so on. Let A_I be the set of input actions in all state machines and AV_I be the set of their input variables. A *queue* is a sequence of triples (a_i, val, p), where $a_i \in A_I$, $val \in Val^s[AV_I']$ with $AV_I' \subseteq AV_I$, and $p \in P^+$. Let the set of all possible queues be denoted by *Queues*. For $q \in Queues$, we assume operations to add: $q' = enq(q, (ai, val, p))$ and to remove: $(q', (ai, val, p)) = deq(q)$ an element from a queue.

A state of an execution of the model structured as illustrated in Figure 1 includes an execution state of its state machine along with execution states of all its parts. To this end, Definition 5 defines a composite symbolic execution state to contain the current execution state of the state machine with the queue and a function that maps parts to their respective execution states. In the definition we use a replacement operator $[s|val]$ for a symbolic state s and a symbolic valuation $val \in Val^s[X]$. It returns a new symbolic state in which all occurrences (in path constraints and in a symbolic valuation of attributes) of a variable $x \in X$ are replaced with its symbolic value $val(x)$. We assume that the operator is also defined for formulas.

Definition 5 (Composite Symbolic State). Let M be a model with a set of parts P and let $SET = (S^s, InVars^s, Tr^s, s_0^s, \prec^s)$ be the symbolic execution tree of its state machine SM. A *composite symbolic state* is a tuple (s, q, sm), where:

- $s \in S^{s*}$ is the current symbolic execution state of SM, which either occurs explicitly in S^s or is a state s^* obtained by replacing some input variables with their received symbolic values. That is, $S^{s*} = S^s \cup \{s^* : s^* = [s|val] \wedge s \in S^s \wedge val \in Val^s[AV_I^s]\}$,
- q is the current contents of the queue for the model M,
- sm is a mapping: $sm : P \rightarrow \mathbb{S}^s$ where $\mathbb{S}^s$ are composite symbolic states defined for models of parts.

For each composite symbolic state (s, q, sm) we assume that the union of the path constraints in s and all states referred to in sm is not contradictory. So the formula $pc(s) \wedge \bigwedge_i pc(sm(SET_i))$ is satisfiable. Additionally, we will define an extended mapping $\bar{sm}(p)$ defined over $P \cup \{\sigma\}$ that returns (s, q, sm) if $p = \sigma$ and $sm(p)$ if $p \in P$.

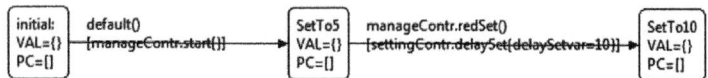

(a) A symbolic execution tree of the state machine of the `Controller` capsule (with two states `SetTo5`, `SetTo10` and the send action on the transition between states, which updates `lightsDelay` of `CarLights`)

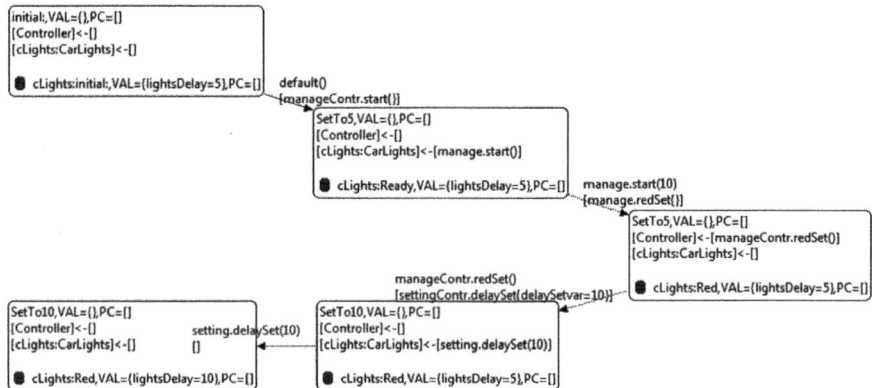

(b) A composite symbolic execution tree for the `Controller` capsule. In each state, the currently active state in $SET_{Controller}$ is given in the first line, lines 2 and 3 show the queues obtained from states, line 4 shows the active state of the included part `cLights`.

Fig. 5. A composition of symbolic execution trees

Example 3. Figure 5(b) shows composite symbolic states for the `Controller` capsule (SET of its state machine is in Figure 5(a)). The second state contains: a current state $s = (\text{SetTo5}, \emptyset, \emptyset)$, the empty queue $q = \emptyset$, and the mapping sm of its only part $sm(\text{cLights}) = ((\text{Ready}, \text{lightsDelay}=5, \emptyset), [(\text{manage.start}(),\emptyset, \text{cLights})], \emptyset)$. In this composite symbolic state both included symbolic states are taken from SETs without any changes.

In a given composite state, when a transition from a tree of some part is taken, it may contain a sequence of output actions. These actions must be placed in appropriate queues. To achieve this, an update operation snd is defined, which takes a composite state along with an output sequence and returns a new composite state, that is: $cs' = \text{snd}(cs, seq)$. The new composite state cs' results from taking each pair (a_o, v_o) in seq and if a_o is connected, i.e., $(a_o, (a_i, p)) \in conn$, updating the receiving queue $q(sm(p)) = \text{enq}(q(sm(p)), (a_i, v_i, p))$. The valuation v_i is obtained from a valuation v_o by mapping output variables to their input counterparts. If there is no connection for a_o, then $cs' = cs$.

Example 4. Consider the transition between the first and the second composite symbolic state in Figure 5(b) (denoted with cs and cs', respectively). Initially, queues in cs are empty. We apply $cs' = \text{snd}(cs, [(\text{manageContr.start}, \emptyset)])$. The action `manageContr.start()` is connected to the input action `manage.start()`, that is, in *conn* there is a pair: $(\texttt{manageContr.start()}, (\texttt{manage.start()}, \texttt{cLights}))$, so the receiving part is `cLights`. Therefore, the new composite state cs' contains a queue for `cLights` with contents: $[(\texttt{manage.start()}, \emptyset, \texttt{cLights})]$.

A symbolic execution tree for communicating state machines has states as in Definition 5 and transitions that are derived from the included trees as given by the rules below.

Definition 6 (Composite Symbolic Execution Tree CSET)
Let $P = \{P_1, ..., P_n\}$ be parts of some model M and let SM be its state machine. The symbolic execution tree of SM is given as SET_0 and symbolic execution trees of parts are $SET_1, ..., SET_n$. Let $SET = \{SET_0, SET_1, ..., SET_n\}$ and *conn* be the connection relation. A *composite symbolic execution tree* of M is a tuple $\mathcal{CSET}(M) = (S^{cs}, SVars^{cs}, Tr^{cs}, sc_0^{cs}, \prec^{cs})$, where:

- S^{cs} is a set of composite symbolic states (see Definition 5),
- $SVars^{cs} \subseteq Val^s[AV_I^s]$ is a set of symbolic valuations assigning symbolic values (received with output actions) to variables, which are used in some SET_i to symbolically represent input values, so $AV_I^s \subseteq \bigcup_i sv(InVars_i^s)$,
- Tr^{cs} is a transition relation $Tr^{cs} \subseteq (S^{cs} \times (A_I \cup Un(A_I)) \times SVars^{cs} \times Seq(A_O \times Val^s[AV_O'])) \times S^{cs}$, where A_I are all input actions, A_O are all output actions, $AV_O' \subseteq AV_O$ and $Un(A_I)$ is a set that contains $un_a \in Un(A_I)$ for each $a \in A_I$ (un_a indicates that the action a has not been received),
- $sc_0^{cs} \in S^{cs}$ is an initial composite symbolic state $sc_0^{cs} = (s_0, q_0, sm_0)$, where s_0 is an initial state in SET_0, q_0 is an empty queue and sm_0 is a function such that $sm_0(P_i) = s_{0i}^s$ for all SET_i,
- $\prec^{cs} \subseteq (S^{cs} \times S^{cs})$ is a subsumption relation between composite symbolic states. Composite symbolic states subsume one another, if they have the same symbolic states and their queues have the same contents. That is, for $cs_1 = (s_1, q_1, sm_1)$ and $cs_2 = (s_2, q_2, sm_2)$ we have $(cs_1, cs_2) \in \prec^{cs}$ iff $s_1 \prec_0^S s_2$ and q_1 is the same as q_2 and $\forall k = 1...n : sm_1(P_k) \prec_k^s sm_2(P_k)$.

The transition relation Tr^{cs} is defined by the following rules:
1) *Matching* - applies if a queue q (or a queue for some part $p \in P$) has the first element (a, v, σ) (or (a, v, p)) and if the currently active state s (or $sm(p)$) has an outgoing transition with the action a and output o. In this case the sequence of actions o is used to update queues and the valuation v received with the input action is used in a replace operation on the target state cs'' of the transition. According to Definition 5 path constraints in cs'' must not be contradictory.

$$\exists p \in P \cup \{\sigma\} : ((q', (a, v, p)) = deq(q(s\bar{m}(p))) \wedge (s_p, q_p, sm_p) = s\bar{m}(p) \wedge$$

$$\exists (s_p, a, iv, o, s_p') \in Tr_p^s : v' = (iv^{-1} \circ v) \wedge s\bar{m}'(p) = ([s_p'|v'], q', sm_p) \wedge s\bar{m}''(p) = snd(s\bar{m}'(p), o)$$

$$(s, q, sm) \xrightarrow{a, v'; o} (s'', q'', sm'')$$

where $iv^{-1} \circ v$ is the function that maps input symbolic variables (from a mapping iv) to the symbolic values (in a valuation v) received with an output action.

2) *Unreceived* - applies if a queue q (or a queue for some part p) has the first element (a, v, σ) (or (a, v, p)) and either there is no outgoing transition with the action a or a new composite state cannot be created due to contradictory path constraints. In this case, the above element is removed from the queue, the state and the state mapping remains the same and the action is marked as unreceived using un_a.

$$\exists p \in P \cup \{\sigma\} : ((q'_p, (a, v, p)) = \text{deq}(q(s\bar{m}(p))) \wedge (s_p, q_p, sm_p) = s\bar{m}(p) \wedge$$
$$\neg \exists (s_p, a, iv, o, s'_p) \in Tr^s_p : (v' = (iv^{-1} \circ v) \wedge s\bar{m}'(p) = ([s'_p|v'], q'_p, sm_p)))$$
$$(s, q, sm) \xrightarrow{un_a, v', \emptyset} (s, q', sm')$$

3) *Open* - applies if for the current state s (or some part p and its current state $sm(p)$) there is an outgoing transition with an input action in the *opn* set as in Definition 3 (in the UML-RT models this means that a signal is received on an external port, such as manageContr in Figure 2). In this case, the state is updated and the queues are updated only to account for output actions.

$$\exists p \in P \cup \{\sigma\} : (s_p, q_p, sm_p) = s\bar{m}(p) \wedge \exists (s_p, a, v, o, s'_p) \in Tr^s_p : (a \in opn \wedge s\bar{m}'(p) = \text{snd}(s\bar{m}(p), o))$$
$$(s, q, sm) \xrightarrow{a_o, v, o} (s', q', sm')$$

The definition of CSET is similar to the definition of non-composite trees, so we can extend the definition such that both types of trees are composed.

Example 5. Figure 5(b) presents an example of a composite symbolic execution tree composed from the trees in Figure 4 and Figure 5(a). It is assumed that the trees have separate queues and all actions are connected (i.e., *opn* is empty).

4   Implementation and Case Studies

4.1   Implementation

The prototype implementation[2] [25] follows Algorithm 1, in which the composition of trees (line 12) is performed according to the method outlined in Section 3.2, while non-composite capsules are symbolically executed using the prototype discussed in [26]. The implementation is an extension of the IBM RSA RTE [2] tool.

The prototype (just like the theory shown in Section 3) makes several assumptions. Among those assumptions the most important one is the exclusion of dynamic bindings and dynamic instantiations of capsules. The first feature allows the dynamic creation and destruction of connectors between ports. To support dynamic bindings in the technique, the relation *conn* must be also made dynamic. Dynamic instantiation enables the creation and destruction of parts of a given capsule at runtime. The support of this feature would require usage of 'create' and 'destroy' signals that would activate and deactivate symbolic execution trees as required. Our symbolic execution inherits also the limitations of

---

[2] Version 0.1.0 available at http://cs.queensu.ca/~mase/software.html

the symbolic execution of non-composite described in [26]. These limitations lie mostly in the supported action language: we currently assume that this language is based on the subset of C++ with assignments, if and while statements. We also require that the UML-RT State Machines do not include history states.

4.2   Using Symbolic Execution Trees for Analysis

The composite symbolic execution tree generated for `TrafficController` introduced in Example 1 has 25 composite states. Besides representing all possible execution paths, the tree has been used to perform the analyses that are similar to the ones possible for non-composite models introduced in [26]:

1. *Invariants checking* checks whether the path constraints provide sufficient restrictions to satisfy invariants that relate attributes of a capsule.
2. *Actions analysis* makes it possible to check which outputs are generated and which values output variables have.
3. *Reachability analysis* is based on the reachable locations of included state machines in the generated symbolic execution tree. Firstly, it can be determined which locations of all state machines are not reachable. Secondly, it can be checked whether a given configuration of parts and their locations is reachable from some other configuration.
4. A symbolic execution tree can be also used to *generate test cases*, which are all paths that lead to leaves in the tree and which have concrete values for variables, such that they satisfy the path constraints.

Besides analyses similar to the ones possible for non-composite symbolic execution trees, there are analyses specific for composite trees. For instance:

1. *Tree projection* allows extracting from the composite tree states and transitions that are from a given part. For instance, the composite tree of the `TrafficController` capsule can be projected to the part `cLights`. Such a projected tree has only 9 states (as opposed to 29 in the unconstrained tree in Figure 4), which illustrates how the communication topology restricts the interactions between capsules to the ones allowed by other capsules.
2. *Unreceived actions* analysis makes it possible to show paths to transitions with actions that are unreceived. In this way it can be checked that all important signals are not dropped.

4.3   Case Studies

In order to evaluate the presented method and its implementation we experimented with 3 UML-RT models using two scenarios. In the first scenario, called "modular", each capsule was executed only once even if it was used as a part more than once. In the second scenario, "non-modular", each capsule was executed every time it has been encountered as a part.

Table 1 gathers the results of the experiments (performed on a standard PC with 4GB of RAM and Intel Core i7 CPU, 2.93 GHz) with 3 UML-RT models.

The first one (traffic lights controller) is presented in Figure 2. The second model (Intersection controller) builds on the first one and combines 2 traffic lights in an intersection of two streets. The last model (Street controller), combines 2 intersections to represent their sequence on a street (e.g., to synchronize green lights). The analysis of other models is reported in [26,25].

Table 1. Performance of generating composite symbolic execution trees (CSET) and test cases (TC)

UML-RT model	number of states in CSET	Generation time in seconds			JVM memory usage in MB	
		modular	non modul.	TCs	modular	non modul.
Traffic lights controller	25	0.38	0.31	0.01	133	134
Intersection controller	287	0.43	0.67	0.01	133	150
Street controller	78338	133.5	157.7	3.9	564	773

The results of the experiments given in Table 1 show that the increase in complexity due to the parallel and hierarchical combination of capsules is substantial. For the street controller model, in which capsules of intersections (each of which contains two traffic controllers) may behave independently, the composite symbolic execution, as well as time to generate it, is quite large.

Table 1 also compares the modular and non-modular scenarios. In the case of the first model all capsules are used only once so there is no gain from the modular approach. However, in the last model there are a total of 15 parts and only 5 different capsules and in this case there is 18% less time required. The difference is not larger, because even in the modular case symbolic execution trees must be copied (line 11 in Algorithm 1). The memory usage in both cases is comparable and is only the result of more complex processing of the non-modular approach.

5   Related Work

Following the traditional approach [17], there has recently been much interest in the area of symbolic execution of programs, usually aiming at test case generation [19]. In some of these works the emphasis is put on the generation of test cases that cover all reachable code, especially for complex data structures in Java [6]. In another popular approach, dynamic (concrete) testing is combined with static (symbolic) execution [16].

Symbolic execution has also been explored as the analysis method for models. For example, symbolic execution trees have been used for model checking of reactive models IOSTSs (Input Output Transition Systems) [20]. In another approach, symbolic execution forms the basis for proving correctness for Statecharts [23] and UML State Machines [8]. Symbolic execution for models with timing constraints has been proposed for Modecharts [18]. An approach incorporating ideas from [15] has been presented also for Simulink/Stateflow models [5]. As opposed to our work the above approaches deal with non-composite (atomic) models only.

Compositional methods to symbolic execution have been proposed for source code to deal with module calls. For instance in [14,7] functions are represented using summaries that are logical formulas. Compositions are also considered for parallel systems. In [9,12] it is shown how to formally prove the correctness of parallel programs, but symbolic execution is performed globally for all components. The inclusion of communication enables a more modular approach. In [13], which is the most similar to the work presented here, symbolic execution of hierarchically composed modules is proposed. Modules are represented with reusable designs, which communicate through typed ports. Although close, there are several differences to our work. Firstly, we use state machines to specify behavior as opposed to the high-level language used in [13]. Secondly, we support asynchronous communication, whereas in [13] synchronous communication is assumed. Finally, our work reports also on a prototype implementation.

The work presented in this paper does not aim at formalization of the UML-RT language, which is treated in other works (e.g. [11,10]). Finally, we note that the current model analysis capabilities of the IBM RSA RTE [2] are quite limited and do not support the analyses enabled by the symbolic execution.

6   Conclusions

This paper presents a modular technique to symbolically execute UML-RT models. The technique combines symbolic execution trees of components contained in such models based on their communication topology. In this way it is possible to analyze complex and hierarchical structures of UML-RT capsules. The method is formally introduced and the algorithm to build composite symbolic execution trees is presented. The paper also includes some details of the prototype implementation. Using a running example we show, how the composite trees can be used to verify some properties of UML-RT models.

In future work we will use abstraction techniques during the computation of symbolic execution tree to improve the applicability of our approach to large models, without compromising the usability of the trees for analysis. Moreover, we are investigating more general reasoning techniques for symbolic execution trees based on ,e.g., temporal logic.

Acknowledgment. Authors wish to acknowledge the support of NSERC, IBM Canada, and Malina Software.

References

1. IBM Rational Rhapsody Architect, Version 7.5.5,
   http://www-01.ibm.com/software/rational/products/rhapsody/swarchitect/
2. IBM Rational Software Architect, RealTime Edition, Version 7.5.5,
   http://publib.boulder.ibm.com/infocenter/rsarthlp/v7r5m1/index.jsp
3. Scade Suite, http://www.esterel-technologies.com/products/scade-suite/

 4. Unified Modeling Language (UML 2.0) Superstructure, http://www.uml.org/
 5. Alur, R., Kanade, A., Ramesh, S., Shashidhar, K.C.: Symbolic analysis for improving simulation coverage of Simulink/Stateflow models. In: EMSOFT 2008 (2008)
 6. Anand, S., Păsăreanu, C., Visser, W.: Symbolic execution with abstraction. Journ. on Soft. Tools for Techn. Transfer 11(1), 53–67 (2009)
 7. Anand, S., Godefroid, P., Tillmann, N.: Demand-Driven Compositional Symbolic Execution. In: Ramakrishnan, C.R., Rehof, J. (eds.) TACAS 2008. LNCS, vol. 4963, pp. 367–381. Springer, Heidelberg (2008)
 8. Balser, M., Bäumler, S., Knapp, A., Reif, W., Thums, A.: Interactive Verification of UML State Machines. In: Davies, J., Schulte, W., Barnett, M. (eds.) ICFEM 2004. LNCS, vol. 3308, pp. 434–448. Springer, Heidelberg (2004)
 9. Balser, M., Duelli, C., Reif, W., Schellhorn, G.: Verifying concurrent systems with symbolic execution. Journal of Logic and Computation 12(4), 549 (2002)
10. von der Beeck, M.: A Formal Semantics of UML-RT. In: Wang, J., Whittle, J., Harel, D., Reggio, G. (eds.) MoDELS 2006. LNCS, vol. 4199, pp. 768–782. Springer, Heidelberg (2006)
11. Burmester, S., Giese, H., Hirsch, M., Schilling, D.: Incremental design and formal verification with UML/RT in the FUJABA real-time tool suite. In: SVERTS 2004 (part of UML 2004) (2004)
12. Dillon, L.: Verifying General Safety Properties of Ada Tasking Programs. IEEE Trans. on Soft. Eng. 16 (1990)
13. Gaston, C., Aiguier, M., Bahrami, D., Lapitre, A.: Symbolic execution techniques extended to systems. In: ICSEA 2009 (2009)
14. Godefroid, P.: Compositional dynamic test generation. In: POPL 2007 (2007)
15. Godefroid, P., Klarlund, N., Sen, K.: DART: directed automated random testing. SIGPLAN Notices 40(6), 213–223 (2005)
16. Godefroid, P., Levin, M., Molnar, D.: Automated whitebox fuzz testing. In: NDSS 2008 (2008)
17. King, J.: Symbolic execution and program testing. Communications of the ACM 19(7), 385–394 (1976)
18. Lee, N.H., Cha, S.D.: Generating test sequence using symbolic execution for event-driven real-time systems. Microproc. and Microsys. 27, 523–531 (2003)
19. Păsăreanu, C., Visser, W.: A survey of new trends in symbolic execution for software testing and analysis. J. on Software Tools for Technology Transfer 11 (2009)
20. Rapin, N.: Symbolic Execution Based Model Checking of Open Systems with Unbounded Variables. In: Dubois, C. (ed.) TAP 2009. LNCS, vol. 5668, pp. 137–152. Springer, Heidelberg (2009)
21. Selic, B.: Using UML for Modeling Complex Real-Time Systems. In: Müller, F., Bestavros, A. (eds.) LCTES 1998. LNCS, vol. 1474, pp. 250–260. Springer, Heidelberg (1998)
22. Sperschneider, V., Antoniou, G.: Logic: a foundation for computer science. Addison-Wesley (1991)
23. Thums, A., Schellhorn, G., Ortmeier, F., Reif, W.: Interactive Verification of Statecharts. In: Ehrig, H., Damm, W., Desel, J., Große-Rhode, M., Reif, W., Schnieder, E., Westkämper, E. (eds.) INT 2004. LNCS, vol. 3147, pp. 355–373. Springer, Heidelberg (2004)
24. Tomb, A., Brat, G., Visser, W.: Variably interprocedural program analysis for runtime error detection. In: ISSTA 2007 (2007)
25. Zurowska, K., Dingel, J.: SAUML: a Tool for Symbolic Execution of UML-RT Models. In: ASE 2011 - Tool Demonstrations (2011) (to appear)
26. Zurowska, K., Dingel, J.: Symbolic execution of UML-RT State Machines. In: SAC - Software Verification and Testing 2012 (to appear, 2012)

Inferring Definite Counterexamples through Under-Approximation

Jörg Brauer[1,*] and Axel Simon[2,**]

[1] Embedded Software Laboratory, RWTH Aachen University, Germany
[2] Informatik 2, Technical University Munich, Germany

Abstract. Abstract interpretation for proving safety properties summarizes concrete traces into abstract states, thereby trading the ability to distinguish traces for tractability. Given a violation of a safety property, it is thus unclear which trace led to the violation. Moreover, since part of the abstract state is over-approximate, such a trace may not exist at all. We propose a novel backward analysis that is based on abduction of propositional Boolean logic and that only generates legitimate traces that reveal actual defects. The key to tractability lies in modifying an existing projection algorithm to stop prematurely with an under-approximation and by combining various algorithmic techniques to handle loops finitely.

1 Introduction

Model checking has the attractive property that, once a specification cannot be verified, a trace illustrating a counterexample is returned which can be inspected by the user. These traces have been highlighted as invaluable for fixing the defect [9]. In contrast, abstract interpretation for asserting safety properties typically summarizes traces into abstract states, thereby trading the ability to distinguish traces for computational tractability. Upon encountering a violation of the specification, it is then unclear which trace led to the violation. Moreover, since the abstract state is an over-approximation of the set of actually reachable states, a trace leading to an erroneous abstract state may not exist at all.

Given a safety property that cannot be proved correct, a trace to the beginning of the program would be similarly instructive to the user as in model checking. However, obtaining such a trace is hard as this trace needs to be constructed by going backwards step-by-step, starting at the property violation. One approach is to apply the abstract transfer functions that were used in the forward analysis in reverse [28]. However, these transfer functions over-approximate. Thus, a counterexample computed using this approach may therefore be spurious, too. However, spurious warnings are the major hinderance of many static analyses, except those crafted for a specific application domain [11]. It has even been noted that unsound static analyses might be preferable over sound ones because

---

* Supported partly by DFG GRK 1298 and DFG EXC 89.
** Supported by DFG Emmy Noether SI 1579/1.

A. Goodloe and S. Person (Eds.): NFM 2012, LNCS 7226, pp. 54–69, 2012.

the number of false positives can be traded off against missed bugs, thereby delivering tools that find defects rather than prove their absence [3].

Rather than giving up on soundness, we propose a practical technique to find legitimate traces that reveal actual defects, thereby turning sound static analyses into practical bug-finding tools. We use the results of an approximate forward analysis to guide a backward analysis that builds up a trace from the violation of the property to the beginning of the program. At its core, it uses a novel SAT-based projection algorithm [6] that has been adapted to deliver an under-approximation of the transition relation in case the exact solution would be too expensive to compute. Furthermore, assuming that the projection is exact, if the intersection between a backward propagated state and the states of the forward analysis is empty on all paths, the analysis has identified a warning as spurious. Hence our analysis has the ability to both, find true counterexamples and to identify warnings as spurious. To our knowledge, our work is the first to remove spurious warnings without refining or enhancing the abstract domain.

One challenge to the inference of backward traces is the judicious treatment of loops. Given a state s' after a loop, it is non-trivial to infer a state s that is valid prior to entering the loop. In particular, it is necessary to assess how often the loop body needs to be executed to reach the exit state s'. This problem is exacerbated whenever analyzing several loops that are nested or appear in sequence. Our solution to this issue is to summarize multiple loop iterations in a closed Boolean formula and to use iterative deepening in the number of loop executions across all loops until a feasible path between s to s' is found.

The practicality of our approach is based on the following technical contributions:

- We use an over-approximating affine analysis between the backward propagated state s' after the loop and the precondition s of the loop inferred by the forward analysis to estimate the number of loop iterations. If an affine relationship exists, we derive a minimum number of loop iterations that the state s' has to be transformed by the loop.
- We synthesize a relational Boolean loop transformer f^{2^i}, which expresses 2^i executions of a loop, given $f^{2^{i-1}}$. These loop transformers are then used to construct f^n for arbitrary n, thereby providing the transfer function to calculate an input state from the given output state of the loop in $\log_2(n)$ steps for n iterations. This approach can also be applied to nested loops.
- We provide a summarization technique, which describes $0, \ldots, 2^n$ iterations of a loop as one input/output relation. This method combines the Boolean transfer functions f^{2^n} with a SAT-based existential elimination algorithm. The force of this combination is that we can modify the elimination algorithm to generate under-approximate state descriptions — any approximated result thus still describes states which are possible in a concrete execution.

The remainder of the paper is structured as follows: After the next section details our overall analysis strategy, Sect. 3 illustrates the three contributions in turn. Section 4 details the modifications to the projection algorithm to allow under-approximations which Sect. 5 evaluates in our implementation. Section 6 presents related work before Sect. 7 discusses possible future work and concludes.

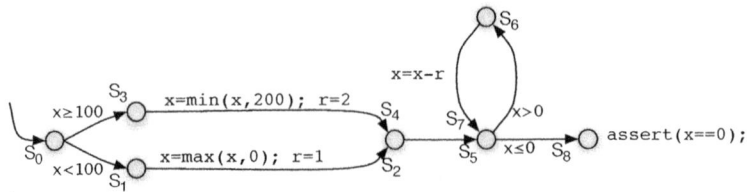

Fig. 1. Backward propagation past a loop

Table 1. Abstract states in the analyzer, presented as ranges for conciseness

i		0	1	2	3	4	5	6	7	8
$S_i =$	x	any	≤ 99	$[0, 99]$	≥ 100	$[100, 200]$	$[0, 200]$	$[1, 200]$	$[-1, 199]$	$[-1, 0]$
	r	any	any	$[1, 1]$	any	$[2, 2]$	$[1, 2]$	$[1, 2]$	$[1, 2]$	$[1, 2]$
$S_i' =$	x			$\perp$		$\perp$	1	$[0, 1]$	-1	-1
	r						2	$[1, 2]$	$[1, 2]$	$[1, 2]$
$S_i'' =$	x	$[101, 125]$		$\perp$		$[101, 125]$	$[101, 125]$	$[1, 125]$		
	r	2				2	2	2		

2   Backward Analysis Using Under-Approximation

The various SAT-based algorithms that constitute our backwards analysis are orchestrated by a strategy that tries to find a path to the beginning of the program with minimal effort. Specifically, the idea is to perform iterative deepening when unrolling loops until either a counterexample is found or a proof that the alarm was spurious. We illustrate this strategy using Fig. 1 which shows a program that limits some signed input variable x to $0 \leq x \leq 200$ and then iteratively decreases x by one if the original input was $x < 100$ and by two otherwise. The abstract states $S_0, \ldots S_8$ inferred by a forward analysis (here based on intervals) are stored for each tip of an edge where an edge represents either a guard or some assignments. The resulting states of the forward analysis are listed in Table 1. Since S_8 violates the assertion $x = 0$, we propagate the negated assertion $x \leq -1 \vee x \geq 1$ backwards as an assumption. As the forward analysis constitutes a sound over-approximation of the reachable states, we may intersect the assumption with S_8, leading to the refined assumption S_8' in Table 1. We follow the flow backwards by applying the guard $x \leq 0$ which has no effect on S_8'.

At this point, we try to continue backwards without entering the loop. This strategy ensures that the simplest counterexample is found first. However, in this case S_8' conjoined with S_5 yields an empty state, indicating that the chosen path cannot represent a counterexample. The only feasible trace is therefore one that passes through the loops that we have skipped so far. In the example, only one loop exists on the path, and we calculate the effect of this loop having executed $0, \ldots, 2^i$ times, beginning with $i = 0$. At the bit-level, the effect of executing a loop body backwards can be modelled as a Boolean function f which one can compose with itself to express the effect of executing the body twice: $f^2 = f \circ f$.

For the sake of presentation, let $f \vee f^2$ denote the effect of executing the loop once or twice. We then pre-compute f^{2^i} and express the semantics of $0, \ldots, 2^i$ iterations as $\varphi_{i+1} = \varphi_i \vee \varphi_i \circ f^{2^i}$ with φ_0 being defined as the identity. For each $i = 1, 2, \ldots$, we unroll all loops that we have encountered so far until we manage to propagate the resulting state further backwards. For instance, in the example we unroll the loop once by propagating the state S_8' backwards through the loop, yielding $S_7' = S_8'$ and $S_6' = \{x \mapsto r - 1 \in [0, 1], r \mapsto [1, 2]\}$. Applying the guard $x > 0$ yields a non-empty $S_5' = \{x \mapsto [1, 1], r \mapsto [2, 2]\}$. However, $S_5' \sqcap S_2 = \emptyset$ and $S_5' \sqcap S_4 = \emptyset$ and hence the loop must be unrolled further. After five more iterations, we find $S_5'' = \varphi_5(S_5') = \{x \mapsto 2n - 1 \wedge n \in [1, 63], r \mapsto 2\}$ which has a non-empty intersection with S_4, leading to $S_4'' = \{x \mapsto 2n - 1 \wedge n \in [51, 63], r \mapsto 2\} = S_3'' = S_0''$, thereby providing a counterexample that violates the assertion.

Interestingly, the above construction can also be used to identify a warning as false positive: If during the unrolling of a loop $\varphi_{i+1}(S) \models \varphi_i(S)$ then further unrolling does not add any new states. If propagating this fixpoint beyond a certain point p in the program is impossible (it drops to bottom) then the warning is spurious and the forward analysis lost precision between p and the assertion.

However, calculating φ_i can become very expensive and a fixpoint might be impossible to obtain. The source of the complexity is the elimination of existentially quantified variables that tie the input of a function to the output. For instance, the Boolean formula $(o = f^2(i)) \equiv \exists t : (o = f(t) \wedge t = f(i))$ introduces fresh variables t that must be removed in order to avoid exponential growth of the formula when calculating $f^{2n} = f^n \circ f^n$. Further intermediate variables are required in φ_i to express that the result is either o or t.

In order to reduce the cost of the calculation, we employ a simple pre-analysis that infers a minimal number of loop iterations 2^m that are required to proceed past the initialization in the loop header. In case $m > 0$, we calculate the formula $f^m \circ \varphi_{i-m}$ that does not consider cases in which the loop exits in the first 2^m iterations and which is cheaper to calculate than φ_i for $i \geq m$. Moreover, rather than examining all φ_i with $i \geq m+1$ at once, we fix $i = 0$ in $f^m \circ \varphi_{i-m}$ for all loops in the program. If no trace can be found, we retry for each i. The two heuristics square with the observation that, usually, an error trace through a loop exists for small k, unless a loop iterates m times where m is constant, which is addressed by composing φ_i with f^m. If calculating $f^n \circ \varphi_{i-m}$ is still too expensive, we apply an algorithm that can under-approximate the elimination of the existentially quantified variables. Once under-approximation is used, traces may be missed and an inferred error cannot be shown to be a false positive. However, any trace found using under-approximation is still a valid counterexample.

In summary, if loops must be unrolled, our approach uses an iterative deepening approach where in each step the number of iterations that are considered is doubled. It also applies a heuristic that unrolls a loop by n iterations if it is clear that the loop cannot exit earlier. These techniques are applied in the next section to eliminate false positives. For complex loops and many iterations, we under-approximate the existential elimination in a well-motivated fashion. This approach is detailed in the context of inferring counterexamples in Sect. 4.

```
1   unsigned int log2(unsigned char c) {
2        unsigned char i = 0;
3        if (c==0) return 0; else c--;
4        while (c > 0) {
5            i = i + 1;
6            c = c >> 1;
7        }
8        assert(i <= 7);
9        return i;
10  }
```

Fig. 2. Computation of the log2 of an unsigned integer c; even though the code is correct, abstract interpreters based on domains such as convex polyhedra emit a warning

3   Eliminating False Positives

Consider the program in Fig. 2, which computes the logarithm to the base 2 of an unsigned character c (a bit-vector of length 8) and stores the result in i. Clearly, i should hold a value less than 8, which is formulated in terms of an assertion. The assertion is valid, yet most abstract interpreters emit a warning; typical domains fail to capture the relation between i, which is used in the assertion, and c, which specifies the termination condition. We build towards our technique, which proves the non-existence of a defective path, in three steps.

3.1   Concrete Relational Semantics in Boolean Logic

To mark the warning as spurious, our analysis thus attempts to exclude all paths that lead to a state satisfying the invariant $\iota = (0 \leq i \leq 255 \land 0 \leq c \leq 0)$ produced by the forward analysis for line 8, and at the same time violates $0 \leq i \leq 7$. We express the concrete relational semantics of each block in the program in Boolean logic. The values of i on entry and exit of each basic block are represented using bit-vectors i and i', respectively. Likewise, use bit-vectors c and c' to represent c. In the following, let $\langle x \rangle = \sum_{i=0}^{7} 2^i \cdot x[i]$ denote the unsigned value of a bit-vector x, and let $x[j]$ denote the jth bit of x. Let the notation $[\![\cdot]\!]$ encode an arithmetic constraint as Boolean formula. Then, $f_I(V, V') := [\![i' = 0, c \neq 0, c' = c - 1]\!]$ encodes the initialization block of the function over inputs $V = \{c, i\}$ and outputs $V' = \{c', i'\}$. In a similar fashion, $f_L(V, V')$ encodes the loop body:

$$f_I(V, V') = \left\{ \bigwedge_{j=0}^{7} \neg i'[j] \land \bigvee_{j=0}^{7} c'[j] \land ((\bigwedge_{j=0}^{7} c[j]) \leftrightarrow (c'[j] \oplus \bigwedge_{k=0}^{j-1} c'[k]) \right.$$

$$f_L(V, V') = \left\{ \begin{array}{l} (\bigvee_{j=0}^{7} c[j]) \land \neg c'[7] \land (\bigwedge_{j=0}^{6} c'[j] \leftrightarrow c[j+1]) \land \\ (\bigwedge_{j=0}^{7} i'[j] \leftrightarrow (i[j] \oplus \bigwedge_{k=0}^{j-1} i[k]) \end{array} \right.$$

In order to find a path to a state that satisfies ι and violates the assertion, encode ι in Boolean logic as $[\![\iota]\!] = \bigwedge_{j=0}^{7} \neg c[j]$. Furthermore, let $[\![0 \leq \langle i \rangle \leq 7]\!] = \bigwedge_{j=4}^{7} \neg i[j]$ encode the assertion. The error state $g(V)$ after the loop is given as:

$$g(V) = [\![\iota]\!] \wedge \neg [\![0 \leq \langle i \rangle \leq 7]\!]$$
$$= [\![(0 \leq \langle i \rangle \leq 255) \wedge (\langle c \rangle = 0)]\!] \wedge [\![8 \leq \langle i \rangle \leq 255]\!]$$
$$= [\![(8 \leq \langle i \rangle \leq 255) \wedge (\langle c \rangle = 0)]\!]$$
$$= \bigwedge_{j=0}^{7} \neg c[j] \wedge \bigvee_{j=4}^{7} i[j]$$

We commence by testing the shortest trace to the erroneous state $g(V)$, i.e., the trace going through the initialization block followed directly by an assertion violation. This path is feasible if $f_I(V, V') \wedge g(V')$ is satisfiable; since the formula is unsatisfiable, this path cannot be part of a counterexample. A valid trace thus traverses the loop $n \geq 1$ times. One way to discover n, and thus a trace to the loop-entry state $\omega = (\langle i \rangle = 0) \wedge (0 \leq \langle c \rangle \leq 255)$, is to iteratively unroll the loop. However, instead of composing n functions f_L (each representing one iteration), we infer an $m \leq n$ using affine abstraction and derive f_L^m in $\log_2(m)$ steps.

3.2   Lower Bounds on the Number of Loop Iterations

The first step of computing a lower bound on the number of loop iterations is to abstract $f_L(V, V')$ using a conjunction of affine equalities [16], which relate symbolic bounds $V_{\ell,u} = \{c_\ell, c_u, i_\ell, i_u\}$ on entry of the block to symbolic bounds $V'_{\ell,u} = \{c'_\ell, c'_u, i'_\ell, i'_u\}$ on exit. Here, c_ℓ, c_u, c'_ℓ, and c'_u are bit-vectors representing the lower and upper bounds of c, respectively; similarly for i. Applying the abstraction scheme from [5, Sect. 3] yields the following system of affine equations:

$$F = \left\{ \begin{array}{llll} \langle c_\ell \rangle = 0 & \wedge & \langle c_u \rangle = 2 \cdot (\langle c'_u \rangle + 1) - 1 & \wedge \\ \langle i_\ell \rangle = \langle i'_\ell \rangle - 1 & \wedge & \langle i_u \rangle = \langle i'_u \rangle - 1 & \end{array} \right\}$$

We transform $g(V)$ to express affine constraints on the outputs $V'_{\ell,u}$ by automatically lifting the characterization over program variables to relations over range variables (see [5, Sect. 3.2] for further details of this operation):

$$g_{\text{aff}}(V') = \left\{ \begin{array}{llll} \langle i'_\ell \rangle = 8 & \wedge & \langle i'_u \rangle = 255 & \wedge \\ \langle c'_\ell \rangle = 0 & \wedge & \langle c'_u \rangle = 0 & \end{array} \right\}$$

Then, applying F to $g_{\text{aff}}(V')$ yields

$$F(g_{\text{aff}}(V')) = \left\{ \begin{array}{llll} \langle i_\ell \rangle = 7 & \wedge & \langle i_u \rangle = 254 & \wedge \\ \langle c_\ell \rangle = 0 & \wedge & \langle c_u \rangle = 1) & \end{array} \right\}$$

which, in turn, gives $(7 \leq \langle i \rangle \leq 254) \wedge 0 \leq \langle c \rangle \leq 1)$. The intersection with the precondition $[\![\omega]\!]$ yields $\bot$; thus a single loop iteration does not suffice. In the next iteration, we summarize two executions of the loop using relational composition $\circ_{\text{Lin}}$ of two affine systems F_1 and F_2. This amounts to renaming the outputs of F_1 and the inputs of F_2 to the same temporary variables, and eliminating these from the conjunction of both systems using projection [19,27]. The projection, in turn, has a straightforward implementation using Gauss elimination. In the example, two iterations of the loop are characterized by:

$$F^2 = F \circ_{\mathsf{Lin}} F = \left\{ \begin{array}{llll} \langle c_\ell \rangle = 0 & \wedge & \langle c_u \rangle = 4 \cdot (\langle c'_u \rangle + 1) - 1 & \wedge \\ \langle i_\ell \rangle = \langle i'_\ell \rangle - 2 & \wedge & \langle i_u \rangle = \langle i'_u \rangle - 2 \end{array} \right\}$$

Again, we get $F^2(g_{\mathsf{aff}}(\boldsymbol{V})) \sqcap \llbracket \omega \rrbracket = \bot$. Likewise, compute:

$$F^4 = F^2 \circ_{\mathsf{Lin}} F^2 = \left\{ \begin{array}{llll} \langle c_\ell \rangle = 0 & \wedge & \langle c_u \rangle = 16 \cdot (\langle c'_u \rangle + 1) - 1 & \wedge \\ \langle i_\ell \rangle = \langle i'_\ell \rangle - 4 & \wedge & \langle i_u \rangle = \langle i'_u \rangle - 4 \end{array} \right\}$$

$$F^8 = F^4 \circ_{\mathsf{Lin}} F^4 = \left\{ \begin{array}{llll} \langle c_\ell \rangle = 0 & \wedge & \langle c_u \rangle = 256 \cdot (\langle c'_u \rangle + 1) - 1 & \wedge \\ \langle i_\ell \rangle = \langle i'_\ell \rangle - 8 & \wedge & \langle i_u \rangle = \langle i'_u \rangle - 8 \end{array} \right\}$$

Observe that

$$F^8(g_{\mathsf{aff}}(\boldsymbol{V}')) = \left\{ \begin{array}{llll} \langle c_\ell \rangle = 0 & \wedge & \langle c_u \rangle = 255 & \wedge \\ \langle i_\ell \rangle = 0 & \wedge & \langle i_u \rangle = 247 \end{array} \right\}$$

describes states that satisfy the invariant $\llbracket \omega \rrbracket$ prior to the loop. Thus, the minimum number of loop iterations is $5 \le m \le 8$. Using binary search, we determine which F^m is the first to satisfy $\llbracket \omega \rrbracket$. This gives $m = 8$. Observe that, due to abstraction, this bound is not necessarily exact (though it is in this example) in that any counterexample trace must traverse the loop at least eight times.

3.3   Summarizing a Number of Iterations

We now face the task of efficiently calculating input-output behavior of eight loop iterations as a Boolean formula f_L^8 which is later used to compute the preimage of f_L^8 subject to $g(\boldsymbol{V})$. Analogous to the construction of composing affine transformers, we incrementally double the number of iterations summarized in a single formula, thus finessing the need to unroll the loop. Specifically, put:

$$\begin{array}{ll} f_L^0(\boldsymbol{V}, \boldsymbol{V}') = \mathsf{id}(\boldsymbol{V}, \boldsymbol{V}') & f_L^2(\boldsymbol{V}, \boldsymbol{V}') = \exists \boldsymbol{V}'' : f^1(\boldsymbol{V}, \boldsymbol{V}'') \wedge f^1(\boldsymbol{V}'', \boldsymbol{V}') \\ f_L^1(\boldsymbol{V}, \boldsymbol{V}') = f_L(\boldsymbol{V}, \boldsymbol{V}') & f_L^4(\boldsymbol{V}, \boldsymbol{V}') = \exists \boldsymbol{V}'' : f^2(\boldsymbol{V}, \boldsymbol{V}'') \wedge f^2(\boldsymbol{V}'', \boldsymbol{V}') \end{array}$$

Each f_L^i describes an input-output relation for exactly i applications of f_L. To eliminate $\boldsymbol{V}''$ from the formulae, we apply a SAT-based projection algorithm [6]. This construction suffices to test $\llbracket \omega \rrbracket \wedge f_L^8(\boldsymbol{V}, \boldsymbol{V}') \wedge g(\boldsymbol{V}')$ for satisfiability, i.e., to check if the erroneous state $g(\boldsymbol{V}')$ can be reached with exactly eight iterations starting in $\llbracket \omega \rrbracket$. If unsatisfiable, it is necessary to unroll the loop further. Again, we construct a summary φ_i of $m + 2^i$ iterations, where m is the lower bound on the number of iterations. Then, φ_i describes all states reachable after $m, \ldots, m + 2^i$ iterations whereas $f_L^{m+2^i}$ describes exactly $m + 2^i$ iterations.

3.4   Summarizing a Range of Iterations

Formally, let $\varphi_i(\boldsymbol{V}) = \exists \boldsymbol{V}' : \exists \boldsymbol{V}'' : (\bigvee_{j=0}^{2^i} f_L^j(\boldsymbol{V}, \boldsymbol{V}')) \wedge f_L^m(\boldsymbol{V}', \boldsymbol{V}'') \wedge g(\boldsymbol{V}'')$, that is, erroneous states expressed over $\boldsymbol{V}''$ being backpropagated m times around

```
 1  unsigned int hamDist(int x, int y) {
 2        unsigned int d = 0;
 3        unsigned int v = x ^ y;
 4        while (v != 0) {
 5            d = d + 1;
 6            v = v & (v - 1);
 7        }
 8        assert(d < 32);
 9        return d;
10  }
```

Fig. 3. Erroneous hamming distance calculation; the assertion in line 8 does not hold

the loop giving V', which are, in turn, $j = 0, \ldots, 2^i$ times transformed into constraints over V. Rather than recalculating each $\varphi_i(V)$ from scratch, we compute $\varphi_i(V)$ based on the following inductive definition, allowing us to reuse $\varphi_{i-1}(V)$ to compute $\varphi_i(V)$ and requiring only i instead of 2^i steps:

$$\varphi_i(V) = \begin{cases} \exists V' : f_L^m(V, V') \land g(V') & : i = 0 \\ \varphi_{i-1}(V) \lor (\exists V' : f^{2^i}(V, V') \land \varphi_{i-1}(V')) : \text{otherwise} \end{cases}$$

Note that, due to monotonicity, there exists an $i \geq 0$ with $\varphi_i(V) \models \varphi_{i-1}(V)$. In the example, since $\varphi_4(V) \models \varphi_3(V)$ and $\varphi_3(V) \land [\![\omega]\!]$ is unsatisfiable, we deduce that no trace from $[\![\omega]\!]$ to the erroneous state $g(V)$ exists that iterates more than eight times. Hence, the warning emitted by the forward analysis is spurious. In certain cases, calculating φ_i can become too costly, which is addressed next.

4   Finding Counterexamples

Although the iterative deepening heuristic reduces the complexity of the generated formulae, exact state spaces cannot always be computed since calculating $\exists V' : f^{2^i}(V, V') \land \varphi_{i-1}(V')$ may result in an exponentially sized formula. However, if the aim is to only find a counterexample rather than eliminating false positives, an under-approximation of the projection $\exists V' : \psi$ suffices. In order to illustrate the idea, consider Fig. 3 which presents a function to calculate the Hamming distance of two integers x and y. Once more, we bit-blast the concrete semantics of both, loop body and loop pre-condition. Here, $\oplus$ denotes the Boolean exclusive-or and u is an auxiliary bit-vector that captures the intermediate value of v-1:

$$f_I(V, V') = \begin{cases} \bigwedge_{j=0}^{31} \neg d'[j] \land \bigwedge_{j=0}^{31} v'[i] \leftrightarrow x[j] \oplus y[j] \end{cases}$$

$$f_L(V, V') = \begin{cases} (\bigwedge_{j=0}^{31} d'[j] \leftrightarrow (d[j] \oplus \bigwedge_{k=0}^{j-1} d[k])) \land (\bigvee_{j=0}^{31} v[j]) \land \\ (\bigwedge_{j=0}^{31} v[j] \leftrightarrow (u[j] \oplus \bigwedge_{k=0}^{j-1} u[k])) \land (\bigwedge_{j=0}^{31} v'[j] \leftrightarrow (v[j] \land u[j])) \end{cases}$$

As before, let $\iota = (\langle v \rangle = 0 \land \langle d \rangle = \top)$ describe the invariant derived at the assertion which was inferred during the forward analysis and let $\omega = ((\langle d \rangle =$

$0 \wedge \langle \boldsymbol{v} \rangle = \langle \boldsymbol{x} \rangle \oplus \langle \boldsymbol{y} \rangle)$ represent the state at loop entry. The erroneous state after the loop is thus characterized as $g(\boldsymbol{V}) = [\![\iota]\!] \wedge \bigvee_{j=5}^{31} \boldsymbol{d}[j]$ in Boolean logic.

4.1   Lower Bounds on the Number of Loop Iterations

Again, to compute a lower bound on the number of loop iterations required to reach the erroneous state $g(\boldsymbol{V})$ from the pre-condition $\boldsymbol{V}'$ defined by $f_I(\boldsymbol{V}, \boldsymbol{V}')$, we derive an abstraction of the loop transfer function $f_L(\boldsymbol{V}, \boldsymbol{V}')$ in terms of a conjunction of affine equalities. This operation gives:

$$F = \left\{ \langle \boldsymbol{d}_\ell \rangle = \langle \boldsymbol{d}'_\ell \rangle - 1 \quad \wedge \quad \langle \boldsymbol{d}_u \rangle = \langle \boldsymbol{d}'_u \rangle - 1 \right\}$$

Note that v = v & (v - 1) is non-affine, hence the lack of an affine constraint over v. We transform $g(\boldsymbol{V})$ to express affine constraints on the outputs as per [5]:

$$g_{\text{aff}}(\boldsymbol{V}') = \left\{ \langle \boldsymbol{d}'_\ell \rangle = 32 \quad \wedge \quad \langle \boldsymbol{d}'_u \rangle = 2^{32} - 1 \quad \right\}$$

Applying F to $g_{\text{aff}}(\boldsymbol{V}')$ yields $F(g_{\text{aff}}(\boldsymbol{V}')) = \{ \langle \boldsymbol{d}_\ell \rangle = 31 \wedge \langle \boldsymbol{d}_u \rangle = 2^{32} - 2 \}$ which, in turn, gives $31 \leq \langle \boldsymbol{d} \rangle \leq 2^{32} - 2$. The intersection with the state $\omega = (\langle \boldsymbol{d} \rangle = 0)$ that the forward analysis inferred for the loop entry yields $[\![31 \leq \langle \boldsymbol{d} \rangle \leq 2^{32} - 2]\!] \sqcap [\![\omega]\!] = \bot$; thus a single loop iteration does not suffice. Following the strategy discussed in Sect. 3.2, we compute $F^2(g_{\text{aff}}(\boldsymbol{V}'))$, $F^4(g_{\text{aff}}(\boldsymbol{V}'))$, $F^8(g_{\text{aff}}(\boldsymbol{V}'))$, $F^{16}(g_{\text{aff}}(\boldsymbol{V}'))$, and $F^{32}(g_{\text{aff}}(\boldsymbol{V}'))$. It is only $F^{32}(g_{\text{aff}}(\boldsymbol{V}'))$ that satisfies $F^{32}(g_{\text{aff}}(\boldsymbol{V}')) \sqcap [\![\omega]\!] \neq \bot$. Consequently, the minimum number of loop iterations is $17 \leq m \leq 32$. Using binary search, we determine which F^m is the first to satisfy $[\![\omega]\!]$. This gives $m = 32$ as the minimum number of iterations.

4.2   Under-Approximating a Range of Iterations

As in Sect. 3.3, we face the task of summarizing the execution of 32 consecutive loop iterations. To find a backward trace, compute $f_L^0(\boldsymbol{V}, \boldsymbol{V}')$ and $f_L^1(\boldsymbol{V}, \boldsymbol{V}')$ as before. Rather than computing $f_L^i(\boldsymbol{V}, \boldsymbol{V}')$ exactly by enumerating all of the projection space, we preempt the computation of $f_L^2(\boldsymbol{V}, \boldsymbol{V}') = \exists \boldsymbol{V}'' : f^1(\boldsymbol{V}, \boldsymbol{V}'') \wedge f^1(\boldsymbol{V}'', \boldsymbol{V}')$ prematurely after, say, 100 models have been enumerated (though in our implementation, we have used a heuristic based on the structure of the erroneous goal state $g(\boldsymbol{V})$ rather than one that is based solely on the number of models, see Sect. 5). This tactic yields a formula $h_L^2(\boldsymbol{V}, \boldsymbol{V}')$ in CNF that entails $f_L^2(\boldsymbol{V}, \boldsymbol{V}')$. In other words, every of model $h_L^2(\boldsymbol{V}, \boldsymbol{V}')$ is also a model of $f_L^2(\boldsymbol{V}, \boldsymbol{V}')$, i.e., the formula is easier to compute and under-approximates $f_L^2(\boldsymbol{V}, \boldsymbol{V}')$. Based on $h_L^2(\boldsymbol{V}, \boldsymbol{V}')$, we compute $h^4(\boldsymbol{V}, \boldsymbol{V}'') = \exists \boldsymbol{V}'' : h_L^2(\boldsymbol{V}, \boldsymbol{V}'') \wedge h_L^2(\boldsymbol{V}'', \boldsymbol{V}')$. Likewise compute $h_L^8(\boldsymbol{V}, \boldsymbol{V}')$, $h_L^{16}(\boldsymbol{V}, \boldsymbol{V}')$ and $h_L^{32}(\boldsymbol{V}, \boldsymbol{V}')$. This under-approximating strategy may decrease the size of the formulae exponentially.

4.3   Failing to Derive a Counterexample Trace

Suppose now that the summary of states $\varphi_0(\boldsymbol{V}) = \exists \boldsymbol{V}' : h_L^{32}(\boldsymbol{V}, \boldsymbol{V}') \wedge g(\boldsymbol{V}')$ yields a state description such that $\varphi_0(\boldsymbol{V}) \wedge [\![\omega]\!]$ is unsatisfiable and that

$\varphi_i(V) \models \varphi_0(V)$ for any $i \geq 1$. Then the under-approximated transfer function $h_L^{32}(V, V')$ is insufficient to reach a loop-entry state. Hence, it is necessary to compute a greater under-approximation $\hat{h}_L^{32}(V, V')$ such that $h_L^{32}(V, V') \models \hat{h}_L^{32}(V, V')$. Doing so necessitates computing $\hat{h}_L^2(V, V')$ such that $h_L^2(V, V') \models \hat{h}_L^2(V, V')$; likewise for $\hat{h}_L^4(V, V')$, $\hat{h}_L^8(V, V')$, $\hat{h}_L^{16}(V, V')$, and $\hat{h}_L^{32}(V, V')$. Based on the enlarged under-approximation $\hat{h}_L^{32}(V, V')$, we compute $\hat{\varphi}_0(V) = \exists V' : \hat{h}_L^{32}(V, V') \wedge g(V')$, which by monotonicity satisfies $\varphi_0(V) \models \hat{\varphi}_0(V)$. Suppose that $\hat{\varphi}_0(V) \wedge [\![\omega]\!]$ is satisfiable, producing a model $\mathbf{m} \models \hat{\varphi}_0(V) \wedge [\![\omega]\!]$ defined as follows:

$$\mathbf{m} = \left\{ \begin{array}{l} \boldsymbol{d}[0] \mapsto 0, \, \boldsymbol{d}[1] \mapsto 0, \, \boldsymbol{d}[2] \mapsto 0, \, \boldsymbol{d}[3] \mapsto 0, \, \ldots, \, \boldsymbol{d}[31] \mapsto 0 \\ \boldsymbol{x}[0] \mapsto 0, \, \boldsymbol{x}[1] \mapsto 1, \, \boldsymbol{x}[2] \mapsto 0, \, \boldsymbol{x}[3] \mapsto 1, \, \ldots, \, \boldsymbol{x}[31] \mapsto 1 \\ \boldsymbol{y}[0] \mapsto 1, \, \boldsymbol{y}[1] \mapsto 0, \, \boldsymbol{y}[2] \mapsto 1, \, \boldsymbol{y}[3] \mapsto 0, \, \ldots, \, \boldsymbol{y}[31] \mapsto 0 \end{array} \right\}$$

This model entails that we successfully applied an under-approximate loop transformer to find a trace that executes 32 iterations. Bit-vectors $\boldsymbol{x} = \langle 0101 \ldots 01 \rangle$ and $\boldsymbol{y} = \langle 1010 \ldots 10 \rangle$ then indicate values that give a Hamming distance of 32 and therefore violate the assertion. We have thus computed a definite counterexample.

5   Experiments

We have integrated the techniques described in this paper into the [MC]SQUARE framework, which is written in JAVA. Several programs have been analyzed that contain at least one loop each. The benchmarks shown in Tab. 2 include Wegner's bit-counting bit-cnt, the algorithm in Fig. 3 ham-dist, consecutive loops that shift and add inc-1shift, the algorithm in Fig. 2 log, parity calculation parity, parity_mit, bit-reversal randerson, swapping of bytes swap and two interdependent, nested loops. The running times were obtained on a 2.4 GHz MACBOOK PRO equipped with 4 GB of RAM. The programs are written in Instruction List, a language used in Programmable Logic Controllers. The semantics of these programs are translated into bit-vector relations, similarly to the examples in Sect. 3 and Sect. 4. This translation and the calculation of the affine loop transformers is written in JAVA using SAT4J. In none of the benchmark do these calculations take more than 0.1s of the runtimes. The Boolean summarization of loop iterations and the counterexample generation are implemented in C++ using MINISAT and CUDD. MINISAT frequently outperforms SAT4J by a factor of 5-10 and was thus chosen for the more demanding transfer function synthesis.

Table 2 presents the timings for different analysis strategies. In the simplest strategy, the post-condition state that violates the assertion g is propagated through φ_n, the fixpoint of the input/output behavior of a loop. The times to calculate the loop transfer function φ_n is given in column "Runtime (Full) / TF" whereas propagating the state g through φ_n is given in column "Runtime (Full) / CE". Note that these two phases are interleaved and that the table presents the accumulated times spent in each phase. The next sections discuss the impact of pre-computing a minimal number of unrollings of a loop using affine abstractions and of restricting the post-condition g to find a counterexample quicker.

5.1   Affine Estimation of Iterations

Inferring a lower bound on the number of loop iterations follows the algorithm presented in [4, Sect. 3.2]. The affine relationships on the bounds of the variables are inferred by asking for an initial solution to the loop transfer function f, yielding an assignment for the input and output variables. These assignments form a linear equation system. By using cheap incremental SAT solving, different assignments are queried and joined into the equation system by calculating the affine hull which, in turn, reduces to Gauss elimination. This process will terminate after at most $n + 1$ queries to the SAT solver for n input/output variables. Each query is rather trivial by current standards. If an affine relationship exists, a minimal number of loop iterations can be calculated by composing the affine transfer function repeatedly with itself using the $\circ_{\mathsf{Lin}}$ operation which, again, reduces to cheap Gauss elimination. Indeed, these steps contribute less than 0.1s for each benchmark and we therefore omitted this phase from the table. All our examples contain at least one variable that increases with each loop iteration such that the estimated minimal number of iterations is exactly the number of iterations it takes to exit the loop. This minimum number of iterations n allows us to reduce the size of the formula by those conjuncts that model that the loop may be exited after $i < n$ unrollings, thereby alleviating the SAT solver from proving this fact at the binary level. The speedup due to unrolling is minor (and thus omitted from the table). Still, it shows that proving the exit condition in MiniSat is more costly than Gauss elimination in Java and querying Sat4J.

5.2   Focussing the Search for Counterexamples

According to Table 2, the dominant part of the backwards analysis is the phase of calculating and composing loop transformers, which hinges on the performance of projection. In our experiments, we used model enumeration [6] and combined it with BDDs so as to derive a quantifier-free CNF formula [22]. Cudd v2.4.2 was used since it offers direct support for enumerating a compact CNF formula. Although this combination of data structures for representing Boolean formulae during projection is the best we could find, there naturally exist problems that result in large (intermediate) formulae. Indeed, McMillan [26], amongst others, has observed that no Boolean structure (such as BDDs or CNF) exists which is suitably small for all kinds of inputs; indeed, some problems exist where the CNF is exponential in the size of the respective BDD, and vice versa. However, the projection algorithm of [6] enumerates *prime implicants*, i.e., a Boolean formula that contains a minimum number of literals, thereby covering as many models as possible. This observation is relevant for the task of inferring counterexamples (rather than eliminating false positives where the state space has to be enumerated exhaustively): The algorithm enumerates prime implicants, which entailed that stopping the projection early means that a maximum number of models of the Boolean formula is propagated backwards. In principle, this means that the largest number of states, which also has the simplest representation, is tried first when resorting to under-approximation. Unfortunately, not every model of the

Table 2. Experimental results for PLC benchmarks

Benchmark	# Instr.	Runtime (Full)		Runtime (Simp.)	
		TF	CE	TF	CE
bit–cnt	26	4.1s	0.9s	0.4s	0.4s
ham–dist	19	4.8s	1.7s	0.8s	0.3s
inc–lshift	14	3.2s	2.7s	0.8s	0.6s
log	22	1.9s	1.3s	0.3s	0.3s
parity	28	8.3s	1.2s	1.2s	0.4s
parity_mit	17	6.2s	2.6s	1.5s	1.2s
randerson	23	8.0s	2.4s	4.2s	0.6s
swap	15	5.9s	1.8s	0.9s	0.5s
loops	207	43.6s	8.0s	13.1s	5.8s

formula has the same probability to constitute a counterexample trace. Consider the erroneous target $g(V) = [\![\iota]\!] \wedge \bigvee_{j=5}^{31} d[j]$ from Sect. 4. The prime implicant that captures the maximum number of states is $d[31]$, i.e., the formula stating that the most significant bit of d is set. This choice is in contrast to the intuition that many errors are off-by-one errors and thus happened close to those numeric values $d \in [0, 31]$ that do not violate the assertion.

Hence, we employ a heuristic that constrains $g(V)$ so that a sub-range of target values are considered that lie close to the feasible state, extending the sub-range to the next power of two iff the given under-approximation is insufficient for finding a counterexample. This is a straightforward extension considering the bit-level encodings of integer values. For the example in Sect. 4, this strategy is applied as follows: The goal-state requires $2^5 \leq \langle d \rangle \leq 2^{32} - 1$. In the first iteration, our strategy tries to find values that satisfy $2^5 \leq \langle d \rangle \leq 2^6$. If no counterexample is found, we proceed with $2^5 \leq \langle d \rangle \leq 2^7$, and so forth. This focusses model enumeration to regions that are more likely to contain an actual trace. These simpler models also reduce the runtime of computing projection. The difference is shown in the columns "Runtime (Simp.)", showing significant speed-ups to find counterexamples compared to the "Full" column where g is used without restrictions. Depending on the problems, counterexamples can be found up to 10 times faster by searching near states that do not violate the assertion.

5.3   Discussion

Using Boolean functions to represent a program state has obvious limits. However, when trading the ability to remove false positives for the aspiration of finding backwards traces, under-approximation can yield useful results, even on complex loops. Interestingly, each prime implicant and each sub-range can be tested for feasibility in parallel, which squares with the advent of multi-core processors and may allow the search for counterexamples on larger computer clusters.

6   Related Work

A sound static analysis, usually expressed using the abstract interpretation framework [10], is bound to calculate an over-approximate result to elude undecidability.

Due to over-approximation, a safety property may not be verifiable even though it holds. In this case, the emitted warning is a so-called false positive [3] which cannot a priori be distinguished from an actual defect in the software. While an analysis with zero false positives is possible [11], it is crucial to understand the origin of each alarm in order to either refine the analysis or to fix the defect. Thus, analyzing warnings which are emitted poses two related questions: firstly, *is the warning legitimate?*, and if so, *how can the error state be reached in terms of a concrete execution?* The difficulty of answering the first has led to approaches that rank warnings based on the likelihood of being actual defects. Statistical classifications have been based on error correlation [20] or bayesian filtering [15]. Recent work [23] clusters defects, allowing to eliminate dependent defects if a master defect is shown to be spurious (defects can be proven legitimate, too).

An exact answer to both questions is required in counterexample-guided abstraction refinement (CEGAR) in model checking [8]. However, deciding if a warning is legitimate is strictly easier in the context of CEGAR than in a general static analysis as the model checker produces an abstract counterexample. A concrete counterexample may then be inferable by replaying the trace in the concrete program [21]. If successful, the concrete trace can be used afterwards for, e.g., error localization [2]. If constructing the trace fails at a certain program point, a new predicate can be introduced to refine the abstract model [1]. In the context of numeric analysis, Gulavani and Rajamani [14] propose to refine a pre-analysis, based on a fixed point computation with widening, by introducing predicates using so-called hints. Later, they extended their technique to combine widening with interpolants between verification conditions and the inferred state [13]. Yet, neither work is concerned with computing the backward trace but assumes that it has been inferred by a theorem prover. Our approach can infer a set of traces that could provide additional hints as to what new predicates are needed, thus extending their work [13,14]. Finitization, as performed by our techniques, also appears in bounded model checking [7]. There, the state space of the system is explored in a breadth-first fashion in forward direction, up to a given depth k. By way of comparison, our approach unrolls the program back-to-front, implementing strategies to minimize the unrolling depth k during the generation of a counterexample on an under-approximate description of each block.

For static analyses that operate on the semantics of the actual program, no model program exists in which the trace can be inferred, and backward reasoning from the warning to the program entry is required [28]. Backward reasoning, in turn, amounts to solving the following abduction problem: Given B and C, compute a non-empty A in $(A \wedge B) \Rightarrow C$. Here, A and C can be thought of as states before and after a guard B, respectively. When A, B and C are elements of an abstract domain then $A \neq \bot$ is called the pseudo-complement of B relative to C if it is the largest unique element with $(A \sqcap B) \models C$. A domain in which each pair of elements has a pseudo-complement is called a Heyting domain. Few classes of linear constraints allow abduction [24] and no single numeric domain commonly used in forward analyses is Heyting, nor is the combination of Heyting domains necessarily a Heyting domain [25]. As an example, consider the intervals $B = [0, 0]$,

$C = [-5, 5]$ for which two incomparable A can be found, namely $A = [-5, -1]$ and $A = [1, 5]$. One way out of this dilemma is to lift a non-Heyting domain to its power-set domain [18], which yields a Boolean domain. A Boolean domain $\mathcal{B}$ is always Heyting since for each $b \in \mathcal{B}$ there exists a "full" complement $\bar{b} \in \mathcal{B}$ with $b \sqcup \bar{b} = \top$ and $b \sqcap \bar{b} = \bot$. Boolean functions naturally form a Boolean domain which motivates our choice for inferring backward traces. Given their expressiveness and the recent advances in SAT solving, it is sufficient to only use Boolean functions which also forestalls potential difficulties of combining this domain with other (Heyting) domains. Rival [28] sidesteps the abduction problem by calculating an A' with $A \models A'$ using the same domains as in forward analysis. To cap the over-approximation of the backward transformer, backward states are intersected with the forward invariants. Over-approximation makes it unlikely that an empty state is ever observed. Then, a warning cannot be identified as a false positive. Indeed, Rival's analysis merely informs a tool-users about inputs in which a counterexample might lie. In contrast, Erez [12] aims at reducing the number of false positives by performing a bounded search for backward traces using theorem proving. Further afield is the work of Kim et al. [17] who, after a fast but imprecise forward analysis, slice the program for a property violation before running a more expensive forward analysis based on SMT solving.

7   Conclusion

This paper advocates integrating under-approximate abduction using SAT into forward abstract interpretation frameworks. The motivation is to generate a definitive counterexample once a property violation has been detected or to identify a warning as spurious. Using Boolean formulae as abstract domain is theoretically motivated as many domains used in verification cannot express abduction. Moreover, the domain benefits from the progress in SAT solving, specifically the recent advances in computing under-approximate projections.

Acknowledgements. The authors want to thank Andy King for interesting discussions and Stefan Kowalewski for his support in this line of scientific enquiry.

References

1. Ball, T., Cook, B., Lahiri, S.K., Zhang, L.: ZAPATO: Automatic Theorem Proving for Predicate Abstraction Refinement. In: Alur, R., Peled, D.A. (eds.) CAV 2004. LNCS, vol. 3114, pp. 457–461. Springer, Heidelberg (2004)
2. Ball, T., Naik, M., Rajamani, S.K.: From symptom to cause: localizing errors in counterexample traces. In: POPL, pp. 97–105. ACM (2003)
3. Bessey, A., Block, K., Chelf, B., Chou, A., Fulton, B., Hallem, S., Henri-Gros, C., Kamsky, A., McPeak, S., Engler, D.R.: A few billion lines of code later: using static analysis to find bugs in the real world. Commun. ACM 53(2), 66–75 (2010)
4. Brauer, J., King, A.: Automatic Abstraction for Intervals Using Boolean Formulae. In: Cousot, R., Martel, M. (eds.) SAS 2010. LNCS, vol. 6337, pp. 167–183. Springer, Heidelberg (2010)

5. Brauer, J., King, A.: Transfer Function Synthesis without Quantifier Elimination. In: Barthe, G. (ed.) ESOP 2011. LNCS, vol. 6602, pp. 97–115. Springer, Heidelberg (2011)
6. Brauer, J., King, A., Kriener, J.: Existential Quantification as Incremental SAT. In: Gopalakrishnan, G., Qadeer, S. (eds.) CAV 2011. LNCS, vol. 6806, pp. 191–207. Springer, Heidelberg (2011)
7. Clarke, E., Biere, A., Raimi, R., Zhu, Y.: Bounded model checking using satisfiability solving. Formal Methods in System Design 19(1), 7–34 (2001)
8. Clarke, E., Grumberg, O., Jha, S., Lu, Y., Veith, H.: Counterexample-Guided Abstraction Refinement. In: Emerson, E.A., Sistla, A.P. (eds.) CAV 2000. LNCS, vol. 1855, pp. 154–169. Springer, Heidelberg (2000)
9. Clarke, E., Veith, H.: Counterexamples Revisited: Principles, Algorithms, Applications. In: Dershowitz, N. (ed.) Verification: Theory and Practice. LNCS, vol. 2772, pp. 208–224. Springer, Heidelberg (2004)
10. Cousot, P., Cousot, R.: Abstract Interpretation: A Unified Lattice model for Static Analysis of Programs by Construction or Approximation of Fixpoints. In: POPL, pp. 238–252. ACM (1977)
11. Cousot, P., Cousot, R., Feret, J., Mauborgne, L., Miné, A., Monniaux, D., Rival, X.: The ASTREÉ Analyzer. In: Sagiv, M. (ed.) ESOP 2005. LNCS, vol. 3444, pp. 21–30. Springer, Heidelberg (2005)
12. Erez, G.: Generating concrete counterexamples for sound abstract interpretation. Master's thesis, School of Computer Science, Tel-Aviv University, Israel (2004)
13. Gulavani, B.S., Chakraborty, S., Nori, A.V., Rajamani, S.K.: Automatically Refining Abstract Interpretations. In: Ramakrishnan, C.R., Rehof, J. (eds.) TACAS 2008. LNCS, vol. 4963, pp. 443–458. Springer, Heidelberg (2008)
14. Gulavani, B.S., Rajamani, S.K.: Counterexample Driven Refinement for Abstract Interpretation. In: Hermanns, H. (ed.) TACAS 2006. LNCS, vol. 3920, pp. 474–488. Springer, Heidelberg (2006)
15. Jung, Y., Kim, J., Shin, J., Yi, K.: Taming False Alarms from a Domain-Unaware C Analyzer by a Bayesian Statistical Post Analysis. In: Hankin, C., Siveroni, I. (eds.) SAS 2005. LNCS, vol. 3672, pp. 203–217. Springer, Heidelberg (2005)
16. Karr, M.: Affine Relationships among Variables of a Program. Acta Informatica 6, 133–151 (1976)
17. Kim, Y., Lee, J., Han, H., Choe, K.-M.: Filtering false alarms of buffer overflow analysis using SMT solvers. Inform. & Softw. Techn. 52(2), 210–219 (2010)
18. King, A., Lu, L.: Forward versus Backward Verification of Logic Programs. In: Palamidessi, C. (ed.) ICLP 2003. LNCS, vol. 2916, pp. 315–330. Springer, Heidelberg (2003)
19. King, A., Søndergaard, H.: Inferring Congruence Equations Using SAT. In: Gupta, A., Malik, S. (eds.) CAV 2008. LNCS, vol. 5123, pp. 281–293. Springer, Heidelberg (2008)
20. Kremenek, T., Engler, D.R.: Z-ranking: Using Statistical Analysis to Counter the Impact of Static Analysis Approximations. In: Cousot, R. (ed.) SAS 2003. LNCS, vol. 2694, pp. 295–315. Springer, Heidelberg (2003)
21. Kroning, D., Groce, A., Clarke, E.: Counterexample Guided Abstraction Refinement Via Program Execution. In: Davies, J., Schulte, W., Barnett, M. (eds.) ICFEM 2004. LNCS, vol. 3308, pp. 224–238. Springer, Heidelberg (2004)
22. Lahiri, S.K., Bryant, R.E., Cook, B.: A Symbolic Approach to Predicate Abstraction. In: Hunt Jr., W.A., Somenzi, F. (eds.) CAV 2003. LNCS, vol. 2725, pp. 141–153. Springer, Heidelberg (2003)

23. Lee, W., Lee, W., Yi, K.: Sound Non-statistical Clustering of Static Analysis Alarms. In: Kuncak, V., Rybalchenko, A. (eds.) VMCAI 2012. LNCS, vol. 7148, pp. 299–314. Springer, Heidelberg (2012)

24. Maher, M.J.: Abduction of Linear Arithmetic Constraints. In: Gabbrielli, M., Gupta, G. (eds.) ICLP 2005. LNCS, vol. 3668, pp. 174–188. Springer, Heidelberg (2005)

25. Maher, M.J., Huang, G.: On Computing Constraint Abduction Answers. In: Cervesato, I., Veith, H., Voronkov, A. (eds.) LPAR 2008. LNCS (LNAI), vol. 5330, pp. 421–435. Springer, Heidelberg (2008)

26. McMillan, K.L.: Applying SAT Methods in Unbounded Symbolic Model Checking. In: Brinksma, E., Larsen, K.G. (eds.) CAV 2002. LNCS, vol. 2404, pp. 250–264. Springer, Heidelberg (2002)

27. Müller-Olm, M., Seidl, H.: Analysis of Modular Arithmetic. ACM Trans. Program. Lang. Syst. 29(5) (August 2007)

28. Rival, X.: Understanding the Origin of Alarms in ASTRÉE. In: Hankin, C., Siveroni, I. (eds.) SAS 2005. LNCS, vol. 3672, pp. 303–319. Springer, Heidelberg (2005)

Modifying Test Suite Composition to Enable Effective Predicate-Level Statistical Debugging

Ross Gore and Paul F. Reynolds

University of Virginia, Department of Computer Science
P.O. Box 400740 Charlottesville, VA 22904 USA
{rjg7v,reynolds}@viriginia.edu

Abstract. In order to effectively deal with increased complexity and production pressures for the development of safety-critical systems, organizations need automated assistance in program analysis and testing. This need is intensified for systems that make heavy use of floating-point computations. Challenges related to the use of floating-point computations exist in the fields of testing, formal verification and debugging. While testing and formal verification provide mechanisms to identify possible failures within safety-critical systems, debugging techniques are employed to automatically isolate the cause of the failure. Recent advances in predicate-level statistical debugging have addressed localizing faults due to floating-point computations. Here, we present a methodology to modify the composition of a test suite to enable predicate-level statistical debuggers to more effectively isolate the causes of failures in safety-critical systems. Our methodology makes test suites significantly more effective for a class of debuggers, including those built to address faults due to floating-point computations.

Keywords: causal model, matching, debugging, safety-critical systems.

1   Introduction

The success of experiments involving safety-critical systems including autonomous robots, Next Generation Air Transportation (NextGen), and fly-by-wire spacecraft depends on the correctness of software [1, 2]. Achieving correctness in these systems is significantly more difficult when floating-point computations are used because the desire to employ efficient floating-point computations increases the likelihood of numerical analysis errors [3].

Floating-point correctness creates challenges within the fields of testing, formal verification and fault localization. Formal verification is difficult because the semantics of floating-point computations may change according to factors beyond source-code level, such as choices made by compilers. Testing is difficult because non-deterministic numerical analysis errors can result in difficult to replicate failures. Fault localization is difficult because the values of variables associated with a fault rarely are exactly equal to one another or a predetermined value.

A. Goodloe and S. Person (Eds.): NFM 2012, LNCS 7226, pp. 70–84, 2012.
© Springer-Verlag Berlin Heidelberg 2012

Numerous studies have shown that among verification, testing and fault localization, fault localization (debugging) takes up the most time in the development process [4, 5]. Recently, there has been considerable research on using statistical approaches for debugging [6-11]. Statistical debuggers require a test suite, execution profiles, and a labeling of the test executions as either succeeding or failing. The execution profiles reflect coverage of program elements. Program elements refer to individual statements or other inserted predicates. The approaches employ an estimate of the suspiciousness of the program elements. Then developers examine program elements in decreasing order of suspiciousness until the fault is discovered.

Here, we are concerned with predicate-level statistical debuggers. All predicate-level statistical debuggers share a common structure. Each debugger uses a set of conditional propositions, or predicates, which are inserted into a program and tested at particular points. A single predicate can be thought of as partitioning the space of all test cases into two subspaces: those satisfying the predicate and those not. Better predicates create partitions that more closely match where the fault is expressed.

In the canonical predicate-level statistical debugger Cooperative Bug Isolation (CBI), three predicates are inserted and tested for each variable x within a program statement: $(x>0)$, $(x=0)$ and $(x<0)$ [6]. In the statistical debugger Exploratory Software Predictor (ESP), these three predicates are complemented with *elastic predicates*. *Elastic predicates* use profiling to compute the mean, μ_x, and standard deviation, σ_x, of the values of variable x. Then, the CBI predicates are complemented with elastic predicates: $(x > \mu_x + \sigma_x)$, $(\mu_x + \sigma_x > x > \mu_x - \sigma_x)$ and $(x < \mu_x - \sigma_x)$ [8].

Elastic predicate debuggers, such as ESP, are the only fault localization techniques, which are designed to target faults due to floating-point computations. Elastic predicates are effective for such faults because the predicates: (1) expand or contract based on observed variable values and (2) do not employ a rigid notion of equality.

The standard suspiciousness metric for a predicate (elastic or otherwise) is the probability of a program Q failing given that a predicate p is true. This probability, $\Pr(Q\ fails \mid p=true)$, indicates if predicate p was true during an execution of Q at least once. Given the execution of a test suite, $\Pr(Q\ fails \mid p=true)$ is typically estimated by the sample ratio $(f_p\ /\ (f_p+s_p))$, where f_p is the number of tests for which p is true and the system fails and where s_p is the number of tests for which p is true and the system succeeds (does not fail). However, this estimate and other similarly derived estimates of suspiciousness are susceptible to at least two types of confounding bias: *control-flow dependency bias* and *failure-flow bias*. *Control-flow dependency bias* occurs when the conditions specified by a predicate corresponding to a fault cause other predicates to be evaluated during system failures [7]. *Failure-flow bias* occurs when a triggered fault causes the probability of reaching a subsequent statement where a predicate p is evaluated to be the same as the probability of p being true [6, 9].

In previous work, we introduced a causal model that accounts for these biases, to estimate the suspiciousness of a predicate by considering two groups of executions: those where predicate p is true at least once (the treatment group) and those where predicate p is not true (the control group) [9]. The estimate resulting from this model is more accurate than existing suspiciousness estimates, because it accounts for the possible confounding influences of other predicates on a given predicate p.

When estimating the suspiciousness of a predicate p, executions in the treatment group should have the same pattern of control flow dependences (control-flow dependencies bias) and statement coverage (failure-flow bias) as executions in the control group. When this condition is met the groups are balanced. When it is not met, there is a lack-of-balance and suspiciousness estimates become unreliable.

Lack-of-balance issues are avoided in controlled experiments by the random assignment of subjects to different groups. Unfortunately, in practice the set of test cases is given and we cannot assume that the subsets of test cases where p is true (treatment group) and where p is not true (control group) are balanced. Furthermore, generating a random test suite with respect to a set of predicates is a non-trivial task that is made even more difficult when elastic predicates are employed. Here, instead of generating such a test suite, we modify the existing suite to create a suite that mitigates lack-of-balance issues for predicates. First we show that for effective fault localization within safety-critical systems, the composition of a test suite should exhibit balance. Second, we present new methods to modify the existing test suite to overcome lack-of-balance problems using statistical matching techniques. Finally we present empirical evidence that our genetic matching technique modifies the existing test suite to improve the effectiveness of predicate-level statistical debuggers. This is a particularly significant contribution since these improvements affect elastic-predicate debuggers, the only class of fault localization tools that target faults due to floating-point computations. Ultimately, our work makes the development of safety-critical systems a more effective and efficient process.

2   Background

Predicate-level statistical debuggers can be improved by estimating the suspiciousness of predicates with *causal models* [9]. In this section we review causal models and discuss their application to predicate-level statistical debuggers.

2.1   Causal Models

Causal graphs represent assumptions about causality that permit statistical techniques to be used with observational data. A causal graph is a directed acyclic graph G. Within G, nodes represent random variables and edges represent cause-effect relationships. An edge $X \rightarrow Y$ indicates that X causes Y. Each random variable X has a probability distribution $P(x)$, which may not be known. The values of random variables are denoted by the corresponding lowercase letter of the random variable.

All causal effects associated with the causal model $M=(G, P)$ can be estimated if M is *Markovian*. *Markovian* means that each random variable X_i is conditionally independent of all its nondescendants, given the values of its parents (immediately preceding nodes) PA_i in G [12]. If M is *Markovian* then the joint distribution of the random variables is factored as:

$$p(x_1, x_2, \ldots, x_n) = \prod p(x_i \mid pa_i) . \tag{1}$$

M will satisfy the *Markovian* condition if for each node X_i in G, the relationship between X_i and its parents can be described by the structural equation [13]:

$$x_i = f_i(pa_i, u_i) .$$ (2)

Thus M is Markovian if it represents functional relationships (f_i) among a set of random variables and any external sources of error (u_i) are mutually independent. Markovian models are a powerful and concise formalism to combine causal estimation together with causal graphs. For a binary cause, there are two states to which each member of the population can be exposed: *treatment* and *control*. These states correspond to the values of a causal treatment variable T where: $T = 1$ for treated population members and $T = 0$ for controlled population members. Given an outcome variable Y over the population, there are two potential outcome random variables: Y^1 and Y^0. The average treatment effect, τ, in the population is [12]:

$$\tau = E[Y^1] - E[Y^0] .$$ (3)

Many problems call for estimating the average treatment effect in a population from a sample S. Let S_1 be the subset of S consisting of the treatment sample members, and let S_0 be the subset of S consisting of control sample members. The estimator of τ, $\hat{\tau}_{re}$, is the difference of the sample means of the outcomes for those in the treatment group (S_1) and those in the control (S_0) group [14]:

$$\hat{\tau}_{re} = \frac{1}{|S_1|}\sum_{i \in S_1} y_i - \frac{1}{|S_0|}\sum_{i \in S_0} y_i .$$ (4)

In an ideal randomized experiment, sample members are assigned to the treatment group or the control group randomly. Thus the treatment indicator variable T is independent of the potential outcomes Y^1 and Y^0 and $\hat{\tau}_{re}$ is an unbiased. However, ideal randomized experiments are rare. Instead, data often comes from an observational study. In an observational study the effects are not under the control of the investigator and occur in the past. Treatment selection is not random, so the treatment indicator variable T is not independent of potential outcomes Y^1 and Y^0 and $\hat{\tau}_{re}$ is likely to be biased and a better estimate of τ is needed.

Often one can characterize a better estimate of τ in terms of one or more variables that are suspected of influencing selection. These variables are covariates of the treatment indicator T. If a set X of covariates accounts for which members received the treatment and which did not, then the confounding bias, when estimating the average treatment effect τ on an outcome Y, can be reduced by conditioning on X. One way of conditioning on X when estimating an average treatment effect is to include it as a predictor in a linear regression model, such as in Eq. 5 [12]:

$$Y = \alpha + \tau T + \beta X + \varepsilon.$$ (5)

In Eq. 5, Y is the outcome variable, T is the treatment variable, X is a vector of covariates and ε is a random error term. The fitted value $\hat{\tau}$ of the coefficient τ is an unbiased estimate of the average-treatment effect.

2.2   Control Flow Dependencies

Recall that control-flow dependencies can be covariates of a predicate's influence on system failure. Here we review predicate control-flow dependencies and show how to incorporate them into our causal model for estimating predicate suspiciousness.

Defining predicate control flow dependencies requires first defining statement control flow dependencies. A program's dependence graph is a directed graph where nodes correspond to program statements and edges represent data and control dependences between statements [15]. Node Y is *control dependent* on node X if X has two outgoing edges and the traversal of one edge always leads to the execution of Y while the traversal of the other edge does not necessarily execute Y. Node X *dominates* node Y in a control flow graph if every path from the entry node to Y contains X. Node Y is *forward control dependent* on node X if Y is control dependent on X and Y does not dominate X [15]. Forward control dependences are control dependences that can be realized during execution without necessarily executing the dependent node more than once. Node X is a forward control flow predecessor of Node Y if Y is forward control dependent on X and X immediately precedes Y in the dependence graph. The statement corresponding to node X is the *forward control flow predecessor statement* of the statement corresponding to node Y.

Defining forward control flow predecessor predicates, as opposed to forward control flow predecessor statements, requires an additional step. For each variable x within a statement, all variables y_1, y_2,..., y_n referenced in the forward control flow predecessor statement are identified. Each pair of variables (x, y_i) induces additional predicate instrumentation that partitions the value of x and y_i with compound predicates. Thus, for a given test case and predicate p, the *forward control flow predecessor predicate* is the set of predicates that correspond to the *control flow predecessor statement* for p and are true when combined with p via a compound predicate. For CBI, the nine compound predicates are shown in Table 1. The elastic compound predicates employed in ESP are formed in the same manner [9].

Table 1. Compound CBI predicates for (x, y_i)

	$x = 0$	$x < 0$	$x > 0$
$y_i = 0$	$x = 0 \wedge y_i = 0$	$x < 0 \wedge y_i = 0$	$x > 0 \wedge y_i = 0$
$y_i < 0$	$x = 0 \wedge y_i > 0$	$x < 0 \wedge y_i < 0$	$x > 0 \wedge y_i < 0$
$y_i > 0$	$x = 0 \wedge y_i < 0$	$x < 0 \wedge y_i > 0$	$x > 0 \wedge y_i > 0$

2.3   Predicate-Level Statistical Debugging with Causal Models

In our previous work we presented the following linear regression model for estimating the suspiciousness of a predicate p [9]:

$$Y' = \alpha'_p + \tau'_p T'_p + \beta'_p C'_p + \omega'_p D'_p + \varepsilon'_p \tag{6}$$

The model in Eq. 6 is fit separately for each predicate in the subject program, using execution profiles of the instrumented predicates from a set of passing and failing test cases. In the model, Y is a binary variable that is 1 for a given execution if the program failed, T'_p is a treatment variable that is 1 if predicate p was true at least once during the execution, C'_p is a binary covariate that is 1 if the dynamic forward control-flow dependence predecessor control-flow ($cfp(p)$) of p was true at least once during the execution, D'_p is a binary covariate that is 1 if the statement corresponding to predicate p was covered during execution, α'_p is a constant intercept and ε'_p is a random error term that does not depend on the values of T'_p, C'_p and D'_p. The average treatment effect of T'_p upon Y is estimated by the fitted value $\hat{\tau}'_p$ of the coefficient τ'_p, which is the estimate of suspiciousness for p. The role of C'_p, the control-flow predecessor covariate and D'_p, the statement-coverage covariate is to account for confounding bias in the suspiciousness estimate for p, due to coverage of other predicates. Conditioning on C'_p and D'_p reduces confounding because they reflect the most immediate causes of p being true or not true during a particular execution [9].

3   Modifying Test Suites with Matching

We illustrate that lack-of-balance can create unreliable predicate suspiciousness estimates using an implementation of the Traffic Collision Avoidance System (TCAS). TCAS monitors an aircraft's airspace and warns pilots of possible collisions with other aircraft [16]. The code snippet in Fig. 1 comes from the TCAS implementation in the Software-artifact Infrastructure Repository (SIR) [17]. Statement 126 contains a fault, which causes some system test cases to fail.

```
124    if (enabled && ((tcas_equipped && intent_not_known) || !tcas_equipped)
125    {
126        need_upward_RA = Non_Crossing_Biased_Climb(); // BUG: && Own_Below_Threat();
127        need_downward_RA = Non_Crossing_Biased_Descend() && Own_Above_Threat();
128        if (need_upward_RA && need_downward_RA)
129            alt_sep = UNRESOLVED;
130        else if (need_upward_RA)
131            alt_sep = UPWARD_RA;
132        else if (need_downward_RA)
133            alt_sep = DOWNWARD_RA;
134        else
135            alt_sep = UNRESOLVED;
136    }
```

Fig. 1. A code snippet from a faulty implementation of **tcas**

Table 2 summarizes execution data gathered for the predicate **need_upward_RA > 0** in statement 126 in Fig. 1. An entry of "1" in Table 2 denotes that the characteristic was observed in the specified number of execution traces and an entry of "0" denotes the characteristic was not observed in the specified number of execution traces. Combinations of characteristics not listed in Table 2 do not occur.

Table 2. Execution data for predicate **need_upward_RA > 0** at Statement 126

# of Tests	Pred. Truth	Forward Control-flow Predecessor Pred. (*cfp(p)*)	Statement Coverage	Failure
294	1	1	1	1
721	1	1	1	0
32	0	1	1	0
8	0	0	1	0
516	0	0	0	0

The tests for this predicate are not balanced. In all the tests where **need_upward_RA > 0** is true (treatment group), the forward control flow predecessor predicate (*cfp(p)*) for the predicate is also true and the tests cover statement 126. However, in only 32 of the tests where **need_upward_RA > 0** is not true (control group), is the *cfp(p)* true. Furthermore, there are another 8 test cases where the *cfp(p)* is not true but the statement is covered. The remaining 516 tests branch away before reaching the *cfp(p)* or statement 126. These differences represent a lack-of-balance. Next, we'll illustrate how this lack-of-balance can make a predicate suspiciousness estimate unreliable and how matching can mitigate this effect.

3.1   Exact Matching

In *exact matching*, each treatment unit is matched with one or more control units that have exactly the same covariate values as the treatment unit [14]. In the context of predicate-level statistical debugging, exact matching excludes tests where predicate *p* is not true, *cfp(p)* is not true and the statement *s* corresponding to *p* is not covered. The only tests from the control group that are used in fitting the causal model are those where *cfp(p)* is true and the statement *s* corresponding to *p* is covered. This ensures that the modified control and treatment groups are balanced with respect to the covariates (truth of *cfp(p)* and the coverage of *s*) because each of these covariates will be present in all entries in the treatment and control groups.

An example helps elucidate the purpose of exact matching. We consider the predicate **need_upward_RA > 0** in statement 126. Exact matching excludes 524 test cases from the test suite that is used to fit the causal model for this predicate. These tests are reflected in the bottom two rows of Table 2. If we do not employ matching while estimating the suspiciousness of **need_upward_RA > 0** in statement 126, then the predicate suspiciousness estimate is 0.24. However, if we exclude the 524 unmatched test cases the predicate suspiciousness estimate becomes 0.29. Furthermore, the suspiciousness of the predicate **(enabled && ((tcas_equipped && intent_not_known) || !tcas_equipped) > 0** in statement 124 is 0.27 regardless of the use of matching. Thus without matching the predicate **(enabled && ((tcas_equipped && intent_not_known) || !tcas_equipped) > 0** in statement 124 will be considered more suspicious than the predicate **need_upward_RA > 0** in statement 126. This is incorrect. The fault in Fig. 1 lies in statement 126. By excluding the unmatched test cases we are able to more accurately estimate the suspiciousness of the predicates in Fig. 1 and identify the fault.

Unfortunately, as the number of covariates increases, it is more difficult to find exact matches between treatment and control units. This results in more discarded test cases, which makes suspiciousness estimates less reliable. In the next section we present more advanced techniques, which reduce the number of discarded test cases.

3.2   More Advanced Matching Techniques

To obtain more flexibility in matching treatment and control units, more advanced matching techniques, such as *Mahalanobis Distance* (MD) matching are employed [18]. The MD metric, $d_M(a,b)$, measures the similarity between two vectors. Eq. 7 shows the computation of the MD between two random vectors a and b where T is the vector transpose and S^{-1} is the inverse of the covariance matrix for a and b.

$$d_M(a,b) = \sqrt{(a-b)^T S^{-1} (a-b)}. \tag{7}$$

In the context of predicate-level statistical debugging, a reflects one covariate vector from a test case where predicate p is true and b reflects one covariate vector from a test case where predicate p is not true. The matrix S is the sample covariance matrix for all the test cases. In MD matching a treatment unit a is matched with a control unit b if and only if $d_M(a,b)$ is minimal.

One property of MD matching is that all the components of the covariate vector are given equal weight in the MD metric. In other words, the truth of $cfp(p)$ is given the same weight as the statement coverage of s in the computation of $d_M(a,b)$ [18]. This property seems reasonable but it does not always yield optimal covariate balance. To address this issue a generalization of MD Matching, *genetic matching* has been proposed [18].

Genetic Matching (GM) considers many different distance metrics, rather than just one, and employs the metric that optimizes covariate balance. Each potential distance metric considered in GM corresponds to a particular assignment of weights to covariates. The algorithm weights each covariate according to its relative importance for achieving the best overall balance. Eq. 8 shows the general equation for the computation of the GM distance, $d_{GM}(a,b)$, between two random vectors a and b given a weight matrix W. In Eq. 8, T denotes the vector transpose and $S^{-1/2}$ is the Cholesky Decomposition of the covariance matrix for a and b [18]:

$$d_{GM}(a,b,W) = \sqrt{(a-b)^T \left(S^{-1/2}\right)^T W S^{-1/2} (a-b)}. \tag{8}$$

GM uses a genetic search algorithm to choose the weights, W. The algorithm moves towards the W that maximizes overall balance by iteratively minimizing the largest observed covariate discrepancy. Since the largest observed covariate discrepancy is iteratively minimized, GM achieves optimal balance in the limit. When W converges to the identity matrix GM is equivalent to MD matching [18].

Our test suite modification employing GM for each predicate is shown in Alg. 1. T is the set containing the test cases and execution profiles for those test cases where predicate p is true (treatment covariate data), C is the set containing the test cases and execution profiles for those test cases where predicate p is not true (control covariate data). GM is the genetic matching function. The algorithm returns the balanced set of test cases, *Tests*, which is used to fit the causal model specified in Eq. 6 and compute the suspiciousness estimate for the predicate. In the next section we will see how Alg. 1 significantly improves the effectiveness of predicate-level statistical debuggers for a variety of faulty software applications including several safety-critical systems.

Algorithm 1. Test Suite Modification Algorithm: TEST SUITE MOD

TEST SUITE MOD(T, C)

1	$Tests \leftarrow NULL$
2	$w \quad \leftarrow \mathrm{GM}(T, C)$
3	$index \leftarrow 0$
4	**for** (each $T_i \in T$)
5	$\quad min \leftarrow \infty$
6	$\quad$ **for** (each $C_j \in C$)
7	$\qquad distance \leftarrow d_{GM}(T_i,\ C_j, w)$
8	$\qquad$ **if** ($distance < min$)
9	$\qquad\quad min \leftarrow distance$
10	$\qquad\quad index \leftarrow j$
11	$\quad$ add C_{index} to *Tests*
12	$\quad$ add $T_i \quad$ to *Tests*
13	**return** *Tests*

4   Empirical Evaluation

The evaluation includes programs used to evaluate statistical debuggers (print-tokens, print-tokens2, replace, totinfo, sed, space, gzip, bc) and safety-critical programs (schedule, schedule2, tcas). Recall tcas is an air traffic collision avoidance system. schedule and schedule2 are different implementations of priority scheduling algorithms found in engine controllers and nuclear power plants [19]. These three programs make up approximately one-third of our evaluation subject programs. Furthermore approximately one-half of our evaluation programs employ floating-point computations. Table 3 shows the characteristics of all the subject programs.

Each test suite modification algorithm is implemented in R, which is a statistical computation language and runtime environment [20]. For matching on control flow predicate dependences and statement coverage data we used the R package *Matching*, which implements MD matching and GM [21]. We perform MD matching and GM with replacement. In matching with replacement, the control unit is retained so it can be matched with other treatment units. We use replacement because it has been shown to be more effective for similar matching applications [18]. Matching introduces uncertainty into the suspiciousness estimate for a predicate. In GM the optimization that identifies the weight matrix includes a random component and in MD matching ties for minimality are broken randomly. In both cases, the resulting predicate suspiciousness estimate can vary. To reduce the uncertainty we compute a predicate's suspiciousness 100 times and take the mean as the suspiciousness estimate.

Table 3. Subject programs used in our evaluation

Name	LOC	Vers.	# Tests	Description
tcas	138	41	1608	altitude separation
totinfo	396	23	1052	information measure
schedule	299	9	2650	priority scheduler
schedule2	297	9	2710	priority scheduler
print-tokens	472	5	4130	lexical analyzer
print-tokens2	399	10	4115	lexical analyzer
replace	512	31	5542	pattern recognition
sed	6,092	7	363	stream editing utility
space	14,382	35	157	ADL interpreter
bc (1.06)	14,288	1	4,000	basic calculator
gzip	7,266	9	217	compression utility

4.1   Effectiveness and Efficiency Studies

To measure the effectiveness of the debugging techniques we use an established cost-measuring function (*Cost*) [7-9, 11]. *Cost* measures the percentage of predicates a developer must examine before the fault is found, assuming the predicates are sorted in descending order of suspiciousness. To compare two techniques A and B for effectiveness, we subtract the *Cost* value for A from the *Cost* value for B. If A performs better than B, then the *Cost* is positive and if B performs better than A, the *Cost* is negative. For example, if for a given program the Cost of A is 30% and the *Cost* of B is 40%, then A is a 10% absolute improvement over B.

Table 4 and Table 5 summarize the results of comparing the test suite modification techniques in the statistical debuggers CBI and ESP respectively. Our new genetic matching statistical debugging technique is GM. A variant of the new technique that employs MD matching instead of genetic matching is MDM. Our previous technique, which does not employ matching is NM. The first column of each table shows the two techniques being compared. The second column (Positive %) shows the percentage of faulty versions where the first technique performed better, the third column (Neutral %) shows the percentage of faulty versions where there was no improvement and the fourth column (Negative %) shows the percentage of faulty versions where the second technique performed better. Table 4 and Table 5 also show the minimum (Min), mean (Mean) and maximum (Max) improvement or degradation for the Positive % column and Negative % column. For example in Table 4, comparing GM vs NM, for the 33.88% of faulty versions with positive improvement, the minimum improvement was 0.03%, the mean improvement was 6.03% and maximum improvement was 14.23%.

Table 4. Comparison of Matching Techniques within CBI

Comparing	Positive %			Neutral %	Negative %		
	Min	Mean	Max		Min	Mean	Max
GM vs NM		33.88		62.79		3.33	
	0.03	6.03	14.23		0.03	2.26	9.08
MDM vs NM		21.11		73.89		5.00	
	0.03	5.09	10.21		0.03	3.16	12.90
GM vs MDM		12.77		85.56		1.67	
	0.02	1.82	7.88		0.02	0.08	0.21

Table 5. Comparison of Matching Techniques within ESP

Comparing	Positive %			Neutral %	Negative %		
	Min	Mean	Max		Min	Mean	Max
GM vs NM		39.44		57.78		2.77	
	0.02	4.45	10.23		0.02	1.24	6.12
MDM vs NM		24.44		71.05		4.51	
	0.02	3.06	8.39		0.02	2.92	8.84
GM vs MDM		15.00		83.89		1.11	
	0.02	1.26	5.05		0.02	0.11	0.21

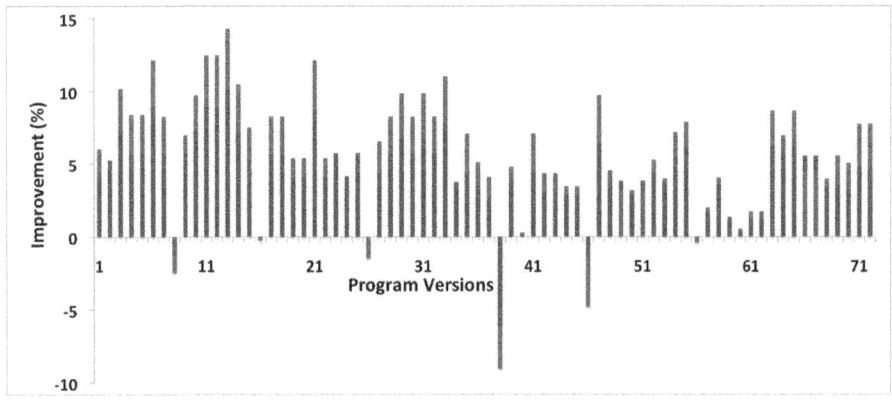

Fig. 2. Effectiveness of GM vs NM in ESP (blue) and CBI (red)

Fig. 2 summarizes the comparison of GM to NM over all program versions where there was some change in effectiveness for the statistical debuggers CBI (red) and ESP (blue). For each program version, the absolute improvement within each predicate-level debugger provided by GM is represented with a bi-colored bar. The height of the colored portion of the bar closest to the x-axis reflects the improvement for the matching debugger. The total height of both portions reflects the improvement for the debugger matching the colored portion of the bar furthest from the x-axis. For example, for program version 21 on the graph, using GM in ESP resulted in examining 8.62% fewer predicates. For the same version on the graph, using GM in CBI resulted in examining 12.11% fewer predicates. Overall Fig. 2 shows that GM performed better than NM for both CBI and ESP. Table 4 shows that within CBI, GM performed better than NM on 33.88% of the faulty versions, worse on 3.33% of the faulty versions, and showed no improvement on 62.79% of the versions. Table 5 shows that within ESP, GM performed better than NM on 39.44% of the faulty versions, worse on 2.77% of the faulty versions, and showed no improvement on 57.78% of the faulty versions. Furthermore, the improvements in the ESP and CBI with GM are significantly greater than the degradations.

Although GM performed the best of the three test suite modification techniques, we explored how much of the improvement resulted from the choice to use genetic matching as opposed to MD matching. To do this, we compared GM to MDM. Table 4 shows that within CBI, GM performed better than MDM on 12.77% of the faulty versions and performed worse on only 1.67% of the versions. Similarly for ESP, GM performed better than MDM on 15.00% of the versions and performed worse on only 1.11% of the versions. Thus, finding the optimal covariate balance in a test suite as opposed to simply improving the balance, significantly improves the effectiveness of a predicate-level statistical debugger. Overall, the results show that by using genetic matching to find optimal covariate balance, confounding bias can be significantly reduced when estimating the suspiciousness of a predicate.

Table 6. Mean efficiency results for each subject program

Name	# of Test	CBI Mins. Spent Matching	CBI Total Mins.	ESP Mins. Spent Matching	ESP Total Mins.
tcas	1608	30.9	31.2	62.6	64.0
totinfo	1052	15.3	15.9	31.4	33.0
schedule	2650	90.7	91.1	183.4	184.7
schedule2	2710	100.5	100.8	204.1	205.4
print-tokens	4130	480.2	480.6	966.8	968.1
print-tokens2	4115	470.4	470.8	945.3	947.1
replace	5542	885.5	885.9	1,774.2	1,775.8
sed	363	4.8	5.4	10.1	12.9
space	157	1.8	2.7	3.9	7.5
bc (1.06)	4000	443.6	444.3	899.3	902.2
gzip	217	2.1	2.9	4.3	8.4

Table 6 shows the mean absolute computation time for each of the evaluation subject programs. The computation time is largely dependent on the size of the test suite. The GM and MD Matching algorithms within the Matching package take considerable time to invert large covariance matrices resulting from large test suites [21]. The time required to match the test cases for ESP is roughly double that of CBI because ESP contains twice as many predicates as CBI.

While matching makes CBI and ESP less efficient, it only requires machine time not developer time. If developers can remain productive while matching is performed, overall efficiency will be improved because the developer is given a more effective list of ranked predicates to identify faults. This rationale has made formal verification methods useful despite execution times measured in days and hours [22].

4.2   Discussion

Although GM and MDM performed well in our evaluation, each was not as effective as our previous causal model (NM) for some program versions. The faults in these versions are not triggered by the coverage of the statement corresponding to a predicate. Instead the faults reflect missing statements and the predicates corresponding to statements adjacent to the missing code qualify as the fault. While these faults demonstrate a limitation of our approach, they have not been effectively localized by other automated debugging approaches either [6-11].

The effectiveness of GM and MDM relative to CBI and ESP is also important to discuss. For the preponderance of the program versions CBI offers more improvement than ESP. However, for most of the program versions ESP incurs less overall *Cost* for developers. This paradox can be explained. ESP has been shown to be more effective than CBI when biased suspiciousness metrics or no test suite modification is employed [8, 9]. Thus while GM and MDM improve the effectiveness of each statistical debugger, ESP appears to improve less because of its superior effectiveness. Similarly, for most of the versions where negative improvement is observed, the effectiveness of ESP degrades less than CBI. Furthermore, given a faulty program version, ESP is more likely than CBI to offer some improvement. This is because the additional predicates employed in ESP create more situations where there is a lack-of-balance. As a result matching has a better chance of improving the effectiveness.

4.3   Threats to Validity

Validity threats affect our evaluation. Threats to internal validity concern factors that might affect dependent variables without the researcher's knowledge. The implementations of the matching algorithms we used in our studies could contain errors. However, the Matching package we used is open source and is widely used. Threats to external validity occur when the results of our evaluation cannot be generalized. While we performed our evaluations on nine programs with a total of 180 versions and two different predicate-level statistical debuggers (CBI and ESP), we cannot claim that the effectiveness observed in our evaluation can be generalized to other faults in other programs. Threats to construct validity concern the appropriateness of the evaluation metrics used. More studies into how useful developers find ranked predicates need to be performed. However, the more accurate ranking methods are the more meaningful such studies will become.

5   Related Work

The debugging approach that is most closely related to our work is the statement-level statistical debugging work of Baah et. al [7]. Baah et. al have also developed a causal model that employs matching to balance test cases. However, our work differs from their work in three important ways. First, our work addresses debugging at the predicate-level while Baah et. al's approach operates on the statement-level. Debugging at the predicate-level allows developers to identify more faults than debugging at the statement-level [8]. Second, while both our predicate-level approach and Baah et al.'s statement-level approach reduce confounding bias due to control-flow dependencies, our approach is capable of further reducing bias by considering failure-flow. Third, our test suite modification algorithm uses genetic matching to balance the test suite while Baah et al.'s work employs MD matching. Recall, MD matching does not ensure optimal balance of covariates and did not perform as well as genetic matching in our evaluation [18, 21].

Other statistical debugging techniques exist but they employ biased suspiciousness estimates [6, 10]. State-altering debugging approaches such as Delta Debugging [23] and IVMP [11] attempt to find the cause of program failure by altering program states and re-executing the program. However, the experiments they perform on altered programs can be time consuming and require an oracle to determine the success or failure of each altered program. Other approaches use slicing [24-26] to compute the statements that potentially affect the values of a given program point. However, these techniques do not provide any guidance, such as rankings, to the developer. Thus, it is difficult to compare them with our work.

6   Conclusion

The development of safety-critical systems presents challenges in the fields of testing, formal verification and debugging. Numerous studies have shown that debugging takes up the most time of these three tasks [4, 5]. We have presented a novel statistical debugging technique that modifies the program's test suite based on information about dynamic control-flow dependences and statement coverage to obtain more accurate estimates of an instrumented predicate's effect on the occurrences of safety critical system failures. Our evaluation shows that our technique provides more effective debugging for a variety of software applications including safety-critical systems. The result is a more effective development process.

References

1. Swenson, H., Barhydt, R., Landis, M.: Next Generation Air Transportation System Air Traffic Management, Technical report, NASA Ames Research Center (2006)
2. Stanton, N.A., Marsden, P.: From Fly-by-Wire to Drive-by-Wire: Safety Implications of Automation in Vehicles. Safety Sci. 24(1), 35–49 (1996)
3. Monniaux, D.: The Pitfalls of Verifying Floating-point Computations. ACM Trans. Program. Lang. Syst. 30(3), 1–41 (2008)

4. Hallam, P.: What do programmers really do anyway? In: Microsoft Developer Network (MSDN) C# Compiler (2006)
5. Pigoski, T.M.: Practical Software Maintenance: Best Practices for Managing Your Software Investments. John Wiley & Sons, New York (1996)
6. Liblit, B., Naik, M., Zheng, A.X., Aiken, A., Jordan, M.I.: Scalable Statistical Bug Isolation. In: Conf. on Programming Language Design and Implementation, pp. 15–26. ACM Press, New York (2005)
7. Baah, G.K., Podgurski, A., Harrold, M.J.: Mitigating the Confounding Effects of Program Dependences for Effective Fault Localization. In: Symp. on Foundations of Software Engineering, pp. 146–156. ACM Press, New York (2011)
8. Gore, R., Reynolds, P.F., Kamensky, D.: Statistical Debugging with Elastic Predicates. In: Conf. on Automated Software Engineering, pp. 492–495. ACM Press, New York (2011)
9. Gore, R., Reynolds, P.F.: Reducing Confounding Bias in Predicate-level Statistical Debuggers. Submitted to: Conf. on Software Engineering. ACM Press, New York (2012)
10. Renieris, M., Reiss, S.: Fault Localization with Nearest Neighbor Queries. In: Conf. on Automated Software Engineering, pp. 30–39. ACM Press, New York (2003)
11. Jeffery, D., Gupta, N., Gupta, R.: Fault Localization using Value Replacement. In: Symp. on Software Testing and Analysis, pp. 167–177. ACM Press, New York (2008)
12. Pearl, J.: Causality: Models, Reasoning, and Inference. Cambridge University Press, San Francisco (2000)
13. Pearl, J., Verma, T.: A Theory of Inferred Causation. In: Conf. on Principles of Knowledge Representation and Reasoning, pp. 441–452. AAAI Press, Palo Alto (1991)
14. Morgan, S.L., Winship, C.: Counterfactuals and Causal Inference: Methods and Principles of Social Research. Cambridge University Press, Cambridge (2007)
15. Ferrante, J., Ottenstein, K.J., Warren, J.D.: The Program dependence graph and its use in optimization. ACM Trans. on Program Lang. Sys. 9(30), 319–349 (1987)
16. Harman, W.H.: TCAS: A System for Preventing Midair Collisions. Lincoln Laboratory Journal 2, 437–458 (1989)
17. SIR: Software-artifact Infrastructure Repository, http://sir.unl.edu/portal/index.html
18. Diamond, A., Sekhon, J.S.: Genetic Matching for Estimating Causal Effects: A General Multivariate Matching Method for Achieving Balance in Observational Studies. Technical report (2006)
19. Bate, I., Burns, A.: An Integrated Approach to Scheduling in Safety-critical Embedded Control Systems. Real-Time Syst. 25(1), 5–37 (2003)
20. R Development Core Team. R: A Language and Environment for Statistical Computing. R Foundation for Statistical Computing, Vienna, Austria (2008)
21. Sekhon, J.S.: Multivariate and Propensity Score Matching Software with Automated Balance Optimization: The Matching package for R. J. Stats. Soft. 42(7), 1–52 (2011)
22. D'Silva, V., Kroening, D., Weissenbacher, G.: A Survey of Automated Techniques for Formal Software Verification. IEEE Trans. Comp. Aided Design 27(7), 1165–1178 (2008)
23. Cleve, H., Zeller, A.: Locating Causes of Program Failures. In: Conf. on Software Engineering, pp. 342–351. ACM Press, New York (2005)
24. Gupta, N., He, H., Zhang, X., Gupta, R.: Locating Faulty Code Using Failure-inducing Chops. In: Conf. on Automated Software Engineering, pp. 263–272. ACM Press, New York (2005)
25. Tip, F.: A Survey of Program Slicing Techniques. J. Program. Lang. 3, 121–189 (1995)
26. Weiser, M.: Program Slicing. In: Conf. on Software Engineering, pp. 439–449. ACM Press, New York (1981)

Rigorous Polynomial Approximation Using Taylor Models in Coq[*]

Nicolas Brisebarre[1], Mioara Joldeş[4], Érik Martin-Dorel[1],
Micaela Mayero[1,2], Jean-Michel Muller[1], Ioana Paşca[1],
Laurence Rideau[3], and Laurent Théry[3]

[1] LIP, CNRS UMR 5668, ENS de Lyon, INRIA Grenoble - Rhône-Alpes, UCBL,
Arénaire, Lyon, F-69364
[2] LIPN, UMR 7030, Université Paris 13, LCR, Villetaneuse, F-93430
[3] Marelle, INRIA Sophia Antipolis - Méditerranée, Sophia Antipolis, F-06902
[4] CAPA, Dpt. of Mathematics, Uppsala Univ., Box S-524, 75120, Uppsala, Sweden

Abstract. One of the most common and practical ways of representing a real function on machines is by using a polynomial approximation. It is then important to properly handle the error introduced by such an approximation. The purpose of this work is to offer guaranteed error bounds for a specific kind of rigorous polynomial approximation called Taylor model. We carry out this work in the Coq proof assistant, with a special focus on genericity and efficiency for our implementation. We give an abstract interface for rigorous polynomial approximations, parameterized by the type of coefficients and the implementation of polynomials, and we instantiate this interface to the case of Taylor models with interval coefficients, while providing all the machinery for computing them. We compare the performances of our implementation in Coq with those of the Sollya tool, which contains an implementation of Taylor models written in C. This is a milestone in our long-term goal of providing fully formally proved and efficient Taylor models.

Keywords: certified error bounds, Taylor models, Coq proof assistant, rigorous polynomial approximation.

1 Rigorous Approximation of Functions by Polynomials

It is frequently useful to replace a given function of a real variable by a simpler function, such as a polynomial, chosen to have values close to those of the given function, since such an approximation may be more compact to represent and store but also more efficient to evaluate and manipulate. As long as evaluation is concerned, polynomial approximations are especially important. In general the basic functions that are implemented in hardware on a processor are limited to addition, subtraction, multiplication, and sometimes division. Moreover, division is significantly slower than multiplication. The only functions of one variable

---

[*] This research was supported by the TaMaDi project of the French ANR (ref. ANR-2010-BLAN-0203-01).

A. Goodloe and S. Person (Eds.): NFM 2012, LNCS 7226, pp. 85–99, 2012.

that one may evaluate using a bounded number of additions/subtractions, multiplications and comparisons are piecewise polynomials: hence, on such systems, polynomial approximations are not only a good choice for implementing more complex functions, they are frequently the only one that makes sense.

Polynomial approximations for widely used functions used to be tabulated in handbooks [1]. Nowadays, most computer algebra systems provide routines for obtaining polynomial approximations of commonly used functions. However, when bounds for the approximation errors are available, they are not guaranteed to be accurate and are sometimes unreliable.

Our goal is to provide efficient and quickly computable *rigorous polynomial approximations*, i.e., polynomial approximations for which (i) the provided error bound is tight and not underestimated, (ii) the framework is suitable for formal proof (indeed, the computations are done in a formal proof checker), while requiring computation times similar to those of a conventional C implementation.

1.1 Motivations

Most numerical systems depend on standard functions like exp, sin, etc., which are implemented in libraries called `libm`s. These `libm`s must offer guarantees regarding the provided accuracy: they are heavily tested before being published, but for precisions higher that single precision, an exhaustive test is impossible [16]. Hence a proof of the behavior of the program that implements a standard function should come with it, whenever possible. One of the key elements of such a proof would be the guarantee that the used polynomial approximation is within some threshold from the function. This requirement is even more important when *correct rounding* is at stake. Most `libm`s do not provide correctly rounded functions, although the IEEE 754-2008 Standard for Floating-Point (FP) Arithmetic [22] recommends it for a set of basic functions. Implementing a correctly rounded function requires rigorous polynomial approximations at two steps: when actually implementing the function in a given precision, and—before that—when trying to solve the *table maker's dilemma* for that precision.

The 1985 version of the IEEE Standard for FP Arithmetic requires that the basic operations ($+$, $-$, $\times$, $\div$, and $\sqrt{\cdot}$) should produce *correctly rounded results*, as if the operations were first carried out in infinite precision and these intermediate results were then rounded. This contributed to a certain level of portability and provability of FP algorithms. Until 2008, there was no such analogous requirement for standard functions. The main impediment for this was the *table maker's dilemma*, which can be stated as follows: consider a function f and a FP number x. In most cases, $y = f(x)$ cannot be represented exactly. The correctly rounded result is the FP number that is closest to y. Using a finite precision environment, only an approximation $\hat{y}$ to y can be computed. If that approximation is not accurate enough, one cannot decide the correct rounding of y from $\hat{y}$. Ziv [41] suggested to improve the accuracy of the approximation until the correctly rounded value can be decided. A first improvement over that approach derives from the availability of tight bounds on the worst-case accuracy required to compute some functions [25], which made it possible to write a `libm` with correctly rounded functions, where correct rounding is obtained at modest additional costs [37]. The TaMaDi project [32] aims at computing the worst-case accuracy for the most common functions and formats. Doing this requires very accurate polynomial approximations that are formally verified.

Beside the Table Maker's Dilemma, the implementation of correctly rounded elementary functions is a complex process, which includes finding polynomial approximations for the considered function that are accurate enough. Obtaining good polynomial approximations is detailed in [10,9,12]. In the same time, the approximation error between the function and the polynomial is very important since one must make sure that the approximation is good enough. The description of a fast, automatic and verifiable process was given in [23].

In the context of implementing a standard function, we are interested in finding polynomial approximations for which, given a degree n, the maximum error between the function and the polynomial is minimum: this "minimax approximation" has been broadly developed in the literature and its application to function implementation is discussed in detail in [12,33]. Usually this approximation is computed numerically [38], so an a posteriori error bound is needed. Obtaining a tight bound for the approximation error reduces to computing a tight bound for the supremum norm of the error function over the considered interval. Absolute error as well as relative errors can be considered. For the sake of simplicity, in this paper, we consider absolute errors only (relative errors would be handled similarly). Our problem can be seen as a univariate rigorous global optimization problem, however, obtaining a tight and formally verified interval bound for the supremum norm of the error function presents issues unsuspected at a first sight [14], so that techniques like interval arithmetic and Taylor models are needed. An introduction to these concepts is given below.

Interval arithmetic and Taylor models. The usual arithmetic operations and functions are straightforwardly extended to handle intervals. One use of interval arithmetic is bounding the image of a function over an interval. Interval calculations frequently overestimate the image of a function. This phenomenon is in general proportional to the width of the input interval. We are therefore interested in using thin input intervals in order to get a tight bound on the image of the function. While subdivision methods are successfully used in general, when trying to solve this problem, one is faced with what is known as a "dependency phenomenon": since function f and its approximating polynomial p are highly correlated, branch and bound methods based on using intervals of smaller width to obtain less overestimation, end up with an unreasonably high number of small intervals. To reduce the dependency, *Taylor models* are used. They are a basic tool for replacing functions with a polynomial and an interval remainder bound, on which basic arithmetic operations or bounding methods are easier.

1.2   Related Work

Taylor models [27,35,28] are used for solving rigorous global optimization problems [27,5,14,6] and obtaining validated solutions of ODEs [34] with applications to critical systems like particle accelerators [6] or robust space mission design [26]. Freely available implementations are scarce. One such implementation is available in SOLLYA [13]. It handles univariate functions only, but provides multiple-precision support for the coefficients. It was used for proving the correctness of supremum norms of approximation errors in [14], and so far it is the only freely available tool that provides such routines. However, this remains a C implementation that does not provide formally proved Taylor models, although this would be necessary for having a completely formally verified algorithm.

There have been several attempts to formalize Taylor models (TMs) in proof assistants. An implementation of multivariate TMs is presented in [42]. They are implemented on top of a library of exact real arithmetic, which is more costly than FP arithmetic. The purpose of that work is different than ours. It is appropriate for multivariate polynomials with small degrees, while we want univariate polynomials and high degrees. There are no formal proofs for that implementation. An implementation of univariate TMs in PVS is presented in [11]. Though formally proved, it contains ad-hoc models for a few functions only, and it is not efficient enough for our needs, as it is unable to produce Taylor models of degree higher than 6. Another formalization of Taylor models in COQ is presented in [15]. It uses polynomials with FP coefficients. However, the coefficients are axiomatized, so we cannot compute the actual Taylor model in that implementation. We can only talk about the properties of the involved algorithms.

Our purpose is to provide a modular implementation of univariate Taylor models in COQ, which is efficient enough to produce very accurate approximations of elementary real functions. We start by presenting in Section 2 the mathematical definitions of Taylor models as well as efficient algorithms used in their implementation. We then present in Section 3 the COQ implementation. Finally we evaluate in Section 4 the quality of our implementation, both from the point of view of efficient computation and of numerical accuracy of the results.

2   Presentation of the Taylor Models

2.1   Definition, Arithmetic

A Taylor model (TM) of order n for a function f which is supposed to be $n+1$ times differentiable over an interval $[a, b]$, is a pair (T, Δ) formed by a polynomial T of degree n, and an interval part Δ, such that $f(x) - T(x) \in \Delta, \forall x \in [a, b]$. The polynomial can be seen as a Taylor expansion of the function at a given point. The *interval remainder* Δ provides an enclosure of the approximation errors encountered (truncation, roundings).

For usual functions, the polynomial coefficients and the error bounds are computed using the Taylor-Lagrange formula and recurrence relations satisfied by successive derivatives of the functions. When using the same approach for composite functions, the error we get for the remainder is too pessimistic [14]. Hence, an arithmetic for TMs was introduced: simple algebraic rules like addition, multiplication and composition with TMs are applied recursively on the structure of function f, so that the final model obtained is a TM for f over

$[a, b]$. Usually, the use of these operations with TMs offers a much tighter error bound than the one directly computed for the whole function [14]. For example, addition is defined as follows: let two TMs of order n for f_1 and f_2, over $[a, b]$: (P_1, Δ_1) and (P_2, Δ_2). Their sum is an order n TM for $f_1 + f_2$ over $[a, b]$ and is obtained by adding the two polynomials and the remainder bounds: $(P_1, \Delta_1) + (P_2, \Delta_2) = (P_1 + P_2, \Delta_1 + \Delta_2)$. For multiplication and composition, similar rules are defined.

We follow the definitions in [23,14], and represent the polynomial T with *tight interval coefficients*. This choice is motivated by the ease of programming (rounding errors are directly handled by the interval arithmetic) and also by the fact that we want to ensure that the true coefficients of the Taylor polynomial lie inside the corresponding intervals. This is essential for applications that need to handle removable discontinuities [14]. For our formalization purpose, we recall and explain briefly in what follows the definition of valid Taylor models [23, Def. 2.1.3], and refer to [23, Chap. 2] for detailed algorithms regarding operations with Taylor models for univariate functions.

2.2 Valid Taylor Models

A Taylor model for a function f is a pair (T, Δ). The relation between f and (T, Δ) can be rigorously formalized as follows.

Definition 1. *Let* $f : I \to \mathbb{R}$ *be a function,* x_0 *be a small interval around an expansion point* x_0. *Let* T *be a polynomial with interval coefficients* $a_0, \dots, a_n$ *and* Δ *an interval. We say that* (T, Δ) *is a Taylor model of* f *at* x_0 *on* I *when*

$$\begin{cases} x_0 \subseteq I \text{ and } 0 \in \Delta, \\ \forall \zeta_0 \in x_0, \exists \alpha_0 \in a_0, \dots, \alpha_n \in a_n, \forall x \subset I, \exists \delta \subset \Delta, f(x) \sum_{i=0}^{n} \alpha_i (x \quad \xi_0)^i = \delta. \end{cases}$$

Informally, this definition says that there is always a way to pick some values α_i in the intervals a_i so that the difference between the resulting polynomial and f around x_0 is contained in Δ. This validity is the invariant that is preserved when performing operations on Taylor models. Obviously, once a Taylor model (T, Δ) is computed, if needed, one can get rid of the interval coefficients a_i in T by picking arbitrary α_i and accumulating in Δ the resulting errors.

2.3 Computing the Coefficients and the Remainder

We are now interested in an automatic way of providing the terms $a_0, \dots, a_n$ and Δ of Definition 1 for basic functions. It is classical to use the following.

Lemma 1 (Taylor-Lagrange Formula). *If* f *is* $n + 1$ *times differentiable on a domain* I, *then we can expand* f *in its Taylor series around any point* $x_0 \in I$ *and we have:* $\forall x \in I, \exists \xi$ *between* x_0 *and* x *such that*

$$f(x) = \underbrace{\left(\sum_{i=0}^{n} \frac{f^{(i)}(x_0)}{i!} (x - x_0)^i \right)}_{T(x)} + \underbrace{\frac{f^{(n+1)}(\xi)}{(n+1)!} (x - x_0)^{n+1}}_{\Delta(x, \xi)}.$$

Computing interval enclosures $a_0, \ldots, a_n$, for the coefficients of T, reduces to finding enclosures of the first n derivatives of f at x_0 in an efficient way. The same applies for computing Δ based on an interval enclosure of the $n+1$ derivative of f over I. However, the expressions for successive derivatives of practical functions typically become very involved with increasing n. Fortunately, it is not necessary to generate these expressions for obtaining values of $\{f^{(i)}(x_0), i = 0, \ldots, n\}$. For basic functions, formulas are available since Moore [31] (see also [21]). There one finds either recurrence relations between successive derivatives of f, or a simple closed formula for them. And yet, this is a case-by-case approach, and we would like to use a more generic process, which would allow us to deal with a broader class of functions in a more uniform way suitable to formalization.

Recurrence Relations for D-finite Functions. An algorithmic approach exists for finding recurrence relations between the Taylor coefficients for a class of functions that are solutions of linear ordinary differential equations (LODE) with polynomial coefficients, called *D-finite functions*. The Taylor coefficients of these functions satisfy a linear recurrence with polynomial coefficients [40]. Most common functions are *D-finite*, while a simple counter-example is tan. For any D-finite function one can generate the recurrence relation directly from the differential equation that defines the function, see, e.g., the Gfun module in Maple [39]. From the recurrence relation, the computation of the first n coefficients is done in linear time. Let us take a simple example and consider $f = \exp$. It satisfies the LODE $f' = f$, $f(0) = 1$, which gives the following recurrence for the Taylor coefficients $(c_n)_{n \in \mathbb{N}}$: $(n + 1)c_{n+1} - c_n = 0$, $c_0 = 1$, whose solution is $c_n = 1/n!$.

This property lets us include in the class of *basic functions* all the D-finite functions. We will see in Section 3.2 that this allows us to provide a uniform and efficient approach for computing Taylor coefficients, suitable for formalization. We note that our data structure for that is *recurrence relation + initial conditions* and that the formalization of the isomorphic transformation from the *LODE + initial conditions*, used as input in Gfun is subject of future research.

3   Formalization of Taylor Models in CoQ

We provide an implementation[1] of TMs that is efficient enough to produce very accurate approximating polynomials in a reasonable amount of time. The work is carried out in the CoQ proof assistant, which provides a formal setting where we will be able to formally verify our implementation. We wish to be as generic as possible. A TM is just an instance of a more general object called *rigorous polynomial approximation* (RPA). For a function f, a RPA is a pair (T, Δ) where T is a polynomial and Δ an interval containing the approximation error between f and T. We can choose Taylor polynomials for T and get TMs but other types of approximation are also available like Chebyshev models, based on Chebyshev polynomials. This generic RPA structure will look like:

---

[1] It is available at http://tamadi.gforge.inria.fr/CoqApprox/

```
Structure rpa := { approx: polynomial; error: interval }
```

In this structure, we want genericity not only for **polynomial** with respect to the type of its coefficients and to its physical implementation but also for the type for intervals. Users can then experiment with different combinations of datatypes. Also, this genericity lets us factorize our implementation and will hopefully facilitate the proofs of correctness. We implement Taylor models as an instance of a generic RPA following what is presented in Section 2. Before describing our modular implementation, we present the CoQ proof assistant, the libraries we have been using and how computation is handled.

3.1   The CoQ Proof Assistant

CoQ [4] is an interactive theorem prover that combines a higher-order logic and a richly-typed functional programming language. Thus, it provides an expressive language for defining not only mathematical objects but also datatypes and algorithms and for stating and proving their properties. The user builds proofs in CoQ in an interactive manner. In our development, we use the SSREFLECT [19] extension that provides its own tactic language and libraries.

There are two main formalizations of real numbers in CoQ: an axiomatic one [29] and a constructive one [18]. For effective computations, several implementations of computable real numbers exist. A library for multiple-precision FP arithmetic is described in [8]. Based on this library, an interval arithmetic library is defined in [30]. It implements intervals with FP bounds. Also, the libraries [36] and [24] provide an arbitrary precision real arithmetic. All these libraries are proved correct by deriving a formal link between the computational reals and one of the formalizations of real numbers. We follow the same idea: implement a computable TM for a given function and formally prove its correctness with respect to the abstract formalization of that function in CoQ. This is done by using Definition 1 and the functions defined in the axiomatic formalization.

The logic of CoQ is computational: it is possible to write programs in CoQ that can be directly executed within the logic. This is why the result of a computation with a correct algorithm can always be trusted. Thanks to recent progress in the evaluation mechanism [7], a program in CoQ runs as fast as an equivalent version written directly and compiled in OCAML. There are some restrictions to the programs that can be executed in CoQ: they must always terminate and be purely functional, i.e., no side-effects are allowed. This is the case for the above mentioned computable real libraries. Moreover, they are defined within CoQ on top of the multiple-precision arithmetic library based on binary tree described in [20]. So only the machine modular arithmetic (32 or 64 bits depending on the machine) is used in the computations in CoQ.

For our development of Taylor models we use polynomials with coefficients being some kind of computable reals. Following the description in Section 2, we use intervals with FP bounds given by [30] as coefficients. Since the interval and FP libraries are proved correct, so is the arithmetic on our coefficients. By choosing a functional implementation for polynomials (e.g., lists), we then obtain

TMs that are directly executable within COQ. Now, we describe in detail this modular implementation.

3.2   A Modular Implementation of Taylor Models

COQ provides three mechanisms for modularization: *type classes*, *structures*, and *modules*. Modules are less generic than the other two (that are first-class citizens) but they have a better computational behavior: module applications are performed statically, so the code that is executed is often more compact. Since our generic implementation only requires simple parametricity, we have been using modules.

First, abstract interfaces called Module Types are defined. Then concrete "instances" of these abstract interfaces are created by providing an implementation for all the fields of the Module Type. The definition of Modules can be parameterized by other Modules. These parameterized modules are crucial to factorize code in our structures.

Abstract Polynomials, Coefficients and Intervals. We describe abstract interfaces for *polynomials* and for their *coefficients* using COQ's Module Type. The interface for coefficients contains the common base of all the computable real numbers we may want to use. Usually coefficients of a polynomial are taken in a ring. We cannot do this here. For example, addition of two intervals is not associative. Therefore, the abstract interface for coefficients contains the required operations (addition, multiplication, etc.) only, where some basic properties (associativity, distributivity, etc.) are ruled out. The case of abstract polynomials is similar. They are also a Module Type but this time parameterized by the coefficients. The interface contains only the operations on polynomials (addition, evaluation, iterator, etc.) with the properties that are satisfied by all common instantiations of polynomials. For intervals, we directly use the abstract interface provided by the Coq.Interval library [30].

Rigorous Polynomial Approximations. We are now able to give the definition of our rigorous polynomial approximation.

```
Module RigPolyApprox (C : BaseOps)(P : PolyOps C)(I : IntervalOps).
Structure rpa : Type := RPA { approx : P.T;   error : I.type }.
```

The module is parameterized by C (the coefficients), by P (the polynomials with coefficients in C), and by I (the intervals).

Generic Taylor Polynomials. Before implementing our Taylor models, we use the abstract coefficients and polynomials to implement generic Taylor polynomials. These polynomials are computed using an algorithm based on recurrence relations as described in Section 2.3. This algorithm can be implemented in a generic way. It takes as argument the relation between successive coefficients, the initial conditions and outputs the Taylor polynomial.

We detail the example of the exponential, which was also presented in Section 2.3. The Taylor coefficients $(c_n)_{n \in \mathbb{N}}$ satisfy $(n+1)c_{n+1} - c_n = 0$. The corresponding COQ code is

```
Definition exp_rec (n : nat) u := tdiv u (tnat n).
```

where `tdiv` is the division on our coefficients and `tnat` is an injection of integers to our type of coefficients. We then implement the generic Taylor polynomial for the exponential around a point x_0 with the following definition.

```
Definition T_exp n x0 := trec1 exp_rec (texp x0) n.
```

In this definition, `trec1` is the function in the polynomial interface that is in charge of producing a polynomial of size **n** from a recurrence relation of order 1 (here, `exp_rec`) and an initial condition (here, `texp x0`, the value of the exponential at `x0`). The interface also contains `trec2` and `trecN` for producing polynomials from recurrences of order 2 and order N with the appropriate number of initial conditions. Having specific functions for recurrences of order 1 and 2 makes it possible to have optimized implementations for these frequent recurrences. All the functions we currently dispose of in our library are in fact defined with `trec1` and `trec2`. We provide generic Taylor polynomials for constant functions, identity, $x \mapsto \frac{1}{x}$, $\sqrt{\cdot}$, $\frac{1}{\sqrt{\cdot}}$, exp, ln, sin, cos, arcsin, arccos, arctan.

Taylor Models. We implement TMs on top of the RPA structure by using polynomials with coefficients that are intervals with FP bounds, according to Section 2. Yet we are still generic with respect to the effective implementation of polynomials. For the remainder, we also use *intervals with FP bounds*. This datatype is provided by the Coq.Interval library [30], whose design is also based on modules, in such a way that it is possible to plug all the machinery on the desired kind of COQ integers (i.e., Z or BigZ).

In a TM for a basic function (e.g., exp), polynomials are instances of the generic Taylor polynomials implemented with the help of the recurrence relations. The remainder is computed with the help of the Taylor-Lagrange formula in Lemma 1. For this computation, thanks to the parameterized module, we reuse the generic recurrence relations. The order-n Taylor model for the exponential on interval X expanded at the small interval X0 is as follows:

```
Definition TM_exp (n : nat) X X0 :=
  RPA (T_exp n X0) (Trem T_exp n X X0).
```

We implement Taylor models for the addition, multiplication, and composition of two functions by arithmetic manipulations on the Taylor models of the two functions, as described in Section 2. Here is the example of addition:

```
Definition TM_add (Mf Mg : rpa) :=
  RPA (P.tadd (approx Mf) (approx Mg))
      (I.add (error Mf) (error Mg)).
```

The polynomial approximation is just the sum of the two approximations and the interval error is the sum of the two errors. Multiplication is almost as intuitive. We consider the truncated multiplication of the two polynomials and we make sure that the error interval takes into account the remaining parts of the truncated multiplication. Composition is more complex. It uses addition and multiplication of Taylor polynomials. Division of Taylor models is implemented in term of multiplication and composition with the inverse function $x \mapsto 1/x$. The corresponding algorithms are fully described in [23].

Discussion on the Formal Verification of Taylor Models. The Taylor model `Module` also contains a version of Taylor polynomials defined with axiomatic real number coefficients. These polynomials are meant to be used only in the formal verification when linking the computable Taylor models to the corresponding functions on axiomatic real numbers. This link is given by Definition 1. The definition can be easily formalized in the form of a predicate `validTM`.

```
Definition validTM X X0 M f :=
  I.subset X0 X /\
  contains (error M) 0 /\
  let N := tsize (approx M) in
  forall x0, contains X0 x0 -> exists P, tsize P = N /\
    ( forall k, (k < N) ->
      contains (tnth (approx M) k) (tnth P k) ) /\
    forall x, contains X x ->
      contains (error M) (f x - teval P (x - x0)).
```

The theorem of correctness for the Taylor model of the exponential `TM_exp` then establishes the link between the model and the exponential function `Rexp` that is defined in the real library.

```
Lemma TM_exp_correct :
  forall X X0 n, validTM X X0 (TM_exp n X X0) Rexp.
```

Our goal is to formally prove the correctness of our implementation of Taylor models. We want proofs that are generic, so a new instantiation of the polynomials would not require changing the proofs. In a previous version of our COQ development we had managed to prove correct Taylor models for some elementary functions and addition. No proofs are available yet for the version presented here but adapting the proofs to this new setting should be possible.

4   Benchmarks

We want to evaluate the performances of our COQ implementation of Taylor models. For this we compare them to those of SOLLYA [13], a tool specially designed to handle such numerical approximation problems.

The Coq Taylor models we use for our tests are implemented with polynomials represented as simple lists with a linear access to their coefficients. The coefficients of the approximating polynomial in our instantiation of Taylor models as well as the interval errors are implemented by intervals with multiple-precision FP bounds as available in the Coq.Interval library described in [30]. Since we need to evaluate the initial conditions for recurrences, only the basic functions already implemented in Coq.Interval can have their corresponding Taylor models.

In SOLLYA, polynomials have interval coefficients and are represented by a (coefficient) array of intervals with multiple-precision FP bounds. SOLLYA's `autodiff()` function computes interval enclosures of the successive derivatives of a function at a point or over an interval, relying on interval arithmetic computations and recurrence relations similar to the ones we use in our Coq development. Thus, we use it to compute the Taylor models we are interested in.

Timings, Accuracy and Comparisons

We compare the Coq and the SOLLYA implementations presented above on a selection of several benchmarks. Table 1 gives the timings as well as the tightness obtained for the remainders. These benchmarks have been computed on a 4-core computer, Intel(R) Xeon(R) CPU X5482 @ 3.20GHz.

Each cell of the first column of Table 1 contains a target function, the precision in bits used for the computations, the order of the TM, and the interval under consideration. When "split" is mentioned, the interval has been subdivided into a specified amount of intervals of equal length (1024 subintervals for instance in line 3) and a TM has been computed over each subinterval. Each TM is expanded at the middle of the interval. The symbols $RD_t(\ln 4)$, resp. $RU_t(\ln 2)$, denote $\ln(4)$ rounded toward $-\infty$, resp. $\ln(2)$ rounded toward $+\infty$, using precision t.

Columns 2 and 3 give the total duration of the computations (for instance, the total time for computing the 1024 TMs of the third line) in Coq and SOLLYA respectively. Columns 4 and 5 present an approximation error obtained using Coq and SOLLYA, while the last column gives, as a reference, the true approximation error, computed by ad-hoc means (symbolically for instance), of the function by its Taylor polynomial. Note that when "split" is mentioned, the error presented corresponds to the one computed over the last subinterval (for instance, $[2 - 1/256, 2]$ for the arctan example). For simplicity, the errors are given using three significant digits.

In terms of accuracy, the Coq and SOLLYA results are close. We have done other similar checks and obtained the same encouraging results (the error bounds returned by Coq and SOLLYA have the same orders of magnitude). This does not prove anything but is nevertheless very reassuring. Proving the correctness of an implementation that produces too large bounds would be meaningless.

Coq is 6 to 10 times slower than SOLLYA, which is reasonable. This factor gets larger when composition is used. One possible explanation is that composition implies lots of polynomial manipulations and the implementation of polynomials as simple lists in Coq maybe too naive. An interesting alternative could be to use persistent arrays [2] to have more efficient polynomials. Another

Table 1. Benchmarks and timings for our implementation in CoQ

	Execution time		Approximation error		
	CoQ	SOLLYA	CoQ	SOLLYA	Mathematical
exp prec=120, deg=20 I=$[1, \mathrm{RD}_{53}(\ln 4)]$ no split	7.40s	0.01s	7.90×10^{-35}	7.90×10^{-35}	6.57×10^{-35}
exp prec=120, deg=8 I=$[1, \mathrm{RD}_{53}(\ln 4)]$ split in 1024	20.41s	3.77s	3.34×10^{-39}	3.34×10^{-39}	3.34×10^{-39}
exp prec=600, deg=40 I=$[\mathrm{RU}_{113}(\ln 2), 1]$ split in 256	38.10s	16.39s	6.23×10^{-182}	6.22×10^{-182}	6.22×10^{-182}
arctan prec=120, deg=8 I=$[1, 2]$ split in 256	11.45s	1.03s	7.43×10^{-29}	2.93×10^{-29}	2.85×10^{-29}
exp × sin prec=200, deg=10 I=$[1/2, 1]$ split in 2048	1m22s	12.05s	6.92×10^{-50}	6.10×10^{-50}	5.89×10^{-50}
exp/sin prec=200, deg=10 I=$[1/2, 1]$ split in 2048	3m41s	13.29s	4.01×10^{-43}	9.33×10^{-44}	8.97×10^{-44}
exp ∘ sin prec=200, deg=10 I=$[1/2, 1]$ split in 2048	3m24s	12.19s	4.90×10^{-47}	4.92×10^{-47}	4.90×10^{-47}

possible improvement is at algorithmic level: while faster algorithms for polynomial multiplication exist [17], currently in all TMs related works $O(n^2)$ naive multiplication is used. We could improve that by using a Karatsuba-based approach, for instance.

5   Conclusion and Future Works

We have described an implementation of Taylor models in the CoQ proof assistant. Two main issues have been addressed. The first one is genericity. We wanted our implementation to be applicable to a large class of problems. This motivates our use of modules in order to get this flexibility. The second issue is efficiency. Working in a formal setting has some impact in terms of efficiency. Before starting to prove anything, it was then crucial to evaluate if the computational power provided by CoQ was sufficient for our needs. The results given in Section 4 clearly indicate that what we have is worth proving formally.

We are in the process of proving the correctness of our implementation. Our main goal is to prove the validity theorem given in Section 2 formally. This is tedious work but we believe it should be completed in a couple of months. As we aim at a complete formalization, a more subtle issue concerns the Taylor models for the basic functions and in particular how the model and its corresponding function can be formally related. This can be done in an ad-hoc way, deriving the recurrence relation from the formal definition. An interesting future work would be to investigate a more generic approach, trying to mimic what is provided by the Dynamic Dictionary of Mathematical Functions [3] in a formal setting.

Having Taylor models is an initial step in our overall goal of getting formally proved worst-case accuracy for common functions and formats. A natural next step is to couple our models with some positivity test for polynomials, for example some sums-of-squares technique. This would give us an automatic way of verifying polynomial approximations formally. It would also provide another way of evaluating the quality of our Taylor approximations. If they turned out to be not accurate enough for our needs, we could always switch to better kinds of approximations such as Chebyshev truncated series, thanks to our generic setting.

References

1. Abramowitz, M., Stegun, I.A.: Handbook of mathematical functions with formulas, graphs, and mathematical tables. National Bureau of Standards Applied Mathematics Series, vol. 55. For sale by the Superintendent of Documents, U.S. Government Printing Office, Washington, D.C (1964)
2. Armand, M., Grégoire, B., Spiwack, A., Théry, L.: Extending Coq with Imperative Features and Its Application to SAT Verification. In: Kaufmann, M., Paulson, L.C. (eds.) ITP 2010. LNCS, vol. 6172, pp. 83–98. Springer, Heidelberg (2010)
3. Benoit, A., Chyzak, F., Darrasse, A., Gerhold, S., Mezzarobba, M., Salvy, B.: The Dynamic Dictionary of Mathematical Functions (DDMF). In: Fukuda, K., van der Hoeven, J., Joswig, M., Takayama, N. (eds.) ICMS 2010. LNCS, vol. 6327, pp. 35–41. Springer, Heidelberg (2010)
4. Bertot, Y., Castéran, P.: Interactive Theorem Proving and Program Development. Coq'Art: The Calculus of Inductive Constructions. Texts in Theoretical Computer Science. Springer, Heidelberg (2004)
5. Berz, M., Makino, K.: Rigorous global search using Taylor models. In: SNC 2009: Proceedings of the 2009 Conference on Symbolic Numeric Computation, pp. 11–20. ACM, New York (2009)
6. Berz, M., Makino, K., Kim, Y.K.: Long-term stability of the tevatron by verified global optimization. Nuclear Instruments and Methods in Physics Research Section A: Accelerators, Spectrometers, Detectors and Associated Equipment 558(1), 1–10 (2006); Proceedings of the 8th International Computational Accelerator Physics Conference - ICAP 2004
7. Boespflug, M., Dénès, M., Grégoire, B.: Full Reduction at Full Throttle. In: Jouannaud, J.-P., Shao, Z. (eds.) CPP 2011. LNCS, vol. 7086, pp. 362–377. Springer, Heidelberg (2011)
8. Boldo, S., Melquiond, G.: Flocq: A Unified Library for Proving Floating-point Algorithms in Coq. In: Proceedings of the 20th IEEE Symposium on Computer Arithmetic, Tübingen, Germany, pp. 243–252 (2011)

9. Brisebarre, N., Chevillard, S.: Efficient polynomial L^∞-approximations. In: Kornerup, P., Muller, J.M. (eds.) 18th IEEE Symposium on Computer Arithmetic, pp. 169–176. IEEE Computer Society, Los Alamitos (2007)

10. Brisebarre, N., Muller, J.M., Tisserand, A.: Computing Machine-efficient Polynomial Approximations. ACM Trans. Math. Software 32(2), 236–256 (2006)

11. Cháves, F.: Utilisation et certification de l'arithmétique d'intervalles dans un assistant de preuves. Thèse, École normale supérieure de Lyon - ENS LYON (September 2007), http://tel.archives-ouvertes.fr/tel-00177109/en/

12. Chevillard, S.: Évaluation efficace de fonctions numériques. Outils et exemples. Ph.D. thesis, École Normale Supérieure de Lyon, Lyon, France (2009), http://tel.archives-ouvertes.fr/tel-00460776/fr/

13. Chevillard, S.. Joldeş, M., Lauter, C.: Sollya: An Environment for the Development of Numerical Codes. In: Fukuda, K., van der Hoeven, J., Joswig, M., Takayama, N. (eds.) ICMS 2010. LNCS, vol. 6327, pp. 28–31. Springer, Heidelberg (2010)

14. Chevillard, S., Harrison, J., Joldeş, M., Lauter, C.: Efficient and accurate computation of upper bounds of approximation errors. Theoretical Computer Science 16(412), 1523–1543 (2011)

15. Collins, P., Niqui, M., Revol, N.: A Taylor Function Calculus for Hybrid System Analysis: Validation in Coq. In: NSV-3: Third International Workshop on Numerical Software Verification (2010)

16. de Dinechin, F., Lauter, C., Melquiond, G.: Assisted verification of elementary functions using Gappa. In: Proceedings of the 2006 ACM Symposium on Applied Computing, Dijon, France, pp. 1318–1322 (2006), http://www.lri.fr/~melquion/doc/06-mcms-article.pdf

17. von zur Gathen, J., Gerhard, J.: Modern computer algebra, 2nd edn. Cambridge University Press, New York (2003)

18. Geuvers, H., Niqui, M.: Constructive Reals in Coq: Axioms and Categoricity. In: Callaghan, P., Luo, Z., McKinna, J., Pollack, R. (eds.) TYPES 2000. LNCS, vol. 2277, pp. 79–95. Springer, Heidelberg (2002)

19. Gonthier, G., Mahboubi, A., Tassi, E.: A Small Scale Reflection Extension for the Coq system. Rapport de recherche RR-6455, INRIA (2008)

20. Grégoire, B., Théry, L.: A Purely Functional Library for Modular Arithmetic and Its Application to Certifying Large Prime Numbers. In: Furbach, U., Shankar, N. (eds.) IJCAR 2006. LNCS (LNAI), vol. 4130, pp. 423–437. Springer, Heidelberg (2006)

21. Griewank, A.: Evaluating Derivatives - Principles and Techniques of Algorithmic Differentiation. SIAM (2000)

22. IEEE Computer Society: IEEE Standard for Floating-Point Arithmetic. IEEE Std 754^{TM}-2008 (August 2008)

23. Joldeş, M.: Rigourous Polynomial Approximations and Applications. Ph.D. dissertation, École Normale Supérieure de Lyon, Lyon, France (2011), http://perso.ens-lyon.fr/mioara.joldes/these/theseJoldes.pdf

24. Krebbers, R., Spitters, B.: Computer Certified Efficient Exact Reals in Coq. In: Davenport, J.H., Farmer, W.M., Urban, J., Rabe, F. (eds.) Calculemus/MKM 2011. LNCS, vol. 6824, pp. 90–106. Springer, Heidelberg (2011)

25. Lefèvre, V., Muller, J.M.: Worst cases for correct rounding of the elementary functions in double precision. In: Burgess, N., Ciminiera, L. (eds.) Proceedings of the 15th IEEE Symposium on Computer Arithmetic (ARITH-16), Vail, CO (June 2001)
26. Lizia, P.D.: Robust Space Trajectory and Space System Design using Differential Algebra. Ph.D. thesis, Politecnico di Milano, Milano, Italy (2008)
27. Makino, K.: Rigorous Analysis of Nonlinear Motion in Particle Accelerators. Ph.D. thesis, Michigan State University, East Lansing, Michigan, USA (1998)
28. Makino, K., Berz, M.: Taylor models and other validated functional inclusion methods. International Journal of Pure and Applied Mathematics 4(4), 379–456 (2003), http://bt.pa.msu.edu/pub/papers/TMIJPAMO3/TMIJPAMO3.pdf
29. Mayero, M.: Formalisation et automatisation de preuves en analyses réelle et numérique. Ph.D. thesis, Université Paris VI (2001)
30. Melquiond, G.: Proving Bounds on Real-Valued Functions with Computations. In: Armando, A., Baumgartner, P., Dowek, G. (eds.) IJCAR 2008. LNCS (LNAI), vol. 5195, pp. 2–17. Springer, Heidelberg (2008)
31. Moore, R.E.: Methods and Applications of Interval Analysis. Society for Industrial and Applied Mathematics (1979)
32. Muller, J.M.: Projet ANR TaMaDi – Dilemme du fabricant de tables – Table Maker's Dilemma (ref. ANR 2010 BLAN 0203 01), http://tamadiwiki.ens-lyon.fr/tamadiwiki/
33. Muller, J.M.: Elementary Functions, Algorithms and Implementation, 2nd edn. Birkhäuser, Boston (2006)
34. Neher, M., Jackson, K.R., Nedialkov, N.S.: On Taylor model based integration of ODEs. SIAM J. Numer. Anal. 45, 236–262 (2007)
35. Neumaier, A.: Taylor forms – use and limits. Reliable Computing 9(1), 43–79 (2003)
36. O'Connor, R.: Certified Exact Transcendental Real Number Computation in Coq. In: Mohamed, O.A., Muñoz, C., Tahar, S. (eds.) TPHOLs 2008. LNCS, vol. 5170, pp. 246–261. Springer, Heidelberg (2008)
37. The Arénaire Project: CRlibm, Correctly Rounded mathematical library (July 2006), http://lipforge.ens-lyon.fr/www/crlibm/
38. Remez, E.: Sur un procédé convergent d'approximations successives pour déterminer les polynômes d'approximation. C.R. Académie des Sciences 198, 2063–2065 (1934) (in French)
39. Salvy, B., Zimmermann, P.: Gfun: a Maple package for the manipulation of generating and holonomic functions in one variable. ACM Trans. Math. Software 20(2), 163–177 (1994)
40. Stanley, R.P.: Differentiably finite power series. European Journal of Combinatorics 1(2), 175–188 (1980)
41. Ziv, A.: Fast evaluation of elementary mathematical functions with correctly rounded last bit. ACM Trans. Math. Software 17(3), 410–423 (1991)
42. Zumkeller, R.: Formal Global Optimisation with Taylor Models. In: Furbach, U., Shankar, N. (eds.) IJCAR 2006. LNCS (LNAI), vol. 4130, pp. 408–422. Springer, Heidelberg (2006)

Enhancing the Inverse Method
with State Merging

Étienne André[1], Laurent Fribourg[2], and Romain Soulat[2]

[1] LIPN, CNRS UMR 7030, Université Paris 13, France
[2] LSV, ENS Cachan & CNRS, France

Abstract. Keeping the state space small is essential when verifying real-time systems using Timed Automata (TA). In the model-checker Uppaal, the merging operation has been used extensively in order to reduce the number of states. Actually, Uppaal's merging technique applies within the more general setting of Parametric Timed Automata (PTA). The *Inverse Method* (*IM*) for a PTA $\mathcal{A}$ is a procedure that synthesizes a zone around a given point π^0 (parameter valuation) over which $\mathcal{A}$ is guaranteed to behave in an equivalent time-abstract manner. We show that the integration of merging into *IM* leads to the synthesis of larger zones around π^0. It also often improves the performance of *IM*, both in terms of computational space and time, as shown by our experimental results.

1 Introduction

A fundamental problem in the exploration of the reachability space in Timed Automata (TA) is to compact as much as possible the generated space of symbolic states. In [11], the authors show that, in a network of TAs, all the successor states can be merged together when all the interleavings of actions are possible. In [7,8], A. David proposed to replace the union of two states by a unique state when this union is convex. More precisely, if the union of two states is included into their convex hull, then one can replace the two states by their hull. This technique is applied to timed constraints represented under the form of "Difference Bound Matrices" (DBMs). Actually, such a merging technique applies as well in the more general setting of *parametric* timed automata (PTA), where parameters can be used instead of constants, and timed constraints are represented under the form of polyhedra.

The *Inverse Method* (*IM*) for a PTA $\mathcal{A}$ is a procedure that synthesizes a zone around a given point π^0 (parameter valuation) over which $\mathcal{A}$ is guaranteed to behave in an equivalent time-abstract manner [2]. We show that the integration of merging into *IM* often leads to the synthesis of larger zones around π^0. More surprisingly, our experiments show that even a simple implementation of merging often improves the performance of *IM*, not only in terms of computational space but also in time.

A. Goodloe and S. Person (Eds.): NFM 2012, LNCS 7226, pp. 100–105, 2012.

2   Background and Definition

2.1   Timed Automata

Given a finite set X of n non-negative real-valued variables (called "clocks"), a *timed constraint* is a conjunction of linear inequalities of the form $x_i \prec c$, $-x_i \prec c$ or $x_i - x_j \prec c$ with $\prec \in \{<, \leq\}$, $x_i, x_j \in X$ and $c \in \mathbb{Z}$.

A Timed Automaton (TA) is a tuple $(\Sigma, Q, l_0, X, I, \rightarrow)$, with Σ a finite set of *actions*, Q a finite set of *locations*, $l_0 \in Q$ the *initial location*, X a set of *clocks*, I the *invariant* assigning to every $l \in Q$ a constraint over X, and $\rightarrow$ a *step relation* consisting of elements (l, g, a, ρ, l'), where $l, l' \in Q$, $a \in \Sigma$, g is a timed constraint (*guard*) and ρ is a subset of X (set of clocks reset to 0).

A *state* is a couple (l, v) where l is a location of Q and v a valuation of X.

The operational semantics of TA is informally given as follows: given two states $s = (l, v)$ and $s' = (l', v')$ with $l, l' \in Q$, v, v' two valuations of X, the step $s \xrightarrow{a} s'$ means that, for some $(l, g, a, \rho, l') \in \rightarrow$ and some $\delta \in \mathbb{R}_+$:

$$(l, v) \xRightarrow{g, a, \rho} (l', v') \xrightarrow{\delta} (l', v' + \delta),$$

where $(l, v) \xRightarrow{g, a, \rho} (l', v')$ means that discrete transition $(l, g, a, \rho 0, l')$ can take place (i.e. v satisfies g, and v' is obtained from v by resetting the clocks of ρ to zero), and $(l', v') \xrightarrow{\delta} (l', v' + \delta)$ means that time can pass during δ units in location l' (i.e., $v' + \delta'$ satisfies the invariant $I(l')$ for all $0 \leq \delta' \leq \delta$).

A *run* is a sequence of the form $(l_0, v_0) \xrightarrow{a_1} (l_1, v_1) \xrightarrow{a_2} \cdots \xrightarrow{a_n} (l_n, v_n)$. A *trace* (or time-abstracted run) associated to a run is a sequence of the form $l_0 \xRightarrow{a_1} l_1 \xRightarrow{a_2} \cdots \xRightarrow{a_n} l_n$. A trace can be seen as an alternating sequence of locations and actions. Given a TA $\mathcal{A}$, we denote by $Tr(\mathcal{A})$ the set of traces associated to all possible runs of $\mathcal{A}$. When two TAs have the same set of traces, we say that they behave in an equivalent time-abstract manner.

Given a set of states S, one defines $Post_{\mathcal{A}}(S)$ as the set of states reachable from S in one step, i.e.:

$$Post_{\mathcal{A}}(S) = \{s' = (l', v') \in Q \times \mathbb{R}_+^n \mid s = (l, v) \xrightarrow{a} s',$$
$$\text{for some } s \in S,\ l \in Q,\ v \in \mathbb{R}_+^n \text{ and } (l, g, a, \rho, l') \in \rightarrow \}.$$

Likewise, $Post_{\mathcal{A}}^i(S)$ is the set of states reachable from S in exactly i steps and let $Post_{\mathcal{A}}^*(S) = \bigcup_{i \geq 0} Post_{\mathcal{A}}^i(S)$.

2.2   Parametric Timed Automata

We assume now given a finite set P of symbols (called "parameters"). A *parametric term* is a linear combination of parameters and integer constants. A *parametric timed constraint* is a conjunction of linear inequalities of the form $x_i \prec e$, $-x_i \prec e$ or $x_i - x_j \prec e$ where $\prec \in \{<, \leq\}$, $x_i, x_j \in X$ and e is a parametric term. A *constraint over* P is a conjunction of inequalities of the form $e_1 \prec e_2$ with $\prec \in \{<, \leq\}$ and e_1, e_2 two parametric terms. Given a parametric timed constraint C, the expression $(\exists X : C)$ denotes the constraint over P obtained from C by eliminating the variables of X. Given a parametric timed constraint

C and a valuation π over P (i.e. a function from P to $\mathbb{N}$), we denote by $C[\pi]$ the result of replacing every parameter in C by its π-valuation. We write $\pi \models C$ to express that $\exists X : C[\pi]$ is true. A *(symbolic parametric) state* is a couple (l, C) where l is in Q and C is a parametric timed constraint. A Parametric Timed Automaton (PTA) is a TA where some constants appearing in the guard and invariant inequalities have been replaced by parameters. Given a PTA $\mathcal{A}$, we denote by $\mathcal{A}[\pi]$ the TA obtained from $\mathcal{A}$ by replacing the parameters by their π-valuations. Given a parametric constraint K, we denote by $\mathcal{A}(K)$ the PTA where the parameters are assumed to satisfy K.

2.3   Inverse Method

Given a PTA $\mathcal{A}$ and a valuation π^0 over P, the goal of *IM* introduced in [2] is to synthesize a constraint K^0 over P such that: $\pi^0 \models K^0$ and $Tr(\mathcal{A}[\pi^0]) = Tr(\mathcal{A}[\pi])$, for all $\pi \models K^0$. This implies that for every $\pi \models K^0$, $\mathcal{A}[\pi]$ and $\mathcal{A}[\pi^0]$ have the same time-abstracted behavior. The size of K^0 gives us a measure of the "robustness" (see [10]) of the behavior of $\mathcal{A}$ around π^0. The larger K^0 is, the more robust $\mathcal{A}$ is guaranteed to be. The algorithm *IM* is given below (where s_0 denotes the set of states of location l_0 whose clocks are equal and satisfy the invariant $I(l_0)$). The idea of the procedure is to refine iteratively a current constraint K over P by adding inequalities J in order to eliminate all the generated π^0-incompatible states (i.e., states (l, C) such that $\pi^0 \not\models (\exists X : C)$).

Algorithm 1. Algorithm $IM(\mathcal{A}, \pi^0)$

---

    input : PTA $\mathcal{A}$ of initial state s_0, parameter valuation π^0
    output: Constraint K^0 on the parameters

    $i \leftarrow 0$; $K \leftarrow \mathbf{true}$; $S \leftarrow \{s_0\}$
    while true do
        while *there are π^0-incompatible states in S* **do**
            Select a π^0-incompatible state (l, C) of S (i.e., s.t. $\pi^0 \not\models (\exists X : C)$) ;
            Select a π^0-incompatible inequality J in $(\exists X : C)$ (i.e., s.t. $\pi^0 \not\models J$) ;
            $K \leftarrow K \wedge \neg J$; $S \leftarrow \bigcup_{j=0}^{i} Post_{\mathcal{A}(K)}^{j}(\{s_0\})$;
        if $Post_{\mathcal{A}(K)}(S) \sqsubseteq S$ **then return** $K^0 = \bigcap_{(l,C) \in S}(\exists X : C)$
        $i \leftarrow i + 1$; $S \leftarrow S \cup Post_{\mathcal{A}(K)}(S)$;   // $S = \bigcup_{j=0}^{i} Post_{\mathcal{A}(K)}^{j}(\{s_0\})$

---

3   Enhancement of *IM* with Merging

Let us recall the notion of merging, following the lines of [7].

Definition 1 (Merging). *We say that two states $s = (l, C)$ and $s' = (l', C')$ are mergeable iff $l = l'$ and $C \cup C'$ is convex; then, $(l, C \cup C')$ is their merging.*

In [7], the main technique for merging two timed constraints C, C' consists of comparing their convex hull H with their union. If the hull and the union are equal (or alternatively, if $(H \setminus C) \setminus C' = \emptyset$ where $\setminus$ is the operation of *convex*

Table 1. Comparison between IM and IM_{merge}

PTA	X	P	IM				IM_{merge}				$K^0 \subseteq K^0_{merge}$
			t	States	Trans.	M	t	States	Trans.	M	
AndOr	4	12	0.112	16	17	1,262	0.101	13	14	1,187	$=$
Flip-Flop	5	12	0.183	14	13	1,692	0.227	14	13	1,762	$=$
Latch	8	13	1.18	18	68	3,686	0.621	12	40	2,662	$\subsetneq$
BRP	7	6	4.29	428	474	25,483	7.015	426	473	25,845	$=$
WLAN	2	8	220.157	7,038	11,052	733,044	286.141	6,020	9,538	1,408,702	$=$
SPSMALL$_1$	10	26	1.578	31	35	5,098	1.642	31	35	5,442	$=$
SPSMALL$_2$	28	62	-	-	-	overflow	593	397	499	180,888	-
SIMOP	8	7	18.959	1,108	1,404	43,333	5.179	239	347	14,371	$\subsetneq$
CSMA/CD	3	3	0.801	240	383	6,580	0.947	240	383	7,049	$=$
Jobshop	3	8	1.865	253	387	10,658	1.147	118	179	5,221	$\subsetneq$
Mutex 3	3	2	0.802	307	1,060	14,598	0.671	241	811	11,934	$=$
Mutex 4	4	2	22.373	4,769	19,873	373,900	22.03	3,287	13,459	260,962	$=$

difference), then C and C' are mergeable into H. In [7,8], this technique is specialized to the case where the timed constraints are represented as DBMs. Actually, as mentioned in the introduction, such a merging technique based on convex difference is more general and still applies in the setting of PTA, where parametric timed constraints are represented under the form of polyhedra.

Given a set of (symbolic parametric) states S, let $Merging(S)$ denote the result of applying iteratively the merging of a pair of states of S (using convex difference) until no further merging applies. We define $Post_{merge}(S)$ as $Merging(Post(S))$. Let us denote by IM_{merge} the algorithm obtained from IM (see algorithm 1) by replacing the $Post$ operator by $Post_{merge}$. Let $K^0_{merge} = IM_{merge}(\mathcal{A}, \pi^0)$ and $K^0 = IM(\mathcal{A}, \pi^0)$. It is easy to see that we have always

$$K^0 \subseteq K^0_{merge}.$$

Informally, this is because the merging of a π^0-incompatible state with a π^0-compatible state gives a π^0-compatible state. Therefore, there are less π^0-incompatible states generated. Accordingly, the current set K is less often refined with inequalities J in IM_{merge}. The property of trace preservation still holds with IM_{merge}:

Proposition 1. *Given a PTA $\mathcal{A}$ and a valuation π^0, let $K^0_{merge} = IM_{merge}(\mathcal{A}, \pi^0)$. We have: $\pi^0 \models K^0_{merge}$; furthermore: $\forall \pi \models K^0_{merge}, Tr(\mathcal{A}[\pi]) = Tr(\mathcal{A}[\pi^0])$.*

The proof of this proposition is similar to its counterpart in [2].

We have implemented IM_{merge} through a simple extension of the tool IMITATOR [4] using the operation of convex difference on polyhedra from the Parma Polyhedra Library (PPL) [6]. We give in Table 1 some experimental results obtained with IM_{merge} compared with those obtained with IM. The experiments have been done on a 2.4 GHz Intel single-core processor with 4 GB of RAM memory.

The models of PTA in Table 1 are described in [5], except Jobshop which corresponds to the jobshop scheduling problem with 2 jobs and 4 tasks of [1] (Table 1), and the Mutex i model ($i = 3, 4$) which corresponds to Fisher's mutual exclusion protocol with i tasks of [9]. In Table 1, column X (resp. P) denotes the number of clocks (resp. parameters) of the PTA. Column t (resp. M) denotes the computational time in seconds (resp. the memory used in KB), column States (resp. Trans.) the number of states (resp. transitions) of the generated reachability graph. The last column indicates if $K^0 = K^0_{merge}$ or $K^0 \subsetneq K^0_{merge}$.

We can see that K^0_{merge} is strictly larger than K^0 on 3 examples. Furthermore, the reachability graphs produced with IM_{merge} are always smaller than the corresponding graphs produced with IM (as illustrated in Appendix). Let us also point out that IM_{merge}, unlike IM, is able to treat the SPSMALL$_2$ example (which contains no less than 62 parameters). Finally, the experiments are often faster with IM_{merge}, in spite of the simplicity of our implementation.

4   Final Remarks

We have shown that the integration of a general technique of state merging into IM often increases the size of the synthesized constraint while reducing the computation space. Surprisingly, in spite of our simple implementation of merging, the extended procedure is often faster than the basic procedure on our experiments. We presently study the combined integration into IM of the general technique of state merging with specific improvements presented in [3].

Acknowledgment. We are grateful to T. Chatain for helpful discussions.

References

1. Abdeddaïm, Y., Maler, O.: Job-Shop Scheduling Using Timed Automata. In: Berry, G., Comon, H., Finkel, A. (eds.) CAV 2001. LNCS, vol. 2102, pp. 478–492. Springer, Heidelberg (2001)
2. André, É., Chatain, T., Encrenaz, E., Fribourg, L.: An inverse method for parametric timed automata. IJFCS 20(5), 819–836 (2009)
3. André, É., Soulat, R.: Synthesis of Timing Parameters Satisfying Safety Properties. In: Delzanno, G., Potapov, I. (eds.) RP 2011. LNCS, vol. 6945, pp. 31–44. Springer, Heidelberg (2011)
4. André, É.: IMITATOR II: A tool for solving the good parameters problem in timed automata. In: INFINITY, pp. 91–99 (2010)
5. André, É.: An Inverse Method for the Synthesis of Timing Parameters in Concurrent Systems. Thèse de doctorat, ENS Cachan, France (2010)
6. Bagnara, R., Hill, P.M., Zaffanella, E.: The Parma Polyhedra Library: Toward a complete set of numerical abstractions for the analysis and verification of hardware and software systems. Science of Computer Programming 72(1–2), 3–21 (2008)
7. David, A.: Merging DBMs efficiently. In: 17th Nordic Workshop on Programming Theory, pp. 54–56. DIKU, University of Copenhagen (2005)
8. David, A.: Uppaal DBM Library Programmer's Reference (2006), http://people.cs.aau.dk/~adavid/UDBM/manual-061023.pdf
9. Henzinger, T.A., Ho, P.H., Wong-Toi, H.: A user guide to HyTech. In: Brinksma, E., Steffen, B., Cleaveland, W.R., Larsen, K.G., Margaria, T. (eds.) TACAS 1995. LNCS, vol. 1019, pp. 41–71. Springer, Heidelberg (1995)
10. Markey, N.: Robustness in real-time systems. In: SIES 2011, Sweden, pp. 28–34. IEEE Computer Society Press (2011)
11. Salah, R.B., Bozga, M., Maler, O.: On Interleaving in Timed Automata. In: Baier, C., Hermanns, H. (eds.) CONCUR 2006. LNCS, vol. 4137, pp. 465–476. Springer, Heidelberg (2006)

Appendix: Compared Reachability Graphs for Jobshop and SIMOP Examples

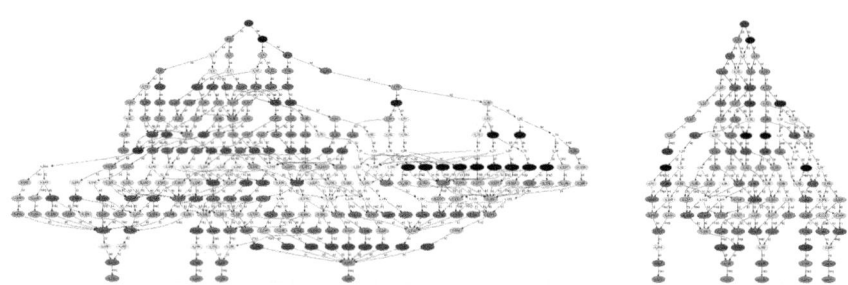

Fig. 1. Reachability graph of the jobshop example with *IM* (left) and *IM*_{merge} (right)

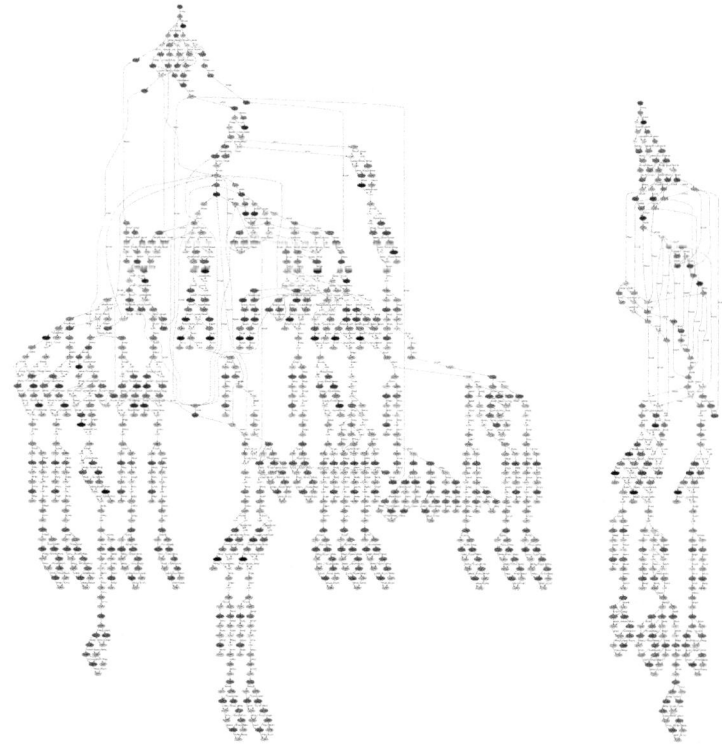

Fig. 2. Reachability graph of the SIMOP example with *IM* (left) and *IM*_{merge} (right)

Class-Modular, Class-Escape and Points-to Analysis for Object-Oriented Languages

Alexander Herz and Kalmer Apinis

Lehrstuhl für Informatik II, Technische Universität München
Boltzmannstraße 3, D-85748 Garching b. München, Germany
{herz,apinis}@in.tum.de

Abstract. We present a combined class-modular points-to and class-escape analysis that allows to analyze class declarations even if no information about the code that invokes the class's methods is available as is the case for e.g. shared libraries. Any standard whole-program or summary-based points-to analysis can be plugged into our framework and thus be transformed into a class-modular, class-escape and points-to analysis. The analysis framework uses the flow restrictions imposed by the access modifiers (e.g. *private*, *public* and *protected* in Java) to find all fields that may be modified by code that is not part of the class declaration. These fields escape the class. Unlike method-based summaries instantiated with an unknown context, our analysis framework can give detailed points-to information for non-escaping fields. In addition, the knowledge of which fields belong to the region that does not escape a class can be exploited to perform other analysis like class-modular object in-lining [6] more efficiently or enable the automatic inference of class invariants [10]. We prove the soundness of the analysis and present a set of benchmarks showing that the analysis is suitable to analyze real world code and that more than 75% of the fields from the benchmarked classes are identified as non-escaping.

1 Introduction

Accessibility of an object from different program parts is important information for optimizing compilers and verification tools. This information can be inferred using may points-to analyses. Often, not all code that uses a class declaration is available to analysis because program modules are compiled independently and linked dynamically at run time. In order to apply optimizations or verification in this scenario, points-to information for a class must be inferred in isolation from the code that uses the class. Such class-modular points-to information cannot be obtained from existing whole-program points-to analyses as these expect the complete program as input. In contrast to whole-program analyses, modular analyses can abstract different program parts independently. Commonly, individual methods are abstracted without calling-context. Later, these method-summaries are instantiated with calling-context information, so that eventually the context information and the summaries of the whole program are

A. Goodloe and S. Person (Eds.): NFM 2012, LNCS 7226, pp. 106–119, 2012.

combined. Instantiating method-summaries with unknown context information (e.g. for class methods without the code that calls the method) yields imprecise results for the *this*-pointer and all other method parameters. Therefore, it is not enough to use method-summary based analyses.

To solve this problem, we present a framework that transforms a common whole-program or summary-based points-to analysis into a class-modular points-to analysis. Given a sound plug-in analysis, the transformed analysis is sound and may be useful even if the whole program is available. The analysis time can be reduced by analyzing class declarations independently or in parallel, mostly without loosing much precision.

As a side-effect, the transformed analysis produces class-escape information. Escape analysis as presented by e.g. Blanchet [2] determines which local objects escape from a method as only local objects that do not escape can be considered truly local to that method and may be stack allocated. In contrast, our framework extends the scope from methods to classes. Local variables, *private* fields and locally used heap objects are considered class-local if and only if they can never become accessible from outside the class. If a local variable, a *private* field or a locally used heap object can possibly become accessible from outside the class (e.g. a pointer to it escapes through one of the *public* class methods) then it is considered *class-escaped*. This class-escape information can be used to improve other analysis (e.g. object in-lining), which depend on object accessibility information but commonly rely on a whole-program or summary-based analysis.

We have implemented an instance of the analysis in the Goblint [20] framework showing that it can handle large classes from industrial and open source C++ code in seconds. Our contributions in this paper are

- we present the combined class modular, class escape and points-to analysis based on encapsulation that is fully independent from the code that uses the analyzed class,
- we present a framework that allows to transform common points-to analyses into a class-modular, class-escape and points-to analysis,
- we prove the soundness of the transformed analysis,
- we present an implementation and a set of benchmarks applying an instance of the analysis to large, real world code in seconds.

Related Work. Many of the points-to analyses are based on work from Steensgaard [17] and Andersen [1] which are neither class-modular nor deal with class-escape information. Abstract interpretation based modular analyses in general is described by Cousot and Cousot [5], where program modules can be analyzed independently but a completely unknown (worst case) context is assumed and access modifiers are not taken into account.

Rountev [13], Cheng and Hwu [4], Horwitz and Shapiro [8] present pointer analyses that are modular on the function level but require additional information from the function's calling context. Whaley and Rinard [21] present a compositional pointer and (method-)escape analysis. They need information from the analyzed methods' calling context as their analysis is not on the class-level.

The precision of non-escaped objects cannot be as good if the class methods are analyzed separately, neglecting the object state information available through the access modifiers. Rountev and Ryder [14] present a similar approach, assuming worst case information on the function level.

The partitioning of fields into escaping and non-escaping fields performed by our analysis relates our analysis to region analysis where mutually unreachable heap regions are identified. The non-escaping field set represents a heap region that is unreachable from outside the class declaration and the resulting points-to information directly represents which of these fields can reach outside the region. Region analysis for C programs was recently investigated by Seidl and Vojdani [16]. These region analyses are not class-modular and often the pointer information is undirected and less precise.

Boyapati, Liskov, and Shrira [3] have applied ownership type-systems to verify encapsulation and alias protection properties of object-oriented programs. These type-systems heavily rely on annotations and restrict the programming language they can be applied to, as e.g. iterators are not easily incorporated. The ownership property verified in these systems is more restrictive than our class-escape property.

To the best of our knowledge, this is the first presentation of a class-modular, points-to and class-escape analysis. Class-level modular static analysis of classes and class methods that automatically infer class invariants have been proposed by Logozzo[10]. For the analysis to be sound it is required that accessibility of object internal state (to code that is outside the analyzed class) is detected by another static analysis. The suggested whole-program escape analysis from Blanchet[2] uses the different notion of method-escaping rather than class-escaping and cannot be applied when only the class declaration is given. As such, it does not provide the accessibility information required for Logozzo's analysis. Instead, the class-escape information provided by our analysis can be used.

Porat et al. [12] present a mutability analysis for Java that can handle missing class definitions and utilizes access modifiers. They give no details on their state accessibility analysis. It is not clear whether their state accessibility analysis is sound nor how it works.

Based on our analysis class-modular object in-lining[6] optimizations for garbage collected languages can be implemented. The life-time of fields which do not escape a class is limited to the life-time of the enclosing object. Since non-escaping fields are guaranteed to be non-accessible from outside a class it is sufficient to modify the code of the class itself to in-line an object, so method cloning is not required and the optimization is modular. JIT compilers could benefit from similar improvements [22].

Structure. First, we present a working example in Section 2. After describing the abstract semantics of our analysis in Section 3, we present our implementation of the analysis and benchmarks in Section 4. In Section 5 we summarize our findings. In the corresponding technical report [7] we define the language our analysis operates on and proof the soundness of the analysis.

2   Example

In this Section a C++ example is shown where pointer assignments in one method of a class have a non-local effect which is visible in another method and how the analysis handles this information.

The *private* fields of a class can become accessible from external code if a pointer to such a field escapes the class, for example as a return value as shown in line 27 in the example in Fig.1.

In the constructor of *Rect*, an instance of *Point* which we denote as Pt_b for convenience, is assigned to *lr*. Then the address of *lr* is assigned to *e* and *e*'s address is assigned to *p* in turn. This is denoted as edges leading from *p* to *e* to *lr* and finally to Pt_b in Fig.2, where an edge from node a to node b denotes that b is in the points-to set of a.

Within DoEscape from line 22 to line 26 Pt_b is escaped through various routes as noted in the comments of Fig.1. Especially interesting is line 23, here the pointer *pr* may be equal to the current instance *this* or another instance of the class. So Pt_b may be assigned to a field from *this* if *pr* equals *this*. Otherwise, Pt_b escapes because it is assigned to an external variable. In line 27 *e* is returned, so the content of *e* and everything reachable thereof may escape. Generally, an object escapes when its address may be stored in an externally accessible object.

Analysis Overview. In order to collect all escaped pointers in the points-to set of the variable *a_ext*, our analysis proceeds in three steps. First an instance *a_this* of the class is created and all public fields of the class are considered escaped. The created instance serves as representative object for this class. Then the effect of calling *any* constructor is over-approximated on the representative object *a_this*. Finally, all possible combinations of *public* method calls on *a_this* are simulated.

When applied to our example class *Rect*, the first step produces points-to information telling us that *a_ext* may point to itself, *a_this*, or *pub* — these are considered class-escaped. In the next step we need to apply the effect of any constructor to our abstract state. As our example only has one constructor, only the effect of that constructor is applied. Finally, we simulate all possible *public* method calls on *Rect*. Class *Rect* only has one (*public*) method DoEscape, but this method could be applied several times on the same object (while changing the escaped objects between each call). Therefore, the effect of *a_ext* = *a_this* → DoEscape(*a_ext*, *a_ext*) is computed until the smallest fix-point for *a_ext* and the fields of *a_this* is reached. In the first iteration, Pt_b class-escapes on line 22, 23 and 25, because a pointer to Pt_b is assigned to a potentially externally accessible variable. In line 24, Pt_b class-escapes because it is given as an argument to an unknown function. Eventually, *lr* and Pt_b class-escape in line 27, because they are returned. In the second iteration, Pt_b also escapes in line 26 because it is assigned to *lr*, which has escaped in the previous iteration. This last step reaches the fix-point and the result is shown in Fig. 2. At the end of each iteration step, all escaped pointers are modified so that any escaped object may point to any other escaped object.

```
1  class Point{ public: int x,y; };
2  extern void unknown(Rect* pr);
3  class Rect
4  {
5  private:   Point *ul,*lr;
6             Point **e,**l;
7             Point ***p;
8             Point *priv;
9  public:    Point *pub;//escapes
10
11 Rect(int x1,int y1, int x2, int y2)
12 {
13    ul=new Point();//Pt_a
14    lr=new Point();//Pt_b
15    p=&e;e=&lr;l=&ul;
16    ul->x=x1;ul->y=y1;
17    lr->x=x2;lr->y=y2;
18 }
19
20 Point** DoEscape(Point**v,Rect* pr)
21 { //pt_b escapes in the following:
22    pub=lr;//copied into public var
23    pr->priv=lr;//maybe copied into
                  other instance
24    unknown(lr);//passed to unknown fun
25    *v=*e;//copied into external var
26    **p=lr;//copied into lr, becomes
                  external in next line
27    return e; //lr, Pt_b escape (
                  returned from public fun)
28 }
29 };
```

externally accessible objects

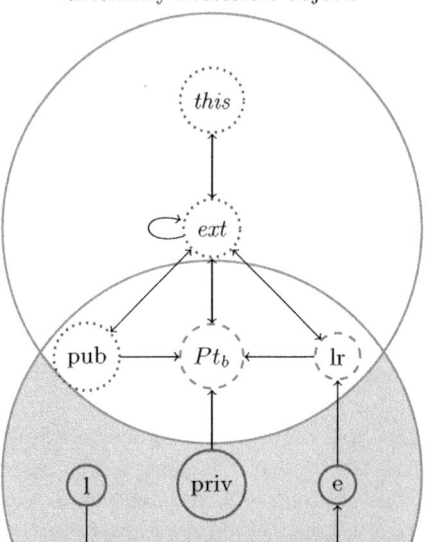

Fig. 1. C++ Example Code. Sound points-to information for class *Rect* in the absence of the code that uses class *Rect* is generated. The field e is assigned the address of lr in line 15, so when the content of e is escaped in line 27 then lr and the instance of Point which lr is pointing to are escaped. An object escapes when its address is stored in an externally accessible object. The points-to information of fields from *Rect* is a global property, points-to relations set up in one method are retained when another method is called.

Fig. 2. Graph based representation of the final domain state for class *Rect* for the program on the left after the analysis has finished, dashed nodes represent escaped objects. Dotted nodes are externally accessible before any method from *Rect* is called. An edge from symbol a to symbol b means that b is in the points-to set of a. All nodes reachable from ext are also connected with each other, this is not shown for clarity. All addresses except *this* that are neither declared nor allocated inside Rect are abstracted to *ext*. *this* is an external instance of *Rect* for which the class-invariant is generated.

3   Abstract Semantics

Our framework transforms a given plug-in points-to analysis from whole-program or summary-based analysis to a class-modular class-escape analysis that can analyze a given class without any context information on how the class may be used. To achieve this, the domain of the plug-in analysis is extended by our own global domain $\mathbb{G}'$. The semantics of our analysis is defined by lifting the plug-in semantics to be able to handle the extended domain and the unknown context in which the class may be used in. A special address *ext* that abstracts all addresses which may exist outside the analyzed class definition is introduced. Furthermore, an address *this* is created that together with our global domain abstracts all possible instances of the analyzed class.

The concrete language our analysis operates on and its semantics is given in the corresponding technical report [7]. Any points-to analyses which adhere to the set of requirements given in this section can be used as plug-in analysis for our framework. Eventually, our analysis inherits the properties from the plug-in analysis while making the analysis class-modular and calculating sets of maybe class-escaping objects which are collected in the points-to set of *ext*.

This section is structured as follows. First, we list the requirements for the plug-in points-to analysis to be suitable for this framework. Afterwards, we define the necessary functions to lift the plug-in semantics $[\![s]\!]^{\sharp}$ to the abstract semantics $[\![s]\!]^{\sharp'}$ of our analysis. Finally, we give a set of initialization steps and a set of constraints that must be solved in order to perform the analysis.

Let $val^{\sharp}$ be the abstract values and $addr^{\sharp} \subseteq val^{\sharp}$ the abstract addresses used by the plug-in analysis.

Let $\mathcal{A} : \mathbb{D} \to lval \to \mathcal{P}(addr^{\sharp})$ be a plug-in provided function that calculates the set of possible abstract addresses of a l-value given an abstract domain state $\rho^{\sharp}$.

Let $[\![s]\!]^{\sharp} : \mathbb{D} \to \mathbb{D}$ be the abstract semantics of the plug-in points-to analysis for the statement s. The complete lattice $\mathbb{D}$ is the abstract domain used by the points-to analysis. From that we construct $\mathbb{D}' = \mathbb{D} \times \mathbb{G}'$ — the domain of our analysis, where $\mathbb{G}' : addr^{\sharp} \to \mathcal{P}(val^{\sharp})$ extends the global (flow-insensitive) domain of the plug-in analysis.

Furthermore, let $q : \mathbb{D} \to addr^{\sharp} \to \mathcal{P}(val^{\sharp})$ be a plug-in provided function that calculates the set of possible abstract values that may be contained by the memory at the given address when provided an abstract domain state and $\forall \rho^{\sharp} \in \mathbb{D} : q_{\rho^{\sharp}}$ **null** $= \emptyset$.

Intuitively, the plug-in analysis maintains some kind of mapping from abstract addresses to sets of abstract values for each program point abstracting the stack and the heap. How exactly this information is encoded inside the plug-in domain $\mathbb{D}$ is not relevant for our analysis. All addresses are initialized to **null**, so if an address a has not yet been written to in $\rho^{\sharp}$ then $q_{\rho^{\sharp}}$ $a = \{$**null**$\}$.

Finally, $\rho_1^\sharp = \rho_0^\sharp[x \to Y]$ denotes the weak update of $\rho_0^\sharp \in \mathbb{D}$ such that:

$$\forall z \in addr^\sharp : q_{\rho_1^\sharp}\ z \sqsupseteq \begin{cases} q_{\rho_0^\sharp}\ z \cup Y & : z = x \\ q_{\rho_0^\sharp}\ z & : else \end{cases}$$

Using this notation we can perform weak updates on the plug-in domain without knowing the details of $\mathbb{D}$.

For example the summary-based points-to and escape analysis from Whaley and Rinard [21], like virtually all other sound points-to analyses, fulfills all our requirements and can be plugged into our framework and thus become a class-modular, points-to and class-escape analysis.

To shorten the notation we also define a function $Q : \mathbb{D} \to \mathcal{P}(addr^\sharp) \to \mathcal{P}(val^\sharp)$ for sets of addresses.

$$Q_{\rho^\sharp}\ S \triangleq \bigcup_{x \in S} q_{\rho^\sharp}\ x$$

The function $Q^*_{\rho^\sharp} : \mathbb{D} \to \mathcal{P}(addr^\sharp) \to \mathcal{P}(val^\sharp)$ defines the abstract reachability using $Q_{\rho^\sharp}$:

$$Q^*_{\rho^\sharp}\ S \triangleq F \cup Q^*_{\rho^\sharp}(F)$$

$$\textbf{where } F = Q_{\rho^\sharp}\ S \cup \bigcup_{\substack{c \in Q_{\rho^\sharp} S \\ f_i \in public\ fields\ of\ c}} \mathcal{A}_{\rho^\sharp}(c \to f_i)$$

The analysis is performed on a given class which we will call *Class*. Before starting the analysis an instance of *Class* is allocated and stored in the global variable *a_this* which we assume is not used in the analyzed code. Also, a global variable called *a_ext* of the most general pointer type (e.g. Object for Java or void* for C++) is created using the plug-in semantics (*a_ext* is also assumed not to be used in the analyzed code):

$$\rho_0^\sharp = ([\![a\_this := new\ Class]\!]^\sharp \circ [\![a\_ext := \textbf{null}]\!]^\sharp)\ d_0^\sharp$$

where $d_0^\sharp$ is the initial state of the plug-in domain, before any code has been analyzed. At this stage of the analysis *new* does not execute any constructors. As both *a_this* and *a_ext* are not used within the analyzed code, they do not change the semantics of the analyzed code and our lifted semantics can use these variables to communicate with the plug-in analysis and store special information as explained in the following.

The set *this* which contains all possible addresses of the allocated *Class* is defined as:

$$this = Q_{\rho_0^\sharp}(\mathcal{A}_{\rho_0^\sharp}(a\_this)).$$

The value *this* is meant to abstract all instances of *Class* that can exist (for the plug-in, *this* is an instance of *Class* that cannot be accessed from the program unless our analysis provides its address). The plug-in analysis should be field sensitive at least for the *Class* instance addressed by *this* in order to exceed the precision of other points-to analyses when an unknown context is used.

The set *fields* contains all addresses of the *public* fields from the *Class* instance *a_this*:

$$fields = \bigcup_{f_i \in public\ fields\ of\ Class} \mathcal{A}_{\rho_0^\sharp}(a\_this \to f_i).$$

Since the analysis is class-modular, only class declarations are analyzed. Hence, most of the program code is hidden from the analysis. We differentiate program segments which are visible to the analysis and the rest by defining external code:

Definition 1 (External Code)
External code with respect to a class C denotes all code that is not part of the class declaration of C. If no class C is stated explicitly, then class Class is assumed.

The points-to set of *ext* abstracts all addresses accessible from external code.

$$ext = \mathcal{A}_{\rho_0^\sharp}(a\_ext)$$

Initially, only *ext* itself, *this* and the *public* fields from *this* are reachable from external code, so *ext* must point to itself, *this* and the *public* fields, as an instance of *Class* may be allocated in code external to *Class*.

$$\rho_1^\sharp = \rho_0^\sharp[a \to ext \cup this \cup fields \mid a \in ext] \tag{1}$$

As the points-to set of *ext* contains multiple distinct objects, only weak updates can be performed on *ext* by the plug-in analysis.

Fields from *a_this* and their content become member of $Q_{\rho^\sharp}$ *ext* during the analysis if they may escape the *Class*. So after the analysis has finished, all possibly escaped memory locations are contained in $Q_{\rho^\sharp}$ *ext*, all other memory locations do not escape the *Class* and are inaccessible from external code.

In the following we describe how the plug-in semantics $[\![s]\!]^\sharp$ is lifted to produce the abstract semantics $[\![s]\!]^{\sharp'}$ of our class-modular class-escape analysis:

The global addresses are constituted by *ext* and the fields of *this* since modifications of these addresses' values are observable inside different member methods of *this*, even if these methods do not call each other. For example, a method from *Class* may return an address to external code which was not previously accessible by external code. Later, external code may invoke a method from *Class* passing the newly accessible address (or something reachable thereof) as parameter to the method.

$$global : \mathcal{P}(val^\sharp)$$

$$global \triangleq ext \cup fields$$

Given the state of the plug-in domain, $globals : \mathbb{G}' \to \mathbb{D} \to \mathbb{G}'$ calculates the new state of the global domain $\mathbb{G}'$:

$$globals\ g^\sharp\ \rho^\sharp\ x \triangleq \begin{cases} q_{\rho^\sharp}\ x \cup g^\sharp\ x & : x \in global \\ \emptyset & : else \end{cases}$$

The global domain state tracks modifications to fields of a_this between different invocations of *Class*-methods from external code.

modify over-approximates the effects of code external to *Class*. In a single-threaded setting, these effects cannot occur inside code of *Class* so *modify* is applied when leaving code from *Class*. This happens either when returning from a *public* method to external code or when calling an unknown method. We assume that all methods from *Class* are executed sequentially. If other threads (that do not call methods from *Class*) exist, then *modify* must be applied after every atomic step a statement is composed of, as the other threads my perform modifications on escaped objects at any time. If no additional threads exit, then modifications of escaped objects can happen only before a method from *Class* is entered, when an external function is called and after a method from *Class* is exited. As external code may modify all values from addresses it can access to all values it can access, *modify* ensures all possible modifications are performed.

$$modify \quad : \mathbb{D} \to \mathbb{D}$$
$$modify \ \rho^\sharp \triangleq \rho^\sharp[x \to Q^*_{\rho^\sharp} \ ext \mid x \in Q^*_{\rho^\sharp} \ ext]$$

The following semantic equation is inserted into the semantics $[\![s]\!]^\sharp : \mathbb{D} \to \mathbb{D}$ of the plug-in analysis (or it replaces the existing version).

$$[\![l := e_0 \to m^{\text{extern}}(e_1, \dots, e_n)]\!]^\sharp_{\rho^\sharp} \triangleq ([\![l := a\_ext]\!]^\sharp \circ modify \ \circ$$
$$[\![\mathbf{deref}(a\_ext) := e_0]\!]^\sharp \circ \dots \circ [\![\mathbf{deref}(a\_ext) := e_n]\!]^\sharp) \ \rho^\sharp$$

Methods m^{extern} are called from within *Class* but are not analyzed (e.g. because the code is not available). This makes the analysis modular with respect to missing methods in addition to its class-modularity. The procedure *unknown*, which is called in line 24 from our example in Fig. 1, represents such an external method where the above rule applies.

As shown in the proof [7], reading from a_ext and writing to $\mathbf{deref}(a\_ext)$ correctly over-approximates reads from non-class-local r-values and writes to non-class-local l-values.

Finally, we give the abstract semantics $[\![s]\!]^{\sharp'} : \mathbb{D} \times \mathbb{G}' \to \mathbb{D} \times \mathbb{G}'$ of our analysis for a statement s.

$$[\![e_0 \to m(e_1, \dots, e_n)]\!]^{\sharp'}_{(\rho^\sharp, g^\sharp)} \triangleq (\rho^\sharp_2, globals \ g^\sharp \ \rho^\sharp_2)$$
$$\mathbf{where} \ \rho^\sharp_2 = modify([\![\mathbf{deref}(a\_ext) := e_0 \to m(e_1, \dots, e_n)]\!]^\sharp_{\rho^\sharp_1})$$
$$\mathbf{and} \ \rho^\sharp_1 = \rho^\sharp[x \to g^\sharp \ x \mid x \in global]$$

Our transfer function is invoked only for top-level methods when solving the constraint systems for the analysis (see Eq. 3,4). First, the current state of the flow-insensitive fields and ext is joined into the plug-in domain. Then the plug-in semantics (which now contains our patched rule for m^{extern}) is applied and stores the return value of the method in *a_ext*. Afterwards, *modify* is applied to over-approximate the effects of external code that may execute after the top-level method is finished. Finally, the new global domain state is calculated using *globals*.

Before starting the actual analysis, the effects of external code that might have executed before a constructor from *Class* is called are simulated by applying *modify*:

$$(\rho_i^\sharp, g_i^\sharp) = modify(\rho_1^\sharp, globals \; g_0^\sharp \; \rho_1^\sharp) \tag{2}$$

Here, $g_0^\sharp$ is the bottom state of the global domain.

Then, the constructors are analyzed. Since we know that only one constructor is executed when a new object is created, it is sufficient to calculate the least upper bound of the effects of all available constructors:

$$(\rho_c^\sharp, g_c^\sharp) = \bigsqcup_{m \in public \; constructor \; of \; Class} [\![a\_this \to m(a\_ext, ..., a\_ext)]\!]^{\sharp'} (\rho_i^\sharp, g_i^\sharp) \tag{3}$$

Afterwards, the *public* methods from *Class* with all possible arguments and in all possible orders of execution are analyzed by calculating the solution [15] of the following constraint system,

$$(\rho_f^\sharp, g_f^\sharp) \sqsupseteq (\rho_c^\sharp, g_c^\sharp) \tag{4}$$

$$\forall m \in public \; method \; of \; Class:$$
$$(\rho_f^\sharp, g_f^\sharp) \sqsupseteq [\![a\_this \to m(a\_cxt, ..., a\_cxt)]\!]^{\sharp'} (\rho_f^\sharp, g_f^\sharp)$$

in order to collect all local and non-local effects on the *Class* until the global solution is reached. *a_ext* is passed for all parameters of the method as it contains all values that might be passed into the method. For non pointer-type arguments the plug-in's top value $\top_{type}$ for the respective argument type must be passed as argument to the top-level methods. If the target language supports function-pointers then all *private* methods for which function-pointers exist must be analyzed like *public* methods, if the corresponding function-pointer may escape the class.

When inheritance and *protected* fields are of interest, the complete class hierarchy must be analyzed. If the language allows to break encapsulation then additional measures must be taken to detect this. For example, C++ allows friends and *reinterpret_cast* to bypass access modifiers [18]. Friend declarations are part of the class declaration and as such easily detected. Usage of *reinterpret_casts* on the analyzed *Class* can be performed outside the class declaration, so additional code must be checked. Still, finding such casts is cheaper than doing a whole-program pointer analysis. In other languages, e.g. Java, such operations are not allowed and no additional verification is required.

4   Experimental Results

In this Section we present an implementation of the analysis and give a set of benchmark results that show how the analysis performs when it is applied to large sets of C++ code. As plug-in analysis a custom points-to analysis was implemented using the Goblint[20] framework.

The C++ code is transformed to semantically equivalent C using the LLVM [9] as the Goblint front-end is limited to C. Inheritance and access modifier information is also extracted and passed into the analysis. During this transformation, we verify that the access modifiers are not circumvented by casts or friends. No circumvention of access modifiers was found for the benchmarked code. Better analysis times can be expected from an analyzer that works directly with OO code as the LLVM introduces many temporary variables that have to be analyzed as well.

As additional input to the analysis a list of commonly used methods from the STL that were verified by hand was provided. Without this information more fields are flagged as escaping incorrectly by our implementation, because they are passed to an STL method which is considered external. This additional input is not required when using a different plug-in analysis that does not treat the STL methods as external.

The analysis is performed on two code-sets — the **Industrial** code is a collection of finite state machines that handle communication protocols in an embedded real-time setting whereas **Ogre**[11] is an open source 3d-engine. The results are given in Table 1.

Table 1. Benchmark results

Code	Classes	C++[loc]	C[loc]	Time[s]	Ext[%]	σ
Industrial	44	28566	1368112	282	23	24
Ogre	134	71886	1998910	42	62	36

The **C++** and **C** columns describe the size of the original code and its C code equivalent, respectively. More complex C++ code lines generate more C lines of code, so the ratio of code size is a measure for the complexity of the code that needs to be analyzed. The table shows that the industrial code is on average more complex than the Ogre code, requiring more time to analyze per line of code.

The time column represents the total time to analyze all the classes. The last two columns in the table show the mean and the standard deviation of the percentage of fields that the analysis identified as escaping. Fields that are identified as escaping limit the precision of subsequent analysis passes since they can be modified by external code. For the rest of the fields detailed information can be generated. A field f_i is escaping if and only if $A_{\rho_f^\sharp}(a\_this \rightarrow f_i) \cap Q_{\rho_f^\sharp} ext \neq \emptyset$.

Only 23% of the fields are escaping for the industrial code. This is the case because most of the code processes the fields directly rather than passing the fields to methods outside the analyzed class, yielding good precision.

Inside the Ogre code most fields are classes themselves, so many operations on fields are not performed by code belonging to the class containing the field but by the class that corresponds to the field's type. Our plug-in analysis implementation handles the methods of these fields as unknown methods and assumes that the field escapes. By using a plug-in analysis that analyses into these methods from other classes (e.g. Whaley and Rinard [21]) the precision of the analysis for the Ogre code can be improved to about 25% escaping fields, as indicated by preliminary results. Our analysis provides an initial context to analyze deeper into code outside of the class declaration. Especially for libraries it is necessary to generate an initial context if the library is analyzed in isolation.

So for the presented examples, for more than 75% of the fields detailed information can be extracted without analyzing the code that instantiates and uses the initial class. The benchmark times are obtained by analyzing all classes sequentially on a 2.8 Ghz Intel Core I7 with 8GB RAM. Since the results for each class are independent from the results for all other classes, all classes could be analyzed in parallel.

5   Conclusion

We have presented a sound class-modular, class-escape and points-to analysis based on the encapsulation mechanisms available in OO-languages. The analysis can be applied to a set of classes independently without analyzing the code that uses the class thus reducing the amount of code that needs to be analyzed compared to whole-program and summary-based analysis.

In addition, we have presented an easy to apply, yet powerful transformation of non-class-modular points-to analyses into class-modular, points-to and class-escape analyses. Since our framework has very weak requirements on potential plug-in points-to analyses, it can be applied to virtually all existing points-to analyses. We have shown, that the transformation will produce a sound analysis, given that the whole-program plug-in analysis was sound. Moreover, the resulting class-modular, class-escape and points-to analysis will inherit the properties of the plug-in and therefore benefits from previous and future work on points-to analyses.

The presented benchmarks show that the analysis can be applied to large, real world code yielding good precision. Due to the modularity of the analysis, flow sensitive pointer analysis becomes viable for compiler optimization passes. Class files can be analyzed and optimized independently before they are linked to form a complete program. Hence, various compiler optimizations and static verifiers can benefit from a fast class-modular class-escape and pointer analysis. Especially in large OO software projects that enforce common coding standards[19] the usage of non-private fields is rare, so good results can be expected.

Acknowledgements. We would like to thank Prof. Helmut Seidl for his valuable input and support. In addition, we would like to thank Axel Simon for contributing his time and expertise on pointer analyses. Finally, we would like to thank Nokia Siemens Networks for providing their source code and additional funding.

References

1. Andersen, L.: Program analysis and specialization for the C programming language. Tech. rep., 94-19, University of Copenhagen (1994)
2. Blanchet, B.: Escape analysis for object-oriented languages: application to Java. SIGPLAN Not. 34, 20–34 (1999)
3. Boyapati, C., Liskov, B., Shrira, L.: Ownership types for object encapsulation. SIGPLAN Not. 38, 213–223 (2003)
4. Cheng, B.C., Hwu, W.M.W.: Modular interprocedural pointer analysis using access paths: design, implementation, and evaluation. In: Proceedings of the ACM SIGPLAN 2000 Conference on Programming Language Design and Implementation, PLDI 2000, pp. 57–69. ACM, New York (2000)
5. Cousot, P., Cousot, R.: Modular Static Program Analysis. In: Horspool, R.N. (ed.) CC 2002. LNCS, vol. 2304, pp. 159–179. Springer, Heidelberg (2002)
6. Dolby, J., Chien, A.: An automatic object inlining optimization and its evaluation. SIGPLAN Not 35(5), 345–357 (2000), aCM ID: 349344
7. Herz, A., Apinis, K.: Class-Modular, Class-Escape and Points-to Analysis (Proof). Tech. rep., TUM-I1202, Technische Universität München (2012)
8. Horwitz, S., Shapiro, M.: Modular Pointer Analysis. Tech. rep., 98-1378, University of Wisconsin–Madison (1998)
9. Lattner, C., Adve, V.: Llvm: a compilation framework for lifelong program analysis transformation. In: International Symposium on Code Generation and Optimization, CGO 2004, pp. 75–86 (2004)
10. Logozzo, F.: Automatic Inference of Class Invariants. In: Steffen, B., Levi, G. (eds.) VMCAI 2004. LNCS, vol. 2937, pp. 211–222. Springer, Heidelberg (2004)
11. Open Source 3D Graphics Engine OGRE, http://www.ogre3d.org/
12. Porat, S., Biberstein, M., Koved, L., Mendelson, B.: Automatic detection of immutable fields in Java. In: Proceedings of the 2000 Conference of the Centre for Advanced Studies on Collaborative Research, CASCON 2000, p. 10. IBM Press (2000)
13. Rountev, A.: Component-Level Dataflow Analysis. In: Heineman, G.T., Crnković, I., Schmidt, H.W., Stafford, J.A., Ren, X.-M., Wallnau, K. (eds.) CBSE 2005. LNCS, vol. 3489, pp. 82–89. Springer, Heidelberg (2005)
14. Rountev, A., Ryder, B.G.: Points-to and Side-Effect Analyses for Programs Built with Precompiled Libraries. In: Wilhelm, R. (ed.) CC 2001. LNCS, vol. 2027, pp. 20–36. Springer, Heidelberg (2001)
15. Seidl, H., Vene, V., Müller-Olm, M.: Global invariants for analyzing multithreaded applications. Proc. of the Estonian Academy of Sciences: Phys., Math. 52(4), 413–436 (2003)
16. Seidl, H., Vojdani, V.: Region Analysis for Race Detection. In: Palsberg, J., Su, Z. (eds.) SAS 2009. LNCS, vol. 5673, pp. 171–187. Springer, Heidelberg (2009)
17. Steensgaard, B.: Points-to analysis in almost linear time. In: Proceedings of the 23rd ACM SIGPLAN-SIGACT Symposium on Principles of Programming Languages, POPL 1996, pp. 32–41. ACM, New York (1996)

18. Stroustrup, B.: The C++ programming language, vol. 3. Addison-Wesley, Reading (1997)
19. Sutter, H., Alexandrescu, A.: C++ coding standards: 101 rules, guidelines, and best practices. Addison-Wesley Professional (2005)
20. Vojdani, V., Vene, V.: Goblint: Path-sensitive data race analysis. Annales Univ. Sci. Budapest., Sect. Comp. 30, 141–155 (2009)
21. Whaley, J., Rinard, M.: Compositional pointer and escape analysis for Java programs. In: Proceedings of the 14th ACM SIGPLAN Conference on Object-Oriented Programming, Systems, Languages, and Applications, OOPSLA 1999, pp. 187–206. ACM, New York (1999)
22. Wimmer, C., Mössenböck, H.: Automatic feedback-directed object inlining in the java hotspot(tm) virtual machine. In: Proceedings of the 3rd International Conference on Virtual Execution Environments, VEE 2007, pp. 12–21. ACM, New York (2007)

Testing Static Analyzers with Randomly Generated Programs*

Pascal Cuoq[1], Benjamin Monate[1], Anne Pacalet[2], Virgile Prevosto[1],
John Regehr[3], Boris Yakobowski[1], and Xuejun Yang[3]

[1] CEA, LIST
[2] INRIA Sophia-Antipolis
[3] University of Utah

Abstract. Static analyzers should be correct. We used the random C-program generator Csmith, initially intended to test C compilers, to test parts of the Frama-C static analysis platform. Although Frama-C was already relatively mature at that point, fifty bugs were found and fixed during the process, in the front-end (AST elaboration and type-checking) and in the value analysis, constant propagation and slicing plug-ins. Several bugs were also found in Csmith, even though it had been extensively tested and had been used to find numerous bugs in compilers.

1 Introduction

A natural place to start for industrial adoption of formal methods is in safety-critical applications [8]. In such a context, it is normal to have to justify that any tool used can fulfill its function, and indeed, various certification standards (DO-178 for aeronautics, EN-50128 for railway, or ISO 26262 for automotive) mention, in one way or another, the need to qualify any tool used. As formal methods make headway in the industry, the question of ensuring the tools work as intended becomes more acute.

This article is concerned with static analysis of software. A static analyzer can be formally correct by construction or it can generate a machine-checkable witness [2]. Both approaches assume that a formal semantics of the analyzed programming language is available. This body of work should not distract us from the fact that the safety-critical program is compiled and run by a compiler with its own version of the semantics of the programming language, and it is the executable code produced by the compiler that actually needs to be safe. Solutions can be imagined here too: the compiler could be a formally verified or verifying compiler based on the same formalization of the programming language as the static analyzer. However, it is not reasonable to expect that any safety-critical software industry is going to move *en masse* to such bundled solutions. Even the most enlightened industrial partners feel reassured if they can substitute one specific, well-delimited step in the existing process with a drop-in replacement based on formal methods [4]. In a context where we assume an early adopter

---

* Part of this work has been conducted during the ANR-funded U3CAT project.

A. Goodloe and S. Person (Eds.): NFM 2012, LNCS 7226, pp. 120–125, 2012.

does not wish to change both compiler and verification process at the same time, how can we ensure the static analyzer being proposed agrees with the compiler already in use? We offer one partial solution to this problem: the compiler and the analyzer can be subjected to automated random testing.

Frama-C is a framework for analysis and transformation of C programs. It enables users to write their own analyses and transformations on top of provided ones. Several such custom analyses are used industrially [3] in addition to various R&D experiments [7]. Csmith [9] is an automatic generator of random C programs, originally intended for differential testing [6] of compilers. It has found bugs in all compilers it has been tried on, for a total of 400+ identified bugs. We used Csmith to test Frama-C. In this article, we report on the experiment's set-up and its results.

2   Testing Frama-C with Random Programs

Csmith generates programs that perform computations and then tally and print a checksum over their global variables. The programs contain no undefined or unspecified behavior, although they contain implementation-defined behavior. They are therefore deterministic for a given set of compilation choices. We defined several oracles for detecting that Frama-C wrongly handled a generated program (subsection 2.1). A few methodological remarks apply regardless of the oracle (subsection 2.2).

2.1   Testable Frama-C Functionalities

Robustness Testing. Random inputs often uncover "crash bugs" in software. Any tool that accepts C programs as input can be tested for crashes with Csmith, discarding the output as long as the tool terminates normally.

Value Analysis as C Interpreter. Frama-C's value analysis computes an over-approximation of the values each variable can take at each point of the program. Another abstract-interpretation-based static analyzer would work in a similar fashion, but make fundamentally different approximations, rendering any naïve application of differential testing difficult. Instead, a first way to test Frama-C's value analysis with Csmith is to turn Frama-C into a C interpreter and check its result against the result computed by the program compiled with a reference compiler.

To allow the value analysis to function as a C interpreter, we needed to make sure that all abstract functions returned a singleton abstract state when applied to a singleton abstract state. This was the hard part: since an abstract interpreter is designed to be imprecise, there are plenty of places where benign approximations are introduced. One example is union types, used in Csmith programs to convert between memory representations of different integer types. We had to improve the treatment of memory accesses to handle all possible partially overwritten values being assembled into any integer type.

Another difference between a plain and abstract interpreter is that a plain interpreter does not join states. The value analysis already had an option for improving precision by postponing joins, so this adaptation was easy. However, postponing joins can make fixpoint detection expensive in time and memory. Noting that when propagating singleton states without ever joining them a fixpoint may be detected if and only if the program fails to terminate, we disabled fixpoint detection. A Csmith program may fail to terminate, but for the sake of efficiency, our script only launches the value analysis on Csmith programs that, once compiled, execute in a few milliseconds (and thus provably terminate).

Printing Internal Invariants in Executable C Form. The previous oracle can identify precision issues that cause the analyzer to lose information where it should not. It is also good at detecting soundness bugs, since the checksum computed by the program during interpretation is compared with the checksum computed in an execution. But a large part of the value analysis plug-in's nominal behavior is not tested in interpreter mode. In order to test the value analysis when used as a static analyzer, we augment it with a function to print as a C assertion everything it thinks it knows about the program variables. Partial information easily translates to C e.g. x >= 3 && x <= 7 or, in the case of a relational abstract interpreter, x - y <= 4. To test, we analyze the program and obtain an assertion that is supposed to hold just before it terminates. The assertion is then inserted at the end of the program, and the program is compiled and run to confirm that the assertion holds.

The weakness of this oracle is that it cannot detect precision bugs. A precision bug makes the printed assertion weaker, but still true. Besides, the natural imprecision of static analysis may hide soundness bugs that would have been caught in interpreter mode. Therefore, both are indeed complementary.

Constant Propagation. A Frama-C program transformation plug-in builds on the results of the value analysis, replacing those expressions that only take one possible value throughout execution with their value. For testing this plug-in, the transformed program is compiled and run, and the computed checksum is compared to the checksum computed by the original program. This oracle tests the recording of value analysis results for further exploitation by other plug-ins, a feature that is not tested by the previous two methods.

Slicing. Frama-C's slicing plug-in removes from an input program everything that does not contribute to the user-defined criterion. We sliced Csmith programs on the criterion "compute and print the final checksum." The slicing plug-in produces slices that can be compiled and executed, so for testing, we did that and compared the computed checksum to the original. This oracle cannot find slicing precision bugs where the plug-in includes unnecessary code, but it finds slicing soundness bugs where the sliced program computes a different checksum.

2.2   Methodological Remarks

Observability. By default, Csmith generates programs with command-line arguments and volatile variables. From the point of view of static analyzers, both command-line arguments and volatile variables are unknown inputs, for which all possible values are considered. Those initial imprecisions can snowball and absorb interesting (faulty) behaviors of the analyzers. We therefore used the Csmith options that disable these two constructs.

When this experiment took place, there was no automatic way to reduce a large Csmith-generated program triggering a bug into a small one triggering the same bug. Manual reduction took too much effort. The obvious solution was to make Csmith work harder to produce smaller test cases. Csmith does not have a single setting for generated program size, but several settings do influence average size, such as maximum number of functions, maximum size of arrays, and maximum expression complexity. We also let automatic scripts run longer, and then picked only the two shortest generated problematic programs out of the twenty found at each wave. Thus filtered, the test cases were 20KB on average, and acceptable for manual reduction. Once the stream of bugs dried up, we increased Csmith's settings again to default values and beyond, but no additional bugs were revealed: all the bugs identified with Csmith were found in the first phase.

Manual Reduction. When a Csmith program produces unexpected results, it may not be obvious where in the program lies the misinterpreted construct. In the case of Frama-C, one way we found to speed up bug identification was to add `printf()` calls at the beginning of each instruction block, so as to observe execution paths. This is not acceptable when manually reducing bugs in an optimizing compiler, because the calls interfere with optimizations. No such considerations apply to Frama-C's value analysis or plug-ins that exploit its results. The tested plug-ins handle statements one by one with little possibility of a bug being affected by interference between original and tracing statements.

Bug Triage. In a 300 kLOC piece of software such as Frama-C, the first step in fixing a bug is to classify it. Frama-C plug-ins build on the results of one another. When trying to make sense of a plug-in bug, it helps to be able to assume that the supporting plug-ins are reliable. We tested the plug-ins in bottom-up order so as to avoid bug reports that would be reclassified one or several times. Some parts of the framework can only be tested indirectly when their results are used by other parts. Thus, despite our efforts, a few bug reports had to be reclassified and looked at by more than one person.

3   Bugs Found

The URL http://j.mp/csmithbugs lists bugs that were reported in Frama-C's bug tracking system. Oracles that use a reference compiler were applied with GCC and Clang on IA32, PowerPC-32 and x86-64 targets, each time configuring the value analysis to simulate the corresponding target. Some bugs were indeed specific to big-endian platforms, or to the LP64 model (bug 785). One crash was identified and fixed (bug 715).

Value-analysis-as-interpreter testing revealed bugs in the front-end and in the value analysis itself, all of which could affect normal use as a static analyzer. An example of a bug in AST elaboration is the normalization into simple assignments of x = long->access[path].bitfield = complex_expr;. One constraint is that long->access[path] may contain arbitrarily complex expressions and should not be duplicated. Before the bug was fixed, this complex statement was normalized as tmp = complex_expr; long->access[path].bitfield = tmp; x = tmp;. Bug 933 explains why this is incorrect.

Testing the value analysis as a static analyzer found bugs related to alarm emission (bugs 715, 718, 1024), which was not tested at all in interpreter mode (no alarm is emitted while interpreting a defined program). Constant propagation testing revealed issues in program pretty-printing (bug 858). It was good to get these out of the way before testing the slicing plug-in. One slicing bug was found that could happen with very simple programs (bug 827). To occur, the other slicing bugs found needed the program to contain jumps in or out of a loop, or from one branch of a conditional to the other.

Unexpectedly, several bugs were also found by Frama-C in Csmith, despite the latter having been put to intensive and productive use finding bugs in compilers. Csmith is intended to generate only defined programs. The bugs found in Csmith involved the generation of programs that, despite being accepted by compilers, were not defined: they could pass around dangling pointers, contain unsequenced assignments to the same memory location, or access uninitialized members of unions. Generated programs could contravene [5, §6.5.16.1:3] either directly (lv1 = lv2; with overlapping lvalues lv1 and lv2) or convolutedly (lv1 = (e, lv2);). Lastly, Csmith could generate the pre-increment ++u.f; with u.f a 31-bit unsigned bitfield containing 0x7fffffff. The bitfield u.f is promoted to int according to [5, §6.3.1.1] (because all possible values of a 31-bit unsigned bitfield fit in a 32-bit int). The increment then causes an undefined signed overflow. This last bug needs just the right conditions to appear. Narrower bitfields do not have the issue. Also, a 32-bit unsigned bitfield would be promoted to unsigned int (because not all its values would fit in type int) and the increment would always be defined [5, §6.2.5:9].

4   Related Work, Future Directions and Conclusion

Another recent application of fuzzing in formal methods is the differential testing of SMT solvers [1]. The originality of our experiment is that differential testing does not apply directly to static analyzers. New functionality had to be implemented in the value analysis both to make it testable as a plain C interpreter and to make it testable as a static analyzer. This was worth it, because the bugs found when misusing the value analysis as a plain interpreter also affected nominal use. Besides, the resulting C interpreter is useful in itself for applications other than testing the value analysis.

It is not shocking that bugs were found in some of the Frama-C plug-ins that are already in operational use. Firstly, qualification of these plug-ins has not taken place yet. Secondly, some of the bugs found can only appear in a specific target configuration or in programs disallowed by industrial coding standards.

Fifty is a large number of bugs to find. It reveals the quantity of dark corners in the C language specification. That bugs were found in Csmith confirms this idea. Another inference is that correctly slicing a real world language such as C into executable slices is hard: it took 22 bug reports on the slicing plug-in itself to converge on an implementation that reliably handles Csmith-generated programs. The Frama-C developers were happy to be informed about every single bug. Almost all the bugs were obscure, which is as it should be, since Frama-C has been in use for some time now.

Csmith generates programs that explore many implementation-defined behaviors. Csmith testing not only uncovers bugs where either the reference compiler or the static analyzer disagree with the standard. It also checks that they agree with each other where the standard allows variations.

Our conclusion is that everyone should be testing their static analyzers with randomly generated programs. The Nitrogen Frama-C release can withstand several CPU-months of automated random testing without any new bugs being found. The development version is continuously tested, so as to find newly-introduced bugs as rapidly as possible.

References

1. Brummayer, R., Biere, A.: Fuzzing and delta-debugging SMT solvers. In: Proceedings of the 7th International Workshop on Satisfiability Modulo Theories, SMT 2009. ACM, New York (2009)
2. Cachera, D., Pichardie, D.: Comparing Techniques for Certified Static Analysis. In: The NASA Formal Methods Symposium, NFM (2009)
3. Delmas, D., Cuoq, P., Moya Lamiel, V., Duprat, S.: Fan-C, a Frama-C plug-in for data flow verification. In: ERTS^2 (to appear, 2012)
4. Delseny, H.: Formal Methods for Avionics Software Verification. Open-DO Conference, presentation (2010), http://www.open-do.org/2010/04/28/formal-versus-agile-survival-of-the-fittest-herve-delseny/
5. International Organization for Standardization: ISO/IEC 9899:TC3: Programming Languages—C (2007), http://www.open-std.org/jtc1/sc22/wg14/www/docs/n1256.pdf
6. McKeeman, W.M.: Differential testing for software. Digital Technical Journal 10(1), 100–107 (1998)
7. Pariente, D., Ledinot, E.: Formal Verification of Industrial C Code using Frama-C: a Case Study. In: FoVeOOS (2010)
8. Woodcock, J., Larsen, P.G., Bicarregui, J., Fitzgerald, J.S.: Formal methods: Practice and experience. ACM Computing Surveys 41(4) (2009)
9. Yang, X., Chen, Y., Eide, E., Regehr, J.: Finding and understanding bugs in C compilers. In: PLDI, San Jose, CA, USA (June 2011)

Compositional Verification of Architectural Models

Darren Cofer[1], Andrew Gacek[1], Steven Miller[1], Michael Whalen[2],
Brian LaValley[3], and Lui Sha[4]

[1] Rockwell Collins Advanced Technology Center
{ddcofer,ajgacek,spmiller}@rockwellcollins.com
[2] University of Minnesota
whalen@cs.umn.edu
[3] WW Technology Group
blavalley@wwtechgroup.com
[4] University of Illinois
lrs@uiuc.edu

Abstract. This paper describes a design flow and supporting tools to significantly improve the design and verification of complex cyber-physical systems. We focus on system architecture models composed from libraries of components and complexity-reducing design patterns having formally verified properties. This allows new system designs to be developed rapidly using patterns that have been shown to reduce unnecessary complexity and coupling between components. Components and patterns are annotated with formal contracts describing their guaranteed behaviors and the contextual assumptions that must be satisfied for their correct operation. We describe the compositional reasoning framework that we have developed for proving the correctness of a system design, and provide a proof of the soundness of our compositional reasoning approach. An example based on an aircraft flight control system is provided to illustrate the method and supporting analysis tools.

Keywords: Cyber-physical systems, design patterns, formal methods, model checking, compositional verification, SysML, AADL, META, DARPA.

1   Introduction

Advanced capabilities being developed for the next generation of commercial and military aircraft will be based on complex new software. These aircraft will incorporate adaptive control algorithms and sophisticated mission software providing enhanced functionality and robustness in the presence of failures and adverse flight conditions. Unmanned aircraft have already displaced manned aircraft in most surveillance missions and are performing many combat missions with increasing levels of autonomy. Manned and unmanned aircraft will be required to coordinate their activities safely and efficiently in both military and commercial airspace.

The *cyber-physical systems* that provide these capabilities are so complex that software development and verification is one of the most costly development tasks and therefore poses the greatest risk to program schedule and budget. Without

A. Goodloe and S. Person (Eds.): NFM 2012, LNCS 7226, pp. 126–140, 2012.

significant changes in current development processes, the cost and time of software development will become the primary barriers to the deployment of the advanced capabilities needed for the next generation of military aircraft.

DARPA's META program was undertaken to significantly improve the design, manufacture, and verification process for complex cyber-physical systems. The work described in this paper directly addresses this goal by allowing the system architecture to be composed from libraries of *complexity-reducing design patterns* with formally guaranteed properties. This allows new system designs to be developed rapidly using patterns that have been shown to reduce unnecessary complexity and coupling between components. This work also deeply embeds formal verification into the design process to enable correct-by-construction development of systems that work the first time. The use of components with formally specified contracts, design patterns that provide formally guaranteed properties, and an architectural modeling language with a well-defined semantics ensures that the system design is known to meet its requirements even before it is implemented. Further details can be found in [1].

In previous work, we have successfully applied model checking to software components that have been created using model-based development (MBD) tools such as Simulink [7]. Our objective in this project was to build on this success and extend the reach of model checking to system design models. Examples of previous work in this area include approaches that essentially flatten the system model by elaborating each component and including its implementation in the same language used for the system [12]. This approach permits accurate modeling of component behaviors and interactions, but suffers from limited scalability. An alternative approach replaces each component with a state machine description that is an abstraction of the component design [11]. This provides better scalability, but can result in the component descriptions that diverge from their implementations and can limit the expressiveness of the overall system model.

The compositional approach we advocate in this paper attempts to exploit the verification effort and artifacts that are already part of our software component verification work. We do this through the use of formal assume-guarantee contracts that correspond to the component requirements for each component. Each component in the system model is annotated with a contract that includes the requirements and constraints that were specified and verified as part of its development process. We then reason about the system-level behavior based on the interaction of the component contracts. The use of contracts is also extended to architectural design patterns that have been formally verified. This approach allows us to leverage our existing MBD process for software components and provides a scalable way to reason about the system as a whole.

Section 2 of this paper presents our architectural modeling framework and describes how we have used the AADL and SysML languages to formally specify system designs. We have developed a mapping between relevant portions of these languages, as well as an automated translation tool and support for contract annotations. Section 3 briefly describes our formalization of architectural design patterns. These patterns encapsulate several fault-tolerance and synchronization mechanisms, increasing the level of design abstraction and supporting verification reuse. Section 4 describes in more detail our compositional verification approach, and Section 5 presents the formulation of our method and a proof sketch of its soundness, the main technical contribution of the paper. Section 6 presents an example based on an aircraft flight control system, and Section 7 briefly describes our tool framework.

2   Architectural Modeling

Our domain of interest is distributed real-time embedded systems (including both hardware and software), such as comprise the critical functionality in commercial and military aircraft. MBD languages and tools are commonly used to implement the components of these systems, but the system-level descriptions of the interactions of distributed components, resource allocation decisions, and communication mechanisms are largely ad hoc. Application of formal analysis methods at the system level requires 1) an abstraction that defines how components will be represented in the system model, and 2) selection of an appropriate formal modeling language.

Assumptions and Guarantees. Many aerospace companies have adopted MBD processes for production of software components. As a result of aircraft certification guidelines, these components must have detailed requirements. We have been successful applying formal methods to software component designs because of our decision to conform (as much as possible) to existing trends in industry. By formalizing the component requirements for verification using a combination of model checking and automated translation of the component models, we have made formal analysis accessible to embedded system developers. Therefore, one of our goals in this project was to create a system modeling methodology that would incorporate existing practices and artifacts and be compatible with tools being used in industry.

In this approach, the architectural model includes interface, interconnections, and specifications for components but not their implementation. It describes the interactions between components and their arrangement in the system, but the components themselves are black boxes. The component implementations are described separately by the existing MBD environment and artifacts (or by traditional programming languages, where applicable). They are represented in the system model by the subset of their specifications that is necessary to describe their system-level interactions.

Assume-guarantee contracts [4] provide an appropriate mechanism for capturing the information needed from other modeling domains to reason about system-level properties. In this formulation, guarantees correspond to the component requirements. These guarantees are verified separately as part of the component development process, either by formal or traditional means. Assumptions correspond to the environmental constraints that were used in verifying the component requirements. For formally verified components, they are the assertions or invariants on the component inputs that were used in the proof process.

A contract specifies precisely the information that is needed to reason about the component's interaction with other parts of the system. Furthermore, contract mechanism supports a hierarchical decomposition of verification process that follows the natural hierarchy in the system model.

SysML and AADL. The two modeling languages that we have worked with in this program are SysML and AADL. These languages were developed for different but related purposes. SysML was designed for modeling the full scope of a system, including its users and the physical world, while AADL was designed for modeling real-time embedded systems. While both SysML and AADL are extensible and can be tailored to support either domain, the fundamental constructs each provides reflect these differences. For example, AADL lacks many of the constructs for eliciting system requirements such as SysML requirement diagrams and use cases, and for specifying the behavior of systems such as SysML activity diagrams. On the other hand, SysML lacks many of the constructs needed to model embedded systems such as processes, threads, processors, buses, and memory.

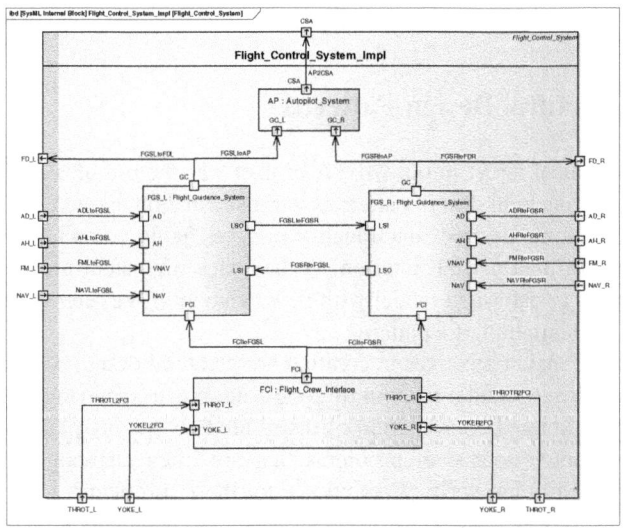

Fig. 1. Flight Control System modeled using SysML

AADL is a good fit for our domain of interest and provides a sufficiently formal notation. However, AADL has yet to gain traction with many industrial users and its lack of a stable graphical environment (at least in the most popular available tool, the Eclipse-based OSATE) has been a barrier to adoption. Consequently, SysML has been adopted by many organizations for system design specification, even though it has no formal semantics and no common textual representation.

Our solution is to allow developers to do at least their initial system development in SysML, and provide support for automatic translation to AADL for analysis. We have built an Eclipse plugin that provides bidirectional translation between SysML and AADL for the domain in which they overlap. We have defined block stereotypes in SysML that correspond to AADL objects, thus effectively mapping the semantics of AADL onto a subset of SysML. The translation is based on the Enterprise Architect SysML tool used by Rockwell Collins. An example system is shown in Fig. 1.

For SysML to be used to model embedded systems in the same way that AADL does, SysML blocks and ports need to be tagged with stereotypes corresponding to AADL constructs such as threads and processors. AADL components are represented using SysML Blocks with stereotypes. If a SysML block is not tagged with one of these stereotypes, the translator treats it as an AADL system. AADL features are represented using SysML flow ports with stereotypes. If a SysML flow port is not tagged with one of these stereotypes, the translator treats it as an AADL port.

The translator also translates the package structure from a SysML model to AADL and vice versa. When translating from AADL to SysML, the translator will create a single SysML block diagram for each AADL package with a SysML block drawn for each AADL component type and implementation. The translator will also create a single internal block diagram for each AADL implementation that has subcomponents showing that implementation, its subcomponents, their features, and connections.

3   Architectural Design Patterns

The second technical thrust in our META project was the use of *architectural design patterns*. An architectural design pattern is a transformation applied to a system model that implements some desired functionality in a verifiably correct way. Each pattern can be thought of as a partial function on the space of system models, mapping an initial model to a transformed model with new behaviors. We refer to the transformed system as the instantiation of a pattern.

We have three main objectives in creating architectural design patterns. The first is the encapsulation and standardization of good solutions to recurring design problems. The synchronization and coordination of distributed computing platforms in avionics system is a common source of problems that are often challenging to implement correctly. By codifying verified solutions to these problems and making them available to developers, we raise level of abstraction and avoid "reinventing the wheel."

Reuse of verification is the second objective. The architectural design patterns are developed in a generic way so that they can be formally verified once, and then reused in many different development projects by changing parameters. Each pattern has a contract associated with it that specifies constraints on the systems to which it can be applied, and specifies the behaviors that can be guaranteed in the transformed system. In this way we are able to amortize the verification effort over many systems.

The final objective is reduction or management of system complexity. An architecture pattern can be said to reduce system complexity if it provides an abstraction that effectively eliminates a type of component interaction in a way that can be syntactically enforced. The PALS (physically asynchronous logically synchronous) architecture pattern is an example in which real time tasks are executed with bounded asynchrony physically but the asynchronous execution is logically equivalent to synchronous execution. This greatly reduces the verification state space.

Four architectural patterns were implemented in this project: PALS, Replication, Leader Selection, and Fusion.

The purpose of the PALS pattern is to make portions of a distributed asynchronous system operate in virtual synchrony. This allows portions of the system logic to be designed and verified as though they will be executed on a synchronous platform, and then deployed in the asynchronous system with the same guaranteed behavior. The pattern relies on certain timing constraints on the delivery and processing of messages that must be enforced by the underlying execution platform. To use the pattern, a group of nodes (systems) is selected that are to execute at approximately the same time at period T. The outputs (ports) of these nodes are to be received by other nodes in the group such that all nodes will receive the same values at each execution step. The pattern does not add any new data connections to the model, but assumes that the required connections already exist.

The purpose of the Replication pattern is to create identical copies of portions of the system. This is typically used to implement fault tolerance by assigning the copies to execute on separate hardware platforms with independent failure modes. To use the pattern, one or more nodes (systems) are selected and the number of copies to create is specified. Optional arguments for each input and output port on the selected systems determine how these ports and their connections are handled in the replication process. Each new system and port created is given a unique name. When multiple outputs are created they may be merged by the addition of a new system block to select, average, or vote the outputs.

The purpose of the Leader Selection pattern is to coordinate a group of nodes so that a single node is agreed upon as the 'leader' at any given time. The nodes typically correspond to replicated computations hosted on distributed computing resources, and are used as part of a fault-tolerance mechanism. If a replicated node fails, this allows a non-failed node to be selected as the one which will interact with the rest of the system. To use the pattern, a group of N nodes (systems or processes) is identified that are to select a leader from among themselves. The leader selection pattern will insert new leader selection threads into each of the systems/processes which are to participate in leader selection. Each thread will have a unique identifier (an integer) to determine its priority in selecting a leader. Connections will be added so that all leader selection threads are able to communicate with each other (N-1 input ports, 1 output port). In addition, each leader selection thread will have an input port from which it determines (from other local systems) if it is failed, and an output port which will say if it is the leader. These two ports are initially left unconnected.

The purpose of the Fusion pattern is to insert a component into the architecture that combines several component interfaces into a single interface. The component supplies properties that define the validation/selection algorithm that is used and its impact on the fault tolerance or performance properties of the interfaces. The fusion algorithm could provide voting through exact or approximate agreement or by mid-value selection. The output could correspond to one of the selected inputs or it could be a computing average. To use the pattern, the user will select from a predefined set of fusion algorithms that are presented in a list. Each option will describe the properties and allow the user to browse these as part of the selection process. The user will select the type of component to be inserted in the model to perform the fusion algorithm. There are three initial choices: System (for abstract system designs),

Thread (for software implemented voting), and Device for hardware implementations. Finally, the user will select the insertion point for the voter by first selecting an existing architecture component that is the current destination of the interfaces to be voted. After component selection the user will be presented with a list of input interfaces that match the constraints required for the voter that was selected. The user can then select the set of interfaces to which the voter will be applied.

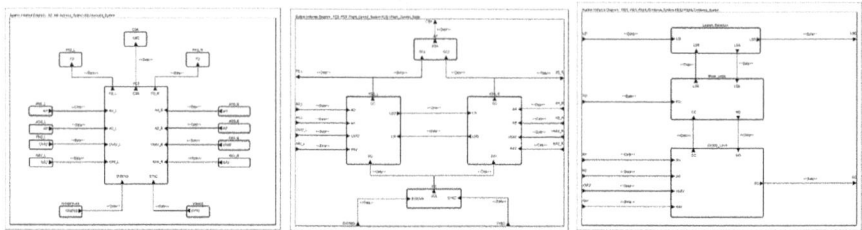

Fig. 2. Avionics System, Flight Control System, and Flight Guidance System models that were used to demonstrate the use of architectural design patterns

We have applied these patterns in an Avionics System modeled in AADL. Three levels of the system architecture are shown in Fig. 2: the Avionics System, the Flight Control System (FCS), and the Flight Guidance System (FGS). The initial system model to which we apply the patterns captures the functionality of the system under the assumption that nothing ever fails. It only has one set of inputs and outputs and has no redundancy in its implementation. We first apply the replication pattern to the FGS component to create two redundant copies. This pattern automatically replicates ports as necessary and applies a property requiring the copies not be hosted on the same hardware. We next apply the Leader Selection pattern to manage the redundant copies of the FGS. This pattern inserts pre-verified leader selection functionality as new threads inside each FGS to determines the current leader. The Leader Selection protocol that we have used requires that the nodes communicate synchronously. To satisfy this assumption, we apply the PALS synchronization pattern. The constraints of the PALS pattern will be verified during implementation to ensure they can actually be satisfied. Finally, the Fusion pattern is used inside the Autopilot component to combine the two outputs produced by the active and standby FGS copies into a single command input.

4   System Verification

The system-level properties that we wish to verify fall into a number of different categories requiring different verification approaches and tools. This is also true for the contracts that are attached to the components and design patterns used in the system model.

– They may be behavioral properties that describe the state of the system as it changes over time. Behavioral properties may be used to describe protocols governing component interactions in the system, or the system response to combinations of triggering events. We will use the Property Specification Language (PSL) [5] to specify most behavioral properties. An example of a behavioral property associated with the Leader Selection pattern is: *A failed node will not be leader in next step*, or G(!device_ok[j]- X(leader[i]!= j)).

- They may be structural properties of the system model to which the pattern is applied (pre-conditions), or of the transformed system model after pattern instantiation (post-conditions). Relationships among timing properties in the model or constraints on the numbers of various objects in the model are in this category.
- Some design patterns rely explicitly on resource allocation properties of the system, including real-time schedulability, memory allocation, and bandwidth allocation. Even after we annotate the model with deadlines and execution times, we must still demonstrate that threads can be scheduled to meet their deadlines. There are many tools available to support verification of these properties, including the ASIIST tool developed by UIUC and Rockwell Collins [8].
- Failure analysis of the system often requires the use of probabilistic methods to demonstrate that the sufficiency of the proposed fault handling mechanisms. The AADL error annex can be used to attach fault behavior models to the system design. As part of the META program we have participated in some demonstrations based on our example AADL model using the PRISM probabilistic model checker [9].

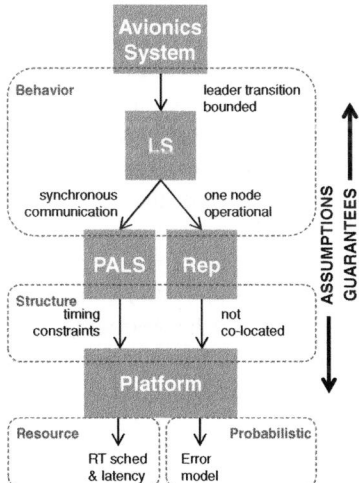

Fig. 3. Contracts between patterns used in the Avionics System example

At the system level, assumptions and guarantees associated with the system components and patterns interact and are composed to achieve desired system properties. For example, the behavior of the avionics system in our example depends upon guarantees provided by the Leader Selection pattern (Fig. 3). Leader Selection includes an assumption of synchronous data exchange which is satisfied by the PALS pattern guarantees. It also includes an assumption that there will be at least one working node, which is satisfied by the Replication pattern. The PALS pattern, in turn, makes assumptions about the timing properties of the underlying execution platform and the Replication pattern requires that copies are not co-located. Finally, the platform can only guarantee these properties if it verified to satisfy its resource allocation constraints, and its probability of failure is sufficiently low.

The focus of our work is the first two categories: behavioral and structural properties. The next section describes our approach to compositional reasoning.

5   Compositional Reasoning

Our idea is to partition the formal analysis of a complex system architecture into a series of verification tasks that correspond to the decomposition of the architecture. By partitioning the verification effort into proofs about each subsystem within the architecture, the analysis will scale to handle large system designs. Additionally, the approach naturally supports an architecture-based notion of requirements refinement: the properties of components necessary to prove a system-level property in effect define the requirements for those components. We call the tool that we have created for managing these proof obligations *AGREE*: Assume Guarantee Reasoning Environment.

There were two goals in creating this verification approach. The first goal was to reuse the verification already performed on components and design patterns. The second goal was to enable distributed development by establishing the formal requirements of subcomponents that are used to assemble a system architecture. If we are able to establish a system property of interest using the contracts of its components, then we have a means for performing *virtual integration* of components. We can use the contract of each of the components as a specification for suppliers and have a great deal of confidence that if all the suppliers meet the specifications, the integrated system will work properly.

In our composition formulation, we use past-time LTL [2]. This logic supports a uniform formulation of composition obligations that can be used for both *liveness* properties and *safety* properties. For the reasoning framework, we use the LTL operator G (globally) supplemented by the past time operators H (historically) and Z (in the previous instant) [3]. They are defined formally over paths σ and time instants t as follows:

$$\sigma, t \models G(f) \equiv \forall (u, t \leq u) : \sigma, u \models f$$
$$\sigma, t \models H(f) \equiv \forall (u, 0 \leq u \leq t) : \sigma, u \models f$$
$$\sigma, t \models Z(f) \equiv (t = 0) \vee (\sigma, (t\text{-}1) \models f)$$

Verification Conditions. Formally, in our framework a component contract is an assume-guarantee pair (A, P), where each element of the pair is a PSL formula. Informally, the meaning of a pair is "if the assumption is true, then the component will ensure that the guarantee is true." To be precise, we need to require a component to meet its guarantee only if its assumptions have been true up to the current instant. We can state this succinctly as a past-time LTL formula $G(H(A) \Rightarrow P)$[1].

Components are organized hierarchically into systems as shown in Fig. 1. We want to be able to compose proofs starting from the leaf components (those whose implementation is specified outside of the architecture model) through several layers of the architecture. Each layer of the architecture is considered to be a system with inputs and outputs and containing a collection of components. A system S can be described by its own contract (A_s, P_s) plus the contracts of its components C_S, so we have $S = (A_S, P_S, C_S)$. Components "communicate" in the sense that their formulas may refer to the same variables. For a given layer, the proof obligation is to demonstrate that the system guarantee P_s is provable given the behavior of its subcomponents C_S and the system assumption A_s.

Our goal is therefore to prove the formula $G(H(A_s) \Rightarrow P_s)$ given the contracted behavior $G(H(A_c) \Rightarrow P_c)$ for each component c within the system. It is conceivable that for a given system instance a sufficiently powerful model checker could prove this goal directly from the system and component assumptions. However, we take a more general approach: we establish generic *verification conditions* which together are sufficient to establish the goal formula. Moreover, we provide verification conditions which have the form of safety properties whenever all the assumptions and guarantees do not contain the G or F LTL operators. In such cases, this allows the verification conditions to be proved even by model checkers which are limited to safety properties, such as k-induction model checkers.

Handling Cycles in the Model. It is often the case that architectural models contain cyclic communication paths among components (e.g., the FGS_L and FGS_R in Fig. 1), and that these components mutually depend on the correctness of one another Therefore, we need to consider circular reasoning among components. To accomplish this, we use a framework similar to the one from Ken McMillan in [4]. We break these cycles using *induction over time*.

Suppose that we have components A and B that mutually refer to each other's guarantees. When trying to establish the assumptions of A *at time t*, we will assume that the guarantees of B are true *only up to time t-1*. Therefore, at time instant t there is no circularity. To accomplish this reasoning, we define a well-founded order $(<)$ between component contracts. If $C_A < C_B$, then B can refer to A's assumptions and guarantees at the current instant, while A can refer to B's assumptions and guarantees only at the previous instant.

Following McMillan, for a contract $c \in C$, we define Θ_c to be the contracts whose assumptions and guarantees are true up to and including time t. We define $c^\wedge$ for a contract c to be $(A_c \wedge P_c)$ and $C^\wedge$ to be $\{c^\wedge \mid c \in C\}$. Every element in Θ_c must be less than c according to the order $<$, so $\Theta_c \subseteq C^\wedge$. Since we are only considering cycles inside system S, its contracts are handled separately and do not need to be included. Therefore, the system assumptions are taken to hold up through the current time, and the system guarantees are proven separately (as shown below):

---

[1] We use "Promises" P in the place of G for guarantees for presentation because G is an LTL operator. Also, in the informal presentation, we represent each of A,P as *sets* of formulas. The formulas here are formed by simple conjunction of the elements of the set.

Theorem 1. Let the following be given:
- $S = (A_S, P_S, C_S)$ with assumption A_s, guarantee P_s and component contracts C_S, with a well-founded order $<$ on C
- Sets $\Theta_c \subseteq C^\wedge$, such that $q \in \Theta_c$ implies $q{<}c$
- For all $c \in C$, $\models G(H(A_c) \Rightarrow P_c)$

Then if for all $c \in C$
$$\models G((H(A_s) \wedge Z(H(C^\wedge)) \wedge \Theta_c) \Rightarrow A_c)$$
and
$$\models G((H(A_s) \wedge H(C^\wedge)) \Rightarrow P_s)$$
then
$$\models G(H(A_s) \Rightarrow P_s).$$

In other words, for a system with n components there are $n{+}1$ verification conditions: one for each component and one for the system as a whole. The component verification conditions establish that the assumptions of each component are implied by the system level assumptions and the properties of its sibling components. These verification conditions are naturally cyclic, but the cycle is broken using the well-founded ordering $<$ and the one-step delay operator Z. The system level verification condition shows that the system guarantees follow from the system assumptions and the properties of each subcomponent. This is essentially an expansion of the original goal, $\models G(H(A_s) \Rightarrow P_s)$, with additional information obtained from each component.

Proof Sketch. It is possible to prove Theorem 1 directly using induction over time. The idea is, at each step, to go through each component (from largest to smallest based on the $<$ ordering) and show that its assumptions hold in the current step. Then we can use the assumption $\models G(H(A_c) \Rightarrow P_c)$ to show that P_c also holds in the current step. Once we have done this for each component we can use the system level verification condition to show that the system level guarantees hold in the current step. Formally, the proof is by induction over time using the strengthened goal formula

$$\models G(H(A_s) \Rightarrow (H(P_s) \wedge H(C^\wedge)))$$

The desired goal formula then follows directly.

Another approach to proving Theorem 1 is to encoding it using McMillan's circular reasoning framework. This is fairly straightforward to do. In fact, the two approaches show many similarities which provides a strong argument for the quality of approach. The details of this equivalence are presented in a companion technical report [6].

6   Flight Control System Example

We have applied our compositional verification approach to the avionics system model. While there is not space to present the entire example, this section provides a summary of the assumptions and guarantees on the flight control system and describes one level of reasoning.

One of the typical requirements levied on a flight control system has to do with transients in the actuator commands. For passenger comfort and safety, a limit is placed on the forces that would be experienced by the passengers during normal operation. For example, the automation should not command a sharp change in the pitch of the aircraft, even in the presence of component failures.

In our system architecture, this property becomes a constraint on the control surface actuator (CSA) output of the system. We would like the commanded pitch to be bounded both in terms of the both the actuator angle and its rate of change. In our notation, we can write these properties as follows:

```
transient_response_1 : assert
   true -> abs(CSA.CSA_Pitch_Delta) < MAX_PITCH_DELTA;
transient_response_2 : assert
   true -> abs(CSA.CSA_Pitch_Delta -
   prev(CSA.CSA_Pitch_Delta, 0.0)) < MAX_PITCH_DELTA_STEP ;
```

The "true ->" portion of each property states the property is initially true. The remainder of the first property states that the absolute value of the commanded pitch (CSA_Pitch_Delta) is less than some constant (MAX_PITCH_DELTA). The second property is similar, but states that the difference between the current pitch and the previously commanded pitch is less than some constant (MAX_PITCH_DELTA_STEP).

Similarly, we have system-level assumptions related to independence of failures:

```
active_assumption: assume (FD_L.mds.active or
   FD_R.mds.active) ;
```

In our model we make assumptions about at least one FGS being active at all times (shown), as well as assumptions about maximum discrepancies between left and right side pitch sensors, and a handful of other assumptions. These assumptions state maximum discrepancies in the pitch inputs in time and between the left and right sides. In order to prove the guarantees for the system, we need to pull in assumptions from the left and right FGSs and the autopilot. In the absence of circularity, the tools automatically compute the dependency order for reasoning: FCI < {FGS_L, FGS_R}, {FGS_L, FGS_R} < AP[2]. For circular dependencies, the user must decide an order (if it is required). In this instance, our proof did not require "same instant" assumptions between FGS_L and FGS_R, so the cycles can be broken and verification conditions produced automatically by our AGREE tool.

The proof relies on the guarantees for the components {FGS_L, FGS_R, AP}, and uses the system level assumptions to discharge the component level assumptions. In addition, the Leader Selection pattern guarantees are brought in as additional *facts* which function as assumptions in the proof. Each set of assumptions and guarantees per component is between ½ page and 1 page of text, and the proof for this layer of the architecture can be discharged in ~5 seconds by the Kind model checker.

---

[2] The set notation is a shorthand: X < {Y, Z} is the same as X <Y and X < Z.

7   Tool Environment

We have produced a prototype implementation of all the tools described in this paper in a single Eclipse environment, shown in Fig. 4 . They have been designed to work with the open source OSATE AADL tool developed by the Software Engineering Institute.

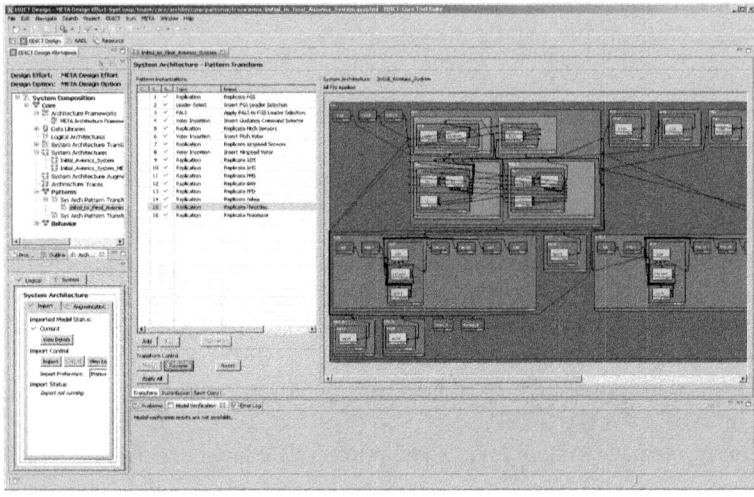

Fig. 4. Eclipse environment for our translation, pattern instantiation, and verification tools

The SysML-AADL translator is implemented as an Eclipse plug-in. It provides a convenient way to import an initial block diagram model created in SysML into OSATE for further development.

The pattern instantiation tool is implemented as an extension to the EDICT tool developed by WW Technology Group. EDICT provides a wide variety analysis capabilities for building dependable systems, and now includes the ability to modify the system design through application of the architectural patterns described above.

We have developed two additional Eclipse plugins to implement the compositional verification approach described in this paper (AGREE), and a static analysis tool called Lute for verifying structural properties of AADL models. These tools are available for download through the AADL wiki page at http://wiki.sei.cmu.edu/aadl/index.php/RC_META.

Complex structural assumptions and guarantees can be verified using the Lute checker. While Lute is similar to the to the REAL verification system [10], it provides several enhancements needed for the META project for specifying and checking complex structural properties. A Lute specification is made up of Lute theorems, which are computational checks over the structure of the model. A typical Lute theorem iterates over a select group of components and aggregates information about each before checking a Boolean condition. For example, a Lute theorem may iterate over each process and verify that the maximum deadline for all threads in the process is less than or equal to the process deadline. The Lute code for this theorem is shown below:

```
theorem Process_Deadline_Greater_Or_Equal_Thread_Deadline
   foreach p in Process_Set do
   Thread_Deadlines := {Property(t, "Deadline")
        for t in Thread_Set | Owner(t) = p};
check Max(Thread_Deadlines) <= Property(p, "Deadline");
   end;
```

Since Lute theorems are purely computational, they can be executed without user interaction. Thus it is feasible to re-verify the Lute specification every time a structural change is made to the model. This enables instant feedback during model development.

The AGREE tool uses the custom AADL property set PSL_Properties to add support for compositional reasoning to AADL. The PSL_Properties property set is currently implemented simply as an AADL string applied as follows:

```
property set PSL_Properties is
   Contract: aadlstring applies to (system, process, thread);
   Facts: aadlstring applies to (system, process, thread);
end PSL_Properties;
```

That is, it supports contracts and facts on systems, processes, and threads specified as AADL strings. Verification of AADL models is performed through the translation of the AADL structure and subcomponent assumptions and guarantees into a form suitable for model checking. Currently the KIND model checker is supported, but it would be straightforward to add support for additional model checkers and theorem provers.

In our initial implementation, subcomponents are assumed to operate synchronously with a one-step communication delay between connected subcomponents. This makes the analysis tractable and creates a sound approximation of the behavior of the system. Any error found during verification corresponds to an error in the actual system. The approximation is complete in the case of synchronous systems (e.g. systems using the PALS pattern), and incomplete in the general case. Incompleteness means that the absence of verification errors does not ensure that the system is correct.

8   Conclusion

The work described here was accomplished under the META program which had a period of performance of only 12 months. Consequently, what we have presented here is just a start in what we consider to be a very important and very interesting research area. There is much important work ahead of us.

First, we plan to extend our compositional verification approach to include more complex models of computation. Synchronous computation platforms are found in many avionics systems, but we also need to provide support for multiple execution rates, variable delays, and asynchronous computation.

We have implemented four architectural design patterns to demonstrate the concept, but there are many more that we have encountered. In particular, there is a great deal of work on standard fault tolerance mechanisms with existing verification artifacts that would fit very well into our design pattern scheme.

The technique we have used to embed contracts in AADL models is expedient but semantically shallow. An improved method for annotation of architecture models with formal contracts would allow much better integration with the system design and more robust tooling. A new AADL annex seems the best way to accomplish this. We would also to provide support for some of the features in SysML that are well-suited for capturing dynamic requirements, such as activity and sequence diagrams.

Acknowledgements. This work was sponsored in part by AFRL under contract FA8650-10-C-7081 in the DARPA META program and NSF grant CNS-1035715.

References

[1]   Cofer, D.D., Miller, S.P., Gacek, A.J., Whalen, M.W., LaValley, B., Sha, L., Al-Nayeem, A.: Complexity-Reducing Design Patterns for Cyber-Physical Systems. Air Force Research Laboratory Technical Report AFRL-RZ-WP-TR-2011-2098 (2011)

[2]   Kamp, J.A.W.: Tense Logic and the Theory of Order. Ph.D. Thesis, UCLA (1968)

[3]   The NuSMV Toolset Users Manual (2005), http://nusmv.irst.itc.it/

[4]   McMillan, K.L.: Circular Compositional Reasoning about Liveness. Technical Report 1999-02, Cadence Berkeley Labs, Berkeley CA (1999)

[5]   IEEE Standard for Property Specification Language (PSL). IEEE Std 1850-2005 (2005)

[6]   Whalen, M., Gacek, A., Cofer, D.: Circular Hierarchical Reasoning using Past Time LTL. Technical Report 2011-1, University of Minnesota Software Engineering Center (2011), http://www.umsec.umn.edu/publications

[7]   The Mathworks Inc. Simulink Product Web Site, http://www.mathworks.com/products/simulink

[8]   Nam, M.-Y., Pellizzoni, R., Sha, L., Bradford, R.M.: ASIIST: Application Specific I/O Integration Support Tool for Real-Time Bus Architecture Designs. In: 2009 14th IEEE International Conference on Engineering of Complex Computer Systems (June 2009)

[9]   Kwiatkowska, M., Norman, G., Parker, D.: Probabilistic Symbolic Model Checking with PRISM: A Hybrid Approach. In: Katoen, J.-P., Stevens, P. (eds.) TACAS 2002. LNCS, vol. 2280, pp. 52–66. Springer, Heidelberg (2002)

[10] Gilles, O., Hugues, J.: Expressing and Enforcing User-Defined Constraints of AADL Models. In: Engineering of Complex Computer Systems (ICECCS), pp. 337–342 (2010)

[11] Ölveczky, P.C., Boronat, A., Meseguer, J.: Formal Semantics and Analysis of Behavioral AADL Models in Real-Time Maude. In: Hatcliff, J., Zucca, E. (eds.) FMOODS/FORTE 2010. LNCS, vol. 6117, pp. 47–62. Springer, Heidelberg (2010)

[12] Jahier, E., Halbwachs, N., Raymond, P., Nicollin, X., Lesens, D.: Virtual Integration of AADL models by a translation into synchronous programs. In: EMSOFT 2007. ACM (2007)

A Safety Case Pattern for Model-Based Development Approach*

Anaheed Ayoub, BaekGyu Kim, Insup Lee, and Oleg Sokolsky

Department of Computer and Information Science
University of Pennsylvania
{anaheed,baekgyu,lee,sokolsky}@seas.upenn.edu

Abstract. In this paper, a safety case pattern is introduced to facilitate the presentation of a correctness argument for a system implemented using formal methods in the development process. We took advantage of our experience in constructing a safety case for the Patient Controlled Analgesic (PCA) infusion pump, to define this safety case pattern. The proposed pattern is appropriate to be instantiated within the safety cases constructed for systems that are developed by applying model-based approaches.

Keywords: safety cases, safety case patterns, model-based development approach, PCA infusion pump.

1 Introduction

A Patient Controlled Analgesic (PCA) infusion pump is one type of infusion pump that primarily delivers pain relievers, and is equipped with a feature that allows for additional limited delivery of medication, called bolus, upon patient demand. We are developing a PCA implementation software by using the model-based approach based on the Generic PCA model [2] and the Generic PCA safety requirements [1] provided by the U.S. Food and Drug Administration (FDA) as shown in [13].

According to FDAs Infusion Pump Improvement Initiative [15], the FDA has received over 56,000 reports of adverse events associated with the use of infusion pumps from 2005 through 2009. The FDA structured 510k guidance document [14] to assist industry in preparing premarket notification submissions for infusion pumps. These recommendations are intended to improve the quality of infusion pumps in order to reduce the number of recalls and infusion pump Medical Device Reports (MDRs). In 510k submissions, the FDA recommends device manufacturers to submit infusion pump information (i.e., what beneficial properties the manufacturer claims for the infusion pump and how those properties are supported by the provided evidences) through a framework known as an assurance case [5]. This recommendation is the main motivation for our work.

* This research was supported in part by NSF CNS-0930647, NSF CNS-1035715, and NSF CNS-1042829.

An assurance case is a way to demonstrate the validity of a claim by providing a convincing argument together with supporting evidence (e.g., testing results, analysis results, etc.). The 510k guidance document specifically mentions the safety case [11] that is a special form of the assurance case that addresses safety. There is often commonality among the structures of the argument used in safety cases. This commonality motivates the definition for the concept of safety case patterns [11], which is an approach to support the reuse of safety arguments between safety cases. For example, patterns extracted from a safety case built for a specific product can be reused in constructing safety cases for other products that are developed via similar processes.

We are constructing a safety case for the PCA implementation software. The term "PCA implementation software" means the software code that is automatically generated from the GPCA reference model, and then extended to interface with the target platform [13]. The ultimate goal of this safety case construction is to show that the PCA implementation software we developed is acceptably safe, with the intention of providing a guiding example of safety cases for other infusion pumps. We are constructing the PCA safety case concurrently with the PCA implementation development. This concurrent development enables assurance needs to drive development decisions [5].

The main contribution of this paper is to define a safety case pattern that allows the incorporation of the belief in the model correctness obtained by using formal methods in the development process. This pattern is appropriate in constructing safety cases for infusion pumps those are developed using the model-based approach. The paper is organized as follows: we start by briefly giving background information in Section 2. Section 3 describes the main contribution of the paper which proposes a safety case pattern. Related work is discussed in Section 4. Finally, conclusions and ongoing work are given in Section 5.

2   Background

Two important concepts are used in this paper: "safety case patterns" and "the model-based development".

Safety case patterns [12] are defined to capture successful (i.e., convincing, sound, etc.) arguments that are used within the safety case. Whenever a safety case pattern is found to be appropriate to apply in a new safety case development, then it is instantiated within this new safety case. Therefore, safety case patterns allow reusing successful arguments among different safety cases. In essence, the patterns concept attempts to encourage best practice in creating and reviewing safety cases [6]. The Goal Structuring Notation (GSN) is one of description techniques that has proven to be useful for constructing safety cases. Details about GSN can be found in [11]. A number of extensions have been made to GSN to define a safety case pattern language. Those extensions are given in [12].

Model-based development is the notion that we can build systems by constructing abstract representations of the system's behavior and translating them into something that executes on a target platform. A typical model-based approach includes the following steps: 1) modeling the system, 2) analyzing and verifying this model against the system requirements, 3) systematic transformation of the model into an implementation, and 4) validating the implementation against the system requirements. We applied such model-based approach in developing the PCA implementation software. Therefore, one of the safety case patterns we suggest (described in the Section 3) is an argument about the correctness of the implementation developed using the model-based approach.

3   The Proposed Safety Case Pattern

We are constructing a safety case for the PCA implementation software we are developing. Due to the page limit, description for the entire PCA safety case is not given. Instead, we concentrate on the safety case pattern extracted from the PCA safety case. The proposed safety case pattern allows one to incorporate the confidence in the model correctness obtained by using formal methods and the confidence in the development process gained by using a well-established development approach. This pattern is appropriate to be used when the system is developed from the formal model using the model-based approach.

Figure 1 shows the GSN structure of the proposed *from_to* pattern. Here, {to} refers to the system implementation and {from} refers to a model of this system. The claim (G1) about the implementation correctness (i.e., satisfaction of some property (referenced in C1.3)) is justified not only by validation (G4 through S1.2) but also by arguing over the model correctness (G2 through S1.1), and the consistency between the model and the implementation created based on it (G3 through S1.1). The model correctness (i.e., further development for G2) is guaranteed through the model verification (i.e., the second step of the model-based approach). The consistency between the model and the implementation (i.e., further development for G3) is supported by the code generation from the verified model (i.e., the third step of the model-based approach). Only part of the property of concern (referenced in C2.1) can be verified at the model level due

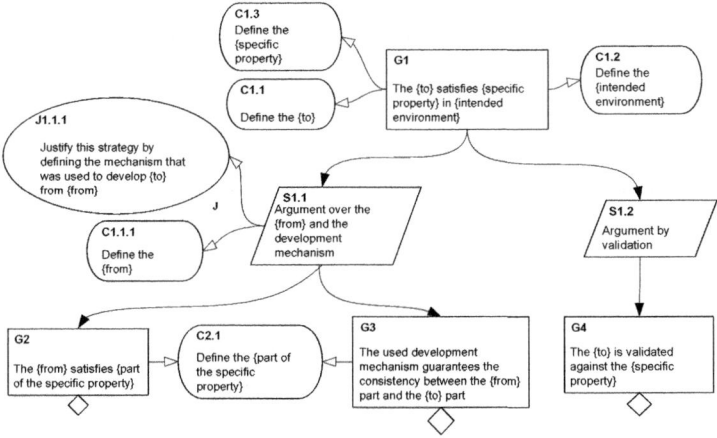

Fig. 1. The proposed *from_to* pattern

to the different abstraction levels between the model and the implementation. However, the validation argument (S1.2) covers the entire property of concern (referenced in C1.3). The additional justification given in (S1.1) increases the assurance in the top-level claim (G1).

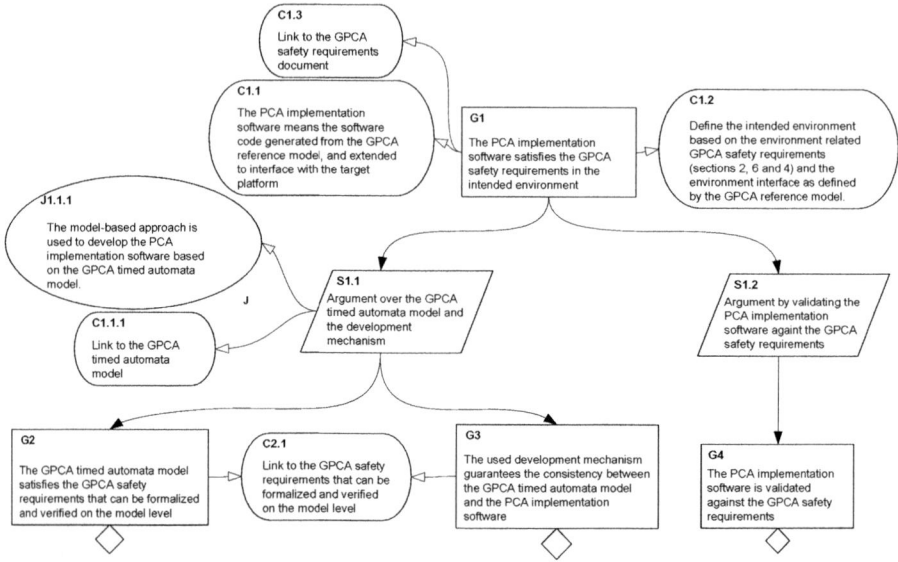

Fig. 2. An instance of the *from_to* pattern

Figure 2 shows an instantiation of this pattern that is part of the PCA safety case. Based on [13], for this pattern instance, the {to} part is the PCA implementation software (referenced in C1.1), the {from} part is the GPCA timed automata model (referenced in C1.1.1) and the GPCA safety requirements (referenced in C1.3) represent the concerned property. In this case, correct PCA implementation means it satisfies the GPCA safety requirements that defined to guarantee the PCA safety. The satisfaction of the GPCA safety requirements in the implementation level (G1) is decomposed by two strategies (S1.1) and (S1.2). The argument in (S1.1) is supported by the correctness of the GPCA timed automata model (G2), and the consistency between the model and the implementation (G3). The correctness of the GPCA timed automata model (i.e., further development for G2) has been proved using the UPPAAL model-checker [4] against the GPCA safety requirements that can be formalized (referenced in C2.1). The consistency between the model and the implementation (i.e., further development for G3) is supported by the code-synthesis from the verified GPCA timed automata model. Not all the GPCA safety requirements (referenced in C1.3) can be verified against the GPCA timed automata model [13]. Only the part referenced in C2.1 can be formalized and verified in the model level (e.g., *"no bolus dose shall be possible during the Power-On Self-Test"*). Other requirements are not formalizable and/or cannot be verification against the model given its level of details (e.g., *"the flow rate for the bolus dose shall be programmable"* cannot be formalized meaningfully and then verified in the model level).

Generally, using safety case patterns does not necessarily guarantee that the constructed safety case will be sufficiently compelling. So when instantiating the *from_to* pattern, it is necessary to be able to provide justification for each taken instantiation decision to guarantee that the constructed safety case is sufficiently compelling. Guidance for justifying such decisions can be found in [8].

4   Related Work

Assurance cases for medical devices have been discussed in [18]. The work in [18] can be used as staring point for the PCA safety case construction. A safety case given in [10] was constructed for a pacemaker was also developed using the model-based approach. This paper takes a step forward by proposing a safety case pattern for the model-based approach. The concept of safety case patterns was defined in [12]. Many safety case patterns were introduced in [3,11,17], but none of them is defined to the model-based approach. Another set of patterns are given in [16]. However, those patterns are introduced only by instantiation examples, limiting their reuse.

The software contribution pattern introduced in [7] is related to the *from_to* pattern. Both concern software development and can be applied iteratively. However, the software contribution pattern is intended to show that the contribution made by the software to the system hazards are acceptably managed. The *from_to* pattern is intended to show the software satisfaction for some concerned property, which can be used to address different aspects. The software contribution pattern is defined to be flexible enough and may be instantiated no matter what development process is used. While the *from_to* pattern is applicable only if the development process guarantees consistency between the developed artifacts. Focusing on a specific development approach (i.e., model-based development) breaks the advantage of the flexibility. The propagation of the correctness between tiers is not part of the software contribution pattern itself [8]. In contrast, the *from_to* pattern argues over the correctness propagation from the {from} artifact to the {to} artifact. This argument strengthens the assurance in the {to} correctness (i.e., the pattern top-level claim).

5   Conclusions

Our ongoing work is constructing a safety case for the PCA infusion pump system that we are developing. In the development of the PCA implementation, we applied the model-based approach starting from the GPCA model. Here, we suggest a safety case pattern that can be instantiated to argue about the correctness of implementations developed using the model-based approach. Where the correctness (i.e., satisfaction of required properties) of the implementation is justified not only by validation but also by arguing over the model correctness and the preservation of this correctness through the development process.

In addition to constructing a safety argument for the PCA infusion pump, we are also working on constructing confidence arguments that are necessary to increase the confidence in the developed safety argument as suggested in [9].

References

1. Safety Requirements for the Generic Patient Controlled Analgesia Pump, http://rtg.cis.upenn.edu/gip.php3
2. The Generic Patient Controlled Analgesia Pump Model, http://rtg.cis.upenn.edu/gip.php3
3. Alexander, R., Kelly, T., Kurd, Z., McDermid, J.: Safety Cases for Advanced Control Software: Safety Case Patterns. Technical report, University of York (2007)
4. Behrmann, G., David, A., Larsen, K.G.: A tutorial on UPPAAL. In: Bernardo, M., Corradini, F. (eds.) SFM-RT 2004. LNCS, vol. 3185, pp. 200–236. Springer, Heidelberg (2004)
5. Graydon, P., Knight, J., Strunk, E.: Assurance Based Development of Critical Systems. In: The 37th Annual IEEE/IFIP International Conference on Dependable Systems and Networks, DSN 2007, Washington, DC, USA, pp. 347–357 (2007)
6. Hawkins, R., Clegg, K., Alexander, R., Kelly, T.: Using a Software Safety Argument Pattern Catalogue: Two Case Studies. In: Flammini, F., Bologna, S., Vittorini, V. (eds.) SAFECOMP 2011. LNCS, vol. 6894, pp. 185–198. Springer, Heidelberg (2011)
7. Hawkins, R., Kelly, T.: A Systematic Approach for Developing Software Safety Arguments. Journal of System Safety 46, 25–33 (2009)
8. Hawkins, R., Kelly, T.: Software Safety Assurance – What is Sufficient? In: The 4th IET International Conference of System Safety, London (2009)
9. Hawkins, R., Kelly, T., Knight, J., Graydon, P.: A New Approach to creating Clear Safety Arguments. In: The 19th Safety Critical Systems Symposium (SSS 2011), pp. 3–23. Springer, London (2011)
10. Jee, E., Lee, I., Sokolsky, O.: Assurance Cases in Model-Driven Development of the Pacemaker Software. In: Margaria, T., Steffen, B. (eds.) ISoLA 2010, Part II. LNCS, vol. 6416, pp. 343–356. Springer, Heidelberg (2010)
11. Kelly, T.: Arguing Safety – A Systematic Approach to Safety Cases Management. PhD thesis, Department of Computer Science, University of York (1999)
12. Kelly, T., McDermid, J.: Safety Case Construction and Reuse using Patterns. In: SAFECOMP, pp. 55–96. Springer, Heidelberg (1997)
13. Kim, B., Ayoub, A., Sokolsky, O., Jones, P., Zhang, Y., Jetley, R., Lee, I.: Safety-Assured Development of the GPCA Infusion Pump Software. In: EMSOFT, Taipei, Taiwan, pp. 155–164 (2011)
14. U.S. Food and Drug Administration, Center for Devices and Radiological Health. Guidance for Industry and FDA Staff - Total Product Life Cycle: Infusion Pump - Premarket Notification (510(k)) Submissions (April 2010)
15. U.S. Food and Drug Administration, Center for Devices and Radiological Health. White Paper: Infusion Pump Improvement Initiative (April 2010)
16. Wagner, S., Schätz, B., Puchner, S., Kock, P.: A Case Study on Safety Cases in the Automotive Domain: Modules, Patterns, and Models. In: ISSRE, pp. 269–278 (2010)
17. Weaver, R.: The Safety of Software - Constructing and Assuring Arguments. PhD thesis, Department of Computer Science, University of York (2003)
18. Weinstock, C., Goodenough, J.: Towards an Assurance Case Practice for Medical Device. Technical report, CMU/SEI-2009-TN-018 (2009)

PVS Linear Algebra Libraries for Verification of Control Software Algorithms in C/ACSL*

Heber Herencia-Zapana[1], Romain Jobredeaux[2], Sam Owre[3],
Pierre-Loïc Garoche[4], Eric Feron[2], Gilberto Perez[5], and Pablo Ascariz[5]

[1] National Institute of Aerospace, Hampton, VA
[2] Georgia Institute of Technology, Atlanta, Georgia
[3] SRI International, Menlo Park, California
[4] ONERA, The French Aerospace Lab, Toulouse, France
[5] University of A Coruña, Coruña, Spain

Abstract. The problem of ensuring control software properties hold on their actual implementation is rarely tackled. While stability proofs are widely used on models, they are never carried to the code. Using program verification techniques requires express these properties at the level of the code but also to have theorem provers that can manipulate the proof elements. We propose to address this challenge by following two phases: first we introduce a way to express stability proofs as C code annotations; second, we propose a PVS linear algebra library that is able to manipulate quadratic invariants, i.e., ellipsoids. Our framework achieves the translation of stability properties expressed on the code to the representation of an associated proof obligation (PO) in PVS. Our library allows us to discharge these POs within PVS.

1 Introduction

Critical computing systems, typically driving machinery or vehicles, are those in which failure may result in unacceptable human losses. Examples of critical systems include fly-by-wire controls on an aircraft or on a manned spacecraft, radiation therapy equipment, and nuclear power plant safety systems. Digital computation facilitates the design and the implementation of complex control algorithms. The software implementation of a control law can be inspected by analysis tools [7,22,24], however these tools are often challenged by issues for which solutions are already available from control theory.

Control theory is a branch of engineering that focuses on the behavior of dynamical systems. The desired output of a system is called the reference point. When one or more output variables of a system need to follow a certain reference

* This work was supported by the National Aeronautics and Space Administration under NASA Cooperative Agreement NCC-1-02043, the National Science Foundation under Grant CNS - 1135955, the Army Research Office under MURI Award W911NF-11-1-0046, the Air Force Research Laboratory as part of the CertaAMOR program, the Dutton/Ducoffe Professorship at Georgia Tech, FNRAE CAVALE project, and Spanish MEC project TIN2009-14562-C05-04.

A. Goodloe and S. Person (Eds.): NFM 2012, LNCS 7226, pp. 147–161, 2012.

over time, a controller manipulates the inputs of the system to obtain the desired effect on its output. The objective of control theory is to calculate a proper action from the controller that will result in stability for the system, that is, the system will hold the reference point and not oscillate around it. Among the different mathematical approaches to prove stability of the controller or the controlled system, Lyapounov based stability relies on ellipsoid characterization and the so-called S-procedure [3,14,15]. These works also address the expression of the proof as C code annotation, but do not give means to automate this expression nor to prove it on C code.

Program verification based on deductive methods uses either automatic decision procedures or proof assistants to ensure the validity of user-provided code annotations. These annotations may express the domain-specific properties of the code. However, formulating annotations correctly (i.e., precisely as the domain expert really intends) is nontrivial in practice [2,12]. By correctly, we mean that the annotations formulate stability properties of an intended mathematical interpretation from control theory.

The challenges of domain-specific code annotation arise along two directions. First, the domain knowledge has its own inherent complexity. When considering control theoretic issues, the annotations need to allow the expression of stability properties using ellipsoids and the S-procedure in the way that was proposed in [3,14,15]. Second, the code annotations are meant to be manipulated by automatic theorem provers. But most of the automatic decision procedures are restricted to decidable logics such as Boolean satisfiability or linear arithmetic, which are generally too weak to express the desired user-defined and domain-specific code annotations.

In order to solve these two challenges this paper proposes an axiomatisation of Lyapunov-based stability as C code annotations, and the implementation of linear algebra and control theory results in PVS [28,26], respectively. The mechanism of theory interpretations [27] enables the translation of POs expressed on the C code as PVS control theory proof obligations. The proof of these obligations can then be discharged using control theory results expressed and proved with our PVS linear algebra library.

Related works. To our knowledge, apart from the work of [3,14,15], no other research endeavor addresses the issue of proving in the C code the high-level correctness properties of control systems such as stability. Some successful attempts have been made at extracting quadratic invariants from the code, in [1] and [13].

Regarding the prover part of our framework, the developments of tools that support the proof of properties in real arithmetic or real linear algebra is a current concern. However these early development do not cover the entire range of mathematics and are often restricted to specific sub-areas. For example a recent project, Coquelicot, develops real functional analysis , Gaussian elimination and basic properties of matrices and determinants for the Coq proof assistant [18]. Generic design patterns were proposed to define algebraic structures[17]. Formalization and instrumentation of Euclidean spaces also appears to be a new concern for Isabelle/HOL [20]. We should also mention automatic decision procedures for floating point arithmetics, such as Gappa [11]. A PVS formalization of multivariate Bernstein polynomials was presented in [25]. In general however, none of these recent extensions of theorem provers are able to deal with the properties of interest in this paper.

Outline. Section 2 reminds the reader of elements of control theory software analysis [3,14,15], i.e., it describes the controller stability proof and its expression as C code annotations with Hoare triples. It also discusses issues to be handled on the theorem prover part, mainly the need for two main theorems: one related to ellipsoids and one on the S-procedure. Section 3 introduces our axiomatisation of stability proofs as C code annotations using Hoare triples. Section 4 presents the implementation of linear algebra and theorems related to ellipsoids in PVS. Finally, Section 5 explains how we plan to map POs generated from the Hoare triple stability annotations to PVS, using theory interpretation in PVS, and how to use the ellipsoid library to discharge these POs. Last, Section 6 concludes the paper and discusses future research.

2   Stability and Correctness

2.1   Expressing and Proving Stability of a Control System

The basic module for the description of a controller can be presented as

$$\xi(k+1) = f(\xi(k), \nu(k)), \ \xi(0) = \xi_0$$
$$\zeta(k) = y(\xi(k), \nu(k))$$

where $\xi \in \mathbb{R}^n$ is the state of the controller, ν is the input of the controller and ζ is the output of the controller. This system is bounded-input, bounded state stable if for every ε there exists a δ such that $||\nu(k)|| \leq \varepsilon$ implies $||\xi(k)|| \leq \delta$, for every positive integer k. If there exists a positive definite function V such that $V(\xi(k)) \leq 1$ implies $V(\xi(k+1)) \leq 1$ then this function can be used to establish the stability of the system; for more details see [8]. This Lyapunov function, V, defines the ellipsoid $\{\xi| \ V(\xi) \leq 1\}$, this ellipsoid plays an important role for the stability preservation at the code level, for more details see [3,14,15].

2.2   Hoare Triple and Deductive Methods

Since their early formalization by Hoare [21] and later by Dijsktra [10], deductive methods from Hoare triple to weakest precondition computation have been widely used on imperative code.

In his initial proposal, Hoare requires a program to be annotated line by line by the invariants that should hold at each program point. He also provides instruction-specific rules that ensure the soundness of the code with respect to the annotation system.

Because it was, in general, not realistic to require a line-based set of annotations, Dijkstra later proposed the weakest precondition computation and the
verification conditions, that automatically generate a PO with respect to a Hoare
style annotation for a block of instructions. Most software analysis tools which
use Hoare Logic are based on this algorithm.

Our current approach does not consist in automating the proof of stability,
but rather, given a stability proof, to check the proof automatically. As was
suggested in [3,14,15], we consider a line-by-line annotation of the code, allowing
a Hoare-like reasoning approach to the program.[1]

In general, Hoare proofs are sound, i.e., the proved property indeed holds, if
and only if the program terminates. They are complete if the underlying logic –
the one used in pre- and post-condition – is complete.

2.3   Application to Controller Stability: An Ellipsoid-Aware Hoare Logic

We present here the two main patterns used in stability proofs. The main concerns are: to relate quadratic invariants and affine or linear combinations of
variables on the one hand – the ellipsoid affine combination theorem; and to
extract one quadratic invariant out of implications between several quadratic
invariants on the other hand – the S-procedure.

Ellipsoid affine combination theorem. The use of ellipsoids to formally specify
bounded input, bounded state stability was proposed in [3,14,15] following prior
work [6]. Stability is then expressed as a predicate stating that the system state
remains in a given ellipsoid. Typically, an instruction S would be annotated in
the following way:

$$\{x \in \mathcal{E}_P\}\, y = Ax + b\, \{y - b \in \mathcal{E}_Q\} \tag{1}$$

where the pre- and post- conditions are predicates expressing that the variables
belong to some ellipsoid, with $\mathcal{E}_p = \{x : \mathbb{R}^n | x^T P^{-1} x \leq 1\}$ and $Q = APA^T$.

The mathematical theorem that guarantees the relations in (1) is now stated:

Theorem 1. *If M, Q are invertible matrices, and $(x - c)^T Q^{-1}(x - c) \leq 1$ and
$y = Mx + b$ then $(y - b - Mc)^T (MQM^T)^{-1}(y - b - Mc) \leq 1$.*

We will refer to it as the *ellipsoid affine combination theorem*. More details about
this result in the context of control theory can be found in [6,23].

The S-procedure. A second common need is to prove the implication between two
quadratic invariants. In the initial Hoare proposal [21], the post-condition of a

---

[1] or equivalently a limited-depth Dijkstra weakest precondition.

statement is exactly the pre-condition of its successor. A *consequence rule* allows to transform a post-condition of a statement into another pre-condition for the following statement. This can be understood as the introduction of a nil statement that contains this translation of predicates as illustrated in Figure 1. This unavoidable step allows software analyzers to manipulate the annotations along the code. The PO associated to this new nil statement is $Q_1 \implies P_2$.

$$\{P_1\}\ S_1\ \{Q_1\}$$
$$\{Q_1\}\ \text{nil}\ \{P_2\}$$
$$\{P_2\}\ S_2\ \{Q_2\}$$

Fig. 1. Conseq. rule

A frequent proof pattern when using ellipsoid-based stability proofs is to show that the inequality $x^T P x \leq 1$ implies $y^T Q y \leq 1$. Such implications are usually difficult to prove. We need to give conditions under which, given symmetric matrices A_0 and A_1, statement 2. implies statement 1. in the following:

1. $\forall x \in \mathbb{R}^n : x^T A_1 x \geq 0 \implies x^T A_0 x \geq 0$
2. $\exists a \in \mathbb{R} : \forall x \in \mathbb{R}^n\, x^T (A_0 - a A_1) x > 0$

From [3,14,15], a typical property for the composition of Hoare triples is to prove that the implication

$$\{\mathbf{x}_c \in \mathcal{E}_P,\ y_c^2 \leq 1\}\ \text{implies}\ \{A_c \mathbf{x}_c + B_c y_c \in \mathcal{E}_P,\ y_c^2 \leq 1\}$$

is a consequence of the inequality

$$(A_c \mathbf{x}_c + B_c y_c)^T P (A_c \mathbf{x}_c + B_c y_c) - \mu \mathbf{x}^T P \mathbf{x} - (1 - \mu) y^2 \leq 0.$$

This type of property may be proved using the following theorem, which the *S-procedure* [6,23] is a by-product of.

Theorem 2. *Let the real valued functionals* $\sigma_k : \mathbb{R}^n \to \mathbb{R}$ *where* $k = 0, 1, 2, \ldots, N$ *and consider the following two conditions:*

1. S_1: $\forall y \in \mathbb{R}^n : (\forall k = 1, 2, \ldots, N : \sigma_k(y) > 0) \implies \sigma_0(y) \geq 0$
2. S_2: *There exists* $\tau_k \geq 0$, $k = 1, 2, \ldots, N$ *such that*

$$\sigma_0(y) - \Sigma_{k=1}^N \tau_k \sigma_k(y) > 0,\ \forall y \in \mathbb{R}^n.$$

Then $S_2 \implies S_1$.

The *S-procedure* is the method of verifying S_1 using S_2.

3   Defining Quadratic Invariants as Code Annotations

Now that we know the annotations that we want to generate on the code, we have to find a concrete way to express them on actual C code. The ANSI/ISO C Specification Language (ACSL) [5] allows its user to specify the properties of a C program within comments, in order to be able to formally verify that the implementation respects these properties. This language was proposed as part of the Frama-C platform [9], which provides a set of tools to reason on both C programs and their ACSL annotations. ACSL offers the means to extend its internal logic with user-defined theory, i.e., types, constructors, functions, predicates and axioms.

We outline the axiomatisation in ACSL to fit our needs, which consist of expressing ellipsoid-based Hoare triples over C code. We first present the axiomatisation of linear algebra elements in ACSL. Then we present the Hoare triple annotations in ACSL and the POs generated by them.

3.1   Linear Algebra in ACSL Predicates

The following abstract types are declared: matrix, vector, integer, and real. With these abstract types, basic matrix operations and properties are introduced : a component of the matrix is a real number accessed using the function mat_select (matrix A, integer i, integer j), total number of rows and columns are integers accessed with mat_row(matrix A), and mat_col(matrix A), respectively. The multiplication of a matrix with a vector is defined with function vect_mult(matrix A, vector x), which returns a vector. The concatenation of vectors x and y, itself a vector, is accessed through Vconcat(vector x, vector y). Addition and multiplication of 2 matrices, multiplication by a scalar, and inverse of a matrix are declared as matrix type as follows:

$$\text{mat\_add(matrix } A, \text{matrix } B), \quad \text{mat\_mult(matrix } A, \text{matrix } B)$$

$$\text{mat\_mult\_scal(matrix } A, \text{real } a), \quad \text{and mat\_inverse}(A).$$

The matrix operations are defined axiomatically, for example the inverse of a matrix A, mat_inverse(A) is defined using the predicate is_invertible(A) as follows:

```
/*@ axiom mat_inv_select_i_eq_j:
@ ∀matrixA, integer i, j;
@ is_invertible(A) && i == j ==>
@ mat_select(mat_mult(A, mat_inverse(A)), i, j) = 1
@
@ axiom mat_inv_select_i_dff_j:
@ ∀matrixA, integer i, j;
@ is_invertible(A) && i! = j ==>
@ mat_select(mat_mult(A, mat_inverse(A)), i, j) = 0
@*/
```
ACSL

In the same axiomatic way, the main matrix operations are declared. Complex constructions or relations can be defined as uninterpreted predicates, i.e., with no associated axiom. The semantics of those predicates are introduced in PVS, as discussed in section 5. The following predicate is meant to express that vector x belongs to $\mathcal{E}_P$:

```
//@ predicate in_ellipsoid(matrix P, vector x);
```
ACSL

And last, a set of typing functions, associated to a set of axioms, such as mat_of_array or vect_of_array, is used to associate an ACSL matrix type to a C array.

```
//@ logic matrix mat_of_array{L}(float *A, integer row, integer col);
```
ACSL

3.2   Linear Algebra Code Annotations

The paramount notion in ACSL is the function contract, [7]. It can be understood as a Hoare triple for a whole function. The key word requires is used to introduce the pre-conditions of the triple, and the key word ensures is used to introduce its post-conditions. Dealing with a low-level language has its disadvantages: we need to deal with memory issues. In general, we want all functions to be called with valid pointers as arguments, i.e., valid array and therefore valid matrices. This is what the built-in ACSL predicate valid does. The followings snippet shows how the contract can be written using mat_select and mat_of_array,

```
ACSL
/*@ requires (valid(a + (0..3)));
@ ensures ∀integer i, j; 0 ≤ i < 2 && 0 ≤ j < 2
@ ==> mat_select(mat_of_array(a, 2, 2), i, j) == 0;
@ */
void zeros_2x2(float* a)
{ a[0]=0; a[1]=0; a[2]=0; a[3]=0; }
```

In the following example the uninterpreted predicate in_ellipsoid is used:

```
ACSL
/*@ requires
@ (valid(xc + (0..1))) && (valid(yc)) && (valid(u)) &&
@ in_ellipsoid(Qmu, Vconcat(vect_of_array(xc, 2), vect_of_array(yc, 1)));
@ ensures
@ in_ellipsoid(Ubound, vect_of_array(u, 1)) &&
@ in_ellipsoid(Qmu, Vconcat(vect_of_array(xc, 2), vect_of_array(yc, 1)));
@ */
void inst2(float* xc, float* yc, float* u)
{ u[0] = 564.48*xc[0] - 1280*yc[0]; }
```

where Q_{mu}, $C = [564.48 \ \ 0 \ \ -1280]$ are matrices and

$$U_{bound} = \mathsf{mat\_inv}(\mathsf{mat\_mult}(\mathsf{mat\_mult}(C, \mathsf{mat\_inv}(Q_{mu})), \mathsf{transpose}(C)))$$

One important assumption which will be made throughout the rest of this article is that all computations in the program yield their exact, real result. Errors due to floating point approximations are thus not taken into account. The Frama-C toolset offers the possibility of making this assumption by including the pragma 'JessieFloatModel(Math)'. Verification conditions are then generated with no concern for floating point computations.

3.3   Generating Proof Obligations

Frama-C tools do not require an annotation at each line as proposed by Hoare. They rather rely on Dijkstra-style weakest precondition calculus to compute the backward semantics of the function code S to the post-condition Q and generate the weakest pre-condition $wp(S, Q)$ that guarantees to obtain Q after executing S. The generated PO is then $P \implies wp(S, Q)$ where P is the pre-condition.

Focusing on single line contract, i.e., the Hoare annotations as described in [21], these tools will generate the following two kinds of POs when used with ellipsoid-based annotations.

First, we have the POs associated with the use of the ellipsoid affine combination theorem, see Equation 1:

```
                                                              ACSL
in_ellipsoid(matrix P, vector x)
IMPLIES
in_ellipsoid(matrix Q, vector (vect_mult(matrix A, vector x))
```

One can remark that both axiom-based and uninterpreted predicates are expressed in the same way. The only difference is that axiom-based predicate definitions appear in the other generated files of the proof obligation generation phase.

Second, we have the POs associated with the use of the S-procedure, cf. Theorem 2:

```
                                                              ACSL
in_ellipsoid($A_1, x$) IMPLIES in_ellipsoid($A_0, x$)
IF AND ONLY IF
in_ellipsoid(mat_add($A_0$, mat_mult_scal($A_1, a$)), x)
```

For both POs, we must first interpret the uninterpreted types and to prove the properties that are defined axiomatically. We must then discharge the verification conditions using the appropriate theorem. This is done by using PVS and a linear algebra extention of it, presented below.

4   Linear Algebra in PVS

First, we define matrices, vectors, etc. in PVS in a way that can be used to interpret in_ellipsoid and S-procedure. Second, we provide the main theorems and basic principles of linear algebra in PVS that are needed to support this interpretation. General linear algebra references include [19,4,16].

4.1   Bases for Linear Algebra in PVS

We first define maps as follows:

```
                                                              PVS
Mapping:TYPE=
    [# dom: posnat, codom: posnat, mp: [Vector[dom]->Vector[codom]] #]
```

This is the set of functions that take a vector and return a vector. A linear map is defined as a map $h \in$ Mapping with the linear property $h(\Sigma_{i=0}^{N}(a(i)x(i))) = \Sigma_{i=0}^{N}(a(i)h(x(i)))$. this property in PVS is expressed as follows:

```
                                                              PVS
linear_map_e?(h,l,n,m): bool = h'dom=n and  h'codom=m and
    ∀(x: Vector[l], F: [below[l]->Vector[n]]):
       h'mp(Σ_{i=0}^{l-1}x(i)*F(i)) = Σ_{i=0}^{l-1}(x(i)*(h'mp(F(i))))
linear_map_e?(n,m)(h): bool = ∀(l): linear_map_e?(h,l,n,m)
Map_linear(n,m): TYPE = {h: Map(n,m) | linear_map_e?(n,m)(h)}
```

The algebra of matrices is the set of matrices together with the operations addition, multiplication and multiplication by scalar, and these operations satisfy the associative and commutative properties. The algebra of linear maps is the set of linear maps with the operations of composition and multiplication and preserving the associative and commutative properties [4,16]. We define the operator L(n,m) from the algebra of linear maps Map_linear(n,m) to the algebra of matrices Mat(m,n) as follows:

```
                                                               PVS
L(n,m)(f) = (# rows:=m, cols:=n, matrix:=λ(j,i): f'mp(e(n)(i))(j) #)
```

where f∈ Map_linear(n,m). We define the operator T(n,m) from Mat(m,n) to Map_linear(n,m) as follows:

```
                                                               PVS
T(n,m)(A) = (# dom:=n, codom:=m,
              mp:=λ(x,j):
                    Σ_{i=0}^{A'cols-1}(λ(i): A'matrix(j,i)*x(i))
              #))
```

With these two operators connecting linear maps and matrices, the following PVS lemmas prove the isomorphism between them:

```
                                                               PVS
Iso   :    LEMMA   bijective?(L(n,m))
Iso_T :    LEMMA   bijective?(T(n,m))
```

Because of the isomorphism between these two operators, the following lemma holds:

```
                                                               PVS
L_inverse: LEMMA inverse(L(n,m))=T(n,m)
```

More practical lemmas for proving properties in PVS are:

```
                                                               PVS
map_matrix_bij: LEMMA  ∀(A:Mat(m,n)): L(n,m)(T(n,m)(A)) = A
iso_map: LEMMA  ∀(f:Map_linear(n,m)): T(n,m)(L(n,m)(f)) = f
```

An important consequence of the isomorphism is the relation between the operations of the isomorphic spaces. For example, the composition of two linear maps is equivalent to the multiplication of their corresponding matrices:

```
                                                               PVS
comp_mult: LEMMA ∀(g: Map_linear(n,m),f:Map_linear(m,p)):
                  L(n,p)(f o g) = L(m,p)(f)*L(n,m)(g)
```

and the addition of two linear maps is equivalent to the addition of their corresponding matrices:

```
                                                               PVS
iso_add: LEMMA ∀(f, g:Map_linear(n,m):
                  L(n,m)(f + g) = L(n,m)(f) + L(n,m)(g)
```

The main reason for the isomorphism is to define the inverse of a matrix; one condition for the existence of the inverse in linear maps is that the linear map needs to be bijective. The space of matrices having inverses is defined in PVS as follows:

```
                                                               PVS
Matrix_inv(n):TYPE =
   {A: Square | squareMat?(n)(A) and bijective?(n)(T(n,n)(A))}
```

where `Square` is the type of matrices having the same number of rows and columns, `squareMat?(n)(A)` is the type of matrices having the same number of rows and column and equal to `n`, and the predicate `bijective?(n)(T(n,n)(A))` expresses that the linear map, `T(n,n)(A)`, associated to a matrix `A` is bijective.

The inverse operator, `inv(n)`, maps `Matrix_inv(n)` to `Matrix_inv(n)` and is defined as follows:

```
                                                                    PVS
 inv(n)(A) = L(n,n)(inverse(n)(T(n,n)(A)))
```

It is important to note that the operators L, T and the isomorphism play an important role in this definition. The main lemmas for the matrix inverse are proved in PVS, such as: the multiplication of the matrix and its inverse is equal to the identity matrix, `I(n)`, the inverse of a transpose matrix is equal to the transpose of its inverse or the distributive property of the inverse over matrix multiplication.

The PVS libraries also have basic lemmas from the matrices theory such as the solution to a matrix equation, the transpose of matrix multiplication, and the multiplication of matrix transpose and vectors.

One important point of this development is that the conditions under which the inverse of a matrix exists is that the linear map associated to the matrix is a bijective map. A common test for the existence of the inverse of a matrix is that the determinant of the matrix be not equal to zero, The equivalence between these two conditions needs to be implemented in PVS for which more mathematical theories such as multi-linear forms, convex spaces, and so forth are currently under development. The two conditions of S-procedure, i.e., Theorem 2, were implemented as follows:

```
                                                                    PVS
 s1_condition?(m)(beta: fun_constraint(m),f: Map_linear(n,1)):
         bool = FORALL (x: Vector[n]):
         pos_constraint_point?(m)(beta,x)
         IMPLIES
         f'mp(x)(0) >= 0
```

```
                                                                    PVS
 s2_condition?(m)(beta: fun_constraint(m),f: Map_linear(n,1)):
         bool = EXISTS (r: pos_scalar_family(m)):
         (FORALL (x: Vector[n]): f'mp(x)(0) -
         sigma(0,m - 1, LAMBDA(i): r(i)*beta(i)'mp(x)(0)) >= 0)
```

We are still working on the proof of the equivalence of these two conditions, one result that is needed for the proof is the Hyperplane theorem, which is a theorem from real analysis currently under development in PVS.

4.2 Ellipsoid Affine Combination Theorem in PVS

The implication associated to Equation 1 can be proved using the following theorem implemented in PVS.

```
ellipsoid_affine_comb: LEMMA                                    [PVS]
  ∀ (n:posnat, Q, M: SquareMat(n), x, y, b, c: Vector[n]):
                  bijective?(n)(T(n,n)(Q)) AND bijective?(n)(T(n,n)(M))
                  AND (x-c)*(inv(n)(Q)*(x-c))≤ 1
                  AND y=M*x + b
              IMPLIES
              (y-b-M*c)*(inv(n)(M*(Q*transpose(M)))*(y-b-M*c))≤ 1
```

This lemma was proved in PVS, the main part of the proof was to show that replacing y by $M * x + b$ in $(y - b - M * c) * (inv(n)(M * (Q * transpose(M))) *$ $(y - b - M * c))$, we obtain $(x - c) * (inv(n)(Q) * (x - c))$. In order to manipulate $(y - b - M * c) * (inv(n)(M * (Q * transpose(M))) * (y - b - M * c))$ the following PVS lemmas *trans_mat_scal*, *prod_inv_oper*, *tran_inv_oper*, *transpose_product* and basic properties of *SigmaV* were used.

5   Mapping ACSL Predicates to PVS Linear Algebra Concepts

On the one hand, using ACSL and the Frama-C framework, we were able to generate POs about the ellipsoid predicate. Frama-C tools even make it possible to express the PO in PVS, along with a complete axiomatisation in PVS of C programs semantics. On the other hand, we have developed a PVS library that is able to reason about these properties.

We now must link these two worlds: ACSL ellipsoids predicate proof obligation in PVS must be connected with with our linear algebra PVS library. We first propose to relate ASCL constructs to PVS Linear Algebra library elements and achieve a proof on the latter. A current ongoing approach, presented at the end of the section, is to automate this mapping using theory interpretations in PVS.

5.1   Mapping ACSL Predicates to PVS Linear Algebra

Frama-C tools automatically generate the proof obligations (POs) associated with a function contract, in our case, a Hoare triple. Depending on the back-end used, the PO can be expressed either to target an automatic decision procedure such as an SMT-solver, or to target a proof assistant, like Coq or PVS.

Using the PVS back-end, both the PO and all the axiomatisation of C semantics and all ACSL defined theories and predicates are expressed in PVS files. We now map PVS-encoded version of ACSL predicates into their PVS linear algebra library equivalent. A few examples of how such a mapping is performed are given in the rest of this section.

The ACSL logic function mat_of_array(ptr, n, m), when put through the PVS back-end, appears with an additional argument, mat_of_array(ptr, n, m, mem), which describes the memory state at the point where the function is used. The mapping for this function and the accessor mat_select are as follows:

```
                                                                    PVS
mat_of_array(ptr, n, m, mem) = A where
  A ∈ Matrix,   A'rows = n,   A'cols = m
  FORALL (i: below(A'rows), j: below(A'cols)):
    A'matrix(i,j) = select[real, floatP](mem, shift[floatP](ptr, i*n+j))

mat_select(A, i, j) = A'matrix(i,j) where A ∈ Matrix
```

The select and shift functions are part of the axiomatisation of C semantics pertaining to memory access.

Function mat_inverse and predicate is_invertible are interpreted as follows:

```
                                                                    PVS
mat_inverse(matrix A) := inv(n)(A)
is_invertible(matrix A)
        := square?(A) AND squareMat?(n)(A) AND bijective?(n)(T(n, n)(A))
```

And the following axiomatic definition of inverse

```
                                                                    ACSL
/* @axiom mat_inv_select_eq: ∀ matrix A, integer: i, j; i=j
@ is_invertible(M)  ⟹  mat_select(mat_mult(A, mat_inverse(A)),i,j) = 1
@*/
```

is mapped to the following lemma:

```
                                                                    PVS
LEMMA squareMat?(n)(M) and bijective?(n)(T(n,n)(M)) and
                i=j and i≤n
                IMPLIES
                (M*inv(n)(M))'(i,j) = 1
```

which was also proved, using the concepts introduced in the linear algebra library and basic properties in PVS.

In the same way we develop the interpretation for the basic matrix operators such as addition, transposition, multiplication by scalars, multiplication by vectors, and so forth.

5.2   Discharging Proof Obligations

We now sketch the typical use of our framework to prove a specific Hoare triple. We consider the following single line function annotated with ellipsoid-based pre- and post-condition. This function corresponds to the definition of the linear combination of matrices as presented in Equation (1).

```
                                                                    ACSL
/*@ requires (valid(xc + (0..1))) && (valid(yc)) && (valid(u)) &&
@ in_ellipsoid(Q_mu, Vconcat(vect_of_array(xc,2), vect_of_array(yc,1)));
@ ensures in_ellipsoid(U_bound, vect_of_array(u,1)) &&
@ in_ellipsoid(Q_mu, Vconcat(vect_of_array(xc,2), vect_of_array(yc,1)));
@ */
void inst2(float* xc, float* yc, float* u) {
u[0] = 564.48*xc[0] - 1280*yc[0];
}
```

Using Frama-C and its PVS back-end on these annotations generate the following PVS PO:

```
                                                                    PVS
FORALL ...   in_ellipsoid(Q_mu,
             Vconcat(vect_of_array(xc, 2, floatP_floatM),
                     vect_of_array(yc, 1, floatP_floatM))) IMPLIES
FORALL (result: real) :
result = select[real, floatP](floatP_floatM, shift[floatP](xc, 0))
IMPLIES
  FORALL (result0: real) :
  result0 = select[real, floatP](floatP_floatM, shift[floatP](yc, 0))
  IMPLIES
    FORALL (floatP_floatM0: memory[floatP, real]) :
    floatP_floatM0 = store[floatP, real]
       (floatP_floatM, u, 564.48 * result - 1280.0 * result0)
    IMPLIES in_ellipsoid(U_bound, vect_of_array(u, 1, floatP_floatM0))
    AND in_ellipsoid(Q_mu,
                     Vconcat(vect_of_array(xc, 2, floatP_floatM0),
                             vect_of_array(yc, 1, floatP_floatM0)))
```

In order to discharge this PO, we first give a meaning to the predicate in_ellipsoid.

```
                                                                    PVS
in_ellipsoid(matrix P, vector x)=x*(inv(n)(P)*x)≤ 1
```

Then, after skolemisation, we can split the conjunction in the consequence and prove the two implications using the ellipsoid affine combination theorem in PVS, presented in paragraph 4.2.

5.3   Theory Interpretations

Theory interpretation is a logical technique for relating one axiomatic theory to another. This technique makes it possible to show that one collection of theories is correctly interpreted by another collection of theories under a user-specified interpretation for the uninterpreted types and constants. PVS supports theory interpretations [27]. A theory instance is generated and imported, while the axiom instances become POs to ensure that the interpretation is valid. Interpretations can be used to show that an implementation is a correct refinement of a specification, or that an axiomatically defined specification is consistent.

We outline here a possible use of theory interpretation to automate this mapping between the two theories. This will be developed as future work.

Jessie to PVS. The Jessie plugin translates obligations to uninterpreted types and constants of PVS. When generating the PVS file associated to an annotated C file, all ACSL definitions and the generated POs are declared under a new theory acsl_theory. This theory contains new types, for example, the uninterpreted type matrix. To provide an interpretation for matrix, we first import the interpreting theory matrices, then we import the uninterpreted theory acsl_theory with mappings for matrix, matselect, mat_mult, etc., as shown below.

```
                                                                    PVS
importing matrices
importing acsl_theory{{ matrix := Matrix,
              mat_select := λ M, i, j: M'matrix(i, j),
              mat_mult := *,
              ... }}
```

This action generates POs corresponding to the axioms of the theory acsl_theory. In a similar fashion, all uninterpreted predicates may be given interpretations, and any axiom instances become POs. In the early stages of development, predicates such as in_ellipsoid may not have axioms provided in Jessie, in which case there is no guarantee of soundness. However, the system still generates POs corresponding to type correctness conditions (TCCs).

6   Conclusion and Future Work

We have described a global approach to validate stability properties of C code implementing controllers. Our approach requires the code to be annotated by Hoare triples, following [3,14,15], proving the stability of the control code using ellipsoid affine combinations and S-procedure.

We have defined an ACSL extension to describe predicates over the code, as well as a PVS library able to manipulate these predicates. This library contains matrices, linear maps, ellipsoid affine combination theorem, isomorphism between matrices and linear maps and theirs basic properties. The PVS libraries can be found at http://shemesh.larc.nasa.gov/fm/ftp/larc/PVS-library/pvslib.html.

We have also outlined an approach based on theory interpretation that maps proof obligations generated from the code to their equivalent in this new PVS library. This mapping allows to discharge POs using the ellipsoid affine combination and S-procedure theorems implemented in PVS.

Currently we are working on the automatic translation, using theory interpretations, of POs in ACSL about matrices properties into POs in PVS and discharging these POs using our linear algebra libraries. We are also working on the proof of the S-procedure in PVS, which involves more mathematical results such as hyperplane theorem, multilinear forms etc. As future research we are going to develop PVS strategies for automatically discharging proof obligation generated from the ACSL annotations of the control code and also to prove the equivalence between $Det(A) \neq 0$ and the inversibility of matrix A.

Acknowledgments. The authors would like to thank Dr. A. Goodloe for his suggestion of the use of the Frama-C toolset and his help in axiomatising of linear algebra in ACSL.

References

1. Adjé, A., Gaubert, S., Goubault, E.: Coupling Policy Iteration with Semi-definite Relaxation to Compute Accurate Numerical Invariants in Static Analysis. In: Gordon, A.D. (ed.) ESOP 2010. LNCS, vol. 6012, pp. 23–42. Springer, Heidelberg (2010)
2. Ahn, K.Y., Denney, E.: Testing first-order logic axioms in program verification
3. Alegre, F., Feron, E., Pande, S.: Using ellipsoidal domains to analyze control systems software. CoRR abs/0909.1977 (2009)

4. Axler, S.: Linear Algebra Done Right, 2nd edn. Springer, Heidelberg (1997)
5. Baudin, P., Filliâtre, J.C., Marché, C., Monate, B., Moy, Y., Prevosto, V.: ACSL: ANSI/ISOC specification language. Preliminary design (version 1.5)
6. Boyd, S., El Ghaoui, L., Feron, E., Balakrishnan, V.: Linear Matrix Inequalities in System and Control Theory. Studies in Applied Mathematics, vol. 15. SIAM (June 1994)
7. Burghardt, J., Gerlach, J., Hartig, K.: ACSL by example towards a verified C standard library version 4.2.0 for Frama-C beryllium 2 (2010)
8. Chen, C.T.: Linear System Theory and Design, 3rd edn. Oxford University Press, USA (1998)
9. Correnson, L., Cuoq, P., Puccetti, A., Signoles, J.: Frama-C user manual
10. Dijkstra, E.: A Discipline of Programming. Prentice-Hall (1976)
11. de Dinechin, F., Quirin Lauter, C., Melquiond, G.: Certifying the floating-point implementation of an elementary function using Gappa. IEEE Trans. Computers 60(2), 242–253 (2011)
12. Eriksson, J., Back, R.-J.: Applying PVS Background Theories and Proof Strategies in Invariant Based Programming. In: Dong, J.S., Zhu, H. (eds.) ICFEM 2010. LNCS, vol. 6447, pp. 24–39. Springer, Heidelberg (2010)
13. Feret, J.: Static Analysis of Digital Filters. In: Schmidt, D. (ed.) ESOP 2004. LNCS, vol. 2986, pp. 33–48. Springer, Heidelberg (2004)
14. Feron, E.: From control systems to control software. IEEE Control Systems 30(6) (2010)
15. Feron, E., Alegre, F.: Control software analysis, part I open-loop properties. CoRR abs/0809.4812 (2008)
16. Friedberg, S., Insel, A., Spence, L.: Linear Algebra, 3rd edn. Prentice-Hall (1997)
17. Garillot, F., Gonthier, G., Mahboubi, A., Rideau, L.: Packaging Mathematical Structures. In: Berghofer, S., Nipkow, T., Urban, C., Wenzel, M. (eds.) TPHOLs 2009. LNCS, vol. 5674, pp. 327–342. Springer, Heidelberg (2009), http://hal.inria.fr/inria-00368403/en/
18. Gonthier, G.: Point-Free, Set-Free Concrete Linear Algebra. In: van Eekelen, M., Geuvers, H., Schmaltz, J., Wiedijk, F. (eds.) ITP 2011. LNCS, vol. 6898, pp. 103–118. Springer, Heidelberg (2011)
19. Halmos, P.: Finite-Dimensional Vector Spaces. Springer, Heidelberg (1974)
20. Harrison, J.: The HOL light formalization of euclidean space. In: AMS Special Session on Formal Mathematics for Mathematicians (2011)
21. Hoare, C.A.R.: An axiomatic basis for computer programming. Comm. ACM 12, 576–580 (1969)
22. Izerrouken, N., Thirioux, X., Pantel, M., Strecker, M.: Certifying an Automated Code Generator Using Formal Tools: Preliminary Experiments in the GeneAuto Project. In: ERTS (2008)
23. Jonsson, U.T.: A lecture on the S-procedure (2001)
24. Moy, Y.: Union and cast in deductive verification
25. Muñoz, C., Narkawicz, A.: Formalization of an efficient representation of Bernstein polynomials and applications to global optimization. J. of Automated Reasoning (2011)
26. Owre, S., Rushby, J.M., Shankar, N.: PVS: A Prototype Verification System. In: Kapur, D. (ed.) CADE 1992. LNCS (LNAI), vol. 607, pp. 748–752. Springer, Heidelberg (1992)
27. Owre, S., Shankar, N.: Theory interpretations in PVS. Tech. Rep. SRI-CSL-01-01, Computer Science Laboratory. SRI International (April 2001)
28. Owre, S., Shankar, N., Rushby, J.M., Stringer-Calvert, D.W.J.: PVS Language Reference. Computer Science Laboratory. SRI International (September 1999)

Temporal Action Language (TAL): A Controlled Language for Consistency Checking of Natural Language Temporal Requirements
(Preliminary Results)

Wenbin Li, Jane Huffman Hayes, and Mirosław Truszczyński

University of Kentucky, USA
wenbin.li@uky.edu, {hayes,mirek}@cs.uky.edu

Abstract. We introduce Temporal Action Language (*TAL*). We design *TAL* as a key component of our approach that aims to semi-automate the process of consistency checking of natural language temporal requirements. Analysts can use *TAL* to express temporal requirements precisely and unambiguously. We describe the syntax and semantics of *TAL* and illustrate how to use *TAL* to represent temporal requirements.

1   Introduction

Requirements such as "*a node should re-identify itself within 10 seconds after making a connection to the server or the server will drop the connection in 2 seconds*" describe temporal dependencies among events. Such *temporal* requirements are common in software projects. Temporal requirements may be inconsistent. Detecting inconsistencies of temporal requirements is essential and should take place before the design phase so that the cost of revisions can be minimized. Automating or partially automating the process is crucial as the task, when performed manually, is time consuming and error-prone.

There has been much research on formal methods for automating the process of requirements analysis [7,9,8,4]. Analyzing temporal constraints automatically requires that they be expressed in a low-level formal language for which good automated reasoning tools are available (such as temporal logic [13] and timed automata [1], which have been used with success to analyze real-time systems [2,12]). Researchers typically assume that formal representations are already given and focus on methods and tools for analyzing them. Our approach is different as it assumes the requirements are stated in *natural language*, and so addresses the needs of the most typical scenario when software requirements are given as a free-flow narrative (cf. the example above).

Our main contribution is a controlled language called *Temporal Action Language* (*TAL*). We propose it as a key component of a process aiming to minimize the time and effort required to check the consistency of temporal requirements specified in natural language. Translating such requirements faithfully into low-level formal languages is difficult due to the significant "distance" between natural language text and formal expressions in logic, ambiguity common in natural

A. Goodloe and S. Person (Eds.): NFM 2012, LNCS 7226, pp. 162–167, 2012.

language descriptions of requirements, importance of implicit information, and insufficient formal method background of analysts. We introduce *TAL* as a bridge between natural language and the low-level target language used for reasoning. This naturally leads to a two-stage process: (1) creating a *TAL* theory that describes the system and (2) detecting conflicts in the *TAL* theory. Each stage can be further decomposed into multiple manageable tasks.

The first stage requires identifying temporal requirements; gathering domain information; making relevant, shared (or commonsense) knowledge explicit; removing ambiguity in requirements; and expressing them in *TAL*. Analyst involvement in the first stage will be necessary. However, we believe natural language processing (NLP) and information retrieval (IR) techniques can effectively assist analysts in the task. The second stage consists of translating *TAL* theory into low-level logic formalism and using its tools to reason about the *TAL* theory. That stage can be fully automated.

We want to use *TAL* as an effective bridge between natural language and a low-level logic, and we designed *TAL* with the following desiderata in mind. First, the syntax of *TAL* must be close to that of natural language because the readability of *TAL* is crucial to the effectiveness and efficiency of the first stage. High readability significantly reduces the time and effort required to verify and validate the *TAL* theories generated in this stage. Second, theories in *TAL* must have a well-defined semantics so that correct automated translations of *TAL* theories into target languages are possible. Third, we want *TAL* to be capable of specifying the temporal constraints that people may find in software requirements. Specifically, we want *TAL* to model the prerequisites and effects of actions, and the time bounds on which actions start and end. Although the overall approach is still under development, we have collected anecdotal evidence suggesting readability of *TAL* theories as well as feasibility of automating translations of *TAL* theories into formal systems.

2 Temporal Action Language *TAL*

Syntax. We use *TAL* to specify temporal constraints on times when events occur. Such events include the start and end of actions and the change of system properties (fluents). We design *TAL* as an extension of *Action Language AL* [3] which allows us to specify actions and fluents, but not temporal information.

A *TAL* theory is a triple (AD, IC, TC) where AD is the set of *action definitions*, IC is the set of *initial constraints*, and TC is the set of *temporal constraints*. The *action definitions* describe the actions by specifying their prerequisites and effects, all expressed as *fluents* (boolean properties). The syntax of AD is that of *AL*. In particular, AD consists of expressions of the following form:

State constraint	L **if** P	(1)
Dynamic causal law	A **causes** L **if** P	(2)
Executability condition	**impossible** $A_1, \ldots, A_k$ **if** P	(3)

where L and P are lists of fluents and their negations, and $A, A_1, \ldots, A_k$ are actions. State constraint (1) says that L holds (every fluent and the negation of a fluent in L holds) in every state in which P holds (in the same sense as L). Dynamic causal law (2) describes the effects of actions. Executability condition (3) specifies the prerequisites of actions. For example:

$connect(serA, nodeA)$ **causes** $connected(nodeA, serA)$ **if** $systemOn$

says that executing the action $connect(serA, nodeA)$ when the system is on results in $nodeA$ and $serA$ being connected, and

impossible $identify(nodeA, serA)$ **if** $\neg\ connected(serA, nodeA)$

specifies the prerequisite of the action $identify(nodeA, serA)$.

The second component of a *TAL* theory, *IC*, consists of *initial constraints* defining the initial state of the system. An *initial constraint* is an expression: **initially** L , where L is a fluent.

The presence of the component *TC* in a *TAL* theory is the key feature that distinguishes *TAL* from *AL*. *TC* specifies temporal information including *temporal constraints* and action durations.

A *duration specification* is an expression: **duration** Act x $unit$, where Act is an action, x is a positive number, and *units* refers to a time unit such as a millisecond, second, or minute.

Temporal constraints describe temporal relationships among the times when events occur. *Temporal conditions* are the basic component of temporal constraints. A temporal condition models the temporal relationship between the occurrence time of two events. In *TAL*, each action Act is associated with two *prompts*: **commence** Act and **terminate** Act, which represent starting and successfully finishing action Act. In *TAL* one can relate two consecutive occurrences of the same action to each other. To distinguish between them, *TAL* provides the keywords **previous** and **next**. A temporal condition is of the form:

$\langle timeReference \rangle$ @ $\langle timeComparator \rangle$ $[\langle timeModifier \rangle]$ $\langle timeReference \rangle$

The expression $\langle timeReference \rangle$ represents the occurrence time of the event. That time can be **startTime** (the time when the system starts), *prompt* (the time when the prompt occurs), a fluent, or its negation (the time when the fluent starts or ceases being true). The expression $\langle timeComparator \rangle$ $[\langle timeModifier \rangle]$ specifies the temporal relationship between these two time moments. In *TAL*, we use $<, \leq, =, \geq$, or $>$ for $\langle timeComparator \rangle$. The parameter *timeModifier* is optional. It modifies the time t given by the second *timeReference* expression as in "x *seconds* **before** t" or "x *milliseconds* **after** t," where $x > 0$. For example, in *TAL*, the temporal condition "*serA drops the connection to nodeB 5 seconds after it establishes a connection to nodeA*" can be written as:

commence $dropConn(serA, nodeB)$ @ $= 5$ *seconds* **after**
terminate $estConn(serA, nodeA)$

The basic form of a *temporal constraint* is:

if A_1 **and** ... **and** A_k, **then** B_1 **or** ... **or** B_m;

where A_1, ..., A_k and B_1, ..., B_m are *temporal conditions* or their negations (temporal conditions can be viewed as special temporal constraints with $k = 0$ and $m = 1$). In *TAL*, one can express "*if a connected node does not re-identify itself to the server within 10 seconds after the connection is established, the server shall drop the connection within 2 seconds*" as:

if not terminate *identify*(*nodeA*, *serA*) @ ≤ 10 *seconds* **after**
terminate *estConn*(*serA*, *nodeA*),
then terminate *dropConn*(*serA*, *nodeA*) @ ≤ 2 *seconds*;

Semantics. We base the semantics of a *TAL* theory (AD, IC, TC) on a transition system T_{AD} of the action description component AD. The use of transition systems as the semantics of action language theories was proposed by Gelfond and Lifschitz [6]. That approach applies also to AL [3]. In an AL transition system, states are combinations of fluents and their negations. Arcs between states are labeled with actions because AL assumes that only actions can cause the system to change its state. Since *TAL*'s action description AD is in the syntax of AL, we create the transition system T_{AD} essentially in the same way as in AL but with two modifications. First, the arcs in T_{AD} are labeled with prompts. This is because the prerequisites and effects of actions specified in AD can be viewed as prerequisites of the corresponding **commence** prompts and effects of the corresponding **terminate** prompts. Second, some arcs are labeled with the term *time* as some fluents in *TAL* can change value simply because of time passing (for instance, a message becomes "old" if it is in the queue for more than 20 seconds – no action is required for that).

A sequence $\langle s_0, pr_0, s_1, pr_1, \ldots, s_{x-1}, pr_{x-1}, s_x \rangle$ is a *path* in a transition system T_{AD} if all s_i are states, all pr_i are prompts or *time*, s_0 satisfies all *initial constraints*, and if for each $i = 0, \ldots, x - 1$, $\langle s_i, pr_i, s_{i+1} \rangle$ is a transition in T_{AD}. A path in T_{AD} represents a scenario, the evolution of the state of the corresponding system as the result of prompts (time) labeling the arcs, assuming we disregard action durations and temporal constraints.

A path does not show when the events occur. We define a *timed path* as a sequence $\langle s_0, pr_0, t_0, s_1, pr_1, t_1, \ldots, s_{x-1}, pr_{x-1}, t_{x-1}, s_x \rangle$, where $\langle s_0, pr_0, s_1, pr_1, \ldots, s_{x-1}, pr_{x-1}, s_x \rangle$ is a path and for every $i = 0, \ldots, x - 1$, $t_i < t_{i+1}$. The times t_i are the times when the system is to progress from state s_i to s_{i+1}. The question of consistency of temporal requirements is that of the existence of arbitrarily long timed paths satisfying all temporal constraints in TC.

Let p be a timed path and t a time in the time range of p (not greater than the time of the last state change). It is straightforward, albeit tedious, to specify when B holds at time t on p. For instance, let B stand for *prompt1* @ $= x$ seconds **after** *prompt2*. If the most recent occurrence of *prompt1* before or at t is at time t' and at time $t' + x$ there is an occurrence of *prompt2*, then we say that the condition holds at B. There are several such cases to cover. We omit details due to space limits. Next, we define a temporal constraint to be satisfied on p at time t if at least one temporal condition in the consequent of C evaluates to true whenever all temporal conditions in the antecedent evaluate to true (interpreting t as "now"). Finally, we say that a temporal constraint C holds on a path p if C holds at every time t within the range of p.

Consistency Checking. Consistency of a TAL theory means the existence of arbitrarily long timed paths. It guarantees that there is no inconsistency in temporal requirements. A weaker notion of *bounded consistency* means the existence of a timed path with a given bound on its time range. It guarantees that no inconsistency in temporal requirements can exhibit itself prior to the bound. The larger the bound, the more accurately the notion approximates that of consistency. Other interesting questions are whether an event can (or will) occur within a given time bound, or whether a system can (will) satisfy a certain property while running. Since TAL is a formal system, a promising approach to decide (bounded) consistency and related questions is to develop translations to low-level logic systems and use automated reasoning tools that are available for them.

3   Validation

To date, we studied bounded consistency and experimented with translations of the TAL representation of the problem of existence of a timed path of bounded length into answer-set programming (ASP) [10,11]. We selected ASP because it is well suited for modeling search problems and has fast solvers [5]. We created an example scenario with multiple temporal requirements and represented it as a TAL theory. We manually translated the TAL theory into an answer-set program. We used a solver, *clingcon* [5], to process it. We found that when the requirements were consistent, *clingcon* returned at least one answer set that represents a valid scenario. Upon modifying the requirements to make them inconsistent, *clingcon* did not return any timed paths. The experiment shows the feasibility of reasoning about consistency of temporal requirements given in TAL by translating them to low-level target logics and then using automated reasoning tools.

We also wrote sixteen natural language temporal requirements and their corresponding TAL statements. We selected four people with various computer science backgrounds, from working in industry for years as a requirements engineer to having a bachelor's degree in computer science. We briefly introduced the syntax of TAL to them and asked them to rate the similarity in meaning of the natural language statements and their formal TAL representations (we used a scale from 0, completely different, to 5, exactly the same). The mean rating for the sixteen pairs was 4.76. The result shows that the participants found TAL statements to be unambiguous and easy to understand.

4   Discussion and Future Work

This paper presents the language TAL for specifying temporal requirements. It extends AL by allowing users to specify temporal dependencies among events using *temporal conditions* and *constraints*. The syntax of *TAL* is close to natural language and, based on anecdotal evidence, easy to follow. The semantics are based on the concepts of transition systems and timed paths. Checking for (in)consistency of temporal requirements is reduced to creating *TAL* expressions from natural language requirements since, once the *TAL* representation is available, it can be processed in a fully automated way.

Future work includes automating the translation from *TAL* to ASP and creating tools based on natural language processing and information retrieval to assist analysts in generating *TAL* theory based on requirements given in natural language. We will also study temporal logics and timed automata as possible target formalisms. Finally, we will perform systematic experiments to validate the scope and feasibility of the approach.

Acknowledgment. This work is funded in part by the National Science Foundation under NSF grant CCF-0811140 and JPL grant 1401954.

References

1. Alur, R., Dill, D.L.: A theory of timed automata. Theoretical Computer Science 126, 183–235 (1994)
2. Baier, C., Katoen, J.P.: Principles of Model Checking. The MIT Press (2008)
3. Baral, C., Gelfond, M.: Reasoning agents in dynamic domains. Logic-Based Artificial Intelligence, 257–279 (2000)
4. Dutertre, B., Stavridou, V.: Formal requirements analysis of an avionics control system. IEEE Transactions on Software Engineering SE 23, 267–278 (1997)
5. Gebser, M., Ostrowski, M., Schaub, T.: Constraint Answer Set Solving. In: Hill, P.M., Warren, D.S. (eds.) ICLP 2009. LNCS, vol. 5649, pp. 235–249. Springer, Heidelberg (2009)
6. Gelfond, M., Lifschitz, V.: Action languages. Electronic Transactions on Artificial Intelligence (ETAI) 2, 193–210 (1998)
7. Heitmeyer, C.: Software cost reduction. In: Marciniak, J.J. (ed.) Encyclopedia of Software Engineering, 2nd edn. John Wiley & Sons (2002)
8. Klein, M.: An exception handling approach to enhancing consistency, completeness and correctness in collaborative requirements capture. Concurrent Engineering Research and Applications 5, 37–46 (1997)
9. Lamsweerde, A.V., Darimont, R., Letier, E.: Managing conflicts in goal-driven requirements engineering. IEEE Transactions on Software Engineering 24(11), 908–926 (1998)
10. Marek, V., Truszczynski, M.: Stable models and an alternative logic programming paradigm. In: The Logic Programming Paradigm: a 25-Year Perspective, pp. 375–398 (1999)
11. Niemela, I.: Logic programs with stable model semantics as a constraint paradigm. Annals of Mathematics and Artificial Intelligence 25, 241–273 (1999)
12. Olderog, E.R., Dierks, H.: Real-Time Systems. CUP (2008)
13. Pnueli, A.: The temporal logic of programs. In: Proceedings of the 18th Annual Symposium on Foundations of Computer Science (FOCS), pp. 46–57 (1977)

Some Steps into Verification of Exact Real Arithmetic[*]

Norbert Th. Müller[1] and Christian Uhrhan[2]

[1] Abteilung Informatik, FB IV, Universität Trier, Germany
[2] Universität Siegen, Faculty IV, Germany

Abstract. The mathematical concept of real numbers is much richer than the double precision numbers widely used as their implementation on a computer. The field of 'exact real arithmetic' tries to combine the elegance and correctness of the mathematical theories with the speed of double precision hardware, as far as possible. In this paper, we describe an ongoing approach using the specification language ACSL, the tool suite Frama-C (with why and jessie) and the proof assistant Coq to verify central aspects of the iRRAM software package, which is known to be a fast C++ implementation of 'exact' reals numbers.

1 Introduction

The verification of programs using double precision numbers often is very complicated: the semantics of this number format does not coincide with the semantics of real numbers, i.e. with the definitions and results found in textbooks on calculus. On the other hand, it is possible to implement 'exact' real numbers in software ([BK08, OS10, Mue01, Les08, Lam07], to name a few), so here verification should be a lot easier and could concentrate on mathematical aspects of the problem and not on the peculiarity of the double precision numbers.

Some of these 'exact' implementations have already been verified themselves: [Les08] used Haskell/PVS, [OS10] used Haskell/Coq, and [BK08, Bau08] used OCaml/Coq. Unfortunately, these implementations are much slower in general than simple computations with double precision, and also much slower than other implementations for real numbers like [Mue01, Lam07]. So what we would like to have is *one* implementation of real numbers that is *exact, fast* and *proven to be correct* at the same time.

This motivates an ongoing project started in 2010 where we try to verify at least central aspects of the iRRAM software package [Mue01], which is known to be a fast C++ implementation of exact reals numbers. Unfortunately, we do not know of any tools for direct verification of C++ programs, so we took the following approach: Using the specification language ACSL, we specify the semantics of core routines of the package, then we use the tool suite Frama-C (with why and jessie) as well as the proof assistant Coq in order to verify versions of these routines that have been manually transformed from C++ to C. Currently, we

---

[*] This work was partially supported by the DFG project 446 CHV 113/240/0-1.

A. Goodloe and S. Person (Eds.): NFM 2012, LNCS 7226, pp. 168–173, 2012.
© Springer-Verlag Berlin Heidelberg 2012

use frama-c-Nitrogen-20111001, why-2.30 and Coq 8.3p12; some proofs have already partly been rewritten to use why3-0.71 instead of why-2.30. Automatic provers like Alt-Ergo or CVC3 could be used to verify some conditions of our test examples, but none of them could do a complete verification.

There are two objectives behind the project: the *internal* goal is just to verify correctness of the iRRAM package, while the *external* goal is to develop verification tools for other users of exact real arithmetic.

Our approach works in 4 levels, that are treated in parallel:

1. *core level*: arbitrarily precise floating-point numbers (mainly internal use)
2. *interval level*: interval arithmetic (mainly internal use)
3. *basic arithmetic level*: basic operations on real numbers (mainly internal use)
4. *application level*: non-basic operations and user tools (mainly external use)

As an example consider the multiplication $x * y$ of real numbers: although level 3 (basic arithmetic) is not yet fully proven, we can already use the multiplication as an exact operation on level 4 (applications). In this paper we will describe parts of this level 4, so basic real operations are assumed to be working correctly.

In section 2, we will briefly describe the background of exact real arithmetic, which will motivate why we emphasize the correct behavior of the verified routines concerning exception handling. Chapters 3 and 4 describe the tools we use, section 5 contains a detailed example and section 6 gives a short summary.

We estimate that until now about 5% of the complete package have been proven: The package consists of about 800 functions and 12000 lines of code, 30 central functions have been considered so far. As the specifications on the different levels are mutually dependent, both specifications and proofs might have to be readjusted later.

2   Exact Real Arithmetic

As the set $\mathbb{R}$ of real numbers is not countable, implementing real numbers must be significantly different from an implementation of countable sets like the natural or even rational numbers. The theoretical background here is usually called 'computable analysis' or 'type-2-theory of effectivity', see [BHW07, We00]: a real number x is represented as a sequence $(r_n)_{n \in \mathbb{N}}$ of rational numbers r_n with a known rate of convergence. Usually this convergence is expressed as a constraint '$\forall n : |x - r_n| \leq 2^{-n}$', i.e. x represented as a converging sequence $(I_n)_{n \in \mathbb{N}}$ of intervals, where $I_n := \{y \in \mathbb{R} : |y - r_n| \leq 2^{-n}\}$. 'Exact real arithmetic' now tries to use similar concepts to implement real numbers on real-world computers.

Functional languages are ideal candidates here, and there exist quite many prototypical implementations based on this programming paradigm: [OS10], [BK08], [Les08], just to name a few. Unfortunately, the performance of these approaches is usually very bad and they can only be used for academic examples. Newer functional based implementations try to improve the performance using 'stateful' functional programming (e.g., [BK08] using monads in OCAML).

Imperative or object-oriented programming languages, as a different paradigm, first have to be enhanced with mechanisms to work with infinite objects like sequences. This is often done by explicit construction of computation diagrams, see e.g. [Lam07]. The performance already increased dramatically, compared to the functional approaches. Unfortunately, the diagrams need a lot of memory.

Already in 1996, the iRRAM package was presented, where computation diagrams were avoided. Instead, iterations of the underlying numerical algorithm are used. This can easily be achieved using the concept of *exceptions* in C++: The algorithm under consideration is executed with interval arithmetic where each real number is represented by a single (initially quite imprecise) interval. If during the computation these intervals grow too large to get satisfactory results, an exception is thrown and the algorithm is executed with smaller intervals. This is repeated until the results are precise enough, i.e. until the algorithm finishes without throwing any exceptions. Although this idea seems to waste computation time, it turned out to be amazingly fast and the memory impact is neglectable compared to approaches previously mentioned: the iRRAM can sometimes perform a billion of dependant operations in a few seconds, where the size of the computation diagrams alone would easily amount to more than 100GB.

In this paper, the main focus is on how we deal with this aspect of exceptions.

3   Verification of C Programs

To verify the iRRAM we use a combination of the proof assistant Coq and the frameworks Frama-C and Why. Coq is a theorem prover which can be used to formulate and proof theories. Frama-C is a static analysis framework for the C programming language which for example provides tools for dead code elimination (Spare Code) but also for formal verification of C programs through a plugin called jessie. The framework Why can be seen as a general verification condition generator. It takes an annotated program as input and is able to generate verification conditions for that input for several proof assistants including Coq.

In order to verify a C program we first have to provide a formal specification of the program. For that we give a formal (predicative) description of the semantics of the C program using the so called *ANSI C Specification Language* (ACSL). As a C program usually consists of a large collection of functions, each of them has to be annotated with a so called 'function contract'.

Next, the annotated C program serves as input to the jessie plugin and then to Why, which is generating the verification conditions in Coq. Having done that we then have to prove that our program in fact is correct with respect to its specification.

4   From C++ to C

First we have to translate the C++ code to C code for the purpose of verification. In fact our specifications and even the resulting correctness proofs should easily be adaptable as soon as there is a verification tool for C++ programs (e.g. as an extension to the jessie plugin) since the semantical description should differ only slightly (e.g. to describe that a function may throw an exception).

We had to consider the following C++ concepts in order to get a reliable translation from C++ to C: (a) classes, (b) constructors and destructors, (c) operator overloading, and (d) exceptions.

(a) Classes can be translated to structs in C. Of course, classes are equipped with a collection of methods for manipulating instances of objects. Additionally, the visibility modifiers like private do not have a counterpart in C. Currently, we simply treat everything to be public.

Taking into account that methods implicitly have access to the 'this'-pointer, every method of n parameters is actually a function with $n+1$ parameters, where the first parameter is a pointer to the object itself:

```
// C++ class for real numbers            // translation to C
class REAL {                             typedef struct REAL { ... } *REAL;
public:                                  ...
double as_double (const int p) const;    double as_double (const REAL this,
...                                                        const int p);
```

(b) The C language does not have constructors and destructors. Fortunately it is easy to detect where constructors are called, so we can replace them by corresponding C functions. Destructors are much harder to handle, as they are almost always called implicitly at the end of a lifetime of objects. Currently we simply ignore the destructors (and rely on a hypothetical garbage collection).

(c) Operator overloading is very useful to keep syntactical structures simple. This especially holds for mathematical software, where writing $x \cdot y \cdot z$ simply as x*y*z instead of mul(mul(x,y),z) can significantly improve readability and reduce errors at the same time. The translation, however, is tedious but quite trivial, as soon as we know the involved classes. Since C does not support overloading we have to define a function with a name of its own:

```
// C++ version                           // translation to C
friend REAL operator
  * (const REAL& x, const REAL& y);      REAL REALREAL_mul (REAL x, REAL y);
friend REAL operator
  * (const REAL& x, const int& y);       REAL REALint_mul (REAL x, int y);
```

(d) The exception mechanism is the most complicated aspect. For JAVA, e.g. the Krakatoa tool[MPMU04] contains a signals construct for the functions contracts to represent exceptions. C itself does not have exceptions, so ACSL does not have any support for exception handling. So currently we have no option for an easy specification of the (vital) exceptions.

We choose to model exceptions by extending the source code: A global pointer exception is introduced in the C version carrying the information about any thrown exceptions. As long as this pointer remains 0, no exception occurred. So using a multiplication z=x*y can be modeled in fact as

```
{REAL tmp = REALREAL_mul (x,y); if(exception != 0) return 0; z=tmp;}
```

On the application level, this is sufficient for verification, as these exceptions are not caught by the application but by the runtime environment of the iRRAM (which is still unverified). To verify this runtime environment, however, we will really need to translate all aspects of the exceptions, maybe using the C functions setjmp and longjmp (that are not yet supported in Frama-C).

5 Example

As an example for a verified function we consider the power function computing x^n with $x \in \mathbb{R}$ and $n \in \mathbb{N}, n \geq 0$. A working implementation in the iRRAM is:

```
REAL power(const REAL& x, int n) {
  REAL y=1;
  for (int k=0; k<n; k=k+1) { y=y*x; }
  return y; }
```

Translated to C we get:

```
REAL REALint_power(const REAL x, int n) {
  REAL y;
  { REAL tmp = REAL_from_int32(1); if (exception != 0) return 0; y=tmp;}
  for (int k=0;k<n;k=k+1)
    { REAL tmp = REALREAL_mul(y,x); if (exception != 0) return 0; y=tmp;}
  return y; }
```

The corresponding function contract in ACSL is build as follows: With requires, we express that the caller has to ensure that x points to a valid (i.e. correctly constructed) data structure for real numbers and that n is non-negative. The assigns part describes the side effects which may happen by calling the function: in our case both the result of the function as well as the exception pointer might be modified. Finally the ensures part expresses that in the program state immediately after returning from the function the result points to a valid real object and represents the n-th power of x, unless an exception was thrown.

```
/*@
requires   valid_REAL(x) && n >= 0;
assigns    \result, exception ;
ensures    exception==0 ==> ( valid_REAL(\result)
   && real_of_iRRAM_REAL (\result) == \pow(real_of_iRRAM_REAL (x),n) );
*/
REAL REALint_power(const REAL x, int n);
```

The last part of the ensures clause is very important here: This is the mathematical statement we want to prove, i.e. that the result is the n-th power of x. The function real_of_iRRAM_REAL is actually a logical defined function mapping iRRAM REALs to the ideal reals Coq knows about, and pow is mapped to Coqs power function. As every interpretation of the value of REAL data will happen via similar mappings, REAL *is* an implementation of real numbers.

What remains to be done for our example is the formal proof in Coq, for which we enhance the source code with the following loop invariant in ACSL:

```
/*@
loop invariant valid_REAL(y) && 0 <= k <= n &&
               real_of_iRRAM_REAL (y) == \pow(real_of_iRRAM_REAL (x),k);
loop variant n-k;
*/
   for (int k=0;k<n;k=k+1)
     { REAL tmp = REALREAL_mul(y,x);  if (exception != 0) return 0; y=tmp;}
```

The `variant` expresses that the (non-negative) value $n - k$ is decreasing, so that we are able to prove that the loop eventually terminates. Having done this we were able to finish the proof of correctness for our example.

6 Summary

As the example showed, the mathematical part of the verification of `iRRAM` algorithms on the application level turns out to be quite easy, as we can rely on the exactness of the operations (unless exceptions occur) and we can use the knowledge already present in `Coqs` libraries on real numbers. Meanwhile, we are also quite certain that the conversion from `C++` to `C` could be done automatically, e.g. by some suitable pre-compilation, at least as far as we need it.

Currently, we would like to concentrate first on other parts of the whole verification: one important task here will be to replace 32- or 64-bit integers almost everywhere by a fast (and verified) datatype for $\mathbb{Z}$ using a similar concept as for real numbers: either operations are correct in the mathematical sense (so without any overflow), or an exception has to be thrown. Then our power operator would not just be correct for 32-bit numbers but for arbitrary $n \in \mathbb{Z}$.

A far goal is to address total correctness, i.e. to identify those cases where no exceptions will be thrown. This will be much harder to do, as equality of real numbers is not decidable. Additionally, out-of-memory errors will be very hard to predict, as they depend on the necessary precision in a computation.

References

[Bau08]   Bauer, A.: Efficient computation with Dedekind reals. In: 5th International Conference on Computability and Complexity in Analysis, CCA 2008, Hagen, Germany, August 21-24 (2008)

[BK08]   Bauer, A., Kavkler, I.: Implementing real numbers with rz. Electron. Notes Theor. Comput. Sci. 202, 365–384 (2008)

[BHW07]   Brattka, V., Hertling, P., Weihrauch, K.: A Tutorial on Computable Analysis. In: Barry Cooper, S., Löwe, B., Sorbi, A. (eds.) New Computational Paradigms: Changing Conceptions of What is Computable, pp. 425–491. Springer, New York (2008)

[Lam07]   Lambov, B.: Reallib: An efficient implementation of exact real arithmetic. Mathematical Structures in Computer Science 17(1), 81–98 (2007)

[Les08]   Lester, D.R.: The world's shortest correct exact real arithmetic program? In: Proc. 8th Conference on Real Numbers and Computers, pp. 103–112 (2008)

[MPMU04]   Marché, C., Paulin-Mohring, C., Urbain, X.: The krakatoa tool for certification of java/javacard programs annotated in jml. J. Log. Algebr. Program. 58(1-2), 89–106 (2004)

[Mue01]   Müller, N.T.: The iRRAM: Exact Arithmetic in C++. In: Blank, J., Brattka, V., Hertling, P. (eds.) CCA 2000. LNCS, vol. 2064, pp. 222–252. Springer, Heidelberg (2001)

[OS10]   O'Connor, R., Spitters, B.: A computer-verified monadic functional implementation of the integral. Theor. Comput. Sci. 411(37), 3386–3402 (2010)

[We00]   Weihrauch, K.: Computable analysis: An introduction. Springer-Verlag New York, Inc. (2000)

Runtime Verification Meets Android Security

Andreas Bauer[1,2], Jan-Christoph Küster[1,2], and Gil Vegliach[1]

[1] NICTA Software Systems Research Group
[2] The Australian National University

Abstract. A dynamic security mechanism for Android-powered devices based on runtime verification is introduced, which lets users monitor the behaviour of installed applications. The general idea and a prototypical implementation are outlined, an application to real-world security threats shown, and the underlying logical foundations, relating to the employed specification formalism, sketched.

1 Introduction

Most mobile platforms, such as Android [8], which is an open-source software stack designed to power tablet PCs and smart phones, offer built-in security mechanisms to protect users from various types of malware, often designed to spy on users or to exert control over (parts of) a mobile device's functionality. An example for the latter consists in the sending of SMS messages secretly, without the user's consent (cf. [11]). However, the existing security mechanisms obviously cannot stop or prevent the rising number of attacks on these platforms: In its Q1/2011 threats report, security firm Kaspersky remarks that "since 2007, the number of new antivirus database records for mobile malware has virtually doubled every year." In case of the Android platform, security firm McAfee asserts in its Q2 threats report that, in fact, "Android OS-based malware became the most popular target for mobile malware developers." In light of these developments various authors have proposed improvements to the built-in security mechanisms of mobile platforms, and in particular to the Android platform which, right now, constitutes the fastest growing platform on the market, and offers researchers the advantage that its source code is freely available.

Arguably, two of the most feature-complete and well-documented security enhancements recently made for Android are TaintDroid [4] and the Saint framework [12]. TaintDroid is an extensive modification of the entire Android stack that tracks the flow of sensitive data through third-party applications at runtime. The modifications allow TaintDroid to detect when sensitive data is leaked in whatever form, e.g., by sending an email or SMS containing the sensitive data, or by uploading a file directly. To cater for all these different scenarios, TaintDroid "taints" sensitive information to keep tracking its use throughout the system. The central components of the Saint framework described in [12] are a modified Android application installer and a so called AppPolicy Provider. The custom installer ensures that at install-time only applications which do not violate policies stored in the AppPolicy Provider can be installed. The authors of Saint have gone to great lengths to check existing applications' permissions for suspicious permission requests and from that derived practically useful policies for that purpose.

While Saint is, more or less, true to Android's own security mechanisms, which are mostly based on assigning permissions statically to third-party applications, thereby

A. Goodloe and S. Person (Eds.): NFM 2012, LNCS 7226, pp. 174–180, 2012.

deciding which operations an application may or not perform at runtime, TaintDroid goes further, in that it controls what happens with data at runtime. The latter, however, comes at a price, in that such a high level of system instrumentation results in up to 27% runtime overhead [4]. Moreover, an expected downside of such a comprehensive system modification is to try and keep up to date with future releases of Android, which is under active development by a world-wide consortium of OEMs. As long as said consortium does not adapt (and thereby maintain) TaintDroid officially, it will be difficult to install and adapt it for off-the-shelf devices.

Our aim therefore, is to introduce a more light-weight, yet dynamic security extension to Android based on a technique known as *runtime verification*. In a nutshell, runtime verification subsumes techniques that aid in showing that an observed system behaviour satisfies or violates a given specification, often given in terms of automata or logic (cf. [1]). The methods developed in this area typically help to automatically generate a *monitor* from a given specification, such that at runtime one must not consider/store the entire behavioural trace, but merely consume observations in a step-wise and therefore efficient manner. That is, the monitor passively observes the system and raises an alarm if a specification is violated, or switches itself off if a specification has been satisfied. While the complexity of monitor generation, depending on the specification formalism at hand, can be very high (sometimes multiple exponents), the runtime complexity is usually constant-time for each new observation. Runtime verification has been employed in safety-critical contexts, but recently also emerged as a generic monitoring and testing methodology for Java (cf. [7,3]). Here, we use it specifically to enhance system security, in that we monitor the individual behaviours of third-party applications, installed on Android devices. To this end, users can specify what constitutes a "suspicious behaviour", e.g., an application starts at boot-time, later checks the device's GPS location, and then connects to the internet (possibly to transmit the location). For each such policy, our implementation automatically creates a monitor that, once active, will raise an alarm when such a sequence of events was produced by any installed application. Admittedly, not every application which queries the GPS and then connects to the internet is malicious, but many so called "spywares" are disguised as seemingly harmless toy (e.g., wallpaper) applications, which have no legitimate reason to behave in the aforementioned way.

Our proposed changes to Android are minimal compared to frameworks such as TaintDroid, yet also target the runtime behaviour of applications rather than Android's static permissions. Unlike TaintDroid, however, our goal was not to trace data at runtime, but to use simple behavioural specifications to detect a whole range of malicious applications. In the area of computer security, this is known as *behavioural detection*, where the aim is not to identify malware by comparing the applications in question with signatures stored in a database, but to detect the behaviour of known and future malware, which is expected to be similar to existing malware (for a survey cf. [10]).

2   Android Security Concepts in a Nutshell

Let us briefly discuss the important Android security concepts that are relevant to this paper. Note that the aim of this section is not to give a comprehensive overview of the Android architecture or its security concepts (cf. [8,9,5] for that).

Firstly, Android applications and most of the Android stack are written in Java, whereas a modified Linux kernel serves as the platform's low-level OS. Applications on Android are "sandboxed", meaning that each executes within its own virtual machine, and, from an OS point of view, as unique user; that is, unlike standard Linux processes, which inherit the UID of the user who started them, Android applications all have a unique UID. In other words, each application is treated as an individual user from the low-level OS's point of view[1]. This strict "sandboxing" basically ensures that one application cannot modify (or even read—unless dedicated inter-process API calls are being made) the data of another installed application. Unfortunately, however, it is generally not true that the harm caused by a malicious application, is therefore restricted to its "sandbox." In fact, as also pointed out above, there are countless ways in which a malicious application could exploit the device's capabilities, or spy on its users.

Whether or not an application is allowed to use a certain functionality that an Android device offers is primarily determined at install-time, when the standard Android installer presents to the user a list of required application permissions. Users cannot revoke individual permissions that they may not feel comfortable with or that they do not understand, rather they need to grant all permissions or cannot install the application. In consequence, many users do not review the permissions at install-time [6]. In fact, Android's permission system is predominantly static, meaning that once an application is installed, users have basically no means of controlling that application's runtime behaviour. For example, once an application has been granted permission to send SMS, it may do so in the background without requesting further user confirmation. According to the official documentation [9], the lack of dynamic security mechanisms is a design principle: "Android has no mechanism for granting permissions dynamically (at runtime) because it complicates the user experience to the detriment of security." Note that the situation on other mobile platforms, like Nokia's Symbian OS, is similar (cf. [2]).

3   Modelling Security Policies

We assume a set of predicate symbols $\mathcal{P} = P \cup R$, such that $P \cap R = \emptyset$. Security policies in our framework are based on the grammar $\varphi ::= p(t_1, \ldots, t_n) | r(t_1, \ldots, t_n) | \neg \varphi | \varphi \wedge \varphi | \mathbf{X}\varphi | \varphi \mathbf{U}\varphi | \forall(x_1, \ldots, x_n) : p.\ \varphi$, where $p \in P$, $r \in R$ are n-ary predicate symbols, t_i terms, and x_i variables. The term structure is determined by variables and function symbols of given arities. For a ground term t, $t \downarrow$ denotes its actual value, e.g., $(2 + 3) \downarrow$ yields 5, assuming the usual arithmetic functions to be part of our language and interpreted accordingly. Variables range over specific domains such as strings, integers, or any finite domain. Hence, in a statement $\forall x : p.\ \varphi$, p's arity, $p : \tau \to \mathbb{B}$, uniquely determines the *sort* of variable x to be τ. We model observed application behaviour in terms of *actions* which, in turn, are represented by ground atoms. Sets of actions are called *events*. An application's behaviour, seen over time, is therefore a finite *trace* of events, e.g., $\{sms(1234)\}\{login(\text{"user"})\} \ldots$ That is, the occurrence of some ground atom $sms(1234)$ in some trace at position $i \in \mathbb{N}_0$ means that at time i it is the case that $sms(1234)$ is true. As is standard, the semantics of this language is defined via

---

[1] There is an exception to this rule, but this is not relevant to this paper: applications which share a developer's signature may run under the same UID.

$$w, i \models p(t_1, \ldots, t_n) \text{ iff } p(t_1 \downarrow, \ldots, t_n \downarrow) \in w(i);$$
$$w, i \models r(t_1, \ldots, t_n) \text{ iff } r(t_1 \downarrow, \ldots, t_n \downarrow) \text{ is true;} \qquad w, i \models \neg\varphi \text{ iff } w, i \not\models \varphi;$$
$$w, i \models \varphi \wedge \psi \text{ iff } w, i \models \varphi \text{ and } w, i \models \psi; \qquad w, i \models \mathbf{X}\varphi \text{ iff } w, i+1 \models \varphi;$$
$$w, i \models \varphi \mathbf{U}\psi \text{ iff there exists } k \geq i \text{ s.t. } w, k \models \psi \text{ and } w, j \models \varphi, \text{ for all } i \leq j < k;$$
$$w, i \models \forall(x_1, \ldots, x_n) : p.\ \varphi \text{ iff } w, i \models \varphi[c_1/x_1, \ldots, c_n/x_n], \text{ for all } p(c_1, \ldots, c_n) \in w(i).$$

Fig. 1. Kripke semantics of the language wrt. infinite trace w and position i therein

infinite traces (see Fig. 1). Note how, unlike symbols from P, symbols from R do not obtain their interpretations via the trace, but by some computational means assumed to be available in the background when evaluating a policy over some trace.

At runtime, a monitor checking φ, will only see a prefix of an infinite trace, denoted u, and therefore return $\top$ if u is a *good prefix* of φ, $\bot$ if u is a *bad prefix*, and ? otherwise. This is akin to the 3-valued finite-trace semantics for LTL introduced in [1], except that our monitor not necessarily reports *minimal* prefixes. For brevity, we cannot give a detailed, step-wise semantics of our monitor, but refer the reader to Sec. 4 for an outline of our algorithm based on the well-known concept of formula progression. Let us now look at example policies, specified in this language and the usual syntactic "sugar".

Recall the promise of our approach and, more generally, of behavioural detection is that it allows not only the detection of specific, known malware, but of new threats as they appear, so long as their damaging behaviour, exhibited on a device, is sufficiently similar to already known malware. Indeed, the databases by security firms such as McAfee, not only list specific malicious applications for Android, but entire evolutions, classified by abstract IDs, such as Android/NickiSpy, to indicate that there exist multiple incarnations of the same malware, realised in differently branded applications. Android/NickiSpy, for example, represents a family of applications which secretly record a user's phone conversation on SD card in the compressed .amr format (adaptive multi-rate). We can detect this family of malware via a simple policy,

$$\mathbf{G}\forall x : sd\_write.\ \neg regcomp(x, \text{“.*}\backslash.\text{amr''}),$$

where $regcomp(x, y)$ is true if the string x, in this case representing a file name, is in the language given by the regular expression that is represented by string y. However, should there be legitimate recording of .amr files to SD card, the user is always able to ignore any reported violations of this policy.

As another example, consider the first ever Android Trojan (Trojan-SMS.Android-OS.FakePlayer.a), disguised as media player, which secretly sent SMS messages to expensive premium numbers [11]. This led us to monitor a more general behaviour, i.e., to be notified if *any* application sends an SMS to a number *not in our contacts*:

$$\mathbf{G}\forall x : sms.\ contact(x).$$

While there may be legitimate violations of this policy, its monitoring at least lets users keep track of which applications exhibit this type of behaviour. It's then up to them to decide to remove an application, if they feel it is not justified.

Finally, a lot of malware is "spyware", meaning that private user data or device details are sent out to remote locations. For example, all applications of type Android/Actrack.

A send GPS location, battery and radio status to a central internet server controlled by the vendor at regular intervals. A policy we may want to monitor in regards to that, more generally, could be "no application should request the GPS location, and later connect to the internet (possibly to transmit said location)", which is captured by the following formula, where $connect(x)$ appears in a trace whenever the application under scrutiny triggers the Linux system call $\texttt{connect}$ to IP address x, and gps whenever it requests the device's current location:

$$\mathbf{G}(\neg((\mathbf{F}\exists x : connect) \wedge gps)).$$

4   Implementation

Currently, our monitors are realised in terms of a stand-alone Android application, written in Java, with a simple GUI that allows users to enter policies. As Android applications are "sandboxed" and therefore unable to monitor each other, we also had to modify the Android stack to facilitate runtime verification in the above sense. To this end, we made some very small, local modifications to exactly two files of the Android system in order to get notified when an application requests permission to perform specific operations or when system events are created, e.g., an application tries to send an SMS message, the system reports low battery status, etc. It is our expectation that this way, our changes will easily carry over

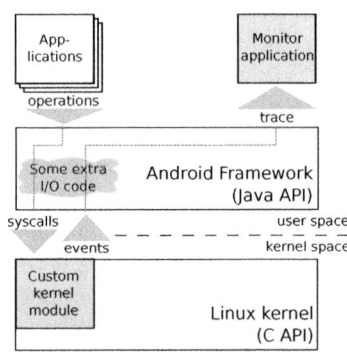

Fig. 2. Architecture

to future releases of the platform. Unfortunately, however, it is not possible to obtain all relevant data by intercepting the high-level Android permission checks. In many cases, Android directly consults the underlying, low-level OS if an application is allowed to perform an operation, e.g., based on the application's UID membership in a Linux group. Examples are the opening of a network socket or the writing to an SD card. But also to extract the actual phone number of an outgoing SMS message, we need to monitor the Linux $\texttt{write}$ system call (or rather, its arguments) as there are no means to obtain this information in user space without having to modify many additional Android files, but then much to the detriment of portability and maintainability of our solution. For reasons of modularity, we "outsourced" this type of information gathering in our own kernel module that dynamically loads during boot[2]. The architecture is sketched in Fig. 2, where the grey areas are constituents of our system, and arrows indicate relevant information flow.

The actual monitoring of a temporal logic formula is currently realised by means of formula progression; that is, for each formula φ and each application, there is a function $prog$, taking a first-order LTL formula and an event as input and returning a first-order LTL formula, such that $\sigma w \models \varphi$ iff $w \models prog(\varphi, \sigma)$. For example, $prog(\mathbf{G}\psi, \sigma) = prog(\psi, \sigma) \wedge \mathbf{G}\psi$, where $prog(\psi, \sigma)$ may return $\top$ or $\bot$ immediately or after expanding

---

[2] Note that there are numerous Android applications, even on the official Market, that also require the installation of custom kernel modules (e.g., DroidWall requires the netfilter module).

to a more complicated formula. The aforementioned 3-valued finite-trace semantics is obtained by mapping all resulting formulae other than $\top$ or $\bot$ to the ?-value.

5   Conclusions and Future Work

Although our work is preliminary, arguably, our results show not only that runtime verification is generally feasible on Android devices, but also that it can improve system security by identifying known and *yet unknown* malware. Due to active development, our code is still unreleased, but we have prepared a system demonstration video: http://baueran.multics.org/droid/. Note also that the performance overhead, when executed on an Android emulator as well as on an actual phone, was negligible even in this preliminary, unoptimised version of the code. However, there can be cases, where the runtime performance of our monitoring procedure necessarily deteriorates over time, i.e., the longer the observed trace gets, the longer the formula becomes that needs to be progressed. Although none of our examples triggers this particular problem, there is a need to characterise fragments of our policy language that lead to monitoring algorithms whose complexity at runtime can be guaranteed to depend only on the size of each new event. One such fragment is obtained by discarding the first rule of both the syntax and the semantics, respectively (see Sec. 3), and assuming predicate symbols from R to be rigid. Additionally, the latter must be either at most unary or, if n-ary, their individual use restricted to at most one variable. However, one of the reasons why we have not adopted this fragment here is due to a symbol like *contact* whose interpretation, arguably, needs to be flexible, i.e., the user can add or delete contacts at any time. Finding useful fragments in the above sense that are also practically relevant is subject of ongoing work.

Acknowledgements. We thank the anonymous referees for their helpful feedback. NICTA is funded by the Australian Government as represented by the Department of Broadband, Communications and the Digital Economy and the Australian Research Council through the ICT Centre of Excellence program.

References

1. Bauer, A., Leucker, M., Schallhart, C.: Runtime verification for LTL and TLTL. ACM Trans. Softw. Eng. Methodol. (TOSEM) 20(4), 14 (2011)
2. Bose, A., Hu, X., Shin, K.G., Park, T.: Behavioral detection of malware on mobile handsets. In: Proc. 6th Int. Conf. Mobile Systems, Applications, and Services (MobiSys), pp. 225–238. ACM (2008)
3. Chen, F., Roşu, G.: Java-MOP: A Monitoring Oriented Programming Environment for Java. In: Halbwachs, N., Zuck, L.D. (eds.) TACAS 2005. LNCS, vol. 3440, pp. 546–550. Springer, Heidelberg (2005)
4. Enck, W., Gilbert, P., Chun, B.-G., Cox, L.P., Jung, J., McDaniel, P., Sheth, A.N.: TaintDroid: an information-flow tracking system for realtime privacy monitoring on smartphones. In: Proc. 9th USENIX Symp. on OS Design and Implementation (OSDI). USENIX (2010)
5. Felt, A.P., Chin, E., Hanna, S., Song, D., Wagner, D.: Android permissions demystified. In: Proc. 18th ACM Conf. Comp. and Comm. Security (CCS), pp. 627–638. ACM (2011)

6. Felt, A.P., Greenwood, K., Wagner, D.: The effectiveness of application permissions. In: Proc. 2nd USENIX Conf. on Web Application Development, pp. 7–19. USENIX (2011)
7. Goldberg, A., Havelund, K., Mcgann, C.: Runtime verification for autonomous spacecraft software. In: IEEE 2005 Aerospace Conference (IEEEAC), pp. 507–516. IEEE (2005)
8. Google Inc., Android development site, `http://developer.android.com/`
9. Google Inc., `http://developer.android.com/guide/topics/security/security.html`
10. Jacob, G., Debar, H., Filiol, E.: Behavioral detection of malware: from a survey towards an established taxonomy. Journal in Computer Virology 4(3), 251–266 (2008)
11. Leyden, J.: First SMS Trojan for Android is in the wild. Web site, The Register (August 2010)
12. Ongtang, M., McLaughlin, S., Enck, W., McDaniel, P.: Semantically rich application-centric security in Android. In: Proc. Annual Comp. Sec. Applications Conference (ACSAC), pp. 340–349. IEEE (2009)

Specification in PDL with Recursion*

Xinxin Liu and Bingtian Xue

State Key Laboratory of Computer Science
Institute of Software, Chinese Academy of Sciences
and Graduate School of Chinese Academy of Sciences
P.O. Box 8718, 100190 Beijing, China
{xinxin,xuebt}@ios.ac.cn

Abstract. By extending regular Propositional Dynamic Logic (PDL) with simple recursive propositions, we obtain a language which has enough expressiveness to allow interesting applications while still enjoying a relatively simple decision procedure. More specifically, it is strictly more expressive than the regular PDL and not more expressive than the single alternation fragment of the modal μ-calculus. We present a decision procedure for satisfiability of a large class of so called simple formulas. The decision procedure has a time complexity which is polynomial in the size of the programs and exponential in the number of the sub-formulas. We show a way to solve process equations of weak bisimulation as an application.

Keywords: specification, labeled transition systems, propositional dynamic logic, modal μ-calculus, fixed point, satisfiability.

1 Introduction

Labeled transition system (LTS) is a widely accepted model for concurrent systems. Many logics use LTS as models, and the formulas of these logics can be used for describing properties of states in LTS. We often call these formulas *specifications* [1] of the concurrent systems represented in the model. Usually, we wish on one hand powerful expressiveness of the language for specification so that more properties can be expressed, and on the other hand simplicity of the language so that the expressions can be easily analyzed. These are contradictory requirements for a language, so we often have to strike a balance based on our needs.

Two well known logics for LTS are Propositional Dynamic Logic (PDL) [2] and the modal μ-calculus [3]. PDL was introduced by Fisher and Ladner [4] in the late 1970s as a formalism for reasoning about programs. Soon afterwards the logic was outdated for that purpose through the introduction of the modal μ-calculus - a much more expressive logic with a little higher complexity. However, there has been a resurgence of interest in PDL in recent years. PDL has by now become a standard logic that is far from being outdated. It can be used

---

* Research supported by ANR-NSFC 61161130530.

A. Goodloe and S. Person (Eds.): NFM 2012, LNCS 7226, pp. 181–194, 2012.

in program verification, to describe the dynamic evolution of agent-based systems, for planning or knowledge engineering, it has links to epistemic logics, it is closely related to description logics, etc. In [5] Lange studied model checking problem for PDL extended with some operators on programs such as `repeat` and `loop`. In [6], instead of introducing recursive definitions for propositions, Leivant proposed PDL with recursive procedures. The resulting logic μPDL is strictly more powerful than the modal μ-calculus. In [7] Löding, Lutz, and Serre studied satisfiability problem for certain non-regular extension of PDL and showed that the problem is still decidable.

PDL uses regular expressions for programs in which limited recursive patterns of traces can be described, while the modal μ-calculus achieves richer expressiveness through a full scale use of recursive properties. Comparing the two logics, PDL is less expressive but the formulas are easy to understand and analyze, while the modal μ-calculus is more expressive but the formulas are often hard to understand and analyze. In particular, some simple properties concerning repetition of traces often have to be encoded into complex recursive properties in μ-calculus.

In this work we introduce limited use of recursive propositions into PDL to obtain a new language, which has a good balance between expressiveness and ease of analysis. As an example, in the extended PDL, we can express a property X which is the weakest one satisfying the equation $X = \langle a^*?\varphi \rangle X$. This X describes a property of a state, which is: there is an infinite trace starting from the state such that the only action occurring in the trace is a, and φ holds infinitely often along the states of the trace. Describing such a property in the modal μ-calculus requires nesting of maximal and minimal fixed-points which leads to complication. We demonstrate that with such extension, the language is strictly more expressive than the original PDL and not more expressive than the single alternation fragment of the modal μ-calculus. And by adding nesting, we can get a language which is more expressive than CTL and CTL* [8], which will be shown in [9]. This language has good decomposition property for a large class of process contexts (this is in another paper [10] because of the limitation of the length of this paper). We identify a class of so called simple formulas which are quite expressive, demonstrate its expressiveness, and present a decision procedure for the satisfiability of such formulas. The worst case time complexity of the decision procedure is polynomial in the size of the programs and exponential in the number of the sub-formulas, which is better than first translating into μ-calculus and then checking the satisfiability of the translated μ-calculus formula [11,12]. Combining the decision procedure and the decomposition property proposed in [10], we can use this language to solve weak (branching) bisimulation equations of processes [13].

In the following section we define the syntax and semantics of the extended language. In section 3 we compare it with modal μ-calculus through translation. In section 4 we study simple formulas and their expressiveness. In section 5 we present a decision procedure for satisfiability of simple formulas, and show a way to solve process equations of weak bisimulation as an application. In the last section we conclude our work, together with some future and related work.

2   PDL with Recursion

This section presents the syntax and semantics of PDL with recursion. It starts with the standard regular PDL. The presentation here is slightly different from but equivalent to that in [2].

The language of regular PDL has expressions of two sorts: *propositions* or *formulas* $\varphi, \psi, \ldots$ and *programs* $\alpha, \beta, \ldots$. There are countably many *atomic* symbols of each sort. Atomic programs are denoted $a, b, \ldots$ which are also called *actions*, and the set of all atomic programs is denoted *Act*. Atomic propositions here are tt and ff. Programs and formulas are built inductively according to the following abstract syntax:

$$\varphi ::= \mathrm{tt} \,\big|\, \mathrm{ff} \,\big|\, \varphi \wedge \psi \,\big|\, \varphi \vee \psi \,\big|\, \langle \alpha \rangle \varphi \,\big|\, [\alpha] \varphi$$
$$\alpha ::= a \,\big|\, ?\varphi \,\big|\, \alpha \cup \beta \,\big|\, \alpha; \beta \,\big|\, \alpha^*$$

The set of all programs is denoted Π and the set of all propositions is denoted Φ. The syntax of programs is essentially regular expressions. That is why it is called regular PDL.

The semantics of regular PDL is interpreted on a labeled transition system $\langle \mathbf{S}, Act, \{ \xrightarrow{a} \mid a \in Act \} \rangle$, where $\mathbf{S}$ is a set of states, each $\xrightarrow{a}$ is a transition relation $\xrightarrow{a} \subseteq \mathbf{S} \times \mathbf{S}$. The satisfaction relation $\models \subseteq \mathbf{S} \times \Phi$ and the transition relation $\Rightarrow \subseteq (\mathbf{S} \times \Pi) \times \mathbf{S}$ for each program α are defined inductively on the structures of formulas and programs as follows:

1. $p \models \mathrm{tt}$ holds for all $p \in \mathbf{S}$;
2. $p \models \mathrm{ff}$ never holds;
3. $p \models \varphi \wedge \psi$ if and only if $p \models \varphi$ and $p \models \psi$;
4. $p \models \varphi \vee \psi$ if and only if $p \models \varphi$ or $p \models \psi$;
5. $p \models \langle \alpha \rangle \varphi$ if and only if there exists $q \in \mathbf{S}$ such that $(p, \alpha) \Rightarrow q$ and $q \models \varphi$;
6. $p \models [\alpha] \varphi$ if and only if for all $q \in \mathbf{S}$ whenever $(p, \alpha) \Rightarrow q$ then $q \models \varphi$.

7. $(p, a) \Rightarrow q$ if and only if $p \xrightarrow{a} q$;
8. $(p, ?\varphi) \Rightarrow q$ if and only if $p = q$ and $p \models \varphi$;
9. $(p, \alpha \cup \beta) \Rightarrow q$ if and only if $(p, \alpha) \Rightarrow q$ or $(p, \beta) \Rightarrow q$;
10. $(p, \alpha; \beta) \Rightarrow q$ if and only if there exists r with $(p, \alpha) \Rightarrow r$ and $(r, \beta) \Rightarrow q$;
11. $(p, \alpha^*) \Rightarrow q$ if and only if there exist $n \geq 0, q_0, \ldots, q_n$ such that $(q_i, \alpha) \Rightarrow q_{i+1}$ for $0 \leq i \leq n - 1$ and $p = q_0, q_n = q$.

We will write $p \not\models \varphi$ when $p \models \varphi$ does not hold.

Although we do not have an explicit negation operation in this presentation of syntax, for each formula φ, there is a dual formula $\overline{\varphi}$ which semantically expresses the negation of φ. Thus tt and ff are dual to each other. For the formula φ with structures, its dual $\overline{\varphi}$ is inductively defined as follows: $\overline{\varphi \wedge \psi} = \overline{\varphi} \vee \overline{\psi}$, $\overline{\varphi \vee \psi} = \overline{\varphi} \wedge \overline{\psi}$, $\overline{\langle \alpha \rangle \varphi} = [\alpha] \overline{\varphi}$, $\overline{[\alpha] \varphi} = \langle \alpha \rangle \overline{\psi}$.

Proposition 1. *Let φ be a PDL formula and $\overline{\varphi}$ be its dual. Then $\forall p \in \mathbf{S}, p \models \varphi$ if and only if $p \not\models \overline{\varphi}$.*

We now present the syntax and semantics of PDL with recursive propositions. We refer to this language as rPDL in this paper. We allow *property identifiers* in rPDL, which are denoted $X, Y, \ldots$ and the set of all property identifiers is denoted $\mathcal{V}$. The syntax of rPDL formulas and programs is as follows:

$$\varphi ::= \mathbf{tt} \,|\, \mathbf{ff} \,|\, X \,|\, \overline{X} \,|\, \varphi \wedge \psi \,|\, \varphi \vee \psi \,|\, \langle \alpha \rangle \varphi \,|\, [\alpha] \varphi$$
$$\alpha ::= a \,|\, ?\varphi \,|\, \alpha \cup \beta \,|\, \alpha; \beta \,|\, \alpha^*$$

The meaning of the property identifiers is determined by a recursive declaration D.

Definition 1. *A declaration is a finite set D with elements of the form $X = \varphi$. No variable is defined more than once in D.*

A simple test is $?X$ or $?\overline{X}$. Since in the declaration the variables are defined as formulas, we can require simple test in the program here, and the restriction of simple tests will not result in any loss of expressiveness.

Definition 2. *For programs, whether it is positive or negative is defined as:*

1. $a, ?X$ *are both positive programs*
2. $?\overline{X}$ *is a negative program*
3. $\alpha \cup \beta$ *and $\alpha; \beta$ are both positive (negative) programs if α and β are positive (negative) programs*
4. α^* *is a positive (negative) program if α is a positive (negative) program*

For formulas, whether it is positive or negative is defined as:

1. $\mathbf{tt}, \mathbf{ff}$ *are positive formulas*
2. X *is a positive formula*
3. $\overline{X}$ *is a negative formula*
4. $\varphi \wedge \psi$ *and $\varphi \vee \psi$ are both positive (negative) formulas if φ and ψ are positive (negative) programs*
5. $\langle \alpha \rangle \varphi$ *is a positive (negative) formula if α is a positive (negative) program and φ is a positive (negative) formula*
6. $[\alpha] \varphi$ *is a positive (negative) formula if α is a negative (positive) program and φ is a positive (negative) formula*

The definition above defines the positive and negative formulas and programs, and it is easy to see that there are formulas and programs which are neither positive nor negative, for example, $X \wedge \overline{Y}$ where $X, Y \in \mathcal{V}$.

A declaration D is well defined if each variable is defined by a positive formula.

The semantics of rPDL formulas and programs under a given environment $\rho : \mathcal{V} \longrightarrow 2^{\mathbf{S}}$, which maps each variable in $\mathcal{V}$ to a set of states in $\mathbf{S}$, are the same as that of PDL formulas and programs under ρ except for the property identifiers:

12. $p \models_\rho X$ if and only if $p \in \rho(X)$;
13. $p \models_\rho \overline{X}$ if and only if $p \in \mathbf{S} - \rho(X)$.

$D = \{X_1 = \varphi_1, \ldots, X_m = \varphi_m\}$ defines an environment for the identifiers $X_1, \ldots, X_m$, which are the weakest properties they satisfy.

Proposition 2. *Let ρ, ρ' be two environments such that $\rho(X) \subseteq \rho'(X)$, φ a positive formula, α a positive program, and β a negative program w.r.t all the X defined in ρ and ρ', then:*

   a. *if $(p, \alpha) \Rightarrow_\rho q$ then $(p, \alpha) \Rightarrow_{\rho'} q$;*
   b. *if $(p, \beta) \Rightarrow_{\rho'} q$ then $(p, \beta) \Rightarrow_\rho q$;*
   c. *if $p \models_\rho \varphi$ then $p \models_{\rho'} \varphi$.*

Since D is well defined and we have the monotonous property (*Proposition 2*), there exists a unique maximal environment $\rho_{\max} = \bigcup\{\rho \mid \rho$ satisfies $D\}$ and $\rho_{\max}$ satisfies D, that is $\forall X = \varphi \in D$, $\rho_{\max}(X) = \{p \in \mathbf{S} \mid p \models_{\rho_{\max}} \varphi\}$.

   Then we have the semantics of rPDL formulas and programs under D:

$$p \models_D \varphi \text{ is defined as } p \models_{\rho_{\max}} \varphi$$

$$(p, \alpha) \Rightarrow_D q \text{ is defined as } (p, \alpha) \Rightarrow_{\rho_{\max}} q$$

For an rPDL formula φ, we can take the same definition as in regular PDL to obtain its dual $\overline{\varphi}$, and in this case $\overline{\varphi}$ is still semantically the negation of φ.

Proposition 3. *Let φ be an rPDL formula and $\overline{\varphi}$ be its dual. Then $\forall p \in \mathbf{S}$, $p \models_D \varphi$ if and only if $p \not\models_D \overline{\varphi}$.*

The proof of this proposition only needs simple induction on the transition rule $\Rightarrow$ and the structure of φ. Just be careful with the test $?\psi$ and modalities $\langle \alpha \rangle$, $[\alpha]$ cases. We will not do that because of the limitation of the length.

Example 1. The following formulas specify the property that the a and b actions always appear alternatively, which is not expressible in PDL.
$X = \langle (\tau.?X)^*.a \rangle Y$
$Y = \langle (\tau.?Y)^*.b \rangle X$

$$\cdot \xrightarrow{\tau} \cdots \xrightarrow{\tau} \cdot \xrightarrow{a} \cdot \xrightarrow{\tau} \cdots \xrightarrow{\tau} \cdot \xrightarrow{b} \cdot \xrightarrow{\tau} \cdots \xrightarrow{\tau} \cdot \xrightarrow{a} \cdot \xrightarrow{\tau} \cdots \xrightarrow{\tau} \cdot \xrightarrow{b} \cdots$$
$$X \quad \cdots \quad X \quad Y \quad \cdots \quad Y \quad X \quad \cdots \quad X \quad Y \quad \cdots \quad Y \qquad \qquad \square$$

3   Expressiveness

It is obviously that rPDL is more expressive than the regular PDL. In this section we will show that any proposition expressible in rPDL can be expressed in the modal μ-calculus.

   To show that any property expressible in rPDL can also be expressed in the modal μ-calculus, we present a translation from rPDL formulas to the μ-calculus formulas. Here we take the version of the μ-calculus in [14] which allows simultaneous mutual recursive definitions. It is clear that this version has the same expressive power as the usual μ-calculus introduced by Kozen [3].

The syntax of this version of μ-calculus is as follows:

$$F ::= \text{tt}\,|\,\text{ff}\,|\,X\,|\,\overline{X}\,|\,F \wedge G\,|\,F \vee G\,|\,\langle a \rangle F\,|\,[a]F$$
$$|\,\text{letmax } D \text{ in } F\,|\,\text{letmin } D \text{ in } F$$
$$D ::= X_1 = F_1, \ldots, X_n = F_n$$

To ensure that the semantics can be well defined, we assume that each F_i in D is positive, i.e. does not have sub-formulas of the form $\overline{X}$.

The semantics is given by two semantic functions $\mathsf{F}[\![\_]\!]\rho$ (from formulas to sets of states) and $\mathsf{D}[\![\_]\!]\rho$ (from declarations to a function on environments) with ρ being a given environment which assigns a subset of $\mathbf{S}$ to each variable X:

$$
\begin{aligned}
\mathsf{F}[\![\text{tt}]\!]\rho &= \mathbf{S} \\
\mathsf{F}[\![\text{ff}]\!]\rho &= \emptyset \\
\mathsf{F}[\![X]\!]\rho &= \rho(X) \\
\mathsf{F}[\![\overline{X}]\!]\rho &= \mathbf{S} - \rho(X) \\
\mathsf{F}[\![F \wedge G]\!]\rho &= \mathsf{F}[\![F]\!]\rho \cap \mathsf{F}[\![G]\!]\rho \\
\mathsf{F}[\![F \vee G]\!]\rho &= \mathsf{F}[\![F]\!]\rho \cup \mathsf{F}[\![G]\!]\rho \\
\mathsf{F}[\![\langle a \rangle F]\!]\rho &= \{p \in \mathbf{S} \mid p \xrightarrow{a} q, q \in \mathsf{F}[\![F]\!]\rho \text{ for some } q \in \mathbf{S}\} \\
\mathsf{F}[\![[a]F]\!]\rho &= \{p \in \mathbf{S} \mid \text{if } p \xrightarrow{a} q \text{ then } q \in \mathsf{F}[\![F]\!]\rho \text{ for all } q \in \mathbf{S}\} \\
\mathsf{F}[\![\text{letmax } D \text{ in } F]\!]\rho &= \mathsf{F}[\![F]\!](\nu\sigma.(\mathsf{D}[\![D]\!]\rho)\sigma) \\
\mathsf{F}[\![\text{letmin } D \text{ in } F]\!]\rho &= \mathsf{F}[\![F]\!](\mu\sigma.(\mathsf{D}[\![D]\!]\rho)\sigma) \\
\mathsf{D}[\![X_1 = F_1, \ldots, X_n = F_n]\!]\rho &= \lambda\sigma.\rho\{\mathsf{F}[\![F_1]\!]\sigma/X_1, \ldots, \mathsf{F}[\![F_n]\!]\sigma/X_n\}
\end{aligned}
$$

Let φ be an rPDL formula. We now define a translation function $\mathcal{T}$ which maps φ to $\mathcal{T}(\varphi)$ as follows:

$$
\begin{aligned}
\mathcal{T}(\varphi) &= \varphi \text{ when } \varphi \text{ is tt}, \text{ff}, X, \overline{X} \\
\mathcal{T}(\varphi \wedge \psi) &= \mathcal{T}(\varphi) \wedge \mathcal{T}(\psi) \\
\mathcal{T}(\varphi \vee \psi) &= \mathcal{T}(\varphi) \vee \mathcal{T}(\psi) \\
\mathcal{T}(\langle a \rangle \varphi) &= \langle a \rangle \mathcal{T}(\varphi) \\
\mathcal{T}(\langle ?\psi \rangle \varphi) &= \mathcal{T}(\psi) \wedge \mathcal{T}(\varphi) \\
\mathcal{T}(\langle \alpha \cup \beta \rangle \varphi) &= \mathcal{T}(\langle \alpha \rangle \varphi) \vee \mathcal{T}(\langle \beta \rangle \varphi) \\
\mathcal{T}(\langle \alpha; \beta \rangle \varphi) &= \mathcal{T}(\langle \alpha \rangle \langle \beta \rangle \varphi) \\
\mathcal{T}(\langle \alpha^* \rangle \varphi) &= \text{letmin } Y = \mathcal{T}(\varphi) \vee \mathcal{T}(\langle \alpha \rangle Y) \text{ in } Y \\
\mathcal{T}([a]\varphi) &= [a]\mathcal{T}(\varphi) \\
\mathcal{T}([?\psi]\varphi) &= \mathcal{T}(\overline{\psi}) \vee \mathcal{T}(\varphi) \\
\mathcal{T}([\alpha \cup \beta]\varphi) &= \mathcal{T}([\alpha]\varphi) \wedge \mathcal{T}([\beta]\varphi) \\
\mathcal{T}([\alpha; \beta]\varphi) &= \mathcal{T}([\alpha][\beta]\varphi) \\
\mathcal{T}([\alpha^*]\varphi) &= \text{letmax } Y = \mathcal{T}(\varphi) \wedge \mathcal{T}([\alpha]Y) \text{ in } Y
\end{aligned}
$$

Lemma 1. *Let φ be an rPDL formula. ρ and ρ' are environments, which satisfy: $p \in \rho(X)$ if and only if $p \in \rho'(X)$, $p \in \mathbf{S}$. Then $p \models_\rho \varphi$ if and only if $p \in F[\![\mathcal{T}(\varphi)]\!]\rho'$.*

Proof. Induction on the structure of φ. And all the cases are quite simple except for the modalities $\langle\alpha\rangle\psi$ and $[\alpha]\psi$ cases. For those, induction on the structure of α. Since the structure of the test cases in α is smaller than the outer formula, we can use the outer inductive hypothesis in the inner proof.

We will not discuss the detail of this proof here because of the limitation of the length.

Theorem 1. *Let φ be an rPDL formula defined with $D = \{X_1 = \varphi_1, \ldots, X_n = \varphi_n\}$, and $p \in \mathbf{S}$. Then $p \models_D \varphi$ if and only if $p \in F[\![letmax\ X_1 = \mathcal{T}(\varphi_1), \ldots, X_n = \mathcal{T}(\varphi_n)$ in $\mathcal{T}(\varphi)]\!]\rho_0$, where ρ_0 is an empty environment.*

Proof. $p \models_D \varphi$ if and only if $p \models_{\rho_{max}} \varphi$;

$F[\![letmax\ D_\mu$ in $\mathcal{T}(\varphi)]\!]\rho_0 = F[\![\mathcal{T}(\varphi)]\!]\rho_\mu$, where $D_\mu = \{X_1 = \mathcal{T}(\varphi_1), \ldots, X_n = \mathcal{T}(\varphi_n)\}$ and $\rho_\mu = \nu\sigma.\rho_0\{F[\![\mathcal{T}(\varphi_1)]\!]\sigma/X_1, \ldots, F[\![\mathcal{T}(\varphi_n)]\!]\sigma/X_n\}$;

Let ρ satisfies: $p \in \rho_\mu(X)$ if and only if $p \in \rho(X)$.

Obviously we have $\rho = \rho_{max}$.

Then by *Lemma 1*, we have $p \models_D F$ if and only if $p \in F[\![letmax\ X_1 = \mathcal{T}(F_1), \ldots, X_n = \mathcal{T}(F_n)$ in $\mathcal{T}(F)]\!]\rho_0$. □

For an rPDL formula φ, it is not hard to see that $\mathcal{T}(\varphi)$ has no free variables except for those defined in D, so $\mathcal{T}(\varphi)$ is free of alternation of fixed-points. However the variables defined in D may occur within some minimal fixed-point definitions in $\mathcal{T}(\varphi)$, the overall result letmax $X_1 = \mathcal{T}(\varphi_1), \ldots, X_n = \mathcal{T}(\varphi_n)$ in $\mathcal{T}(\varphi)$ has one alternation of fixed-points.

Example 2. Fairness property:

$$\cdot \longrightarrow \cdots \cdots \longrightarrow \cdot \longrightarrow \cdots \cdots \longrightarrow \cdot \longrightarrow \cdots \cdots \cdots$$
$$X \qquad\qquad p, X \qquad\qquad p, X$$

CTL*: $EGFp$

μ-calculus: $\nu X.\mu Y.\langle\bullet\rangle((X \wedge p) \vee Y)$

rPDL: $X = \langle\bullet^*.?p\rangle X$ □

By adding nesting to rPDL, we can get a language which is more expressive than CTL and CTL*, and is still no more expressive than modal μ-calculus. This is beyond this paper, and we will present that elsewhere.

4   Simple Formula

An rPDL formula is said to be *simple* if every box modality $[\alpha]$ in it has the simple form $[a]$ for some $a \in Act$. A declaration D is said to be simple if, whenever $X = \varphi \in D$ then φ is simple.

Simple formulas and simple declarations are quite expressive. Here we demonstrate that for a finite state process p, its weak bisimulation [15] equivalence classes are expressible by simple formulas and simple declarations in rPDL.

For that purpose, we associate a proposition identifier X_p for each state p in the (finite) state space. Then we construct a declaration

$$D = \{X_p = \bigwedge_{p \xrightarrow{a} p'} \langle \tau^*.a.\tau^* \rangle X_{p'} \wedge \bigwedge_{a \in Act} [a](\bigvee_{p \xrightarrow{\hat{a}} p'} X_{p'}) \mid p \in \mathbf{S}\}$$

It is easy to see that D is a simple declaration.

Proposition 4. *Let p be a finite state process, X_p and D constructed as above. Then for any process q, $p \approx q$ if and only if $q \models_D X_p$.*

Proof $\Rightarrow$: $p \approx q$ then $q \models_D X_p$

First we construct an environment $\rho_{\approx}$: $q \in \rho_{\approx}(X_p)$ if and only if $p \approx q$;

We know: $q \models_D X_p$ if and only if $p \models_{\rho_{max}} X_p$. That is $\exists \rho$ s.t ρ satisfies D and $q \in \rho(X_p)$;

So if we can prove that $\rho_{\approx}$ satisfies D, then we have $p \in \rho_{\approx}(X_p) \Longrightarrow p \models_D X_p$, and apparently $p \approx q \Longrightarrow q \models_D X_p$ is proved;

That is to prove that: $\forall X_p \in V$, $q \in \rho_{\approx}(X_p)$ then $q \models_{\rho_{\approx}} \bigwedge_{p \xrightarrow{a} p'} \langle \tau^*.a.\tau^* \rangle X_{p'} \wedge \bigwedge_a [a](\bigvee_{p \xrightarrow{\hat{a}} p'} X_{p'})$, which is easy to calculate by the definition of $\approx$.

$\Leftarrow$: $q \models_D X_p$ then $p \approx q$

We construct a binary relation $\mathcal{B}$ first: $\mathcal{B} = \{(p, q) \mid q \models_D X_p\}$;

We show that $\mathcal{B}$ is a weak bisimulation;

For $(p, q) \in \mathcal{B}$, $q \models_D X_p \Longrightarrow q \models_D \varphi_{X_p} \Longrightarrow q \models_D \bigwedge_{p \xrightarrow{a} p'} \langle \tau^*.a.\tau^* \rangle X_{p'} \wedge \bigwedge_a [a](\bigvee_{p \xrightarrow{\hat{a}} p'} X_{p'})$;

- Assume $p \xrightarrow{a} p'$, then $\langle \tau^*.a.\tau^* \rangle X_{p'}$ is a conjunct of φ_{X_p}

  $\Longrightarrow \exists q'$, s.t. $(q, \tau^*.a.\tau^*) \Rightarrow q'$ and $q' \models_D X_{p'}$

  $\Longrightarrow q \xrightarrow{\hat{a}} q'$ and $q' \models_D X_{p'}$

  $\Longrightarrow (p', q') \in \mathcal{B}$

  $\Longrightarrow \exists q'$ s.t. $q \xrightarrow{\hat{a}} q'$ and $(p', q') \in \mathcal{B}$

- Assume $q \xrightarrow{a} q'$, then $[a](\bigvee_{p \xrightarrow{\hat{a}} p'} X_{p'})$ is a conjunct of F_{X_p}

  $\Longrightarrow q' \models \bigvee_{p \xrightarrow{\hat{a}} p'} X_{p'}$

  $\Longrightarrow \exists p'$ s.t. $p \xrightarrow{\hat{a}} p'$ and $q' \models_D X_{p'}$

  $\Longrightarrow (p', q') \in \mathcal{B}$

  $\Longrightarrow \exists q'$ s.t. $q \xrightarrow{\hat{a}} q'$ and $(p', q') \in \mathcal{B}$ $\qquad \square$

In the same way, we can show that branching bisimulation equivalence classes are expressible by simple declarations in rPDL.

Next we will show that in rPDL every positive formula has an equivalent simple formula.

First, we define a translation function $\mathcal{S}$ as follows, which translates a positive formula φ into a simple one. Since φ is a positive formula, the test can only be $?X$ in the diamond modalities and $?\overline{X}$ in the box modalities, and moreover there is no occurrence of $\overline{X}$ in φ besides in the box modalities,.

$$
\begin{aligned}
\mathcal{S}(\varphi) &= \varphi \qquad \text{when } \varphi \text{ is tt, ff, } X \\
\mathcal{S}(\varphi \wedge \psi) &= \mathcal{S}(\varphi) \wedge \mathcal{S}(\psi) \\
\mathcal{S}(\varphi \vee \psi) &= \mathcal{S}(\varphi) \vee \mathcal{S}(\psi) \\
\mathcal{S}(\langle a \rangle \varphi) &= \langle a \rangle \mathcal{S}(\varphi) \\
\mathcal{S}([a]\varphi) &= [a]\mathcal{S}(\varphi) \\
\mathcal{S}([?\overline{X}]\varphi) &= \mathcal{S}(X) \vee \mathcal{S}(\varphi) \\
\mathcal{S}([\alpha \cup \beta]\varphi) &= \mathcal{S}([\alpha]\varphi) \wedge \mathcal{S}([\beta]\varphi) \\
\mathcal{S}([\alpha;\beta]\varphi) &= \mathcal{S}([\alpha][\beta]\varphi) \\
\mathcal{S}([\alpha^*]\varphi) &= X_{[\alpha^*]\varphi}
\end{aligned}
$$

where $X_{[\alpha^*]\varphi}$ is an induced identifier.

Then for a given declaration D, let D_s be the smallest set such that:

1. whenever $X = \varphi \in D$ then $X = \mathcal{S}(\varphi) \in D_s$;
2. whenever $X_{[\alpha^*]\varphi}$ occurs in a right hand side of a definition in D_s, that is $[\alpha^*]\varphi$ occurs in a right hand side of a definition in D, then $X_{[\alpha^*]\varphi} = \mathcal{S}(\varphi) \wedge \mathcal{S}([\alpha]X_{[\alpha^*]\varphi}) \in D_s$.

Theorem 2. *Let D be a declaration, D_s be a simple declaration constructed as above, φ be a positive rPDL formula, and $p \in \mathbf{S}$. Then $p \models_D \varphi$ if and only if $p \models_{D_s} \mathcal{S}(\varphi)$.*

Proof $\Rightarrow$: $p \models_D \varphi$ then $p \models_{D_s} \mathcal{S}(\varphi)$

$$
\text{Define } \rho(X) = \begin{cases} \{p \mid p \in \mathbf{S}, p \models_D X\} & X = \varphi \in D \\ \{p \mid p \in \mathbf{S}, p \models_D [\alpha^*]\varphi\} & X \text{ is } X_{[\alpha^*]\varphi} \end{cases}
$$

If we can prove that $p \models \varphi$ then $p \models_\rho \mathcal{S}(\varphi)$ holds, we can easily prove that ρ satisfies D_s as follows:

- $X = \varphi \in D$, then $X = \mathcal{S}(\varphi) \in D_s$
  $p \in \rho(X) \Longrightarrow p \models_D X \Longrightarrow p \models_D \varphi \Longrightarrow p \models_\rho \mathcal{S}(\varphi)$
- X is $X_{[\alpha^*]\varphi}$, then $X_{[\alpha^*]\varphi} = \mathcal{S}(\varphi) \wedge \mathcal{S}([\alpha]X_{[\alpha^*]\varphi}) \in D_s$
  $p \in \rho(X_{[\alpha^*]\varphi}) \Longrightarrow p \models_D [\alpha^*]\varphi \Longrightarrow p \models_D \varphi \wedge [\alpha][\alpha^*]\varphi \Longrightarrow p \models_\rho \mathcal{S}(\varphi \wedge [\alpha][\alpha^*]\varphi) \Longrightarrow p \models_\rho \mathcal{S}(\varphi) \wedge \mathcal{S}([\alpha]X_{[\alpha^*]\varphi})$

So we have $p \models_\rho \mathcal{S}(\varphi)$ then $p \models_D \mathcal{S}(\varphi)$;

Then $p \models_D \varphi$ then $p \models_{D_s} \mathcal{S}(\varphi)$ is proved;

Then it is easy to prove $p \models_D \varphi$ then $p \models_\rho \mathcal{S}(\varphi)$ by induction on the structure of φ. And for the modalities cases we do the proof by induction on the structure of α.

$\Leftarrow$: $p \models_{D_s} \mathcal{S}(\varphi)$ then $p \models_D \varphi$

Define $\rho(X) = \{p \mid p \in \mathbf{S}, p \models_{D_s} X, X = \varphi \in D\}$

If we can prove that $p \models_{D_s} \mathcal{S}(\varphi)$ then $p \models_\rho \varphi$ holds, we can easily prove that ρ satisfies D as follows:

- $X = \varphi \in D$

$p \in \rho(X) \Longrightarrow p \models_{D_s} X \Longrightarrow p \models_{D_s} \mathcal{S}(\varphi) \Longrightarrow p \models_\rho \varphi$

So we have $p \models_\rho \varphi$ then $p \models_D \varphi$;

Then $p \models_{D_s} \mathcal{S}(\varphi)$ then $p \models_D \varphi$.

Then it is easy to prove $p \models_{D_s} \mathcal{S}(\varphi)$ then $p \models_\rho \varphi$ by induction on the structure of the rPDL formula φ. And for the modalities cases we do the proof by induction on the structure of α.

We will not discuss the detail of this proof here because of the limitation of the length.   □

According to *Theorem 2*, every well defined declaration has an equivalent simple declaration.

5   Deciding Satisfiability of Simple Formula

In this section we present a decision procedure for satisfiability of simple formulas. And since every positive formula has an equivalent simple formula, positive formulas' satisfiability can be decided by translating into simple formula. The key is a syntactic characterization of satisfiability through the notion of *consistency sets* defined as follows.

Definition 3. *A set of formulas Γ is saturated with respect to a declaration D if it satisfies the following:*

1. *whenever $\varphi \wedge \psi \in \Gamma$ then $\varphi \in \Gamma$ and $\psi \in \Gamma$;*
2. *whenever $\varphi \vee \psi \in \Gamma$ then $\varphi \in \Gamma$ or $\psi \in \Gamma$;*
3. *whenever $X \in \Gamma$ and $X = \varphi \in D$ then $\varphi \in \Gamma$.*

Definition 4. *Let $\mathcal{C} \subseteq 2^\Phi$ (i.e. $\mathcal{C}$ is a set of formula sets). Construct a labeled transition system $\langle \mathcal{C}, Act, \{\xrightarrow{a} \mid a \in Act\}\rangle$ with an environment $\rho^{\mathcal{C}}$ defined as $\rho^{\mathcal{C}}(X) = \{\Gamma \in \mathcal{C} \mid X \in \Gamma\}$. $\xrightarrow{a}$ is defined as: $\forall \Gamma, \Gamma' \in \mathcal{C}, \Gamma \xrightarrow{a} \Gamma'$ if whenever $[a]\psi \in \Gamma$ then $\psi \in \Gamma'$. $\mathcal{C}$ is called a consistency set, if for every $\Gamma \in \mathcal{C}$, the following holds:*

1. *Γ is saturated;*
2. *whenever $\varphi \in \Gamma$ then $\Gamma \models_{\rho^{\mathcal{C}}} \varphi$.*

For a simple formula φ, let $\mathsf{sub}(\varphi)$ be the set of sub-formulas of φ, defined by:

$$\begin{aligned}
\mathsf{sub}(\varphi) &= \{\varphi\} \qquad \text{where } \varphi = \mathrm{tt}, \mathrm{ff}, X, \overline{X} \\
\mathsf{sub}(\varphi \wedge \psi) &= \{\psi \wedge \psi\} \cup \mathsf{sub}(\varphi) \cup \mathsf{sub}(\psi) \\
\mathsf{sub}(\varphi \vee \psi) &= \{\varphi \vee \psi\} \cup \mathsf{sub}(\varphi) \cup \mathsf{sub}(\psi) \\
\mathsf{sub}(\langle \alpha \rangle \varphi) &= \{\langle \alpha \rangle \varphi\} \cup \mathsf{sub}(\varphi) \\
\mathsf{sub}([a]\varphi) &= \{[a]\varphi\} \cup \mathsf{sub}(\varphi)
\end{aligned}$$

For a simple declaration D, we write $\mathsf{sub}(D) = \bigcup\{\mathsf{sub}(\varphi) \mid X = \varphi \in D\}$.

Note that the number of sub-formulas in $\mathsf{sub}(\varphi)$ has nothing to do with the size of programs in φ: no matter how big or small a program α is, one occurrence of $\langle \alpha \rangle \psi$ in φ only generates one sub-formula. Therefore, the cardinality of $\mathsf{sub}(\varphi)$ is much smaller than that of $\mathsf{FL}(\varphi)$ (the usual Fischer-Ladner Closure of φ).

Theorem 3. *Let φ be a simple rPDL formula, D be a simple declaration. Then the following two conditions are equivalent:*

1. *there exists a consistency set $\mathcal{C}$ and some $\Gamma \in \mathcal{C}$ such that $\varphi \in \Gamma$;*
2. *there exists an LTS $\langle \mathbf{S}, Act, \{ \overset{a}{\longrightarrow} \mid a \in Act \} \rangle$ such that $p \models_D \varphi$ for some $p \in \mathbf{S}$.*

Proof $\Rightarrow$ To prove 1 $\implies$ 2, let $\mathcal{C}$ be a consistency set.
We construct an LTS $\langle \mathbf{S}, Act, \{ \overset{a}{\longrightarrow} \mid a \in Act \} \rangle$, where: $\mathbf{S} = \mathcal{C}$, and the transition relation $\overset{a}{\longrightarrow}$ is defined such that: $\Gamma \overset{a}{\longrightarrow} \Gamma'$ if and only if, whenever $[a]\psi \in \Gamma$ then $\psi \in \Gamma'$. Let $\rho(X) = \{ \Gamma \mid X \in \Gamma \}$ for $X \in \mathcal{V}$.
With this construction we can prove the following implication by a simple induction on the structure of φ: if $\Gamma \models_{\mathcal{C}} \varphi$ then $\Gamma \models_{\rho} \varphi$.
Then suppose $X = \psi \in D$, the following sequence of implications shows that ρ satisfies D: $\Gamma \in \rho(X) \implies X \in \Gamma$(by construction) $\implies \psi \in \Gamma$ (Γ is saturated) $\implies \Gamma \models_{\mathcal{C}} \psi$ ($\mathcal{C}$ is a consistency set) $\implies \Gamma \models_{\rho} \psi$
Thus $\rho \subseteq \rho_{\max}$.
Then for $\varphi \in \Gamma \in \mathcal{C}$, $\Gamma \models_{\mathcal{C}} \varphi$ ($\mathcal{C}$ is a consistency set) $\implies \Gamma \models_{\rho} \varphi \implies \Gamma \models_{\rho_{\max}} \varphi$ (Proposition 2) $\implies \Gamma \models_D \varphi$.
$\Leftarrow$ To prove 2 $\implies$ 1, construct $\mathcal{C}_0 = \{ \Gamma \subseteq \mathsf{sub}(\varphi) \cup \mathsf{sub}(D) \mid \exists p \in \mathbf{S}, p \models_D \psi$ for all $\psi \in \Gamma \}$.
Now let $\mathcal{C}$ consist of all the maximal elements of $\mathcal{C}_0$, then it is routine to verify that $\mathcal{C}$ is a consistency set. □

The above proof suggests the following simple iterative procedure to decide whether a simple formula φ with a simple declaration D is satisfiable.

Algorithm 4 . *For a given simple formula φ with a simple declaration D, start from $\mathcal{C} = \{ \Gamma \subseteq \mathsf{sub}(\varphi) \cup \mathsf{sub}(D) \mid \}$ and do the following steps.*

1. *For each $\Gamma \in \mathcal{C}$, check whether is saturated, all of which can be checked locally. If not, delete Γ from $\mathcal{C}$.*
2. *Repeat the following until $\mathcal{C}$ does not decrease:*
   If there exists $\Gamma \in \mathcal{C}$, $\exists \psi \in \Gamma$ such that $\Gamma \models_{\mathcal{C}} \psi$ does not hold, delete Γ from $\mathcal{C}$.

The algorithm must terminate, since there are only finitely many states initially, and at least one state must be deleted in each iteration of step 3 in order to continue. Then φ is satisfiable if and only if, upon termination there exists $\Gamma \in \mathcal{C}$ such that $\varphi \in \Gamma$. Obviously $\mathcal{C}$ is a consistency set upon termination. The correctness of this algorithm follows from *Theorem 3.* The 1 $\implies$ 2 direction of the proof guarantees that all formulas in $\mathcal{C}$ are satisfiable. The 2 $\implies$ 1 direction of the proof guarantees that all satisfiable Γ will not be deleted from $\mathcal{C}$.

We now estimate the worst case time complexity of *Algorithm 4*. For a formula, we can simply define its *size* by adding up the number of occurrences of variables, $\vee$'s, $\wedge$'s, and the modalities $\langle\alpha\rangle$'s and $[a]$'s in the formula (in particular note that an occurrence of $\langle\alpha\rangle$ always counts as 1 no matter what complicated structure α has). We can also define the size of programs in a similar way. Now the maximum size of $\mathcal{C}$ is clearly exponential in the number of sub-formulas of D and φ, i.e. the cardinality of $\mathsf{sub}(\varphi)\cup\mathsf{sub}(D)$ which is just the size of φ plus the sum of the size of all right hand side formulas in D. That puts an upper bound on the number of iterations. The size of programs only matters when checking $\Gamma \models_\mathcal{C} \psi$ within an iteration. Then it is not difficult to see that the size of programs only contributes to a polynomial factor. Thus the time complexity of this procedure is exponential in the size of the formulas, but polynomial in the size of the programs. This is better than first translating the formula into a modal μ-calculus formula and then deciding if the translated formula is satisfiable (for example using the decision procedure in [16]), because the size of the translated formula, to which the size of the programs contributes a linear factor, contributes to the exponential blow up of the time complexity.

The construction used here has similarities with the standard Fischer-Ladner construction often used in temporal logics, although as pointed out earlier, we actually used $\mathsf{sub}(\varphi)$ which is much simpler than $\mathsf{FL}(\varphi)$. The way we translate a positive formula into a simple one is quite similar as the way of finding the Fischer-Ladner closure of $[\alpha]\varphi$. So for a positive formula, the time complexity of checking whether it is satisfiable is exponential in the size of the formulas and the programs in the box modalities, but polynomial in the size of the programs in the diamond modalities. The satisfiability of regular PDL can also be decided by our algorithm, and the time complexity is exponential in the size of the formulas and the programs in the box modalities, but polynomial in the size of the programs in the diamond modalities, which is better than the usual algorithm in [2], whose time complexity is exponential in the size of the formulas and programs.

Note that the results of this section only holds for simple formulas. For people familiar with temporal logics, the CTL formula AFp can not be a counter-example here although it fails the Fischer-Ladner technique, because it is not a simple formula when expressed in rPDL.

Now as an application of rPDL, we have a way to solve process equations of the form $C(x) \approx p$ (finding out whether there is a process x which satisfies the equation). Here is how to do it. According to the result of Section 4, there is a simple formula φ_p and a simple declaration D such that $C(x) \approx p$ if and only if $C(x) \models_D \varphi_p$. By the decomposition property of rPDL, which will be shown in [10], there is a simple formula $\mathcal{W}(C,\varphi_p)$ such that $C(x) \models_D \varphi_p$ if and only if $x \models_D \mathcal{W}(C,\varphi_p)$. Thus there is an x with $C(x) \approx p$ if and only if $\mathcal{W}_f(C,\varphi_p)$ is satisfiable under D, and then we can use the decision procedure to check that. In the same way, we can slove the branching bisimulation equations of processes. In [17,18,19], *Disjunctive Modal Transition Systems* (DMTS) were used to solve strong bisimulation equations of the form $C(x) \sim p$. Here by using rPDL, we can do what we could not do with DMTS.

6   Conclusion and Related Works

In this work we propose a specification language for LTS by introducing simple recursive propositions into the regular PDL. This language is strictly more expressive than the regular PDL and not more expressive than the single alternation fragment of the modal μ-calculus. We present a decision procedure for satisfiability of the simple formulas, which are quite expressive. The decision procedure has a time complexity which is polynomial in the size of the programs in the formulas and exponential in the number of the sub-formulas. Many of these are desirable properties as a specification language. We also show an application of rPDL by using it to solve weak (branching) bisimulation equations.

Decomposition [20] is always an interesting issue which relates a specification language to states (systems) with a structure. Informally and in short, a decomposition problem is concerned with reducing a specification required of a combined system into (sufficient and necessary) specifications of the system's components. More precisely, for any property φ expressed in the language and some kind of program context C, the problem of decomposition asks if the property $\mathcal{W}(C, \varphi)$ is always definable as a formula in the language. $\mathcal{W}(C, \varphi)$ is such a property that a state p satisfies it if and only if $C(p)$ satisfies φ. In [10] we show that rPDL enjoys such a good decomposition property for a very large class of process context.

By adding nesting to rPDL, we get a language which is more expressive than CTL and CTL*, and is still no more expressive than modal μ-calculus. We demonstrate that any property expressible in CTL or CTL* can also be expressed in rPDL with nesting. This is beyond this paper, and we will present that in [9], which is in preparation.

Since every positive formula can be transformed into a simple formula, the problem of satisfiability of positive formula is reduced to that of simple formula. According to this, tools for deciding satisfiability of rPDL simple formulas can be built based on the notion of consistency set. The feasibility of such tools is due to the fact that this notion is defined in a way which is effectively computable. With the addition of a proof system, a desirable tool should be able to output a direct proof if a specification is not satisfiable. The work is in progress. Moreover, by combining the tools for deciding satisfiability of rPDL specifications and the tools for decomposing rPDL specifications [10], equation solver (EQ) can be built. EQ can either find a solution or provide a proof to explain why the equation system is not solvable.

However, many interesting problems cannot be reduced to satisfiability of simple formulas. The satisfiability of full rPDL and rPDL with nesting is under consideration. A decision procedure for satisfiability of full rPDL formulas, which runs in deterministic single-exponential time, is promising to be proved to be correct soon. Moreover we also want to provide a tool for deciding satisfiability of full rPDL formulas.

We have not discussed the problem of model checking in rPDL. Although we can always do it through translating into the modal μ-calculus, direct model checking in rPDL may explore features of the language and may lead to improved efficiency. It can be an interesting future research direction.

References

1. Larsen, K.: Modal Specifications. In: Sifakis, J. (ed.) CAV 1989. LNCS, vol. 407, pp. 232–246. Springer, Heidelberg (1990)
2. Harel, D., Kozen, D., Tiuryn, J.: Dynamic Logic. MIT Press (2000)
3. Kozen, D.: Results on the propositional mu–calculus. In: Nielsen, M., Schmidt, E.M. (eds.) ICALP 1982. LNCS, vol. 140, pp. 348–359. Springer, Heidelberg (1982)
4. Fischer, M.J., Ladner, R.E.: Propositional dynamic logic of regular programs. J. Comput. System Sci. 18(2) (1979)
5. Lange, M.: Model checking propositional dynamic logic with all extras. Journal of Applied Logic 4 (2006)
6. Leivant, D.: Propositional Dynamic Logic for Recursive Procedures. In: Shankar, N., Woodcock, J. (eds.) VSTTE 2008. LNCS, vol. 5295, pp. 6–14. Springer, Heidelberg (2008)
7. Löding, C., Lutz, C., Serre, O.: Propositional dynamic logic with recursive programs. Journal Logic and Algebraic Programming 73 (2007)
8. Clarke, E., Grumberg Jr., O., Peled, D.: Model checking. MIT Press (1999)
9. Liu, X., Xue, B.: Recursive pdl with nesting (in preparing)
10. Liu, X., Xue, B.: Decomposition of pdl and its extension. Submitted to International Conference on Computer Science, Hongkong (2012)
11. Streett, R.S., Emerson, E.A.: An automata theoretic decision procedure for the propositional mu-calculus. Information and Computation 81, 249–264 (1989)
12. Kozen, D., Parikh, R.: A decision procedure for the propositional mu–calculus. In: Clarke, E., Kozen, D. (eds.) Logic of Programs 1983. LNCS, vol. 164, pp. 313–325. Springer, Heidelberg (1984)
13. Jonsson, B., Larsen, K.: On the Complexity of Equation Solving in Process Algebra. In: Abramsky, S. (ed.) TAPSOFT 1991. LNCS, vol. 493, pp. 381–396. Springer, Heidelberg (1991)
14. Larsen, K., Liu, X.: Compositionality Through an Operational Semantics of Contexts. In: Paterson, M. (ed.) ICALP 1990. LNCS, vol. 443, pp. 526–539. Springer, Heidelberg (1990)
15. Milner, R.: Communication and Concurrency. Prentice–Hall (1989)
16. Walukiewicz, I.: Notes on the propositional μ-calculus: completeness and related results, brics nots series, Tech. Rep. NS-95-1 (1995)
17. Larsen, K., Liu, X.: Equation solving using modal transition systems. In: Proceedings on Logic in Computer Science (1990)
18. Liu, X.: Specification and decomposition in concurrency, Ph.D. dissertation, University of Aalborg, Fredrik Bajers Vej 7, DK 9220 Aalborg ø, Denmark (1992)
19. Larsen, K., Xinxin, L.: On equation solving. In: Larsen, K., Skou, A. (eds.) 2nd NOrdic Workshop on Program Correctness (1990)
20. Larsen, K.G.: Compositional Theories based on an Operational Semantics of Contexts. In: de Bakker, J.W., de Roever, W.-P., Rozenberg, G. (eds.) REX 1989. LNCS, vol. 430, pp. 487–518. Springer, Heidelberg (1990)

Automatically Proving Thousands of Verification Conditions Using an SMT Solver: An Empirical Study

Aditi Tagore, Diego Zaccai, and Bruce W. Weide

Dept. of Computer Science and Engineering
The Ohio State University
Columbus, Ohio 43210, USA
{tagore.2,zaccai.1,weide.1}@osu.edu

Abstract. Recently it has become possible to verify full functional correctness of certain kinds of software using automated theorem-proving technology. Empirical studies of the difficulty of automatically proving diverse verification conditions (VCs) would be helpful. For example, they could help direct those developing formal specifications toward techniques that tend to simplify VCs. They could also help focus the efforts of those improving automated theorem-proving tools that are targeted to handle VCs. This study explores two specific empirical questions of this sort: How does an SMT solver perform on VCs that involve user-defined mathematical functions and predicates? When it does not perform well, what can be done to improve the prospects for automated proof? Experience using Z3 to prove VCs for a solution to a fully generic sorting benchmark, along with thousands of other VCs generated for both clients and implementations of dozens of RESOLVE software components, suggests that providing the prover with universal algebraic lemmas about user-defined mathematical functions and predicates results in better outcomes than expanding (unfolding) definitions. The importance of such lemmas might not be surprising to those who have tried to carry out such proofs manually or with the help of an interactive prover, but the damage sometimes caused by expanding definitions might be unexpected. A large empirical study of these phenomena in the context of automated software verification has not been previously reported.

1 Introduction

We took the following steps for this study. First, we selected a variety of about 50 RESOLVE [1] software components comprising about 2000 lines of code, including the components involved in an earlier empirical study of different issues [2]. This code includes arithmetic algorithms over integers and natural numbers; sorting of arbitrary items with arbitrary orders; and a variety of client-view manipulations of, and internal data representations for, stacks, queues, lists, sets, etc. These components have specifications based on standard design-by-contract principles with pre- and post-conditions, and the code to be verified includes the

A. Goodloe and S. Person (Eds.): NFM 2012, LNCS 7226, pp. 195–209, 2012.

necessary annotations: loop invariants, progress metrics, representation invariants, and abstraction functions, as appropriate, but very few other assertions. Second, we generated the 4028 VCs needed to prove the full functional correctness of all this code, including termination, by using the RESOLVE VC generator [3]. We believe all these VCs are valid; while we have plenty of code that contains (mostly intentional) bugs, all such code was removed from our library for this study. Third, we machine-translated each VC into Dafny [4] from which it was fed to the SMT solver Z3 [5] for an automated proof attempt, and we recorded what happened. Fourth, we studied carefully the 503 VCs that were *not* proved automatically by Z3 (about 12%; see the red or dark gray region in the middle of the top bar in Figure 1) and tried to determine which changes to specifications, code, or anything else *under the control of the software developer* would improve automated proof success. We emphasize that the intent of this tool-chain is to prove full functional correctness of the code by automated theorem-proving technology, as opposed to abstracting that code to possibly simpler but incomplete finite-state models to be analyzed by a model-checker.

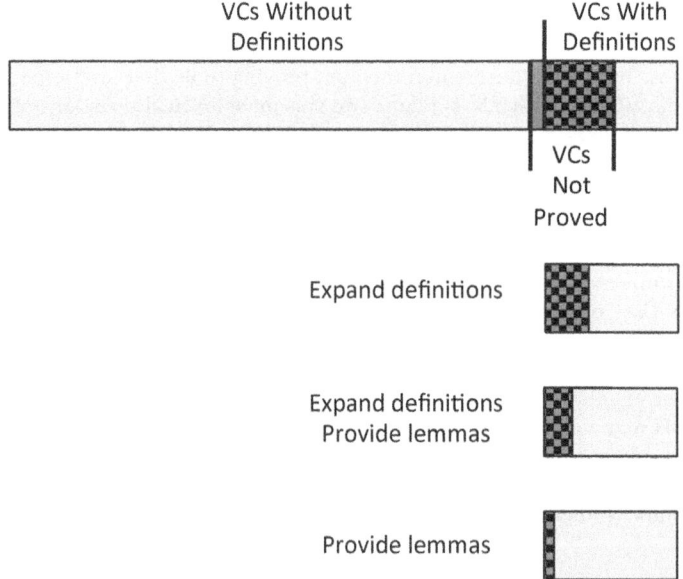

Fig. 1. Results of Proof Attempts on 4028 VCs: For the top bar, user-defined symbols appearing in VCs are treated as uninterpreted function or predicate symbols; for the lower three bars, their definitions are expanded to the prover and/or universal lemmas about those functions and predicates are provided to the prover, as indicated

It turned out that 425 of the 503 VCs *not* proved by Z3 in the first step were among the 817 VCs that involved used-defined mathematical function or predicate symbols. The automated proof success rate for these VCs was so low—about 48%, compared to about 98% for VCs not involving such symbols—that we chose to focus our attention on what one might be able to do about them to improve the automated proof success rate. The rest of this paper is about these 425 VCs (represented in the checkerboard pattern in the top bar of Figure 1) and others like them that might arise in future software verification efforts. It explains how we improved the automated proof success rate for VCs involving user-defined mathematical functions and predicates from 48% to 93%. It is likely that formal specifications of software will often involve some *ad hoc* user-defined mathematical functions and predicates of the sort seen in these VCs, especially when domain-specific software components and application software are involved. So these results are potentially significant.

In considering these 425 VCs, we took the viewpoint of software engineers or mathematical specialists working on the software to be verified. We did not seek to reverse-engineer or second-guess Z3. Any advice provided here is, therefore, intended to be used by software verification researchers (and soon, we believe, by professionals working under a verified software paradigm). That advice is related to issues these "users" of software verification tools should both understand and be able to control, not to internal details of a black-box automated theorem-prover like Z3 that happens to be part of a tool chain. If any results also end up being of interest to those developing better automated provers for VCs, so much the better. We therefore wish to emphasize that none of our results should be interpreted as being critical of Z3. Indeed, both Z3 and the other automated provers we have used in the RESOLVE software verification tool-chain [3] perform far better than their predecessors of the 1990s, when we attempted a similar study on a smaller scale and found that few VCs could be proved automatically. In fact, maybe Dafny/Z3 can do better today on some VCs than the version available at the time of the study: Dafny version 2, which uses Z3 version 2.15. However, we are confident our basic conclusions would remain valid if Z3 were incrementally improved, or even if a different automated prover were substituted for it.

The paper makes two primary contributions. The first contribution is the empirical study itself, including its scale and the conclusions drawn from it: the importance of providing universal algebraic lemmas about user-defined mathematical function and predicates appearing in formal specifications, combined with the importance of not expanding definitions to reveal hidden quantifiers and even more definitions. Overviews of the study process and results are presented here, and full details are available via a companion website that is the first effort to provide such detailed information about so many VCs and the results of attempting to prove them automatically. The second contribution is the first entirely automatic proof of sorting code in which both the type of items being sorted and the total pre-ordering by which they are sorted are client-supplied parameters; this is one of the software verification benchmarks proposed in [6].

2   User-Defined Mathematical Functions and Predicates

Any practical language for writing mathematics must allow new mathematical functions and predicates to be introduced via **user-defined symbols**. This is particularly important for a software specification language because specifications are mathematical statements meant to be read by software developers. Indeed, carefully considered and well-named mathematical functions and predicates can dramatically simplify mathematical statements while providing intuition and understanding for humans.

As a simple example, consider a specification involving odd integers. If the mathematical language used to write this specification has no built-in functions or predicates with which to make direct statements about odd integers, then the specifier might choose to write out, everywhere she needs to say "n is odd", a relatively cumbersome expression such as $\exists k : integer\ (n = 2k + 1)$. The capability to introduce new mathematical symbols and to define them gives her the alternative of saying "$ODD(n)$" in all these places; of course, if she introduces the symbol ODD she must say once and for all that $ODD(n)$ is defined as $\exists k : integer\ (n = 2k + 1)$. We note as an important empirical matter that it is typical for a user-defined mathematical function or predicate symbol such as ODD to hide one or more quantifiers in its definition.

RESOLVE does, therefore, support user-defined mathematical function and predicates. There are a number of mathematical theories and associated operators built-in to the language: booleans, integers, tuples, strings, finite sets, functions, relations, etc. In this study, we deal only with new user-defined mathematical functions and predicates within these theories (i.e., using their mathematical types, or sorts), not with entirely new user-defined mathematical theories.

There are two distinct ways in RESOLVE to introduce a new mathematical function or predicate symbol via a **signature** and provide its definition in a **body**:

- **Explicit definition**: the body is an expression of the result type of the function or predicate.
- **Implicit definition**: the body is an assertion (involving the function or predicate symbol being defined) that uniquely characterizes the function or predicate.

The definition of ODD above uses an explicit definition. An inductive definition is one style of implicit definition. Detailed examples of both explicit and implicit definitions are discussed in Section 5.

It is also possible in RESOLVE for a mathematical function or predicate to be a parameter to a software component. A body does not appear in the software component where this parameter is introduced, but rather in some other component (typically a client program) that completes the definition by binding that parameter to a specific mathematical function or predicate symbol with the same signature. RESOLVE permits the specifier introducing such a parameter to place a **restriction** on the definition ultimately to be bound to it by stating a property that the definition must satisfy—without uniquely characterizing the mathematical function or predicate. A specific example, a client-supplied ordering relation for sorting, is central to the discussion in Section 5.

In a verified software paradigm as envisioned by Hoare [7], mathematical statements in software specifications are seen not only by software developers but also by automated provers as they attempt to prove verification conditions (VCs). If user-defined symbols are simply expanded ("unfolded") into their bodies in the VC proof process, thereby reintroducing the quantifiers and other complexities they were designed to bury for the software developer, then the benefits of user-defined mathematical function and predicate symbols are limited to the human writers and readers of specifications.

Sometimes, no knowledge at all about the definition is needed to prove a VC in which a user-defined mathematical function or predicate symbol appears [2]; it can be treated as an uninterpreted symbol. One assumption in a VC might have a form such as $ODD(n) \implies P$, while another assumption is simply $ODD(n)$. The prover concludes P from these two assumptions and proceeds to use that fact in the proof of the conclusion of the VC—without knowing anything more about ODD. Expanding definitions is not needed here.

Other times, when certain properties associated with a user-defined function or predicate such as ODD are needed in the proof of a VC, they can be stated as universal algebraic lemmas, e.g., $ODD(n) \iff \neg \, ODD(n + 1)$; or, restated without free variables, $\forall n : integer(ODD(n) \iff \neg \, ODD(n + 1))$. Again, expanding definitions is not needed.

In summary, then, when a user-defined mathematical function or predicate symbol appears in a VC, automated proof of that VC may seem to require the prover to have neither, one, or both of (a) the expanded definition, and (b) appropriate lemmas about it. The question studied in this paper is to what extent it is helpful to keep the complexity of expanded definitions hidden from an automated prover, since automated provers (like humans) often have considerable difficulty dealing with quantifiers.[1] The answer is that simply providing the prover with universal algebraic lemmas about user-defined mathematical functions and predicates is generally far better than expanding their definitions.

3   The Tool Chain

Several tools were integrated for this study to process the pipeline from specifications and code to proof of correctness. Our specifications are written in RESOLVE [1] using some of its built-in mathematical theories and its capabilities for making both explicit and implicit definitions and its support for generic parameters. Code to be verified is also written in RESOLVE. VCs are generated by the OSU RESOLVE verification tools [3] using proof rules described in [8,9]. These VCs along with the theories and mathematical definitions used in them are then (by further automatic translation) expressed as assumptions and assertions in Dafny [4,10]. Dafny translates these "VC programs" into the Boogie intermediate language [11], which in turn generates its own VCs to be proved by the SMT solver Z3 [5].

Dafny is a tool for writing programs annotated with various mathematical and specification statements. Dafny's features for making definitions and for writing

---

[1] Not in the case of ODD, which is easy to reason about automatically because it is in Presburger arithmetic, but in general.

assumptions and assertions make it an attractive translation target. We encode each definition as a static function in Dafny, with the definition body encoded as an assumption about it. Each VC generated from the RESOLVE code is encoded in an ordinary method in Dafny that consists of a series of assumptions followed by an assertion: each assumption A_i in the VC generates an `assume` statement, and the conclusion C generates an `assert` statement.

It would have been possible to translate RESOLVE specifications and code for realizations into Dafny or Boogie, but this would have introduced the reference semantics that is inherent in these languages and that RESOLVE avoids. So, we encode RESOLVE VCs themselves into Dafny programs rather than translating them into Z3's input format. The reason for this design choice (rather than providing Z3 with an axiomatic description of, say, RESOLVE's string theory, and relying on its core logic capabilities alone) is that Dafny offers similar mathematical theories into which we can directly translate RESOLVE-generated VCs. So, for instance, Dafny's sequences can be used as a translation target for RESOLVE's strings. Having no direct control over any details in this entire black-box back-end of the tool chain, we trust, of course, that Dafny's interface to Boogie and Boogie's to Z3 appropriately encode these features and that Z3 proves only valid VCs; we have no reason to suspect otherwise.

4   The Study

For the 817 VCs involving user-defined mathematical functions and predicates, we followed a sequence of four steps in an attempt to prove each of them.

In step 1, we attempted to prove each VC without providing Dafny with any information (except the signature) for any user-defined mathematical function or predicate symbol appearing in it. This is also the only step in which we tried to prove VCs that do not have any user-defined symbols in them at all, since providing more information about things not even indirectly appearing in a VC presumably cannot help the prover. In principle, this still allows us to prove any VC whose proof does not require the body or any other knowledge about a user-defined symbol. In Table 1 we see (also shown graphically in Figure 1) that Z3 performed extremely well on VCs without any user-defined mathematical functions or predicate symbols, proving almost 98% of them. It even proved 48% of the VCs that contain such symbols, without any knowledge at all about the underlying mathematical functions and predicates they denote, as illustrated with a similar hypothetical situation involving ODD in Section 2.

In step 2, we provided as an additional assumption of each VC an expanded definition of each user-defined symbol appearing in it. We provided its body for an explicit or implicit definition and its restriction for a parameter. An important point here is that the typical body or restriction involves at least one quantifier and sometimes alternation of quantifiers. This raises the level of complexity of the VCs for the automated prover. Indeed, hiding this complexity from the reader of specifications is one of the primary reasons for introducing a new user-defined mathematical function or predicate in the first place. Yet with expanded definitions available, Z3 nonetheless proved many more of the VCs that involve user-defined mathematical function or predicate symbols: 68% of them, compared to 48% without this additional information.

In step 3, we added to the assumptions of each VC some universal algebraic lemmas about the user-defined mathematical functions and predicates appearing in it, or in the expanded definitions already added as assumptions in step 2. Most of the provided lemmas were from reusable mathematical developments related to the definitions and independent of any VC, as illustrated with *ODD* in Section 2. A few others were not independently identified as mathematically interesting and highly reusable, but rather were directly suggested by human proof attempts on particular VCs that were not proved automatically by Z3. This improved the success rate for Z3 to about 79% of the VCs when *both* expanded definitions and universal algebraic lemmas were available.

In step 4, we removed the bodies of definitions, but left the restrictions on user-defined symbols that are parameters; these restrictions are generally so central to the code in which they appear that they must be kept to have any hope of proving the resulting VCs. Moreover, they tend to be syntactically indistinguishable from universal algebraic lemmas, as seen in Section 5. The *removal* of expanded definitions significantly improved the success rate of Z3 to about 93% of the VCs involving user-defined mathematical functions and predicates.

Table 1. Summary of Empirical Results

Component family	VCs without a user-defined symbol	VCs with a user-defined symbol			
		No expansion; no lemmas	Expansion; no lemmas	Expansion; lemmas	No expansion; lemmas
List	(58/58)	(52/84)	(78/84)	(80/84)	(83/84)
Queue	(48/48)	(108/248)	(155/248)	(222/248)	(240/248)
Sequence	(53/53)	(50/98)	(75/98)	(79/98)	(87/98)
Stack	(6/7)	(7/25)	(17/25)	(25/25)	(21/25)
Integer	(755/788)	(-/-)	(-/-)	(-/-)	(-/-)
Natural	(2147/2190)	(32/48)	(43/48)	(46/48)	(46/48)
Set	(22/23)	(100/242)	(130/242)	(129/242)	(215/242)
Array	(3/3)	(21/31)	(30/31)	(31/31)	(31/31)
BooleanFacility	(23/23)	(-/-)	(-/-)	(-/-)	(-/-)
Others	(18/18)	(22/41)	(30/41)	(35/41)	(38/41)
VCs proved / VCs total	(3133/3211)	(392/817)	(558/817)	(647/817)	(761/817)
% VCs proved	97.6%	48.0%	68.3%	79.2%	93.1%

Table 1 summarizes the empirical study results. Detailed information showing specifications and code for all components, the VCs generated, the user-defined symbols with their bodies and restrictions, and the lemmas are available at http://resolve.cse.ohio-state.edu:8080/archive/nfm2012/ .

5   Example: Fully Generic Sorting

To illuminate issues and results regarding what happens when trying to prove VCs—by expanding the definitions of user-defined mathematical function and predicates and/or by providing universal algebraic lemmas about them—we offer one of the verification benchmark problems proposed in [6]: a generic sorting program in which both the type of the entries to be sorted and the ordering are client-supplied parameters. We specify and then verify an implementation (using merge sort) of an operation that sorts an ordered collection of entries of a user-supplied type according to a user-supplied total pre-order. This benchmark has been addressed by others, e.g., [12], for the case of sorting a fixed type (integers) according to a fixed total pre-order ($\leq$). The fully generic version offered as a benchmark is challenging precisely because it involves an obvious need to introduce some user-defined symbols to simplify the specification. Moreover, the code for the merge sort algorithm is not trivial and involves the standard sequential programming constructs including recursion.

We store the elements to be sorted in a Queue, whose mathematical model is a string of (the mathematical model of) the type of the elements to be sorted. The contract QueueTemplate provides four typical operations: Enqueue, Dequeue, Length, and Is_Empty. The elements inside a Queue can only be obtained by the Dequeue operation in a FIFO order, and there is no way to access the elements inside a Queue other than removing them from it. All code for this example can be found, accompanied with explanations of the relevant RESOLVE language features, on the web site mentioned at the end of Section 4.

Figure 2 shows the contract of an enhancement, or extension, of the Queue-Template contract, called SortExtension. It specifies an operation to Sort a Queue. Note that the contract is parametrized by a binary relation ARE_IN_ORDER that is restricted to be a total pre-order. The ellipsis in this code is where the user-defined mathematical function and predicates in Figure 3 appear. These four definitions (signatures and their bodies) are structured in such a way that the two appearing in the contract of Sort (ARE_PERMUTATIONS and IS_NONDECREASING) are defined in terms of the other two (OCCURS_COUNT and PRECEDES). This style of making definitions is typical. The intent of each definition is as follows:

- OCCURS_COUNT is the number of occurrences of its second argument (an Item) in its first argument (a string of Items). This is an implicit definition, introduced by the keyword **satisfies** followed by an assertion in which OC-CURS_COUNT appears.
- ARE_PERMUTATIONS is true whenever its two arguments (strings of Items) are permutations of one another. This is an explicit definition, introduced by the keyword **is** followed by an expression of the result type.
- PRECEDES is true whenever every entry in its first argument (a string of Items) is "in order with" every entry in its second argument (a string of Items), where the order is based on the relation ARE_IN_ORDER.
- IS_NONDECREASING is true iff its argument (a string of Items) is sorted.

```
contract SortExtension (
    definition ARE_IN_ORDER (x: Item, y: Item): boolean satisfies restriction
        for all z: Item ((ARE_IN_ORDER (x, y) or ARE_IN_ORDER (y, x)) and
            (if (ARE_IN_ORDER (x, y) and ARE_IN_ORDER (y, z)) then ARE_IN_ORDER (x, z)))
    ) enhances QueueTemplate
    ...
    procedure Sort (updates q: Queue)
        ensures
            ARE_PERMUTATIONS (q, #q) and IS_NONDECREASING (q)
end SortExtension
```

Fig. 2. Sort Specification

```
definition OCCURS_COUNT (
    s: string of Item,
    i: Item
    ) : integer satisfies          definition PRECEDES (
if s = <> then                         s1: string of Item,
    OCCURS_COUNT (s, i) = 0             s2: string of Item
else there exists x: Item,             ) : boolean is
                  r: string of Item    for all i, j: Item
(s = <x> * r and                       where (OCCURS_COUNT (s1, i) > 0 and
   (if x = i then OCCURS_COUNT (s, i)         OCCURS_COUNT (s2, j) > 0)
     = OCCURS_COUNT (r, i) + 1        (ARE_IN_ORDER (i, j))
   else OCCURS_COUNT (s, i)
     = OCCURS_COUNT (r, i)))          definition IS_NONDECREASING (
                                          s: string of Item
definition ARE_PERMUTATIONS (             ) : boolean is
    s1: string of Item,                for all a, b: string of Item
    s2: string of Item                     where (s = a * b)
    ) : boolean is                         (PRECEDES (a, b))
for all i: Item
(OCCURS_COUNT (s1, i)
   = OCCURS_COUNT (s2, i))
```

Fig. 3. Mathematical Definitions Used in SortExtension

In Section 6 we use sample VCs from the **MergeSort** realization of the contract in Figure 2 to illustrate the kinds of VCs proved in each step of the study. It is important to notice that this realization is parametrized by a **control**-valued *programming* function **AreInOrder** which returns true whenever its arguments satisfy the *mathematical* relation ARE_IN_ORDER (described previously and arising as a separate parameter to the **SortExtension** contract). The full separation of mathematical and programming functions as illustrated here is an important distinctive feature of RESOLVE. As it is often difficult to select names that convey this distinction, we adopt a typographical convention: all-upper-case identifiers are reserved for mathematical functions and predicates. It is worth noting that since we view the VCs to be putative mathematical theorems, only mathematical symbols (not the names of programming entities) appear in them.

6   VCs from the Fully Generic Sorting Example

In this section, we show some VCs from the **MergeSort** verification that are proved in each step of the study—and some that are not. The intent is to give an indication of some of the difficulties that arise in this example—and some that do not. A total of 62 RESOLVE VCs are generated from the **MergeSort** code, of which 58 mention user-defined mathematical functions or predicates.

Figure 4 shows a VC that is proved in step 1 of the study, i.e., with all user-defined symbols treated as uninterpreted function and predicate symbols. Although this VC involves the predicate **ARE_PERMUTATIONS** in its assumptions and conclusions, it is proved without knowledge of anything more than the signature of **ARE_PERMUTATIONS**. This VC is proved easily due to a contradiction involved in one of the assumptions. Less than half (28 of 58) of the interesting VCs were proved in the first step.

```
var q1_4 , q2_4 , q1_0 : seq<T> ;
var x_6 : T ;
...
assume ARE_PERMUTATIONS ( ( ( ( ( [ x_6 ] + [] )
 + q1_4 ) + q2_4 ) , ( ( q1_0 + [] ) + [] ) );
assume ( ( [ x_6 ] + [] ) == [] );
...
assert ARE_PERMUTATIONS ( ( ( ( [] + ( q1_4
 + [ x_6 ] ) ) + q2_4 ) , ( ( q1_0 + [] ) + [] ) ) );
```

Fig. 4. VC Proved in Step 1

```
var q2_3 , q1_0 : seq<T> ;
var q2Item_3 : T ;
...
assume ( | q1_0 | > 0);
assume ( | ( [ q2Item_3 ] + q2_3 ) | > 0);
...
assert ARE_PERMUTATIONS (
 ( ( ( [] + q1_0 ) + q2_3 ) + [ q2Item_3 ] ) ,
 ( ( ( [] + q1_0 ) + q2_3 ) + [ q2Item_3 ] ) );
```

Fig. 5. VC Proved in Step 2

Figure 5 shows a VC proved in step 2 but not in step 1. Its assumptions include the expanded definitions shown in Figure 3 (but not shown here). The conclusion of this VC states that a particular string is a permutation of itself, which unsurprisingly cannot be proved if **ARE_PERMUTATIONS** is treated as an uninterpreted predicate symbol. With expanded definitions provided, Z3 is able to prove 36 of the 58 VCs containing user-defined function and predicate symbols.

Notice that expanding definitions opens up everything to the prover, including the definitions of **OCCURS_COUNT** and **PRECEDES** that do not directly appear in the specification of **Sort**. In software engineering terms, we might say that expanding definitions breaks encapsulation and flattens out all the underlying mathematical machinery devised to write the specification. There is no information hiding—either from a human reader or a prover—when definitions are expanded.

Figures 6 and 7 show examples of two VCs that are proved in step 3 but not in step 2. In addition to expanding definitions, we now provide the prover with some simple universal algebraic lemmas about the user-defined mathematical functions and predicates. We encode some lemmas involving **ARE_PERMUTATIONS** and **IS_NONDECREASING** as shown in Figures 8 and 9. Z3 proves 57 of the 58 interesting VCs with this additional information.

The VC in Figure 6 contains among its assumptions (not shown here) the expanded definitions of **ARE_PERMUTATIONS** and **OCCURS_COUNT**. It also has several more assumptions, all but one elided as irrelevant to the proof. To a human, the conclusion seems easily provable from the given assumption along with an understanding of concatenation and the meaning of **ARE_PERMUTATIONS**. Some of the lemmas provided as additional assumptions capture this feature.

```
var q1_0 , q2_3 , q1_6 , tmp_4 , q2_4  :  seq<T> ;
var q2Item_3 , q1Item_6 , q2Item_4  :  T ;
...
assume ARE_PERMUTATIONS ( ( ( ( tmp_4 +
([ q1Item_6 ] + q1_6 )) + q2_4 ) + [ q2Item_4 ]) ,
  ( ( ( [] + q1_0 ) + q2_3 ) + [ q2Item_3 ])) ;
...
assert ARE_PERMUTATIONS ( ( ( ( ( tmp_4 +
[ q2Item_4 ]) + q2_4 ) + q1_6 ) + [ q1Item_6 ]) ,
  ( ( ( [] + q1_0 ) + q2_3 ) + [ q2Item_3 ])) ;
```

Fig. 6. VC Proved in Step 3

```
var tmp_4 , q2_4 :  seq<T>;
var q2Item_4 :  T;
...
assume IS_NONDECREASING (
  (( tmp_4 + [ q2Item_4 ]) + q2_4 )) ;
...
assert IS_NONDECREASING (
  ([ q2Item_4 ] + q2_4 )) ;
```

Fig. 7. VC Proved in Step 3

The VC in Figure 7 also includes two expanded definitions (not shown here), and has several other assumptions, all but one elided as irrelevant. To a human, it is obvious that if IS_NONDECREASING holds for a string formed by the concatenation of three strings, then it holds for the concatenation of two of them in the same order. But here it is less clear how long a reasoning path is required for an automated prover to notice this without some simple lemmas to help.

1. **forall** a:**seq**<T>:: ARE_PERMUTATIONS(a,a)

2. **forall** a:**seq**<T>, b:**seq**<T>, c:**seq**<T>:: ARE_PERMUTATIONS(a,b) &&
   ARE_PERMUTATIONS(b,c) ==> ARE_PERMUTATIONS(a,c)

3. **forall** a:**seq**<T>, b:**seq**<T>:: ARE_PERMUTATIONS(a,b) ==> ARE_PERMUTATIONS(b,a)

4. **forall** a:**seq**<T>, b:**seq**<T>, c:**seq**<T>:: ARE_PERMUTATIONS((a + b) + c, a + (b + c))

5. **forall** a:**seq**<T> , b: **seq**<T> :: a == b ==> ARE_PERMUTATIONS(a, b)

6. **forall** a:**seq**<T>, b:**seq**<T>:: ARE_PERMUTATIONS(a,b) ==> |a| == |b|

Fig. 8. ARE_PERMUTATIONS Lemmas

An obvious and important question is, "Which lemmas should be provided to the prover?" The lemmas in Figure 8 for ARE_PERMUTATIONS are the usual equivalence relation properties along with a few others that might be given as problems in an undergraduate math/logic textbook. Given lemma 1, lemma 4 merely restates that concatenation is associative. It turns out to be important for Z3 to have this property separately in order to prove some of the MergeSort VCs involving ARE_PERMUTATIONS. Lemma 5 says the same thing as lemma 1; yet it helps Z3 prove some VCs where lemma 1 does not. In short, none of these lemmas is surprising except possibly for the fact that it helps Z3 when properties are stated in a particular way. This is not a shortcoming of the concept of using universal algebraic lemmas in proofs of VCs, but rather appears to be the expression of a current limitation on how they are processed by the prover.

The lemmas added for IS_NONDECREASING are more extensive. The most basic set are lemmas 1 through 3 in Figure 9. The second set (4 through 6) are of a different nature, relating various concatenations of strings satisfying

1. IS_NONDECREASING ([ ])

2. **forall** x:T :: IS_NONDECREASING ([x])

3. **forall** q:**seq**<T> :: |q| <= 1 ==> IS_NONDECREASING (q)

4. **forall** x:**seq**<T>, y:**seq**<T>:: IS_NONDECREASING (x + y) ==>
    IS_NONDECREASING (x) &&   IS_NONDECREASING (y)

5. **forall** a:**seq**<T>, b:**seq**<T>, c:**seq**<T>:: IS_NONDECREASING (a + b + c) ==>
    IS_NONDECREASING (a + b) && IS_NONDECREASING (b + c) &&
    IS_NONDECREASING (a + c)

6. **forall** a:**seq**<T>, b:**seq**<T>, c:**seq**<T>:: IS_NONDECREASING (a + c) &&
    IS_NONDECREASING (c + b) && c != [ ] ==> IS_NONDECREASING (a + c + b)

7. **forall** a:T, b:T :: ARE_IN_ORDER (a,b) ==> IS_NONDECREASING ([a] + [b])

8. **forall** a:**seq**<T>, b:**seq**<T>, x:T, y:T :: IS_NONDECREASING (a + [x]) &&
    IS_NONDECREASING ([y] + b) && ARE_IN_ORDER (x,y) ==>
    IS_NONDECREASING (a + [x] + [y] + b)

9. **forall** a:**seq**<T>, x:T, y:T :: IS_NONDECREASING ([x] + a + [y]) ==>
    ARE_IN_ORDER (x,y)

Fig. 9. IS_NONDECREASING Lemmas

IS_NONDECREASING. The third set (7 through 9) are different still, relating
ARE_IN_ORDER and IS_NONDECREASING. All these lemmas were proved inter-
actively with the help of Isabelle used as a proof assistant.

```
var  q1_0 , q2_3 , tmp_17 , q2_17  :  seq<T> ;
var  q2Item_3 , q2Item_17  :  T ;

assume  ARE_PERMUTATIONS ( ( ( ( tmp_17 + [ ] ) + q2_17 ) + [ q2Item_17 ] ) ,
            ( ( ( [ ] + q1_0 ) + q2_3 ) + [ q2Item_3 ] ) ) ;
assert  ARE_PERMUTATIONS ( ( ( tmp_17 + [ q2Item_17 ] ) + q2_17 ) , ( q1_0 + ( [ q2Item_3 ] + q2_3 ) ) ) ;
```

Fig. 10. VC Proved in Step 4

Figure 10 shows the lone VC not proved in a previous step but successfully
proved in step 4, where we remove expanded definitions and leave only universal
algebraic lemmas about them. Z3 now proves all 58 interesting lemmas.

It is important to note that in the absence of the expanded definitions, providing universal algebraic lemmas does *not* open up everything to the prover.
In particular, here the very existence of OCCURS_COUNT and PRECEDES remains hidden because they do not directly appear in the specification of Sort. In
software engineering terms, providing universal algebraic lemmas about defined
functions and predicates respects encapsulation and leverages all the underlying
mathematical machinery devised by the software developer to write the specification. Information hiding survives when definitions are not expanded.

7   Discussion

Intuition for believing that otherwise troublesome quantifiers in VCs might be finessed by introducing universally quantified lemmas comes from observing the practice in calculus where, for example, most results are established (by humans) not by appealing to a complex nested quantification like that required to define the concept of a limit, but rather by reusing universal algebraic results proved separately, e.g., $lim\ (f+g) = lim\ f\ +\ lim\ g$. A similar situation characterizes reasoning using big-O notation. The value of universal algebraic lemmas seems clear from such experience with fully manual proofs, as well as from anecdotal evidence involving interactive proofs (e.g., [13,14]). However, it is less clear that such lemmas should be *so helpful* to an automated prover that they should help it produce—fully automatically—proofs of VCs involving complex definitions.

To see why this is plausible, consider a VC in which user-defined symbols appear, but in which no definitions are expanded and no properties about those definitions appear. The form of such a VC as generated by the OSU RESOLVE verification tools [3] is always $\bigwedge A_i \implies C$. When definitions are used to hide *all* quantifiers in software specifications (a recommendation we have followed in the specifications used in this study), all the variables (call them $x_1, ..., x_m$) in the assumptions $A_1, ..., A_n$ and the conclusion C are free variables; equivalently, there is an implicit universal quantifier in front: $\forall x_1, ..., x_m (\bigwedge A_i \implies C)$. Depending on the combination of functions and predicates appearing in the VC, automated provers may do well or not so well on it. Yet if there is trouble then at least it is not the fault of the quantifier structure, because such VCs are in a form to which automated provers are well suited.

On the other hand, if a definition hiding quantifiers is expanded in the VC, then these newly exposed quantifiers might, or might not, cause serious additional trouble for an automated prover. For example, in the lucky special case that one of the assumptions, say A_k, expands to the form $\exists y(P(y))$, then there is no problem at all: x_{m+1} can be introduced as a new free variable and A_k can be replaced with $P(x_{m+1})$, i.e., x_{m+1} is simply treated as a witness to the existence of a value that makes P true. The structure of the revised VC remains the same as the original: it is now $\forall x_1, ..., x_{m+1} (\bigwedge A_i \implies C)$.

A more difficult situation is the equally special case where one of the assumptions, say A_k, expands to the form $\forall y(P(y))$. Now, the prover must instantiate y with term(s) appearing in the VC, so the new instance(s) help in the overall proof. SMT solvers use "triggers" to match terms in such a way that the instantiated version(s) might be useful. Sometimes the prover can find an appropriate match automatically, and sometimes it needs a human-suggested trigger [5,15]. A universal algebraic lemma added as an additional assumption in a VC is exactly of this second, somewhat non-trivial, form. However, this is still far less complex a form for an automated prover to handle than an arbitrary quantified statement. Indeed, in some sense this is *the* non-trivial quantified form for which automated provers are tuned to work particularly well.

Though the general idea of providing universal algebraic lemmas about user-defined mathematical function and predicates has been reported in case studies with interactive proofs [13, 14], it has not previously been systematically and empirically evaluated for use with automated provers for verification conditions.

8   Conclusions

We have presented empirical support for the claim that supplying universal algebraic lemmas about user-defined mathematical functions and predicates is, in general, a better way than expanding definitions to support automated verification of programs; and the approach does not assume that programmers have any knowledge about the intricacies of a back-end theorem prover (such as triggers or proof tactics). We also have demonstrated that the approach can be applied successfully for code whose specifications involve various mathematical theories.

There is clear potential for dependence of these results on specification and programming language features. The VCs we have seen from verification tools for other imperative languages, including Dafny itself and Jahob (for Java), are similar in basic mathematical content to VCs from RESOLVE programs. Yet there is one critical difference as well. RESOLVE programs have value semantics, not reference semantics; hence, there is no possibility for aliasing and no appearance of heap properties in RESOLVE VCs. This makes RESOLVE VCs relatively easier to prove, so our results in one direction should apply across the board: if expanding definitions does not lead to automated proofs of RESOLVE VCs using the same or similar back-end provers as are used for languages with reference semantics, it is unlikely that expanding definitions will lead to successful automated proofs of VCs that involve heap properties *in addition to* the properties of the primary mathematical models used in specifications. In the other direction, as far as we know it remains open to what extent providing universal algebraic lemmas to automated provers rather than expanding definitions has similar value for VCs that also involve heap properties.

Acknowledgment. Jason Kirschenbaum, Ted Pavlic, and Ray McDowell were especially helpful in contributing to this work. The authors are also grateful for the suggestions of Bruce Adcock, Derek Bronish, Paolo Bucci, Harvey M. Friedman, Wayne Heym, Dustin Hoffman, Bill Ogden, and Murali Sitaraman. This material is based upon work supported by the National Science Foundation under Grants No. DMS-0701260, CCF-0811737, and ECCS-0931669. Any opinions, findings, conclusions, or recommendations expressed here are those of the authors and do not necessarily reflect the views of the National Science Foundation.

References

1. Sitaraman, M., Weide, B.: Component-based software using RESOLVE. SIGSOFT Softw. Eng. Notes 19, 21–63 (1994)
2. Kirschenbaum, J., Adcock, B., Bronish, D., Smith, H., Harton, H., Sitaraman, M., Weide, B.W.: Verifying Component-Based Software: Deep Mathematics or Simple Bookkeeping? In: Edwards, S.H., Kulczycki, G. (eds.) ICSR 2009. LNCS, vol. 5791, pp. 31–40. Springer, Heidelberg (2009)

3. Sitaraman, M., et al.: Building a push-button RESOLVE verifier: Progress and challenges. Formal Aspects of Computing 23, 607–626 (2011)
4. Leino, K.R.M.: Dafny: An Automatic Program Verifier for Functional Correctness. In: Clarke, E.M., Voronkov, A. (eds.) LPAR-16 2010. LNCS, vol. 6355, pp. 348–370. Springer, Heidelberg (2010)
5. de Moura, L., Bjørner, N.: Z3: An Efficient SMT Solver. In: Ramakrishnan, C.R., Rehof, J. (eds.) TACAS 2008. LNCS, vol. 4963, pp. 337–340. Springer, Heidelberg (2008)
6. Weide, B.W., Sitaraman, M., Harton, H.K., Adcock, B., Bucci, P., Bronish, D., Heym, W.D., Kirschenbaum, J., Frazier, D.: Incremental Benchmarks for Software Verification Tools and Techniques. In: Shankar, N., Woodcock, J. (eds.) VSTTE 2008. LNCS, vol. 5295, pp. 84–98. Springer, Heidelberg (2008)
7. Hoare, T.: The verifying compiler: A grand challenge for computing research. J. ACM 50, 63–69 (2003)
8. Heym, W.D.: Computer program verification: improvements for human reasoning. PhD thesis, The Ohio State University, Columbus, OH, USA (1995)
9. Sitaraman, M., Atkinson, S., Kulczycki, G., Weide, B.W., Long, T.J., Bucci, P., Heym, W.D., Pike, S.M., Hollingsworth, J.E.: Reasoning about Software-Component Behavior. In: Frakes, W.B. (ed.) ICSR 2000. LNCS, vol. 1844, pp. 266–283. Springer, Heidelberg (2000)
10. Leino, K.R.M.: Specification and verification of object-oriented software. Marktoberdorf International Summer School 2008, lecture notes (2008)
11. Leino, K.R.M.: This is Boogie 2. Manuscript KRML 178 (2008), http://research.microsoft.com/en-us/um/people/leino/papers.html
12. Leino, K.R.M., Monahan, R.: Dafny Meets the Verification Benchmarks Challenge. In: Leavens, G.T., O'Hearn, P., Rajamani, S.K. (eds.) VSTTE 2010. LNCS, vol. 6217, pp. 112–126. Springer, Heidelberg (2010)
13. Nelson, C.C.: Techniques for program verification. PhD thesis, Stanford University, Stanford, CA, USA (1980)
14. Kaufmann, M., Manolios, P., Moore, J.S. (eds.) Computer-Aided Reasoning: ACL2 Case Studies. Kluwer Academic Publishers (2000)
15. Detlefs, D., Nelson, G., Saxe, J.B.: Simplify: a theorem prover for program checking. J. ACM 52, 365–473 (2005)

Sound Formal Verification of Linux's USB BP Keyboard Driver

Willem Penninckx, Jan Tobias Mühlberg, Jan Smans,
Bart Jacobs, and Frank Piessens

IBBT-DistriNet, KU Leuven, 3001 Leuven, Belgium

Abstract. Case studies on formal software verification can be divided
into two categories: while *(i)* unsound approaches may miss errors or
report false-positive alarms due to coarse abstractions, *(ii)* sound approaches typically do not handle certain programming constructs like
concurrency and/or suffer from scalability issues. This paper presents a
case study on successfully verifying the Linux USB BP keyboard driver.
Our verification approach is *(a)* sound, *(b)* takes into account dynamic
memory allocation, complex API rules and concurrency, and *(c)* is applied on a real kernel driver which was not written with verification in
mind. We employ VeriFast, a software verifier based on separation logic.
Besides showing that it is possible to verify this device driver, we identify
the parts where the verification went smoothly and the parts where the
verification approach requires further research to be carried out.

1 Introduction

The safety and security of today's omni-present computer systems critically depends on the reliability of operating systems (OS). Due to their complicated
task of managing a system's physical resources, OSs are difficult to develop and
to debug. As studies show, most defects causing operating systems to crash are
not in the system's kernel but in the large number of OS extensions available
[1,4]. In Windows XP, for example, 85% of reported failures are caused by errors
in device drivers [1]. As explained in [4], the situation is similar for Linux and
FreeBSD: error rates reported for device drivers are up to seven times higher
than error rates stated for the core components of these OSs.

A lot of research aims to prove the correctness of programs. However, not
much work has been carried out to test whether the results of this research is
applicable to complex real-world programs where correctness is important, like
operating systems drivers. To work towards addressing this question, this paper
applies a separation-logic based verifier, VeriFast [11], on a device driver taken
from the Linux kernel.

The driver code subject to verification is Linux's USB Boot Protocol keyboard
driver. While being small, this driver contains a bigger than expected subset of
kernel driver complexity. It involves asynchronous callbacks, dynamic allocated
memory, synchronization and usage of complex APIs. During verification, we
identified and fixed a number of bugs. For these bugs we submitted patches that
have been accepted by the driver's maintainer and are queued for inclusion in
future Linux releases.

A. Goodloe and S. Person (Eds.): NFM 2012, LNCS 7226, pp. 210–215, 2012.

In the remainder of this paper we briefly introduce VeriFast and the device driver. We outline the verification of the driver and elaborate on the challenges involved. Finally, we discuss related work and draw conclusions.

2   Background

The verifier we apply to the USB BP keyboard driver is VeriFast. VeriFast's underlying logic is based on an extension of separation logic. Separation logic [16] builds on Hoare Logic [10] and adds support for the heap by introducing the separating conjunction $*$ and other assertions describing a heap. An assertion $A * B$ expresses that the heap can be divided in two disjoint parts, such that assertion A holds for the first part and B holds for the second part.

Concurrency is supported by associating a real number (called "fraction") from $(0, 1]$ to every heap cell which is regarded as a permission (e.g. to access data) [3]. Multiple threads can obtain different fractions of the same permission. What is allowed with the permission, depends on the fraction, e.g. for an access-data permission, a fraction 1 denotes read-write permission, a fraction of another size denotes read-only permission.

Specifications for (spin)locks are done in a fashion similar to [7]: with a lock a handle and an invariant are associated. A fraction of the handle allows acquiring the lock, which yields (adds to the thread's owned permissions) the invariant which represents the permissions protected by the lock.

VeriFast checks annotated C files. The annotations can contain pre- and post-conditions written in separation logic, ghost data structures and ghost lemmas. The VeriFast tool and its technical documentation, including a tutorial and a formalization of a core subset of VeriFast and its semantics, are available for download at http://www.cs.kuleuven.be/~bartj/verifast/.

3   Overview of How usbkbd Works

The driver subject to verification is Linux's USB Boot Protocol keyboard driver, named usbkbd[1]. This section gives a high-level overview of how the driver works, leaving out details concerning concurrency and the exact API usage.

On loading, usbkbd registers itself with the USB API. When a new keyboard is attached, the API calls the usb_kbd_probe function of usbkbd. usb_kbd_probe checks whether the driver can handle the attached keyboard, and if so initializes a USB Request Block (URB). An URB is an asynchronous request that can be used to send or receive data from a USB device. The purpose of the URB initialized here is to receive key-presses and key-releases. This URB is named the IRQ URB. usb_kbd_probe initializes another URB for updating the LED status (e.g. numlock) named the LED URB. usb_kbd_probe then registers a new input device with the input API to make the keyboard available to applications. When the newly created input device is opened, usbkbd's usb_kbd_open callback

---

[1] The driver's source file, usbkbd.c, is located in drivers/hid/usbhid/ in the Linux kernel distribution available from http://kernel.org/.

is invoked and `usb_kbd_open` submits the IRQ URB. When a key is pressed or released, the URB completion callback `usb_kbd_irq` is called. `usb_kbd_irq` parses the data received from the keyboard and reports key-presses and releases to the input API. It then resubmits the URB. When the input API decides the LED status needs to be changed, the `usb_kbd_event` callback is invoked. This callback checks whether a LED URB is in progress, and if not submits the LED URB with the appropriate data. Otherwise, it stores the new LED info in a buffer. When the LED URB completion callback `usb_kbd_led` is called, this callback checks whether new LED info has appeared while the LED URB was in progress. If so, `usb_kbd_led` resubmits the LED URB with the new LED info.

4   Verifying the USB BP Keyboard Driver

Verification of the driver is against the original API. Wrapper functions are only used in a few cases where API functions return a struct (i.e. not a pointer to a struct) because this is currently not supported by VeriFast. The APIs that usbkbd uses are the USB API, the input API, spinlocks, and some generic functions like `memcpy`. Verification thus consists of (1) writing formal specifications for these APIs, based on official documentation and reading the API implementation for the underspecified or undocumented parts, and (2) of adding annotations to usbkbd. These annotations consists of contracts (pre- and postconditions written in separation logic), predicates to describe data structures, predicate family instances to instantiate callback function contracts, lemmas (i.e. ghost functions), and ghost-code like folding and unfolding predicates.

The verified properties are freedom of data races in the presence of concurrent callbacks, freedom of illegal memory accesses, and correct API usage. This does not include a formal proof of correctness of the hand-written API formalization.

usbkbd is one of the smallest Linux kernel drivers. It consists of 426 lines of C code (including blanks and comments). VeriFast reports 329 lines of actual code and 822 lines of annotations. The API specifications count up to 769 lines of code. VeriFast can be launched for this driver with "`verifast -prover redux -c usbkbd_verified.c`". On an Intel L9400 1.86GHz running the verifier takes about one second. The annotated sources of usbkbd, specifications for the used APIs and the patches submitted to the driver's maintainer are available at `http://people.cs.kuleuven.be/~willem.penninckx/usbkbd/`.

Writing Specifications for the Input API and some generic functions like `kmalloc` was rather straightforward. API rules include forbidding double frees, requiring when registering input devices that the given callbacks are real function pointers with a contract not conflicting with some rules, etc.

Killable URBs were rather tricky to get verified for the LED URB. Because `usb_kbd_event` and `usb_kbd_led` both submit URBs, they are synchronized with a spinlock. A C boolean `led_urb_submitted` represents whether the URB is in progress, and thus also whether the URB data (necessary for URB submitting) is not owned by the lock invariant. After killing the URB, the URB data must be taken out of the lock invariant in order to free it, i.e. VeriFast must be convinced `led_urb_submitted`is false. We used a ghost-counter (associated with a predicate of which a uniqueness-proof must be provided on creation) named `cb_out_count` that yields a ticket on increase and ensures the counter is at least n high if n such tickets are owned. Another counter, `killcount`, keeps track of the number of URB submits. By making sure `killcount` tickets of `cb_out_count` are obtained when killing the URB, we can prove `cb_out_count` is at least as high as `killcount`. Because `cb_out_count` is maximum one less than `cb_out_count`, we know they are equal, which can only happen if the URB is not submitted.

The `usb_kbd_malloc` and `usb_kbd_free`'s Contracts take into account all possible combinations of failed and successful allocation and initialization, which makes their contracts long, and dependent on other parts of the annotations.

Flow Between Callbacks had to be reasoned about: permissions are passed between callbacks by setting up callbacks in other callbacks. Reasoning about flow between multiple callbacks easily gives the impression big parts of the program must be taken into account at the same time.

5   Related Work

Here we discuss related case studies and tools in the context of OS verification. The reader is referred to [11] for a discussion of the related work on VeriFast.

Several automated tools for verifying C programs have been introduced. Notably, CEGAR-based [5] model checkers such as BLAST [9] and SLAM/SDV [1] have been applied to check the conformance of device drivers with a set of API usage rules. In contrast with our work, these tools do not provide support for identifying errors with respect to the inherently concurrent execution environment device drivers are operating in. The tools also assume either that a program "does not have wild pointers" [1] or, as shown in [13], perform poorly when checking OS components for memory safety.

In [18] a model checker with support for pointers, bit-vector operations and concurrency is evaluated on a case study on Linux device drivers. The tool checks for buffer overflows, pointer safety, division by zero and user-written assertions. Yet, it requires a test harness with a fixed number of threads to be generated for each driver. VeriFast, in difference, handles concurrency implicitly and aims at verifying full functional correctness and implements assume-guarantee reasoning using generic API contracts. Therefore, VeriFast can check each function of a driver in isolation, which contributes to the scalability of our approach.

Bounded model checking and symbolic execution have been successfully applied to the source code [15,12] and to the object code [14] of kernel modules. In contrast to the VeriFast approach, these techniques suffer from severe limitations with respect to reasoning about concurrently executing kernel threads.

Shape analysis has been applied to Windows [2] and Linux [19] drivers, and aims to automatically infer, e.g. whether a variable points to a cyclic or acyclic list. Shape analysis can be employed to verify pointer safety, guaranteeing that the shape of data structures is maintained throughout program execution. Ongoing work on VeriFast envisages the use of shape analysis to infer annotations [17].

A competing toolkit to VeriFast is the Verifying C Compiler (VCC) [6]. VCC verifies C programs annotated with contracts in Boogie. The tool generates verification conditions from the annotated program, which are then discharged by an SMT solver. VCC can be expected to require fewer annotations than VeriFast, however, at the expense of a less predictable search times. The toolkit has been employed in a case study on verifying the Microsoft Hypervisor.

Other approaches to OS verification involving modelling and interactive proof. Most notably, the L4.verified [8] project aims at producing a verified OS kernel by establishing refinement relations between several layers of Isabelle/HOL specifications, a prototypic kernel implementation in Haskell and the actual kernel implementation in C and assembly. This differs from our work as we do not employ refinement relations and verification is non-interactive.

6   Conclusions

We report on the successful verification of usbkbd, the USB Boot Protocol keyboard driver distributed with the Linux kernel, using the sound and efficient verification tool VeriFast. The verified properties are crash-freedom, race-freedom, and a set of API usage rules. The usbkbd driver presents a challenging case study as it involves concurrency and employs a complex API.

VeriFast requires the source code to be annotated with method contracts that are typically easy to write. Certain programming constructs that are difficult to annotate are discussed in this paper. During verification, we identified two bugs related to erroneous synchronization and a missing URB kill. Our case study shows that VeriFast is a powerful tool. Yet, the annotation overhead amounts to a total of 4.8 lines of annotations per line of code. About half of these annotations specify API contracts, that can potentially be reused in future case studies.

Verifying functional correctness and unload-safety is left for further work. Unload-safety includes making sure the kernel does not maintain a function-pointer to a callback of a module that is already unloaded. It is hard to tell whether our verification approach will scale for larger device drivers. More automation for writing or generating annotations with a high degree of decomposition might help. From our experience we conclude that execution speed of the verification tool will not impose problems for larger drivers.

Acknowledgments. This research is partially funded by the Interuniversity Attraction Poles Programme Belgian State, Belgian Science Policy, by the Research Fund KU Leuven, and by the EU FP7 projects SecureChange and NESSoS. Jan Smans is a postdoctoral fellow of the Fund for Scientific Research – Flanders (FWO). We acknowledge support from Microsoft Research Cambridge as part of the Verified Software Initiative.

References

1. Ball, T., Bounimova, E., Cook, B., Levin, V., Lichtenberg, J., McGarvey, C., Ondrusek, B., Rajamani, S.K., Ustuner, A.: Thorough static analysis of device drivers. SIGOPS Oper. Syst. Rev. 40(4), 73–85 (2006)

2. Berdine, J., Calcagno, C., Cook, B., Distefano, D., O'Hearn, P., Wies, T., Yang, H.: Shape Analysis for Composite Data Structures. In: Damm, W., Hermanns, H. (eds.) CAV 2007. LNCS, vol. 4590, pp. 178–192. Springer, Heidelberg (2007)
3. Bornat, R., Calcagno, C., O'Hearn, P., Parkinson, M.: Permission accouting in separation logic. In: POPL (2005)
4. Chou, A., Yang, J., Chelf, B., Hallem, S., Engler, D.R.: An empirical study of operating system errors. In: SOSP 2001, pp. 73–88. ACM, New York (2001)
5. Clarke, E., Grumberg, O., Jha, S., Lu, Y., Veith, H.: Counterexample-guided abstraction refinement for symbolic model checking. J. ACM 50(5), 752–794 (2003)
6. Cohen, E., Dahlweid, M., Hillebrand, M., Leinenbach, D., Moskal, M., Santen, T., Schulte, W., Tobies, S.: VCC: A Practical System for Verifying Concurrent C. In: Berghofer, S., Nipkow, T., Urban, C., Wenzel, M. (eds.) TPHOLs 2009. LNCS, vol. 5674, pp. 23–42. Springer, Heidelberg (2009)
7. Gotsman, A., Berdine, J., Cook, B., Rinetzky, N., Sagiv, M.: Local Reasoning for Storable Locks and Threads. In: Shao, Z. (ed.) APLAS 2007. LNCS, vol. 4807, pp. 19–37. Springer, Heidelberg (2007)
8. Heiser, G., Elphinstone, K., Kuz, I., Klein, G., Petters, S.M.: Towards trustworthy computing systems: taking microkernels to the next level. SIGOPS Oper. Syst. Rev. 41, 3–11 (2007)
9. Henzinger, T.A., Jhala, R., Majumdar, R., Necula, G.C., Sutre, G., Weimer, W.: Temporal-Safety Proofs for Systems Code. In: Brinksma, E., Larsen, K.G. (eds.) CAV 2002. LNCS, vol. 2404, pp. 526–538. Springer, Heidelberg (2002)
10. Hoare, C.A.R.: An axiomatic basis for computer programming. Communications of the ACM 12(10), 576–580 and 583 (1969)
11. Jacobs, B., Smans, J., Philippaerts, P., Vogels, F., Penninckx, W., Piessens, F.: VeriFast: A Powerful, Sound, Predictable, Fast Verifier for C and Java. In: Bobaru, M., Havelund, K., Holzmann, G.J., Joshi, R. (eds.) NFM 2011. LNCS, vol. 6617, pp. 41–55. Springer, Heidelberg (2011)
12. Kim, M., Kim, Y.: Concolic Testing of the Multi-sector Read Operation for Flash Memory File System. In: Oliveira, M.V.M., Woodcock, J. (eds.) SBMF 2009. LNCS, vol. 5902, pp. 251–265. Springer, Heidelberg (2009)
13. Mühlberg, J.T., Lüttgen, G.: BLASTing Linux Code. In: Brim, L., Haverkort, B.R., Leucker, M., van de Pol, J. (eds.) FMICS and PDMC 2006. LNCS, vol. 4346, pp. 211–226. Springer, Heidelberg (2007)
14. Mühlberg, J.T., Lüttgen, G.: Verifying Compiled File System Code. In: Oliveira, M.V.M., Woodcock, J. (eds.) SBMF 2009. LNCS, vol. 5902, pp. 306–320. Springer, Heidelberg (2009)
15. Post, H., Sinz, C., Küchlin, W.: Towards automatic software model checking of thousands of Linux modules – a case study with Avinux. Softw. Test. Verif. Reliab. 19, 155–172 (2009)
16. Reynolds, J.C.: Separation logic: A logic for shared mutable data structures. In: LICS 2002, pp. 55–74. IEEE, Washington (2002)
17. Vogels, F., Jacobs, B., Piessens, F., Smans, J.: Annotation Inference for Separation Logic Based Verifiers. In: Bruni, R., Dingel, J. (eds.) FMOODS/FORTE 2011. LNCS, vol. 6722, pp. 319–333. Springer, Heidelberg (2011)
18. Witkowski, T., Blanc, N., Kroening, D., Weissenbacher, G.: Model checking concurrent Linux device drivers. In: ASE 2007, pp. 501–504. ACM, New York (2007)
19. Yang, H., Lee, O., Berdine, J., Calcagno, C., Cook, B., Distefano, D., O'Hearn, P.: Scalable Shape Analysis for Systems Code. In: Gupta, A., Malik, S. (eds.) CAV 2008. LNCS, vol. 5123, pp. 385–398. Springer, Heidelberg (2008)

Learning Markov Models for Stationary System Behaviors

Yingke Chen, Hua Mao, Manfred Jaeger,
Thomas D. Nielsen, Kim G. Larsen, and Brian Nielsen

Dept. of Computer Science, Aalborg University, Denmark
{ykchen,huamao,jaeger,tdn,kgl,bnielsen}@cs.aau.dk

Abstract. Establishing an accurate model for formal verification of an existing hardware or software system is often a manual process that is both time consuming and resource demanding. In order to ease the model construction phase, methods have recently been proposed for automatically learning accurate system models from data in the form of observations of the target system. Common for these approaches is that they assume the data to consist of *multiple* independent observation sequences. However, for certain types of systems, in particular many running embedded systems, one would only have access to a single long observation sequence, and in these situations existing automatic learning methods cannot be applied. In this paper, we adapt algorithms for learning variable order Markov chains from a *single* observation sequence of a target system, so that stationary system properties can be verified using the learned model. Experiments demonstrate that system properties (formulated as stationary probabilities of LTL formulas) can be reliably identified using the learned model.

1 Introduction

Model-driven development (MDD) is increasingly used for the development of complex embedded software systems. An important component in this process is model checking [1], where a formal system model is checked against a specification given by a logical expression. Often, the complexity of a real system and its physical components, unpredictable user interactions, or even the use of randomized algorithms make the use of complete, deterministic system models infeasible. In these cases, probabilistic system models and methods for probabilistic verification are needed.

However, constructing accurate models of industrial systems is hard and time consuming, and is seen by industry as a hindrance to adopt otherwise powerful MDD techniques and tools. Especially, the necessary accurate, updated and detailed documentation rarely exist for legacy software or 3rd party components. We therefore seek an experimental approach where an accurate high-level model can be automatically constructed or *learned* from observations of a given black-box embedded system component.

Sen et al. [12] proposed to learn system models for verification purposes, based on the *Alergia* algorithm for learning finite, deterministic, stochastic automata [2]. In [8] we developed a learning approach related to that of [12], and

A. Goodloe and S. Person (Eds.): NFM 2012, LNCS 7226, pp. 216–230, 2012.

established strong theoretical and experimental consistency results: if a sufficient amount of data, i.e., observed execution runs of the system to be modeled, is available, then the results of model-checking probabilistic linear-time temporal logic (PLTL) properties on the learned model will be good approximations of the results that would be obtained on the true model. Both [12] and [8] assume that learning is based on data consisting of many independent finite execution runs, each starting in a distinguished, unique initial state of the system. In many situations, it will be difficult or impossible to obtain data of this kind: we may not be able to run the system under laboratory conditions where we are free to restart it any number of times, nor may we be able to reset the system to a well-defined unique initial state.

In this paper, therefore, we investigate learning of system models by passively observing a single, ongoing execution of the system, i.e., from data that consists of a single, long observation sequence, which may start at any point in the operation of the system. This scenario calls for different types of models and learning algorithms than used in previous work. The probabilistic system models we are going to construct are *Probabilistic Suffix Automata (PSAs)* [10]. This is a special type of probabilistic finite automaton, in which states can be identified with finite histories of past observations. Since we are constructing models only for the long-run, stationary behavior of a system, we must also limit the model checking of the learned system to such properties as only refer to this long-run behavior, and not to any initial transitions from a distinguished start state. We therefore define *Stationary Probabilistic Linear Time Temporal Logic* (SPLTL) as the specification language for system properties. Roughly speaking, a SPLTL property $S(\varphi)$ specifies the probability that a system run which we start observing at an arbitrary point in time during the stationary, or steady-state, operation of the system satisfies the LTL property φ.

The main contributions of this paper are: we introduce the problem of learning models for stationary system behavior, and adapt an existing learning algorithm for PSAs [10] to this task. We formally define syntax and semantics of SPLTL properties. We conduct experiments which demonstrate that model-checking SPLTL properties on learned models provides good approximations for the results that would be obtained on the true (but in reality unknown) model.

The paper is structured as follows: in Section 2 we introduce the necessary concepts relating to Markov system models, their stationary distributions, and SPLTL. Section 3 describes our method for learning PSAs and Labeled Markov chains (LMCs). Section 4 contains our experimental results. Section 5 includes conclusion and future work.

2 Preliminaries

2.1 Strings and Suffixes

Let Σ denote a (finite) alphabet, and let Σ^* and Σ^ω denote the set of all finite, respectively infinite strings over Σ. The empty string is denoted by e. For any string $s = \sigma_1 \cdots \sigma_i$, where $\sigma_i \in \Sigma$, we use the following notation:

- The longest suffix of s different from s is denoted by suffix$(s) = \sigma_2 \ldots \sigma_i$.
- suffix$^*(s) = \{\sigma_k \ldots \sigma_i | k = 1 \ldots i\} \cup \{e\}$ is the set of all suffixes of s.
- A set of strings S is *suffix free*, if for all $s \in S$, suffix$^*(s) \cap S = \{s\}$.

2.2 Markov System Models

A *Labeled Markov chain (LMC)* is a tuple $M = \langle Q, \Sigma, \pi, \tau, L \rangle$, where

- Q is a finite set of states,
- $\pi : Q \to [0,1]$ is an *initial probability distribution* such that $\sum_{q \in Q} \pi(q) = 1$,
- $\tau : Q \times Q \to [0,1]$ is the *transition probability function* such that for all $q \in Q$, $\sum_{q' \in Q} \tau(q,q') = 1$.
- $L : Q \to \Sigma$ is a *labeling function*

Labeling functions that assign to states a subset of atomic propositions AP can also be accommodated in our framework by assigning $\Sigma = 2^{AP}$. Since a Markov chain defines a probability distribution over sequences of states, an LMC M with alphabet Σ induces a probability distribution P_M^π over Σ^ω through the labeling of the states.

A subset T of Q in LMC M is called *strongly connected* if for each pair (q_i, q_j) of states in T there exists a path $q_0 q_1 \ldots q_n$ such that $q_k \in T$ for $0 \le k \le n$, $\tau(q_k, q_{k+1}) > 0$, $q_0 = q_i$, and $q_n = q_j$. If Q is strongly connected, then M is said to be strongly connected. A distribution π_M^s is a *stationary distribution* for M if it satisfies

$$\pi^s(q) = \sum_{q' \in Q} \pi^s(q')\tau(q',q). \tag{1}$$

We abbreviate $P_M^{\pi^s}$ with P_M^s. If an LMC M is strongly connected, then M defines a unique stationary distribution.

In this paper we focus on so-called *probabilistic suffix automata* (PSA). A PSA is an LMC extended with a labeling function $H : Q \to \Sigma^{\le N}$, which represents the history of the most recent visited states (a string over Σ with length at most N). Given the labeling functions L and H, each state q_i is associated with a string $s_i = H(q_i)L(q_i)$ such that, *i)* the set of strings labeling the states is suffix free, and *ii)* for any two states q_1 and q_2, if $\tau(q_1, q_2) > 0$, then $H(q_2)$ is a suffix of s_1. For example, for the PSA in Figure 1(b) the set of strings associated with the states is suffix free. Furthermore, e.g., by considering the states q_{sa} and q_{aa} we have that $H(q_{sa}) = s$, $H(q_{aa}) = a$, $L(q_{sa}) = a$ and $L(q_{aa}) = a$ which represent past and current information respectively. Here $H(q_{aa}) = a$ is a suffix of the string sa associated with q_{sa}. The latter case implies that aa is sufficient for identifying q_{aa} rather than saa. Similarly, q_s has two incoming transitions with different histories, but s is sufficient for identifying q_s. For a given $N \ge 0$, the collection of PSAs are denoted by N-PSA, where each state are labeled by a string of at length most N. In the special case, where all strings in a N-PSA is of length N, then the N-PSA is also called an N-*order labeled Markov chain*. An LMC is called *PSA-equivalent* if there exists a PSA M', such that they define the same distribution over Σ^ω ($P_M^s = P_{M'}^s$).

Example 1. The LMC M and the PSA M' in Figures 1(a) and (b) are specified over the same alphabet $\Sigma = \{s, a, b\}$ and define the same probability distribution over Σ^ω. The LMC M is therefore PSA-equivalent, but it is not a PSA since the set of strings associated with states can not be suffix free.

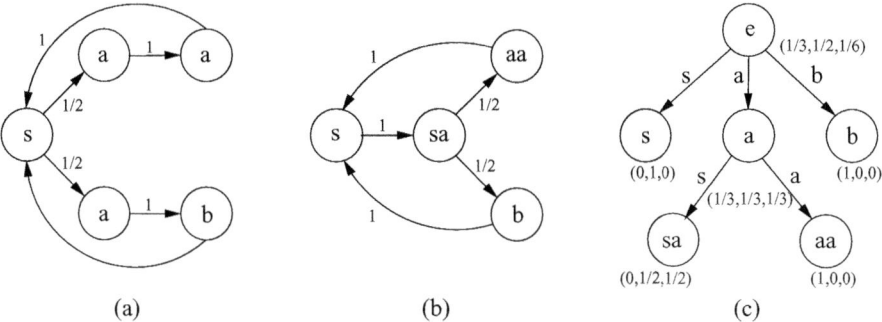

(a)   (b)   (c)

Fig. 1. The LMC in (a) defines the same probability distribution over Σ^ω as the 2-PSA in (b). The probabilistic suffix tree in (c) corresponds to the PSA in (b). The next symbol probabilities associated with the nodes follow the ordering s, a, and b.

The learning algorithm adapted in this paper attempts to find a PSA model that best describes the observed sequence of output symbols generated by a system. However, for the actual learning we will primarily consider an intermediate structure called a *prediction suffix tree* (PST) [10]. A PST over an alphabet Σ is a tree of degree $|\Sigma|$, where each outgoing edge of an internal node is labeled by a symbol in Σ. The nodes of the tree are labeled by pairs (s, γ_s); s is the string defined by labels of the edges on the path from the node in question to the root of the tree. If s' is a descendant of s, then $s \in \text{suffix}^*(s')$. $\gamma_s : \Sigma \to [0, 1]$ is the *next symbol probability function* such that $\sum_{\sigma \in \Sigma} \gamma_s(\sigma) = 1$. The probability that a PST T generates a string $str = \sigma_1 \sigma_2 \cdots \sigma_n \in \Sigma^n$ is $\prod_{i=1}^{n} \gamma_{s_{i-1}}(\sigma_i)$, where $s_0 = e$ and s_i is the label of the deepest node reached by following the links corresponding to $\sigma_i \sigma_{i-1} \cdots \sigma_1$ from the root. The PST T in Figure 1(c) shows a representation of the PSA M in Figure 1(b); the node corresponding to the suffix ba is not shown, since the probability of seeing ba is zero. Based on this PST we can, e.g., calculate the probability of seeing the string $sabsaa$ from the probabilities of the individual symbols in the string. These probabilities can be found as the next symbol probabilities of the deepest nodes in the tree that can be reached by following (in reverse order) the symbols observed so far. For example, at the root node labeled with the empty string e we have that $\gamma_e(s) = 1/3$. After seeing the string s, the probability of seeing an a is encoded at the node labeled s, where we have $\gamma_s(a) = 1$. The probability of seeing the symbol b following the string sa is encoded at the node labeled sa, where $\gamma_{sa}(b) = 1/2$. Given the string sab, the probability of seeing an s is encoded at the node labeled b, which is the deepest node in the tree reached by following the links corresponding to the symbols bas from the root. By following this procedure for each symbol in the string we get $P(sabsaa) = \gamma_e(s) \cdot \gamma_s(a) \cdot \gamma_{sa}(b) \cdot \gamma_b(s) \cdot \gamma_s(a) \cdot \gamma_{sa}(a) = \frac{1}{3} \cdot 1 \cdot \frac{1}{2} \cdot 1 \cdot 1 \cdot \frac{1}{2}$. See also [10] for further discussion about PSAs and PSTs.

2.3 Stationary Probabilistic LTL

Linear time temporal logic (LTL) [1] over the vocabulary Σ is defined as usual by the syntax

$$\varphi ::= true \mid \sigma \mid \varphi_1 \wedge \varphi_2 \mid \neg\varphi \mid \bigcirc\varphi \mid \varphi_1 U \varphi_2 \qquad (\sigma \in \Sigma)$$

For better readability, we also use the derived temporal operators $\Box$ (always) and $\Diamond$ (eventually).

Let φ be an LTL formula over Σ. For $s = \sigma_0\sigma_1\sigma_2\ldots \in \Sigma^\omega$, $s[j\ldots] = \sigma_j\sigma_{j+1}\sigma_{j+2}\ldots$ is the suffix of s starting with the $(j+1)$st symbol σ_j. The LTL semantics for infinite words over Σ are as follows:

- $s \models true$
- $s \models \sigma$, iff $\sigma = \sigma_0$
- $s \models \varphi_1 \wedge \varphi_1$, iff $s \models \varphi_1$ and $s \models \varphi_2$
- $s \models \neg\varphi$, iff $s \not\models \varphi$
- $s \models \bigcirc\varphi$, iff $s[1\ldots] \models \varphi$
- $s \models \varphi_1 U \varphi_2$, iff $\exists j \geq 0.\ s[j\ldots] \models \varphi_2$ and $s[i\ldots] \models \varphi_1$, for all $0 \leq i < j$

The syntax of *stationary probabilistic LTL (SPLTL)* now is defined as by the rule:

$$\phi ::= S_{\bowtie r}(\varphi) \quad (\bowtie \in \geq, \leq, =;\ r \in [0,1];\ \varphi \in LTL)$$

The syntax of SPLTL, thus, is essentially the same as standard probabilistic LTL (PLTL). However, the semantics will be defined in a slightly different manner. Seen as a PLTL formula, $S_{\bowtie r}(\varphi)$ would be satisfied by a LMC if traces of the Markov chain satisfy ϕ with probability $\bowtie r$, when initial states of the system are sampled according to the initial state distribution π. In the SPLTL semantics, the unique initial distribution π is replace with the set of all stationary distributions of the Markov chain, and we define for an LMC M:

$$M \models S_{\bowtie r}(\varphi) \quad \text{iff} \quad \text{for all stationary distributions } \pi^s \text{ for } M:$$
$$P_M^{\pi^s}(\{s \in \Sigma^\omega | s \models \varphi\}) \bowtie r$$

Note, in particular, that the satisfaction relation $\models$ now only depends on the transition probabilities τ of M, but not on the the initial distribution π.

The reason for this design of SPLTL is that we are interesting in analyzing behaviors of systems that are characterized by an open-ended mode of operation, and which we observe during their ongoing operation. Think, for example, of an elevator control program, a network router, or an online web-service. It will then typically be the case that the system (which may originally have started from some special initial configuration) has reached a terminal strongly connected component of states, and has converged to one of its stationary distributions. Starting to observe the system at a random point in time then corresponds to starting the observation at a state sampled from a stationary distribution. The real-world meaning of $M \models S_{\bowtie r}(\varphi)$ then is: assuming that we start observing M at a random point in time, but when M is already past a possible initial "burn-in" phase, then the probability that we see the further execution of M having property ϕ is $\bowtie r$.

3   Learning Labeled Markov Chains

Our algorithm for learning PSAs is a modified version of the method described in [10]. The modifications mostly relate to the fact that the data is to be generated by a strongly connected model. As in [10], for a given sample sequence $Seq = \sigma_1 \sigma_2 \ldots \sigma_n$, we learn a PSA by firstly constructing a PST T, and then translating T into a PSA. The PSA may be considered as an LMC after removing the labeling function $H(q_i)$ (representing past observations), and the resulting model can then directly be used in probabilistic model checker (PRISM [6] here). The translation from a PST to a PSA is performed as described in [10], and will not be discussed further. The key part in the learning process is the construction of T, which takes the form of a top-down tree-growing procedure. At any point in time, the algorithm maintains a current tree, and a set S of strings (representing suffixes) that are candidates for inclusion in the tree. In one iteration, the algorithm

(1) selects a string $s \in S$, and decides whether to add s as a node to T (which may require the addition of intermediate nodes to T that connect s to the leaf in the current T that represents the longest suffix of s contained in T).
(2) (regardless of whether s was added to T) for all $\sigma \in \Sigma$, decide whether to add σs to S.

The crucial question, now, is how exactly to define the decision criteria for (1) and (2). In [10], the decision criteria depend on a parameter as well as a prior specification of both the memory length of the PSA and an upper bound on the number of states of the PSA. The authors prove probably approximately correctness results for the learning algorithm, but the requirement of prior knowledge about model size and memory length is not compatible with our setting. Here we are going to adjust the original criteria by combining parameters and removing prior constraints. As in most statistical learning approaches, a central tool in our learning approach is the *likelihood* of a PST T given the data Seq, i.e., the conditional probability of the data given the model T:

$$L(T \mid Seq) = P(Seq \mid T). \tag{2}$$

We now base both decisions on a single parameter $\epsilon \geq 0$ that is given as input to the PST learning algorithm, and which represents the minimal improvement in likelihood that we want to obtain when adding an extra node to the PST.

For step (1) it is straightforward to compute precisely the improvement in likelihood one will obtain using a tree T containing s as a leaf, compared to the tree T' in which suffix(s) is a leaf (T and T' otherwise being equal), i.e., line 4, in Algorithm 1. We add s to the tree if the improvement is at least ϵ. Exactly the same criterion can not be used in step (2), since here we need to include σs into the candidate set not only when adding σs itself to T leads to a likelihood improvement, but also when this may happen only for some further extension $s' \sigma s$ of σs. However, one can derive a global upper bound on the maximal likelihood improvement obtainable by adding any such $s' \sigma s$, and we add σs to S if this bound is at least ϵ, i.e., line 5, in Algorithm 1. The learning algorithm is described in Algorithm 1, where the empirical (conditional) probabilities $\tilde{P}(\cdot)$ are calculated based on the sample sequence Seq.

Algorithm 1. Learn_PSA

Require:

A sample sequence Seq, and the ϵ

Ensure:

A PST $\bar{T}$

1: Initialize $\bar{T}$ and S: let $\bar{T}$ consist of a single root node (corresponding to e), and let $S = \{\sigma \mid \sigma \in \Sigma \text{ and } \tilde{P}(\sigma) \geq \epsilon\}$

2: **while** $S \neq \emptyset$ **do**

3:     (A) Pick any $s \in S$ and remove s from S

4:     (B) If

$$\tilde{P}(s) \cdot \sum_{\sigma \in \Sigma} \tilde{P}(\sigma|s) \cdot \log \frac{\tilde{P}(\sigma|s)}{\tilde{P}(\sigma|\operatorname{suffix}(s))} \geq \epsilon$$

then add s and all its suffixes which are not in $\bar{T}$ to $\bar{T}$

5:     (C)If $\tilde{P}(s) \geq \epsilon$, then for every $\sigma' \in \Sigma$, if $\tilde{P}(\sigma's) \geq 0$, then add $\sigma's$ to S

6: **end while**

7: Extend $\bar{T}$ by adding all missing sons s of internal nodes if $\tilde{P}(s) > 0$

8: For each $s \in \bar{T}$, let

$$\hat{\gamma}_s(\sigma) = \tilde{P}(\sigma|s')$$

where s' is the longest suffix of s in $\bar{T}$

The learned tree, thus, depends on the value of ϵ. Smaller ϵ lead to the construction of larger trees, and as $\epsilon \to 0$, the size of the tree will typically approach the size of the dataset (because the tree degenerates into a full representation of the data). In Machine Learning terminology, the learned tree then *overfits* the data. In order to avoid overfitting, and to learn an accurate model of the data source, rather than an accurate model of the data itself, one often employs a *penalized likelihood score* to evaluate a model. These scores evaluate candidate models based on likelihood, but subtract a penalty term for the size of the model. Common penalized likelihood scores are *Minimum Description Length* [9] and the *Bayesian Information Criterion (BIC)* [11]. The BIC score of a PSA A relative to data Seq is defined as

$$BIC(A \mid Seq) := log(L(A \mid Seq)) - 1/2\,|A|\,log(|Seq|), \qquad (3)$$

where $|Seq|$ is the length of Seq, and $|A|$ is the number of free parameters which represents the size of model, i.e. $|A| = |Q_A| \cdot (|\Sigma| - 1)$. Using a golden section search [14, Section E.1.1] we systematically search for an ϵ value optimizing the BIC score of the learned model.

4   Experiments

In order to test the proposed algorithm we have generated observation sequences from three different system models. We applied the learning algorithm on each single sampled sequence, and validated resulting models by comparing with the known generating models in terms of their SPLTL properties. For the actual comparison of the models, we considered relevant system properties expressed by LTL formulas as well as a set Φ of randomly generated LTL formulas. Formulas were generated using a stochastic context-free grammar, and each formula was restricted to a maximum length of 30.

In order to avoid generating un-interesting formulas (especially tautologies or unsatisfiable ones), we constructed a dummy model M_d with one state for each symbol in the alphabet, and with uniform transition probabilities. For each generated LTL formula $\varphi \in \Phi$ we tested whether the formula was indistinguishable by the learned model M_l, the generating model M_g, and the dummy model M_d in the sense that $P^s_{M_g}(\varphi) = P^s_{M_l}(\varphi) = P^s_{M_d}(\varphi)$. If that was the case, then φ was removed from Φ.

We compute stationary probabilities of LTL properties using the PRISM model checker [6]; in the experiments performed, all the learned models are strongly connected. PRISM provides algorithms to compute the stationary distribution over the states, and for a given LTL property φ, the probability of φ at any given start state. Combined, this allows us to compute $P^s_M(\varphi)$.

We evaluate the learned models by comparing $P^s_{M_g}(\varphi)$ and $P^s_{M_l}(\varphi)$ for certain properties φ that are of interest for the individual systems, as well as by the mean absolute difference for the random formulas in Φ:

$$D = \frac{1}{|\Phi|} \sum_{\varphi \in \Phi} |P^s_{M_g}(\varphi) - P^s_{M_l}(\varphi)| \qquad (4)$$

The mean absolute difference between M_y and M_d is calculated analogously. It is denoted D_d, and reported as a reference baseline.

We distinguish experiments in which the data was generated by a PSA-equivalent model, and experiments where the generating model is not exactly representable by a PSA.

4.1   Learning Models of PSA-Equivalent Systems

Phone Model. For our first experiment we use a toy model for a telephone. We consider observable state labels i (the phone is idle), r (ringing), t (talking), h (phone is hung up), and p (receiver picked up). The PSA model in Fig. 2 encodes that the probability of a ringing phone being picked up depends on the elapsed time since it has been used last (which can indicate that the phone owner has left in the meantime). To this end, the model has a limited memory for how many time units the phone has been idle since it has last been hung up (or since it has last been ringing without being answered), and, e.g., the probability $P(p|hr)$ is higher than $P(p|hir)$. The model has a memory of histories of at most length 4, but in many cases only a shorter history is relevant for determining the transition probabilities. For example, once the phone is picked up (transition to the state with suffix label p), the previous history becomes irrelevant.

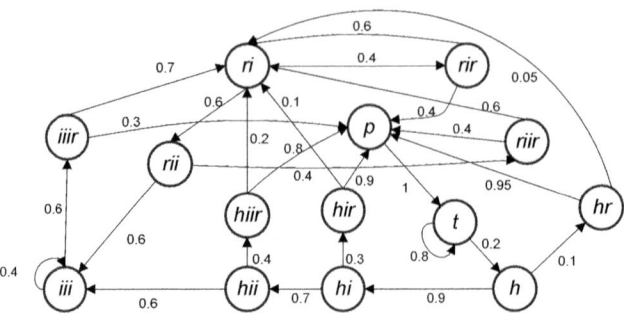

Fig. 2. The phone model. States are labeled by (i)dle,(r)ing, (p)ick-up,(h)ang-up and (t)alk.

Experimental results for this model are summarized in Table 1. The first four columns in the table show: $|Seq|$: the length of the sequence (generated by the model from Figure 2) from which the model was learned; time: time in seconds for the full learning process, including multiple learning runs with different ϵ parameters, and golden section search for optimizing the BIC score; order: the order of the learned model, i.e., the maximal length of a suffix label in the model; $|Q_l|$: number of states in LMC.

The results show that for smaller data sizes a too simple model consisting of a simple Markov chain over the 5 symbols in the alphabet is learned. With more data, the correct structure of the model with its order 4 and 14 states is identified.

Columns 5-10 of Table 1 show the accuracy obtained for checking SPLTL formulas. Column D shows the average error (refer to Equation 4) for 507 random formulas. For comparison: $D_d = 0.1569$. The remaining columns show the stationary probabilities for selected properties of interest. Column t simply contains $P_M^s(t)$, i.e., the long-run frequency of the phone being busy. Column $rp \mid r$ shows the stationary *conditional* probability for the LTL formula $\varphi = r \wedge \bigcirc p$, given that r holds, i.e., the stationary probability that the ringing phone is picked up. Similarly, the next two columns show the probability that the phone is picked up, given that it is ringing, and has been idle for (at least) one, respectively two, time intervals before. Finally, $\Diamond \square\, i$ is the (unbounded) property that eventually the phone will be idle forever. The results show that the learned models provide very good approximations for the SPLTL properties of the generating model.

Randomized Self-stabilizing Protocol. Consider now the randomized self-stabilizing protocol by [4]. This algorithm is designed for ring networks with an odd number of processes, and where each process p_i is equipped with a Boolean variable X_i. The protocol operates synchronously such that if $X_i = X_{i-1}$, then p_i makes a uniform random choice about the next value of X_i; otherwise it sets X_i to the current value of X_{i-1}. Each pair of neighboring processes with the same value assigned to their Boolean variables generates a "token". The network is stable if it only contains a single token. In order to obtain a strongly connected

Table 1. Experimental results for the phone model

$|Seq|$	time(sec)	order	$|Q_l|$	D	t	$rp\|r$	$irp\|ir$	$iirp\|iir$	$\Diamond\Box\, i$
80	9.1	1	5	0.0551	0.253	0.333	0.333	0.333	0
160	4	1	5	0.0096	0.370	0.407	0.407	0.407	0
320	6.2	1	5	0.0281	0.344	0.310	0.309	0.309	0
640	6.13	1	5	0.0094	0.392	0.424	0.424	0.424	0
1280	7.5	1	5	0.0064	0.385	0.446	0.446	0.446	0
2560	11.9	1	5	0.0089	0.366	0.447	0.447	0.447	0
5120	36.9	3	10	0.0020	0.379	0.490	0.490	0.490	0
10240	225.2	4	14	0.0014	0.381	0.506	0.477	0.409	0
20480	456.5	4	14	0.0005	0.378	0.515	0.489	0.414	0
M_g		4	14		0.378	0.512	0.488	0.424	0

model we have modified the original protocol: after reaching a stable state each process will set its Boolean variable to 0, thus returning to an unstable state.

Using the protocol above we have analyzed the behavior of the learning algorithm by varying the number of processes and the length of the observed sample sequence as well as by changing the level of abstraction. In the first experiment, symbols in the sample sequence correspond to a value assignment to all the Boolean variables associated with processes. Thus, with N processes, there are 2^N symbols in Σ. In the second experiment, we replaced the symbols in the sequence with more abstract labels that only represent the number of tokens defined by the value assignments. For N processes, the alphabet Σ then only contains N symbols.

The results of the experiments are given in Figure 3 and Table 2. Figure 3 shows the probability $P_M^s(true\ \mathsf{U}_{\leq L}\ stable|token = N)$ of reaching a stable configuration within L steps conditional on being in a (starting) configuration where all processes assign the same value to the Boolean variables. In general, we observe a very good fit between the probability values computed for the different models (having the same number of processes). One notable difference is the probability values calculated for the full 7 processes model compared to the abstract and the real 7 processes models. We believe that this discrepancy is due to the length of the sample sequence being insufficient for learning an accurate full model. This hypothesis is supported by the results in Table 2. The table lists the learning time (time), the order of the learned PSA (order), the number of iterations performed by the golden section search (iter), the number of states in LMC ($|Q_l|$), and the average difference in probability (D) calculated according to Equation 4 using 503 random LTL formulas. In particular, we see that with 10240 symbols, the learned full model only contains a single state, whereas the abstract model has four states and a lower average difference in probability. Note, however, that with 50000 symbols the algorithm learns the correct order and number of states for the full model and the average difference in probability becomes significantly smaller.

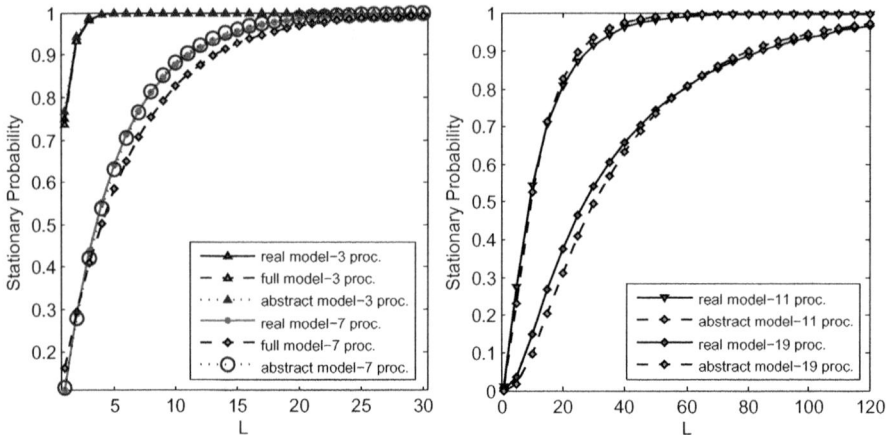

Fig. 3. Left: Experiment results for 3 processes and 7 processes. Right: Experiment results for 11 processes and 19 processes. For 3, 7 and 11 processes, both full and abstract models were learned from 10240 symbols; the abstract 19 process model was learned from 20480 symbols.

From Table 2, we also see (as expected) that the time complexity of learning an abstract model is significantly lower than that of learning a full model. Note that due to time complexity, we have not learned full models for networks with 11 and 21 processes.

Since the abstract models are often significantly smaller than the generating models, the time required for model checking using the abstract models is also expected to be lower. We have analyzed this hypothesis further by measuring the time complexity for evaluating the SPLTL property $P_M^s(true\ U_{\leq L}\ stable|token = N)$ for 19 and 21 processes. For the generating model, the total time is calculated as the time used for compiling the PRISM model description to the internal PRISM representation as well as the time used for the actual model checking. For the abstract model, the total time is calculated as the time used for model learning (which produces a model in the PRISM file format), model compilation, and model checking. Fig. 4 shows the time used by both approaches as a function of L. The time complexity of using the abstract models is close to constant. It consists of a constant time (253 sec. and 284 sec., respectively) for model learning and model compilation, and a negligible additional linear time for model checking.

4.2   Learning Models of Non PSA-Equivalent Systems

Consider the LMC in Figure 5(a), which is a modified version of the model by Knuth and Yao [5] that uses a fair coin to simulate the toss of a six-sided die. For example, *start, H, H, H, T, h2* corresponds to a die toss of 2. Compared to

Table 2. Experimental results for the self-stabilizing protocol with 7 processes. D is based on 503 random LTL formulas. For reference: $D_d = 0.1669$.

	Full model					Abstract model				
$|Seq|$	time(sec)	order	iter	$|Q_l|$	D	time(sec)	order	iter	$|Q_l|$	D
80	73.0	0	30	1	0.0192	1.6	1	38	4	0.0172
160	49.4	0	23	1	0.0325	2.1	1	41	4	0.0079
320	162.9	0	29	1	0.0292	3.3	1	41	4	0.0369
640	34.3	0	19	1	0.0234	2.3	1	23	4	0.0114
1280	37.2	0	19	1	0.0193	4.1	1	32	4	0.0093
2560	42.0	0	19	1	0.0204	5.0	1	23	4	0.0054
5120	47.9	0	19	1	0.0182	8.9	1	23	4	0.0018
10240	59.3	0	19	1	0.0390	16.3	1	23	4	0.0013
20480	80.7	0	19	1	0.0390	31.4	1	23	4	0.0016
50000	1904.4	1	25	128	0.00034	152.42	1	23	4	0.0011
100k	3435.5	1	25	128	0.00071	308.9	1	23	4	0.0007

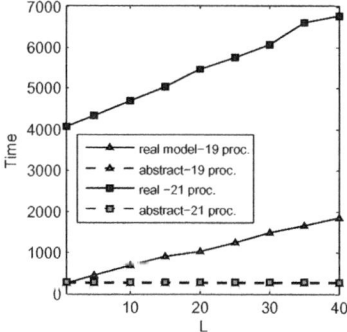

Fig. 4. The time for calculating $P^s_M(true\mathsf{U}_{\leq L}\ stable\,|token = N)$ (N is the number of process in each model) in the generating model and abstract model. Both abstract models for 19 and 21 processes are learned from a single sequence with 20480 symbols.

the original model, the model in Figure 5(a) makes a transition back to the start state after having simulated the outcome of a toss.

In this LMC we see that the next symbol probabilities for the two states labeled H on the top branch differ. Specifically, we have that the next symbol probability depends on whether or not we have seen an even or an odd number of Hs, which implies that the model in Figure 5 cannot be represented by any N-order Markov chain and, in particular, any N-PSA. Note that this also implies that the dice model is not PSA-equivalent. An example of a model that was learned from a sample sequence with 1440 observations can be seen in Figure 5(b).

The results of all the experiments are summarized in Table 3. From the table we see that the learned models provide very good approximations for the stationary probabilities of the randomly generated LTL formulas. For example, for

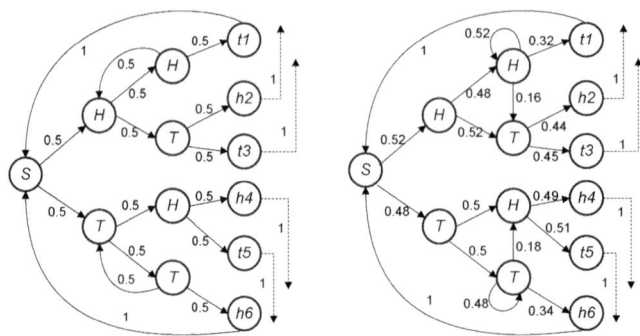

Fig. 5. Dice model. Dash lines will lead to the 'S' state. (a) The generating model. (b) A model learned from a sequence with 1440 symbols.

Table 3. The experiment results for dice model. See table 2 for a description of the columns in the left part of the table. For the right part of the table, D is the mean absolute difference of the learned model and the generating model for stationary probabilities of 501 randomly generated LTL formulas, and $P_M^s(i)$ denotes the stationary probability of getting a i in the next dice toss, and the stationary probability is $1/6$ for each number in the generating model.

$|Seq|$	time(sec)	order	$|Q_l|$	$|D|$	$P_M^s(1)$	$P_M^s(2)$	$P_M^s(3)$	$P_M^s(4)$	$P_M^s(5)$	$P_M^s(6)$
360	11.4	2	13	0.0124	0.137	0.17	0.182	0.103	0.205	0.203
720	14.4	2	13	0.0043	0.188	0.174	0.174	0.149	0.168	0.147
1440	16.9	2	13	0.0023	0.184	0.166	0.169	0.143	0.153	0.185
2880	57.4	4	17	0.0023	0.173	0.166	0.159	0.142	0.176	0.184
5760	90.5	4	17	0.0016	0.173	0.165	0.153	0.161	0.174	0.174
11520	159.4	5	19	0.00094	0.162	0.17	0.176	0.157	0.168	0.167
20000	318.4	6	21	0.00092	0.164	0.173	0.171	0.166	0.164	0.162

the model learned from 20000 observations the mean absolute difference in probability is 0.00092 for 501 random LTL formulas; in comparison, the difference in probability is 0.1014 for the dummy model. A similar behavior is observed for the probability $P^s(i)$ of getting i in the next dice toss.

Finally, we note that the size of the learned model grows as the length of the data sequence increases. This behavior is a consequence of the generating model not being representable by any N-order Markov chain. To illustrate the effect, Figure 6 shows the structure of the model that was learned from 20000 observations. Notice that the differences between the learned model and the generating model relates to the part of the model encoding the number of times we have seen an even number of Hs.

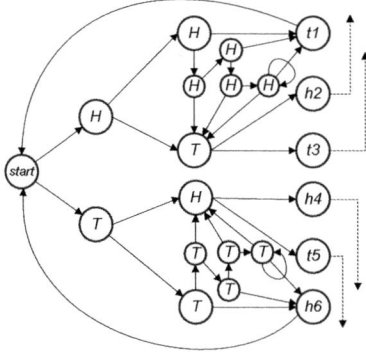

Fig. 6. The dice model learned from a sequence with 20000 symbols. Numbers on edges are omitted.

5   Conclusion

In this paper we proposed to use methods for learning Probabilistic Suffix Automata to learn a formal model for the stationary behavior of a system from a single observation sequence. Compared to previous approaches of learning system models for verification purposes, this extends the scope of applications to scenarios where one can easily obtain data by passively observing a system in its ongoing operation, but where it is difficult to obtain multiple, independent runs under laboratory conditions.

The analysis of the learned model must be restricted to properties that only concern the observed stationary behavior of the system, for which purpose we have introduced SPLTL properties as a suitable specification language. Experimental results show that model-checking SPLTL properties on learned models provides a good approximation to model-checking the true, data-generating system. This can often even be the case when the true system itself is not exactly equivalent to a PSA, in which case the learned model can only be an approximation to the true model.

As in the learning from multiple sequences setting, one could also in the single sequence setting consider methods of *statistical model checking* [15,13,7] as an alternative to model learning. While existing statistical model checking approaches are also based on the assumption that data consists of multiple independent system runs, they could be easily adapted to the single sequence case. As discussed in [8], model learning offers several advantages over statistical model checking: learned models also support the model checking of unbounded properties, whereas statistical model checking is limited to bounded properties. Moreover, learned models can support additional analysis and design processes beyond model-checking.

For learning deterministic stochastic automata from multiple sample strings strong consistency results guarantee that in the large sample limit the learned and the true system agree on the probabilities of LTL formulas [2,8]. Similar results are not yet established for learning PSAs from a single sequence. Existing results on the consistency of the BIC selection criterion for learning variable order Markov chains [3] strongly indicate that such consistency properties also hold for our learning method, but a full analysis is subject of further work.

References

1. Baier, C., Katoen, J.P.: Principles of Model Checking. The MIT Press (2008)
2. Carrasco, R., Oncina, J.: Learning Stochastic Regular Grammars by Means of a State Merging Method. In: Carrasco, R.C., Oncina, J. (eds.) ICGI 1994. LNCS, vol. 862, pp. 139–152. Springer, Heidelberg (1994)
3. Csiszár, I., Talata, Z.: Context tree estimation for not necessarily finite memory processes, via BIC and MDL. IEEE Transactions on Information Theory 52(3), 1007–1016 (2006)
4. Herman, T.: Probabilistic self-stabilization. Information Processing Letters 35(2), 63–67 (1990)
5. Knuth, D., Yao, A.: The complexity of nonuniform random number generation. In: Algorithms and Complexity: New Directions and Recent Results. Academic Press (1976)
6. Kwiatkowska, M., Norman, G., Parker, D.: PRISM 4.0: Verification of Probabilistic Real-Time Systems. In: Gopalakrishnan, G., Qadeer, S. (eds.) CAV 2011. LNCS, vol. 6806, pp. 585–591. Springer, Heidelberg (2011)
7. Legay, A., Delahaye, B.: Statistical model checking: An overview. CoRR, abs/1005.1327 (2010)
8. Mao, H., Chen, Y., Jaeger, M., Nielsen, T., Larsen, K., Nielsen, B.: Learning probabilistic automata for model checking. In: 2011 Eighth International Conference on Quantitative Evaluation of Systems, pp. 111–120 (September 2011)
9. Rissanen, J.: Modeling by shortest data description. Automatica 14(5), 465–471 (1978)
10. Ron, D., Singer, Y., Tishby, N.: Power of amnesia: Learning probabilistic automata with variable memory length. Machine Learning 25, 117–149 (1996)
11. Schwarz, G.: Estimating the dimension of a model. The Annals of Statistics 6(2), 461–464 (1978)
12. Sen, K., Viswanathan, M., Agha, G.: Learning continuous time markov chains from sample executions. In: International Conference on Quantitative Evaluation of Systems, pp. 146–155 (2004)
13. Sen, K., Viswanathan, M., Agha, G.: Statistical Model Checking of Black-Box Probabilistic Systems. In: Alur, R., Peled, D.A. (eds.) CAV 2004. LNCS, vol. 3114, pp. 202–215. Springer, Heidelberg (2004)
14. Tan, P.-N., Steinbach, M., Kumar, V.: Introduction to Data Mining. Addison Wesley (2006)
15. Younes, H.L.S., Simmons, R.G.: Probabilistic Verification of Discrete Event Systems Using Acceptance Sampling. In: Brinksma, E., Larsen, K.G. (eds.) CAV 2002. LNCS, vol. 2404, pp. 223–235. Springer, Heidelberg (2002)

The Use of Rippling to Automate Event-B Invariant Preservation Proofs

Yuhui Lin, Alan Bundy, and Gudmund Grov

School of Informatics, University of Edinburgh, UK

Abstract. Proof automation is a common bottleneck for industrial adoption of formal methods. In Event-B, a significant proportion of proof obligations which require human interaction fall into a family called *invariant preservation*. In this paper we show that a rewriting technique called rippling can increase the automation of proofs in this family, and extend this technique by combining two existing approaches.

Keywords: Event-B, automated reasoning, rippling, lemma conjecture.

1 Introduction

Event-B [2] is a "top-down" formal modelling method, which captures requirements into an abstract formal specification and then stepwise refines the specification into the final product. In each step of development, designers justify the correctness of the specification by proving proof obligations (POs). The generation of the POs is automated by the Rodin platform [3] which also contains support for both automatic and interactive theorem proving. Most of these POs can be discharged automatically, yet still 3% to 10% of them require human interaction [8]. In an industrial sized project, this proportion of interactive proofs can be thousands. For example, 43,610 POs with 3.3% interactive proof in the Roissy Airport Shuttle project, and 27, 800 POs with 8.1% interactive proof in the Paris Metro line 14 project [1]. Moreover, specifications often change frequently, which require users to reprove previous proven POs.

Invariant preservation (INV) POs is a family of Event-B POs which can account for a significant part of all of the POs that require human interaction. To illustrate, 188 out of 317 (59%) undischarged POs in the BepiColombo case study[1] belong to the INV family.

Here we argue that part of the problem is a lack of meta-level reasoning since Rodin provers only work on the object level logic. In [5] it is argued that, to achieve a better understanding of reasoning, both logic and a meta-level understanding should be put into consideration. In this paper we propose to use a meta-level reasoning technique, called *rippling* [6] for Event-B POs. It is a rewriting technique which can be applied when the goal embeds in one of the hypothesis, and details are given in Sect. 2.2. Our hypothesis is

---

[1] The case study can be found in http://deploy-eprints.ecs.soton.ac.uk/136/

A. Goodloe and S. Person (Eds.): NFM 2012, LNCS 7226, pp. 231–236, 2012.
© Springer-Verlag Berlin Heidelberg 2012

By utilising rippling we can increase the automation of Event-B in-
variant preservation proof obligations and make proofs more robust to
changes.

The contributions of this paper are two-fold: (1) In Sect. 2 we illustrate the use of
rippling for INV POs; (2) The key advantage of rippling is that, due to meta-level
reasoning, we can often patch a broken proof, thus making proofs more robust to
changes. In Sect. 3 we describe a novel combination of two existing techniques,
lemma speculation and scheme-based theory exploration, to conjecture lemmas
when proofs are blocked by the absence of lemmas. Finally, we describe further
work and conclude in Sect. 4 and Sect. 5, respectively.

2   INV POs and Rippling

2.1   INV POs

Machines are key components of Event-B specifications. A machine contains
variables, invariant and events. Variables represent the states of the machine,
and invariants describe constraints on the states. INV POs are generated to
guarantee that the invariants are still preserved under changes made by the
events.

any
  s
where
  $s \in Subs \wedge s \notin \text{dom } call$
then
  $call := call \cup \{(s \mapsto (seize \mapsto \varnothing))\}$
end

Notation	Definition
$x \mapsto y$	denotes the pair (x, y)
$\text{dom}(r)$	$\{x \mid \exists y.x \mapsto y \in r\}$
$s \triangleleft r$	$\{x \mapsto y \mid x \mapsto y \in r \wedge x \notin s\}$
$r \triangleright s$	$\{x \mapsto y \mid x \mapsto y \in r \wedge y \in s\}$
$r \triangleleft\!\!\!- s$	$s \cup (\text{dom}(s) \triangleleft r)$
$r \, ; s$	$\{x \mapsto y \mid \exists z.x \mapsto z \in r \wedge z \mapsto y \in s\}$

Fig. 1. An example of events & mathematical notions

To illustrate, let us consider the invariant (1)

$$Callers = \text{dom}((call \, ; st) \triangleright Connected) \tag{1}$$

in which *Callers*, *call*, *st* and *Connected* are variables. Figure 1 defines an event
in which *s* is an argument followed by a *guard* and an *action*, describing how the
state changes.

    An INV PO (2) is generated to ensure (1) holds under changes made in the
event.

$$s \in Subs \wedge s \notin \text{dom } call \wedge Callers = \text{dom}((call \, ; st) \triangleright Connected)$$
$$\vdash \tag{2}$$
$$Callers = \text{dom}((call \cup \{(s \mapsto (seize \mapsto \varnothing)))\} \, ; st) \triangleright Connected)$$

2.2 Rippling

Rippling[2] is a rewriting technique which was developed originally for inductive proofs. Although it was designed to guide the step cases of inductive proofs, it is applicable in any scenario where one of the assumptions can be embedded in the goal. In the case of INV POs, the invariant is embedded in the goal, thus making rippling applicable. The embedding in the goal is called the *skeleton*, while the differences are the *wave-front*.

To illustrate, let us recall the INV PO in Sect. 2.1. The annotated version of the goal (2) with the embedding (1), becomes:

$$Callers = \text{dom}((\boxed{call} \cup \{(s \mapsto (seize \mapsto \varnothing)))\} \ ; st) \triangleright Connected)$$

in which the wave-fronts are shaded by a box. When applying a rewrite rule, the skeleton (i.e. the non-shaded part) in a goal should be preserved, and a *ripple measure* must decrease, e.g. skeletons which are separated by wave-front are moving together. Therefore, by applying

$$(f \cup g) \ ; S = (f \ ; S) \cup (g \ ; S)$$

we have

$$Callers = \text{dom}(\boxed{((call} \ ; st) \cup (\{(s \mapsto (seize \mapsto \varnothing)))\} \ ; st)) \triangleright Connected)$$

With the following two rules

$$(f \sqcup g) \triangleright S = (f \triangleright S) \sqcup (g \triangleright S)$$
$$\text{dom}(f \cup g) = \text{dom} f \cup \text{dom} g$$

we have

$$Callers = \text{dom}(\boxed{(call \ ; st \triangleright Connected)} \cup (\{(s \mapsto (seize \mapsto \varnothing)))\} \ ; st) \triangleright Connected)$$

and finally get the INV PO to

$$Callers = \boxed{\text{dom}((call \ ; st) \triangleright Connected) \cup \text{dom}((\{(s \mapsto (seize, \varnothing))\} \ ; st) \triangleright Connected)}$$

Then the hypothesis (1) can be applied to simplify the proof by substituting $\text{dom}((call \ ; st) \triangleright Connected)$ with $Callers$, which results in

$$Callers = Callers \cup \text{dom}((\{(s \mapsto (seize, \varnothing))\} \ ; st) \triangleright Connected)$$

This step of using the hypothesis is called *fertilisation*. Now as st projects the second first argument (e.g. $(x \mapsto (y \mapsto z))$ into $(x \mapsto y)$) and $Connected =$

---

[2] Rippling has been implemented in *IsaPlanner* which we use to start our experiments. For more details about IsaPlanner, please refer to
http://dream.inf.ed.ac.uk/projects/isaplanner/

$\{ringing, speech, suspended\}$, $dom((\{(s \mapsto (seize, \varnothing))\} \,;\, st) \triangleright Connected)$ can be trivially simplified into $\varnothing$ by using the assumptions in (2). This completes the proof.

The advantage of rippling is the meta-level guidance which can be used to guide the application of rewrite rules. Due to its lack of meta-level guidance, many rules[3] can only be applied manually in Rodin. Let us consider the following distribution rule:

$$(p \cup q) \lhd r = (p \lhd r) \cup (q \lhd r)$$

We can generate two rewrite rules from the left hand side (lhs) to the right hand side (rhs) and the other way round. Rodin does not add both of them into its rewriting system, because the rewriting may not terminate. However, due to the meta-level guidance of rippling, both directions can be added in rippling whilst guaranteeing termination. For example:

$$From\_lhs\_to\_rhs : \boxed{(p \cup q)} \lhd r \Rightarrow \boxed{(p \lhd r)} \cup \boxed{(q \lhd r)}$$

$$From\_rhs\_to\_lhs : \boxed{(p \lhd r)} \cup \boxed{(q \lhd r)} \Rightarrow \boxed{(p \cup q) \lhd r}$$

3   IsaScheme and Lemma Conjecture

The key advantage of rippling is that the strong expectation of how the proof should succeed can help us to build a proof patching mechanism when a proof is blocked, e.g. due to a missing lemma. It can contribute to proof automation and makes proofs more robust to change. This mechanism is known as proof critics [7]. One useful critic in our case is lemma speculation [7]. It's applicable when proofs are blocked due to a missing lemma. Meta-level annotations are used to guide and construct the lemma being conjectured. Consider the following blocked rippling proof:

$$Callers = dom(\; \boxed{(call \lhd x)} \;;\, st \triangleright Connected)$$

we construct the left hand side of the missing lemma with one of our wave-fronts and parts of skeletons, which is $\boxed{(call \lhd x)} \;;\, st$. With the skeleton preservation rule, we can construct the right hand side of the lemma by introducing a meta-variable $?F_1$ to represent the unknown part. Then we have

$$\boxed{(call \lhd x)} \;;\, st = \boxed{?F_1} \, call \,;\, st \,\boxed{x}\, st$$

This meta-variable is stepwise instantiated by unification during the proof. This approach is called middle-out reasoning [9]. However, higher-order unification brings a challenge for this approach.

Therefore, instead of using middle-out reasoning, we propose a new approach by using IsaScheme[10] which is a scheme-based approach to instantiate these

---

[3] For more rules, please refer to
http://wiki.event-b.org/index.php/All_Rewrite_Rules

meta-variables, to generate the missing lemmas. Given a scheme and candidate terms and operations, IsaScheme can instantiate the meta-variables and return lemmas that pass a counter-example checker. The scheme will help constrain the lemmas generated, and we can further filter out those that will not provide valid ripple steps. The following algorithm shows more details about how to construct a scheme and get the potential lemmas with the example showed in the beginning of this section.

Precondition: When no rewriting rules are applicable and fertilisation can not be applied, the following process can be triggered.

1. Construct the lhs of the scheme with a wave-front and part of skeletons, i.e.
   $\boxed{(call \lhd x)}$; $st = ...$
2. Since the skeleton has to be preserved and a ripple measure must decrease, we can partially predict how the term evolves on the rhs.
   i.e. evolve from $\boxed{(call \lhd x)}$; st to $\boxed{call ; st \lhd ...}$
   Also we need to construct a term with meta-variables to specify the new shape of combination of the constants and variables in the wave-fronts, i.e. x, and those next to the wave-fronts in the skeleton, i.e. st. In our example this term would be $(?F_2 \; x \; st)$. Now we introduce meta-variables to combine these terms to compose the rhs of the missing lemma, i.e.
   $... = \boxed{?F_1 (call ; st)(?F_2 \; x \; st)}$.
3. Now we have a scheme to instantiate. i.e.

   $Myscheme \; ?F_1 \; ?F_2 \equiv (call \lhd r) ; st = ?F_1 \; (call ; st) \; (?F_2 \; r \; st)$

   in which $Myscheme$ is the name of our scheme; $?F_1$ and $?F_2$ are meta-variables to be instantiated from a set of given terms. Then we try this scheme in IsaScheme with relevant proof context, including assumptions.
4. IsaScheme returns potential lemmas which we can apply to unblock the current proof. In our example, we get the following lemma which can help to proceed the proof[4]

   $$(call \lhd r) ; st = (call ; st) \lhd (r ; st)$$

4 Further Work

These schemes are currently deduced manually, and next we plan to automate this process. Moreover, we observe that many POs contain quantifiers and are conditional. Rippling is not particularly suited for such POs, and to handle this, we are currently exploring integrating a technique called *piecewise fertilisation* [4] with rippling. Longer term we will develop a rippling plug-in for Rodin, which will be based on the existing Rodin to Isabelle translator by Schmalz[5].

---

[4] This lemma relies on some properties of the specification
[5] See http://wiki.event-b.org/index.php/Export_to_Isabelle for details.

5   Conclusion

We have showed that the use of rippling can improve the automation of proofs in Event-B INV POs and make it more robust to changes. We have combined lemma speculation and scheme-based theory exploration to discovery the missing lemma when proofs are blocked. This has to be done manually in Rodin. Moreover, with meta-level reasoning and its patching mechanism, the robustness of proofs can be improved, as the proof strategy remains the same even if the POs are required to be re-proven when specifications change.

Acknowledgement. This work is supported by EPSRC grant EP/H024204/1 (AI4FM). Thanks to Omar Montano Rivas, Andrew Ireland, Moa Johansson and the AI4FM partners for the useful discussions.

References

1. Abrial, J.R.: Formal methods in industry: achievements, problems, future. In: Proceedings of the 28th International Conference on Software Engineering, pp. 761–768. ACM (2006)
2. Abrial, J.R.: Modeling in Event-B - System and Software Engineering. Cambridge University Press (2010)
3. Abrial, J.R., Butler, M.J., Hallerstede, S., Hoang, T.S., Mehta, F., Voisin, L.: Rodin: an open toolset for modelling and reasoning in event-B. STTT 12(6), 447–466 (2010)
4. Armando, A., Smaill, A., Green, I.: Automatic synthesis of recursive programs: The proof-planning paradigm. Autom. Softw. Eng. 6(4), 329–356 (1999)
5. Bundy, A.: A Science of Reasoning. In: Stickel, M.E. (ed.) CADE 1990. LNCS, vol. 449, pp. 633–640. Springer, Heidelberg (1990)
6. Bundy, A.: Rippling: meta-level guidance for mathematical reasoning, vol. 56. Cambridge Univ. Pr. (2005)
7. Ireland, A.: Productive use of failure in inductive proof. Journal of Automated Reasoning 16(1-2), 79–111 (1996)
8. Jones, C.B., Grov, G., Bundy, A.: Ideas for a high-level proof strategy language. Tech. Rep. CS-TR-1210, School of Computing Science, Newcastle University (2010)
9. Kraan, I., Basin, D., Bundy, A.: Middle-out reasoning for synthesis and induction. Journal of Automated Reasoning 16(1), 113–145 (1996)
10. Montano-Rivas, O., McCasland, R.L., Dixon, L., Bundy, A.: Scheme-Based Synthesis of Inductive Theories. In: Sidorov, G., Hernández Aguirre, A., Reyes García, C.A. (eds.) MICAI 2010, Part I. LNCS, vol. 6437, pp. 348–361. Springer, Heidelberg (2010)

Thread-Modular Model Checking with Iterative Refinement[*]

Wenrui Meng[1,2,3], Fei He[1,2,3], Bow-Yaw Wang[4], and Qiang Liu[1,2,3]

[1] School of Software, Tsinghua University
[2] Tsinghua National Laboratory for Information Science and Technology (TNList)
[3] Key Laboratory for Information System Security, MOE, China
[4] Institute of Information Science, Academia Sinica

Abstract. Thread-modular analysis is an incomplete compositional technique for verifying concurrent systems. The heuristic works rather well when there is limited interaction among system components. In this paper, we develop a refinement algorithm that makes thread-modular model checking complete. Our algorithm refines abstract reachable states by exposing local information through auxiliary variables. The experiments show that our complete thread-modular model checking can outperform other complete compositional reasoning techniques.

1 Introduction

Compositional reasoning is a promising technique to alleviate the state explosion problem in model checking [2,17]. In compositional reasoning, one decomposes a verification problem into simpler subproblems and solves each subproblem one at a time. By the soundness of decomposition, the verification problem is solved if all subproblems are solved. Soundness of decomposition apparently depends on the underlying computation model. In this paper, we are interested in verifying invariant properties on shared-memory interleaving systems.

A shared-memory interleaving system consists of several components. Each component has two types of variables. Global variables are accessible to every component in the system. Local variables, on the other hand, are only accessible to the defining component. At any moment, exactly one component is active. Inactive components do not perform any computation and hence keep their local variables unchanged. Global variables may be updated by the active component nonetheless. In such systems, global variables are used for communication among components. Given a predicate on system states, the invariant checking problem is to verify whether the given predicate holds on every reachable states.

Two compositional techniques for the invariant checking problem on shared-memory interleaving systems are known. In thread-modular reasoning [9,5],

---

[*] This work was supported in part by the Chinese National 973 Plan under grant No. 2010CB328003, the NSF of China under grants No. 60903030, the Tsinghua University Initiative Scientific Research Program.

A. Goodloe and S. Person (Eds.): NFM 2012, LNCS 7226, pp. 237–251, 2012.

one computes an over-approximation of reachable system states by intersecting reachable component states of all components. In order to compute reachable component states of a designated component, one disregards local variables of other components and computes an abstract model of global variables. The designated component is then composed with the abstract model to compute its reachable component states. The effectiveness of thread-modular reasoning depends on the abstraction. If the abstraction of global variables is able to establish the property, one concludes the verification. Otherwise, one reports that the verification is inconclusive.

The (in)effectiveness problem in thread-modular reasoning is solved in the second compositional technique called local proof [4]. In local proof, one still requires reachable states of each component. Reachable component states however are computed by early quantification of reachable system states. Abstract models for global variables are hence not needed. Moreover, techniques have been developed to refine reachable component states. Local proof is hence a complete compositional technique for shared-memory interleaving systems.

Although both techniques compute reachable component states and use the intersection as an over-approximation of reachable system states, we would like to point out a subtle difference between them. In thread-modular reasoning, one constructs an abstract model for global variables. Reachable component states are then computed via the abstract model. In local proof, on the other hand, reachable component states are computed by quantifying out inaccessible local variables during the exploration of reachable system states. Since no abstraction is deployed during the exploration of component states, reachable component states in local proof are more precise than those of thread-modular reasoning. On the other hand, the computation of reachable component states in local proof can be more expensive then thread-modular reasoning due to no abstraction. One wonders whether an efficient yet complete compositional technique exists for such systems.

Inspired by the refinement in local proof, we propose a complete thread-modular model checking algorithm for the invariant checking problem on shared-memory interleaving systems. Our technique contains two phases. At the verification phase, we apply thread-modular reasoning to the verification problem. If the compositional technique suffices to conclude the verification, we are done. Otherwise, our technique moves to the refinement phase. In the other phase, we adopt ideas from local proof and expose information about local variables during refinement. More precisely, we identify local variables that can refine the approximation to reachable system states. Such information is then exposed to other components by adding global variables. When our technique returns to the verification phase, added variables will induce a refined abstract model for global variables. Efficiency of thread-modular reasoning and effectiveness of local proof are thus attained by our proposed technique.

We implement our thread-modular model checking with iterative refinement algorithm on NuSMV, and compare with other algorithms in five examples. Due to its aggressive abstraction, thread-modular reasoning fails to verify all examples but the bakery algorithm. Our new technique performs better than local proof in our examples. In several examples, our compositional technique outperforms monolithic techniques in orders of magnitude. Our preliminary experimental results suggest that an efficient yet complete compositional technique is indeed possible for shared-memory interleaving systems.

1.1   Related Work

In 1976, Owicki and Gries proposed some non-interference proof rules for parallel programs in their work [15]. Chandy and Misra [13] and Jones [9] [10] extended those rules with interference to introduce thread-modular reasoning. To make thread-modular model checking automatic, the environment is automatically generated [5] according to the interactions of the programs. Henzinger et al. [7] [8] improved the original thread-modular model checking and made it complete for safety property verification on finite state systems. In their approaches, each thread is initialized as true and is then iteratively refined by addition of new predicates, and the guarantee of each thread is initialized as false and is successively refined by considering abstract of current thread and guarantees of other threads. Recently, Gu et al. [6] attempted to improve the generation of environment assumptions with horn logic deductive rule. Malkis et al. [12] proposed a technique, called thread-modular counterexample guided abstraction refinement, which computes reachable states with cartesian abstraction. A refinement step was involved to eliminate the infeasible states by excluding them from the cartesian product. But this approach directly computes the reachable states for all processes of concurrent system in an explicit way.

Another interesting branch for concurrent system verification is based on the inductive invariant rule. The invisible invariants method [16] [1] generated quantified invariants for parameterized protocols by analyzing reachable states of a small instance; however, it is incomplete for some protocols. Absorbing the completeness theory of [15] and [11], Namjoshi extended the inductive invariant to non-interference invariant named split invariant [14]. Based on split invariant, Cohen and Namjoshi proposed a local proof algorithm for global safety properties of concurrent systems and used refinement procedure to make the verification complete [4].

The remainder of this paper proceeds as follows. Section 2 gives basic definitions. It is followed by a brief overview of thread-modular reasoning in Section 3. Our technical contribution is presented in Section 4. Section 5 gives our experimental results. We conclude our presentation in Section 6.

2   Preliminary

We assume a fixed set V of typed variables. A *state over* $W \subseteq V$ is a valuation for the variables in W. The set of states over $W \subseteq V$ is denoted by $St[W]$. For $W \subseteq V$ and $s \in St[V]$, the *projection of s on* W (written $s \downarrow_W$) is a state over W that $s \downarrow_W (w) = s(w)$ for every $w \in W$. Let $W \subseteq V$, we write $St[V] \downarrow_W$

to indicate the set $St[W] = \{s \downarrow_W \| \forall s \in St[V]\}$. Given $St[W]$ and $St[X]$, their join is $St[W \cup X] = \{s | s \downarrow_W \in St[W]$ and $s \downarrow_X \in St[X]\}$ which is denoted by $St[W] \bowtie St[X]$. A *predicate over* $St[V]$ is a function from $St[V]$ to the Boolean domain $\mathbb{B}$. Given a state $s \in St[V]$ and a predicate ϕ over $St[V]$, we say s *satisfies* ϕ (written $s \models \phi$) if $\phi(s) = \top$. For any predicate ϕ over $St[V]$, define $\llbracket \phi \rrbracket = \{s \in St[V] : \phi(s)\}$. That is, $\llbracket \phi \rrbracket$ consists of states that satisfy ϕ. For $W \subseteq V$ and a predicate ϕ over $St[V]$, define the predicate $\phi \downarrow_W$ over $St[W]$ to be that for any $t \in St[W]$,

$$\phi \downarrow_W (t) = \top \text{ if and only if there is an } s \in St[V] \text{ with } \phi(s) = \top \text{ and } s \downarrow_W = t.$$

A *process* $P = \langle X, L, I, T \rangle$ is a quadruple where $X \subseteq V$ is the set of *global variables*, $L \subseteq V$ the set of *local variables* disjoint from X, I the *initial predicate* over $St[X \cup L]$, and T the *transition predicate* over $St[X \cup L] \times St[X \cup L]$. Let $s, s' \in St[X \cup L]$. We say s is *initial* if $I(s) = \top$. If $T(s, s') = \top$, we say s is a *predecessor* of s' and s' a *successor* of s. A *trace* τ is a sequence of states $s^0, s^1, \ldots, s^n$ such that $I(s^0) = \top$ and $T(s^i, s^{i+1}) = \top$ for $0 \leq i < n$. The set of traces of P is denoted by $Tr[P]$. A state s is *reachable* in P if there is a trace $\tau = s^0, s^1, \ldots, s^n \in Tr[P]$ such that $s^n = s$. The set of states reachable in P is denoted by $Re[P]$. Let π be a predicate over $St[X \cup L]$. We say P satisfies π (written $P \models \pi$) if $s \models \pi$ for every $s \in Re[P]$.

Let $P_j = \langle X, L_j, I_j, T_j \rangle$ be processes for $j = 0, 1$ where L_0 and L_1 are disjoint. Let $W_j = X \cup L_j \subseteq V$ for $j = 0, 1$. The *composition* of P_0 and P_1 (written $P_0 \| P_1$) is a process $\langle X, L, I, T \rangle$ where

- $L = L_0 \cup L_1$;
- $I(s) = \top$ if $I_0(s \downarrow_{W_0}) = \top$ and $I_1(s \downarrow_{W_1}) = \top$;
- $T(s, s') = \top$ if
  - $T_0(s \downarrow_{W_0}, s' \downarrow_{W_0}) = \top$ and $s \downarrow_{L_1} = s' \downarrow_{L_1}$; or
  - $T_1(s \downarrow_{W_1}, s' \downarrow_{W_1}) = \top$ and $s \downarrow_{L_0} = s' \downarrow_{L_0}$.

That is, exactly one process updates the global variables and its local variables; the other process stutters in a transition of the composition. It is straightforward to see that the composition is associative. $P_1 \| P_2 \| \cdots \| P_N$ is thus well-defined for $N \geq 2$.

3   Thread-Modular Reasoning

Definition 1. *Let* $P = \langle X, L, I, T \rangle$ *be a process. The* guarantee *of* P *is a process* $G(P) = \langle X, \emptyset, I_G, T_G \rangle$ *where* $I_G = I \downarrow_X$ *and* T_G *is a predicate over* $St[X] \times St[X]$ *such that*

$$T_G(t, t') = \top \text{ if } \exists s, s' \in St[X \cup L] \text{ with } T(s, s') = \top, s \downarrow_X = t, \text{ and } s' \downarrow_X = t'.$$

The main process of thread-modular model checking is shown in Algorithm 1. It first computes the reachable component states $\overline{R}_j$ for each process P_j, then computes the reachable system states $\overline{R}$ by joining reachable component states

Input: $P_j = \langle X, L_j, I_j, T_j \rangle$: a process for $1 \le j \le N$; π : a predicate over
$\qquad St[X \cup L_1 \cup \cdots \cup L_N]$
Output: "*PASS*" or "*UNKNOWN*"

```
1  error ← ⟦¬π⟧;
2  foreach j = 1, . . . , N do
3      R̄_j ← Re[G(P₁)‖ ⋯ ‖G(P_{j−1})‖P_j‖G(P_{j+1})‖ ⋯ ‖G(P_N)];
4  end
5  R̄ ← R̄₁ ⋈ R̄₂ ⋈ ⋯ ⋈ R̄_N;
6  if R̄ ∩ error = ∅ then
7      return PASS;
8  else
9      return UNKNOWN;
```

Algorithm 1. Thread-Modular Model Checking

global x: boolean initially $x = 1$
loop forever
$\left[\begin{array}{l} l = 0 : \text{Non-Critical} \\ l = 1 : \text{request } x \\ l = 2 : \text{Critical} \\ l = 3 : \text{release } x \end{array}\right.$

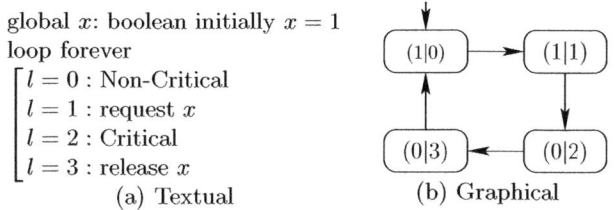

(a) Textual   (b) Graphical

Fig. 1. MUX-SEM$_k$

of all processes. Apparently, $\overline{R}$ is an over-approximation of the reachable system states, so it can report "PASS" when there is no error state in $\overline{R}$. Otherwise, it cannot make any conclusion.

Example 1. Consider a simple solution to the mutual exclusion problem in Fig. 1. In the figure, N processes attain mutual exclusion by the semaphore x. Each process requests x before entering the critical section, and releases x after leaving the critical section. Assume there are two processes P_1 and P_2. We use $P_j.l$ to denote the local variable l in process P_j where $j \in \{1, 2\}$. Each state in Fig. 1 (b) is marked with the corresponding valuation for all variables, where before the separator | is the valuation for global variables, and after the separator | is the valuation for local variables. Mutual exclusion is specified as $\pi : \neg((P_1.l = 2 \vee P_1.l = 3) \wedge (P_2.l = 2 \vee P_2.l = 3))$. We have the guarantee $G(P_j) = \langle \{x\}, \emptyset, I_j, T_j \rangle$ where $j \in \{1, 2\}$, $I_j(s)$ is $s(x) = 1$, and $T_j(s, s')$ is $\top$. Hence

$$\overline{R}_1 = Re[G(P_1)\|P_2] = \{s : s(x) \in \{0,1\} \text{ and } s(P_1.l) \in \{0,1,2,3\}\}$$
$$\overline{R}_2 = Re[P_1\|G(P_2)] = \{s : s(x) \in \{0,1\} \text{ and } s(P_2.l) \in \{0,1,2,3\}\}$$

Thus,

$$\overline{R} = \overline{R}_1 \bowtie \overline{R}_2 = \{s : s(x) \in \{0,1\}, s(P_1.l) \in \{0,1,2,3\}, \text{ and } s(P_2.l) \in \{0,1,2,3\}\}.$$

Since $\overline{R} \cap \llbracket \neg\pi \rrbracket \neq \emptyset$, Algorithm 1 reports "*UNKNOWN*."

4   Iterative Refinement

Let $P = \langle X, L, I, T \rangle$ be a process, $l \in L$, and S a set of states. We say l is an *essential variable* of P with respect to $s \in S$ if there is a $t \in St[X \cup L]$ such that

- $t \notin S$;
- $s(l) \neq t(l)$; and
- $s(v) = t(v)$ for every $v \in (X \cup L) \setminus \{l\}$.

In other words, a local variable is essential with respect to a state set if its value signifies the membership of the given state set.

Let l be an essential variable with respect to $s \in S$. Define the *essential predicate* χ_l^s for l with respect to $s \in S$ by

$$\chi_l^s(t) = \top \text{ if } t(l) = s(l).$$

Two essential predicate χ_l^s and χ_m^t are *distinct* if either l is different from m or $s(l) \neq t(m)$.

Example 2. In Example 1, observe that

$$\overline{R} \cap [\![\neg \pi]\!] = \{s : s(x) \in \{0,1\}, s(P_1.l) \in \{2,3\}, \text{ and } s(P_2.l) \in \{2,3\}\}.$$

Let us consider the state $s_0 \in \overline{R} \cap [\![\neg \pi]\!]$ that $s_0(x) = 0$, $s_0(P_1.l) = s_0(P_2.l) = 2$. Define $t_0 \in St[\{x, P_1.l, P_2.l\}]$ where $t_0(x) = 0$, $t_0(P_1.l) = 1$, and $t_0(P_2.l) = 2$. Then $t_0 \notin \overline{R} \cap [\![\neg \pi]\!]$, $s_0(P_1.l) \neq t_0(P_1.l)$, and $s_0(v) = t_0(v)$ for $v \in \{x, P_2.l\}$. Hence $P_1.l$ is an essential variable of P_1 with respect to s_0. The essential predicate $\chi_{P_1.l}^{s_0}$ for $P_1.l$ is hence

$$\chi_{P_1.l}^{s_0}(t) = \top \text{ if } t(P_1.l) = 2.$$

Similarly, consider the state $s_1 \in \overline{R} \cap [\![\neg \pi]\!]$ that $s_1(x) = 0$, $s_1(P_1.l) = 3$, $s_1(P_2.l) = 2$. Define $t_1(x) = 0$, $t_1(P_1.l) = 1$, and $t_1(P_2.l) = 2$. Then $P_1.l$ is an essential variable of P_1 with respect to s_1. The essential predicate $\chi_{P_1.l}^{s_1}$ for $P_1.l$ is therefore

$$\chi_{P_1.l}^{s_1}(t) = \top \text{ if } t(P_1.l) = 3.$$

Definition 2. *Let $P = \langle X, L, I, T \rangle$ be a process and Ψ a set of predicates. Define $W = X \cup L$. The augmented process $A(P, X_A, X_\Psi, \Psi) = \langle X \cup X_A, L, I_A, T_A \rangle$ of P with Ψ is defined by*

- $X_\Psi = \{u_\chi \in V : \chi \in \Psi\}$ *is the set of auxiliary variables with respect to Ψ;*
- $X_\Psi \subset X_A$ *and $X_A - X_\Psi$ is other processes' auxiliary variables;*
- $I_A(s) = \top$ *if $I(s \downarrow_W) = \top$ and $s(u_\chi) = \chi(s)$ for every $\chi \in \Psi$;*
- $T_A(s, s') = \top$ *if $T(s \downarrow_W, s' \downarrow_W) = \top$, $s(u_\chi) = \chi(s)$, $s'(u_\chi) = \chi(s')$ for every $\chi \in \Psi$ and $s'(v) = s(v)$ for every $v \in X_A - X_\Psi$.*

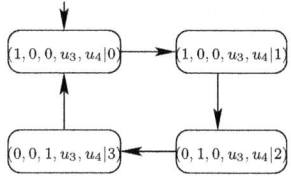

Fig. 2. $A(\text{MUX-SEM}_1, \{\chi^{s_0}_{P_1.l}, \chi^{s_1}_{P_1.l}\})$

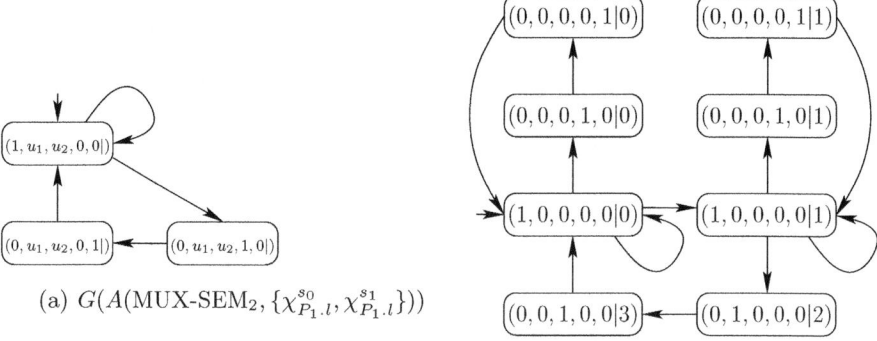

(a) $G(A(\text{MUX-SEM}_2, \{\chi^{s_0}_{P_1.l}, \chi^{s_1}_{P_1.l}\}))$

(b) $A(\text{MUX-SEM}_1, \{\chi^{s_0}_{P_1.l}, \chi^{s_1}_{P_1.l}\}) \|$
$G(A(\text{MUX-SEM}_2, \{\chi^{s_0}_{P_2.l}, \chi^{s_1}_{P_2.l}\}))$

Fig. 3. Example 3

Example 3. Recall the essential predicates $\chi^{s_0}_{P_1.l}$ and $\chi^{s_1}_{P_1.l}$ from Example 2. Let $\Psi = \{\chi^{s_0}_{P_1.l}, \chi^{s_1}_{P_1.l}\}$. Denote the auxiliary variables for $\chi^{s_0}_{P_1.l}$, $\chi^{s_1}_{P_1.l}$, $\chi^{s_0}_{P_2.l}$, and $\chi^{s_1}_{P_2.l}$ as u_1, u_2, u_3, u_4 respectively. Fig. 2 shows the augmented process $A(\text{MUX-SEM}_1, X_A, X_\Psi, \Psi)$, where $X_\Psi = \{u_1, u_2\}$ and $X_A = \{u_1, u_2, u_3, u_4\}$, and Fig. 3 shows its composition with the guarantee of augmented MUX-SEM$_2$. Thus

$$\overline{R}_1 = \left\{ \begin{array}{l} (1,0,0,0,0|0),\ (1,0,0,0,0|1),\ (0,1,0,0,0|2),\ (0,0,1,0,0|3), \\ (0,0,0,1,0|0),\ (0,0,0,0,1|0),\ (0,0,0,1,0|1),\ (0,0,0,0,1|1) \end{array} \right\}$$

Similarly,

$$\overline{R}_2 = \left\{ \begin{array}{l} (1,0,0,0,0|0),\ (1,0,0,0,0|1),\ (0,0,0,1,0|2),\ (0,0,0,0,1|3), \\ (0,1,0,0,0|0),\ (0,0,1,0,0|0),\ (0,1,0,0,0|1),\ (0,0,1,0,0|1) \end{array} \right\}$$

Thus,

$$\overline{R} = \overline{R}_1 \bowtie \overline{R}_2$$
$$= \left\{ \begin{array}{l} (1,0,0,0,0|0,0),\ (1,0,0,0,0|0,1),\ (1,0,0,0,0|1,0),\ (1,0,0,0,0|1,1), \\ (0,1,0,0,0|2,0),\ (0,1,0,0,0|2,1),\ (0,0,1,0,0|3,0),\ (0,0,1,0,0|3,1), \\ (0,0,0,1,0|0,2),\ (0,0,0,0,1|0,3),\ (0,0,0,1,0|1,2),\ (0,0,0,0,1|1,3) \end{array} \right\}$$

Since $\overline{R} \bowtie [\![\neg\pi]\!] = \emptyset$, we conclude that MUX-SEM$_1 \|$MUX-SEM$_2 \models \pi$.

Observe that additional constraints on the initial and transition predicates are non-interfering. They merely update the augmented variables X_A by the values of predicates. The following lemma hence follows from the definition.

Lemma 1. *Let* $P = \langle X, L, I, T \rangle$ *be a process,* Ψ *a set of predicates, and* π *a predicate. Then* $P \models \pi$ *if and only if* $A(P, \Psi) \models \pi$.

Proof. According to Definition 2, given any trace $\alpha = s^0 s^1 \ldots s^k$ in P, the corresponding trace in $A(P, \Psi)$ is $\beta = t^0 t^1 \ldots t^k$, where $t^i(u_\chi) = \chi(s^i)$ for any $\chi \in \Psi$ and $s^i = t^i \downarrow_{X \cup L} (0 \le i \le k)$. It is easy to prove $\alpha \models \pi \iff \beta \models \pi$. So we can conclude that $(P \models \pi) \Leftrightarrow (A \models \pi)$. $\square$

The main process for thread-modular model checking with iterative refinement is shown in Algorithm 2. Given N processes $P_1, P_2, \cdots, P_N$ and a predicate π, the algorithm decides is π satisfied on the whole system or not. In lines 3-8, the algorithm performs the regular thread-modular model checking. Then it analyzes is there any initial state in the reachable set of error states. If so, it reports "FAILURE". Otherwise, it calls a subroutine to refine the model.

Input: $P_j = \langle X, L_j, I_j, T_j \rangle$: a process for $1 \le j \le N$; π : a predicate over
$\quad\quad St[X \cup L_1 \cup \cdots \cup L_N]$
Output: *"PASS"* or *"FAILURE"*

1 $error \leftarrow [\![\neg\pi]\!]$;
2 $\Psi_j \leftarrow \emptyset$, for $j = 1, \ldots, N$; // the essential predicate set for P_j
3 **repeat**
4   **foreach** $j = 1, \ldots, N$ **do**
5          $\overline{R}_j \leftarrow Re[G(P_1)\| \cdots G(P_{j-1})\|P_j\|G(P_{j+1})\| \cdots \|G(P_N)]$;
6   **end**
7      $\overline{R} \leftarrow \overline{R}_1 \bowtie \overline{R}_2 \bowtie \cdots \bowtie \overline{R}_N$;
8   **if** $\overline{R} \bowtie error = \emptyset$ **then**
9   **return** *PASS*;
10   **if** $\overline{R} \bowtie error \bowtie [\![I_1]\!] \bowtie \cdots \bowtie [\![I_N]\!] \ne \emptyset$ **then**
11   **return** *FAILURE*;
    // refine $P_1, P_2, \ldots, P_N$ by $\overline{R}$ and $error$
12     $refinable \leftarrow Refine(\overline{R}, error, P_1, P_2, \ldots, P_N, \Psi_1, \ldots, \Psi_N)$;
13   **if** $\neg refinable$ **then**
14   **return** *PASS*;
15 **until** *forever* ;

Algorithm 2. Thread-Modular Model Checking with Refinement

Algorithm 3 gives the subroutine for refining a model. For each state $s \in \overline{R} \bowtie error$, the algorithm tries to find the distinct essential predicate for each process. If successes, it refines the component model using these found predicates. Otherwise, it adds the predecessors of s into $error$.

Input: $\overline{R}$: a state set; *error* : a state set; $P_j = \langle X, L_j, I_j, T_j \rangle$: a process for
$\qquad 1 \le j \le N$; $\Psi_1, \cdots, \Psi_N$
Output: $\top$ if any of the processes is refined; $\bot$ otherwise
1 *refined* $\leftarrow \bot$;
2 $S \leftarrow \overline{R} \bowtie error$;
3 **while** $S \ne \emptyset$ **do**
4 $\quad$ *predicateAdded* $\leftarrow \bot$;
5 $\quad$ remove an s from S;
6 $\quad$ **foreach** $j = 1, \ldots, N$ **do**
7 $\quad\quad$ $\Psi_j^s \leftarrow \{\chi_i^s : \chi_i^s$ is a distinct essential predicate from all $\chi \in \Psi_j\}$;
8 $\quad\quad$ **if** $\Psi_j^s \ne \emptyset$ **then**
9 $\quad\quad\quad$ $P_j, \Psi_j \leftarrow A(P_j, X_{\Psi_j^s}, X_{\Psi_j^s}, \Psi_j^s), \Psi_j \cup \Psi_j^s$;
10 $\quad\quad\quad$ **foreach** $i \ne j$ **do**
11 $\quad\quad\quad\quad$ $P_i \leftarrow A(P_i, X_{\Psi_j^s}, \emptyset, \emptyset)$;
12 $\quad\quad\quad$ **end**
13 $\quad\quad\quad$ *refined*, *predicateAdded* $\leftarrow \top, \top$;
14 $\quad\quad\quad$ break;
15 $\quad$ **end**
16 $\quad$ **if** $\neg$*predicateAdded* **then**
$\qquad\quad$ // $\langle X, L, I, T \rangle = P_1 \| P_2 \| \cdots \| P_N$, $W = X \cup L$
17 $\quad\quad$ $pre \leftarrow \{'s \downarrow_W : T('s, s) = \top\}$;
18 $\quad\quad$ **if** $pre \setminus error \ne \emptyset$ **then**
19 $\quad\quad\quad$ *refined*, *error* $\leftarrow \top$, *error* $\cup$ *pre*;
20 **end**
21 **return** *refined*;

$$\textbf{Algorithm 3. } \textit{Refine}(\overline{R}, error, P_1, \ldots, P_N)$$

Lemma 2. *Let* $P_j = \langle X, L_j, I_j, T_j \rangle$ *for* $j = 1, \ldots, N$, *and* π *a predicate. For any system state* s *in* $P_1 \| P_2 \| \cdots \| P_N = \langle X, L, I, T \rangle$, *when Algorithm 2 terminates, we have*

1. $s \not\models \pi$ *implies* $s \in error$;
2. $s \in error$ *implies there is a sequence* $s^i = s, s^{i+1}, \ldots, s^n$ *such that* $s^n \not\models \pi$ *and* $T(s^k, s^{k+1}) = \top$ *for every* $i \le k < n$.

Proof. (1) Note all states in $[\![\neg\pi]\!]$ are added to *error* in the begining of the algorithm; (2) Note *error* contains only states in $[\![\neg\pi]\!]$ and their predecessors. $\qquad \square$

Lemma 3. *Let* $P_j = \langle X, L_j, I_j, T_j \rangle$ *for* $j = 1, \ldots, N$, *and* π *a predicate. Then* $Re[P_1 \| P_2 \| \cdots \| P_N] \subseteq \overline{R}$ *at line 2, Algorithm 2.*

Proof. According to Definition 1, $G(P_j)$ simulates P_j for $j = 1, \ldots, N$. Then we can conclude $G(P_1) \| \cdots \| G(P_{j-1}) \| P_j \| G(P_{j+1}) \| \cdots \| G(P_N)$ simulates $P_1 \| \cdots \| P_N$ for $j = 1, \ldots, N$. So $\overline{R}_j$ is an over-approximation of $Re[P_1 \| \cdots \| P_N] \downarrow_{X \cup L_j}$ for $j = 1, \ldots, N$. Then the conclusion holds. $\qquad \square$

Theorem 1. *Let $P_j = \langle X, L_j, I_j, T_j \rangle$ for $j = 1, \ldots, N$, and π a predicate.*

1. *If Algorithm 2 returns "PASS", then $P_1 \| P_2 \| \cdots \| P_N \models \pi$;*
2. *If Algorithm 2 returns "FAILURE", then $P_1 \| P_2 \| \cdots \| P_N \not\models \pi$.*

Proof. (1) If Algorithm 2 returns "PASS" from line 11, with the precondition $\overline{R} \bowtie error = \emptyset$ and Lemma 3, we get the conclusion immediately. Otherwise, if Algorithm 2 returns "PASS" from line 16, the model cannot be refined anymore. Proof by contradiction, suppose $P_1 \| P_2 \| \cdots \| P_N \not\models \pi$, then there must be a state $s \in error$ and it is reachable from an initial system state s^0. Since $s^0 \notin error$ (otherwise the Algorithm 2 returns "FAILURE" from line 13), there must be two adjacent states s^i, s^{i+1} along the trace from s^0 to s, such that $s^i \notin error$ and $s^{i+1} \in error$. According to Algorithm 3, the state s^i should be added into $error$, which means the model is refinable. This is contradictory with the assumption. (2) According to Lemma 2, if $\overline{R} \bowtie error \bowtie \llbracket I_1 \rrbracket \bowtie \cdots \bowtie \llbracket I_N \rrbracket \neq \emptyset$, then $\exists s^0 \in \llbracket I_1 \rrbracket \bowtie \cdots \bowtie \llbracket I_N \rrbracket, \exists \alpha = s^0 \ldots s^k \in Tr[P_1 \| \cdots \| P_N] : s^k \not\models \pi$, which means $(P_1 \| \cdots \| P_N) \not\models \pi$. $\qquad\square$

Theorem 2. *Let $P_j = \langle X, L_j, I_j, T_j \rangle$ for $j = 1, \ldots, N$, and π a predicate. Algorithm 2 always terminates.*

Proof. In each refinement iteration, either some new states are added to the $error$ set, or the system is augmented by some new predicates. Note the state space of the system is finite, the number of possible predicates is also finite (each predicate corresponds to a subset of the states). In the worst case that the algorithm cannot give conclusive answer in all iterations, it finally terminates for no new state or new predicate can be found. $\qquad\square$

Theorem 3. *Let $P_j = \langle X, L_j, I_j, T_j \rangle$ for $j = 1, \ldots, N$, and π a predicate.*

1. *If $P_1 \| P_2 \| \cdots \| P_N \models \pi$, then Algorithm 2 returns "PASS";*
2. *If $P_1 \| P_2 \| \cdots \| P_N \not\models \pi$, then Algorithm 2 returns "FAILURE".*

Proof. (1) According to the second statement of Theorem 1, if $P_1 \| \cdots \| P_N \models \pi$, Algorithm 2 cannot return with "FAILURE". According to Theorem 2, Algorithm 2 always terminates. Thus, if $P_1 \| \cdots \| P_N \models \pi$, the algorithm can only terminate with "PASS". (2) Similarly, according to the first statement of Theorem 1, and Theorem 2, if $P_1 \| \cdots \| P_N \not\models \pi$, the algorithm can only terminate with "FAILURE". $\qquad\square$

5   Experiments

We implemented our thread-modular model checking algorithm with iterative refinement (TMMCIR) in NuSMV. For comparison, several model checking algorithms are implemented as well. They are asynchronous forward reachability (AFR) and thread-modular model checking (TMMC). To compare with SPLIT[3], we configure the tool to use the CUDD package. All benchmarks are downloaded from [18] and conducted on an 2.0GHz Intel T6400 CPU with 2GB memory.

Table 1 shows the experimental results for the simple mutual exclusion protocol MUX-SEM in Fig. 1. In the table, the column "method" shows the name of the model checking algorithm (TMMCIR, SPLIT, AFR, NuSMV, or TMMC). The number of processes instantiated in MUX-SEM is shown in the column "processes." The time needed for verification is indicated by the column "time." The column "BDD's" shows the peak number of BDD nodes required. The number of refinement applied in TTMCIR and SPLIT is shown in the column "refinement." The column "preds" gives the number of essential predicates added during verification. Finally, the column "conclusive?" shows whether the verification result is conclusive.

Table 1. Experimental Results of MUX-SEM

method	processes	time	BDD's	refinement	preds	conclusive?
TMMCIR	20	0.064	45990	1	40	Y
SPLIT	20	0.887	331128	1	38	Y
AFR	20	0.064	45990	na	na	Y
NuSMV	20	0.144	141036	na	na	Y
TMMC	20	0.032	20440	na	na	N
TMMCIR	50	0.580	401646	1	100	Y
SPLIT	50	12.187	4555054	1	98	Y
AFR	50	3.320	1242752	na	na	Y
NuSMV	50	3.412	2444624	na	na	Y
TMMC	50	0.228	203378	na	na	N
TMMCIR	100	5.536	1510344	1	200	Y
SPLIT	100	207.233	57265726	1	198	Y
AFR	100	208.561	3059868	na	na	Y
NuSMV	100	614.806	4762520	na	na	Y
TMMC	100	4.200	2057907	na	na	N
TMMCIR	200	27.966	2439514	1	400	Y
TMMCIR	300	145.093	5767146	1	600	Y

For MUX-SEM (Fig. 1), thread-modular model checking does not give a conclusive verification result due to abstraction. Our algorithm (TTMCIR) clearly outperforms other complete algorithms in large cases. For 100 processes, TTMCIR takes only 5.536 seconds to conclude the verification; other algorithms require more than 200 seconds to give conclusive results. Moreover, our algorithm is able to finish cases with 200 and 300 processes in less than 2.5 minutes. Other complete algorithms fail to finish the verification within an hour.

We now consider a variant of the simple mutual exclusion algorithm called MUX-SEM-LAST (Fig. 4(a)). In the new algorithm, a new global variable *last* is added to record the last process which enters its critical section. Table 2 gives the experimental results.

global x: boolean initially $x = 1$
global $last$: $\mathbb{N}$ initially $last = 0$
loop forever
$$\begin{bmatrix} l = 0 : \text{Non-Critical} \\ l = 1 : \text{request } x \wedge last := j \\ l = 2 : \text{Critical} \\ l = 3 : \text{release } x \end{bmatrix}$$
(a) MUX-SEM-LAST$_j$

global x: boolean initially $x = 1$
local $counter$: initially $counter = 0$
loop forever
$$\begin{bmatrix} l = 0 : \text{Non-Critical} \\ l = 1 : \text{request } x \wedge \\ \qquad counter := (counter + 1)\% M \\ l = 2 : \text{Critical} \\ l = 3 : \text{release } x \end{bmatrix}$$
(b) MUX-SEM-COUNT$_j$

Fig. 4. MUX-SEM-LAST$_j$ and MUX-SEM-COUNT$_j$

Table 2. Experimental Results for MUX-SEM-LAST

method	processes	time	BDD's	refinement	preds	conclusive?
TMMCIR	50	1.548	1263192	1	100	Y
SPLIT	50	4.047	3219300	0	0	Y
AFR	50	87.297	2480394	na	na	Y
NuSMV	50	189.900	3113012	na	na	Y
TMMC	50	0.624	488516	na	na	N
TMMCIR	100	12.945	4143188	1	200	Y
SPLIT	100	36.557	28470876	0	0	Y
AFR	100	>1h	-	na	na	N
NuSMV	100	>1h	-	na	na	N
TMMC	100	6.636	2454844	na	na	N

Thread-modular model checking again fails to verify the property conclusively. Our algorithm still performs better than other complete algorithms in this example. The SPLIT tool also performs reasonably well; it finishes the case with 100 processes in 36.557 seconds whereas forward reachability and NuSMV cannot conclude in an hour. Interestingly, the SPLIT tool is able to prove the result without any refinement. Although TMMCIR requires one refinement and adds 100 essential predicates, the algorithm still concludes the verification with less time and space than SPLIT. This suggests the overhead of the proposed refinement technique is insignificant in this example.

We now consider another variant of the simple mutual exclusion algorithm (Fig. 4(b)). In MUX-SEM-COUNT, a local counter is added to each process. When a process enters its critical section, the local counter is incremented by one (modulo a constant M). Thread-modular model checking fails to give any conclusive result in this example. Thanks to abstraction, our algorithm and SPLIT can verify all cases in seconds. In comparison, forward reachability and NuSMV need more than 20 minutes to finish the case with 20 processes (Table 3).

For the bakery algorithm, thread modular model checking is able to verify the property conclusively (Table 4). It therefore attain the best performance with the larger case with 8 processes. Our algorithm is slightly slower (.466 seconds) than the incomplete algorithm and finishes the verification of the same case in

Table 3. Experimental Results for MUX-SEM-COUNT

method	processes	time	BDD's	refinement	preds	conclusive?
TMMCIR	10	0.020	14308	1	20	Y
SPLIT	10	0.321	100019	1	18	Y
AFR	10	5.996	617288	na	na	Y
NuSMV	10	2.288	1030176	na	na	Y
TMMC	10	0.016	10220	na	na	N
TMMCIR	20	0.104	122640	1	40	Y
SPLIT	20	2.520	930020	1	38	Y
AFR	20	1584.179	2634716	na	na	Y
NuSMV	20	3914.385	159986	na	na	Y
TMMC	20	0.044	47012	na	na	N

less than a half minute. The SPLIT tool is able to prove the same property in less than 1.5 minutes. Conventional forward reachability and NuSMV require more than 6 and 21 minutes to obtain the verification result respectively.

Table 4. Experimental Results for the Bakery Algorithm

method	processes	time	BDD's	refinement	preds	conclusive?
TMMCIR	4	0.100	96068	0	0	Y
SPLIT	4	0.267	218708	0	0	Y
AFR	4	0.084	91980	na	na	Y
NuSMV	4	0.140	106288	na	na	Y
TMMC	4	0.100	96068	na	na	Y
TMMCIR	8	26.246	2389436	0	0	Y
SPLIT	8	75.141	26776400	0	0	Y
AFR	8	240.555	4258674	na	na	Y
NuSMV	8	1282.984	25237268	na	na	Y
TMMC	8	25.780	2389436	na	na	Y

Finally, we consider the dining philosopher problem (Table 5). Thread-modular model checking cannot give conclusive answers. Most interestingly, conventional forward reachability algorithm is most efficient in this example. It takes less than 3 seconds to prove the property in the case with 10 processes. NuSMV is about 1 second slower than forward reachability. In comparison, our algorithm and the SPLIT tool require several refinements to conclude the verification. In the case with 8 processes, TMMCIR adds 20 essential predicates in 3 refinements; SPLIT adds 11 essential predicates in 6 refinement. Subsequently, both are significantly inefficient than conventional algorithms. Our algorithm requires about 16 seconds to finish whereas SPLIT takes more than 80 minutes.

In our experiments, TMMC does not give conclusive results in all examples but the bakery algorithm. If an example needs no refinement, our algorithm and thread-modular model checking have comparable performance. In most

Table 5. Experimental Results for Dining Philosophers

method	processes	time	BDD's	refinement	preds	conclusive?
TMMCIR	6	0.104	85848	3	12	Y
SPLIT	6	0.320	158410	3	6	Y
AFR	6	0.016	14308	na	na	Y
NuSMV	6	0.036	17374	na	na	Y
TMMC	6	0.008	7154	na	na	N
TMMCIR	8	1.028	367920	3	16	Y
SPLIT	8	16.442	1176322	5	10	Y
AFR	8	0.236	243236	na	na	Y
NuSMV	8	0.236	223818	na	na	Y
TMMC	8	0.020	24528	na	na	N
TMMCIR	10	15.981	1475768	3	20	Y
SPLIT	10	5274.488	4193266	6	11	Y
AFR	10	2.592	1815434	na	na	Y
NuSMV	10	3.556	1739444	na	na	Y
TMMC	10	0.052	48034	na	na	N

examples, TMMCIR and SPLIT are faster than conventional forward reachability and and NuSMV. Between our algorithm and SPLIT, ours usually performs better. This is due to the fact that our algorithm computes the reachable states separately with only one process and its environment.

6   Conclusions

This paper uses iterative refinement to make thread-modular model checking complete. Thread-modular model checking computes the reachable states of each process with its environment—the composition of other processes' global information. With limited global information, thread-modular model checking can compute the system reachable states quickly. However, it is incomplete for many protocols, which is what we resolved by the refinement in our approach. In most examples, our approach performs substantially better than other complete verification algorithms. The main reason is that we compute the reachable states separately with only one process and its environment. In MUX-SEM with 200 processes, we only use about 27 seconds, while other approaches use more than 1 hour.

According to our experimental data, the approach about thread-modular model checking cannot give good performance when the global variables are much more than local variables. We will take abstraction for global variables to improve its efficiency in our future work.

Acknowledgment. Thanks to Kedar Namjoshi for introducing more detail about the auxiliary variable refinement.

References

1. Arons, T., Pnueli, A., Ruah, S., Xu, J., Zuck, L.: Parameterized Verification with Automatically Computed Inductive Assertions. In: Berry, G., Comon, H., Finkel, A. (eds.) CAV 2001. LNCS, vol. 2102, pp. 221–234. Springer, Heidelberg (2001)
2. Clarke, E., Emerson, E.: Design and Synthesis of Synchronization Skeletons Using Branching Time Temporal Logic. In: Kozen, D. (ed.) Logic of Programs 1981. LNCS, vol. 131, pp. 52–71. Springer, Heidelberg (1982)
3. Cohen, A., Namjoshi, K.S., Sa'ar, Y.: SPLIT: A Compositional LTL Verifier. In: Touili, T., Cook, B., Jackson, P. (eds.) CAV 2010. LNCS, vol. 6174, pp. 558–561. Springer, Heidelberg (2010)
4. Cohen, A., Namjoshi, K.: Local proofs for global safety properties. Formal Methods in System Design 34(2), 104–125 (2009)
5. Flanagan, C., Qadeer, S.: Thread-Modular Model Checking. In: Ball, T., Rajamani, S.K. (eds.) SPIN 2003. LNCS, vol. 2648, pp. 213–224. Springer, Heidelberg (2003)
6. Gu, M., Liu, Q.: Automatic compositional reasoning for multi-thread programs. In: 15th International Conference on Computer Supported Cooperative Work in Design (CSCWD), pp. 175–182 (2011)
7. Henzinger, T.A., Jhala, R., Majumdar, R., Qadeer, S.: Thread-Modular Abstraction Refinement. In: Hunt Jr., W.A., Somenzi, F. (eds.) CAV 2003. LNCS, vol. 2725, pp. 262–274. Springer, Heidelberg (2003)
8. Henzinger, T., Jhala, R., Majumdar, R.: Race checking by context inference. ACM SIGPLAN Notices 39(6), 1–13 (2004)
9. Jones, C.: Development methods for computer programs including a notion of interference. PhD thesis, Oxford University, June 1981. Printed as: Programming Research Group, Technical Monograph 25 (1981)
10. Jones, C.: Tentative steps toward a development method for interfering programs. ACM Transactions on Programming Languages and Systems (TOPLAS) 5(4), 596–619 (1983)
11. Lamport, L.: Proving the correctness of multiprocess programs. IEEE Transactions on Software Engineering (2), 125–143 (1977)
12. Malkis, A., Podelski, A., Rybalchenko, A.: Thread-Modular Counterexample-Guided Abstraction Refinement. In: Cousot, R., Martel, M. (eds.) SAS 2010. LNCS, vol. 6337, pp. 356–372. Springer, Heidelberg (2010)
13. Misra, J., Chandy, K.: Proofs of networks of processes. IEEE Transactions on Software Engineering (4), 417–426 (1981)
14. Namjoshi, K.S.: Symmetry and Completeness in the Analysis of Parameterized Systems. In: Cook, B., Podelski, A. (eds.) VMCAI 2007. LNCS, vol. 4349, pp. 299–313. Springer, Heidelberg (2007)
15. Owicki, S., Gries, D.: Verifying properties of parallel programs: an axiomatic approach. Communications of the ACM 19(5), 279–285 (1976)
16. Pnueli, A., Ruah, S., Zuck, L.D.: Automatic Deductive Verification with Invisible Invariants. In: Margaria, T., Yi, W. (eds.) TACAS 2001. LNCS, vol. 2031, pp. 82–97. Springer, Heidelberg (2001)
17. Queille, J., Sifakis, J.: Specification and Verification of Concurrent Systems in Cesar. In: Dezani-Ciancaglini, M., Montanari, U. (eds.) Programming 1982. LNCS, vol. 137, pp. 337–351. Springer, Heidelberg (1982)
18. SPLIT, http://split.ysaar.net/

Towards LTL Model Checking of Unmodified Thread-Based C & C++ Programs*

J. Barnat, L. Brim, and P. Ročkai**

Faculty of Informatics, Masaryk University
Brno, Czech Republic
{barnat,brim,xrockai}@fi.muni.cz

Abstract. In this paper we present a new approach to verification of multi-threaded C/C++ programs. Our solution effectively chains the parallel and distributed-memory model checker DiVinE with CLang and the LLVM bitcode interpreter. This combination offers full LTL, distributed-memory model checking of virtually unmodified C/C++ source code and is supported by a newly introduced path-reduction technique. We demonstrate the efficiency of the reduction and also the capacity to produce human-readable counter-examples in two small case studies: a C implementation of the Peterson's mutual exclusion protocol and a C++ implementation of a shared-memory, lock-free FIFO data structure designed for fast inter-thread communication.

1 Introduction

Direct applicability of model checking to unmodified, or at most lightly annotated, software systems is extremely desirable, since it substantially reduces costs associated with this otherwise appealing technique. Tools that bypass the modelling step, i.e. those, that model-check software directly, remove the need for a significant part of the specialist work normally required for model checking. This in turn enables wider applicability of automated formal verification.

A number of advancements have been made in this area. One of the first forays into the territory is the support for combining C code with ProMeLa models in SPIN [11], which can be used, although with a number of caveats and substantial amount of extra work, to verify implementation-level properties. Another early approach to the problem is constituted by automated model extraction [10,12,6]. In a similar spirit, the ZING [1] model checker is shipped with automated model extraction tools. More direct approaches, which are in many cases also easier to apply, are embodied by model checkers based on a particular programming language (or runtime), like CMC [15], JCat [8], MCP [19], Java PathFinder [20] and MoonWalker [7].

---

* This work has been partially supported by the Czech Science Foundation grant No. GAP202/11/0312 and by ARTEMIS-IA iFEST project grant No. 100203.
** Petr Ročkai has been partially supported by Red Hat, Inc. and is a holder of Brno PhD Talent financial aid provided by Brno City Municipality.

A. Goodloe and S. Person (Eds.): NFM 2012, LNCS 7226, pp. 252–266, 2012.
© Springer-Verlag Berlin Heidelberg 2012

While the CEGAR-based software model checkers like SLAM [2], BLAST [9] or Magic [5] have a related set of goals, they in fact live on a somewhat remote branch in relation to previously mentioned explicit-state model checkers. With real programs, the abstraction refinement process allows for processing of larger systems, but also introduces a degree of infidelity into the process. Therefore, the overall result is more in the spirit of static analysis (with a model checker back end) than it is to exhaustive model checking. Even though the CEGAR approach (and predicate abstraction in general) has been quite successful, it does require a significant insight into the input language of the tool. As a direct consequence, there is no C++ support in CEGAR-based tools. Moreover, verification of multi-threaded programs is usually not supported by CEGAR-based tools, since a suitable symbolic representation for control-based, interleaving parallelism is not available.

With the lack of abstraction in direct software model checking the problem of state space explosion is even more poignant. The most successful techniques used in explicit-state model checking to fight state explosion are Partial Order Reduction [16] and distributed-memory processing, the latter of which has not been yet applied to direct model checking of programs. Hash compaction and bitstate hashing techniques can also be used to help overcome a state space explosion, although at a cost of small infidelity in the model checking process. Hash compaction combined with distributed-memory computation can achieve enormous capacity (on the order of 10^{10} states) in an explicit-state model checker [4]. While symbolic approaches can match or even exceed this capacity, this only applies to special classes of models: unfortunately, software systems with interleaving concurrency (in the imperative style) resist significant symbolic reduction.

Another trait of existing approaches to direct software model checking is their focus on verification of safety properties. However, liveness properties, such as the requirement of guaranteed progress or response, are crucial in specifications of parallel programs. Verification of general liveness properties poses a far greater technical challenge compared to state-space exploration (which is sufficient for an important class of safety properties). Nevertheless, even fully generic safety properties are often neglected. For both generic safety and for liveness, we need a mechanism to conveniently describe atomic propositions, which relate the properties of any given instantaneous state of a system (this problem will be described in more detail in Section 3). Additionally, a property automaton (which refers to these atomic propositions to describe undesired program behaviour) needs to be synchronised with the execution of the program.

In this paper we describe an extension of the model checker DiVinE [3] that enables analysis of models written in LLVM assembly language (in the form of LLVM bitcode). Since compilers that produce LLVM bitcode are available for a variety of programming languages, including C, C++, and Java, this extension allows DiVinE to effectively verify unmodified software. As DiVinE is a parallel and distributed-memory model checker, this is, to our best knowledge, the first approach that allows for verification of unmodified C/C++ programs using distributed memory. Additionally, we introduce a path reduction technique

(τ-reduction) that works at the level of LLVM bitcode. This technique significantly reduces the space explosion effect of the very fine detail available in the model structure arising from the LLVM bitcode. We also describe how general atomic propositions can be defined for C/C++ programs so that verification tasks performed by model checking can go beyond the typically delivered state space exploration that allows for deadlock detection and assertion violation checking.

The rest of paper is organised as follows. In Section 2 we describe the LLVM extension to DiVinE, in Section 3 we elaborate on full LTL model checking of C/C++ programs, in Section 4 we describe the principle of the τ-reduction. In Section 5 we give some related work, in Section 6 we demonstrate the technique on two small use cases, and finally, in Section 7 we give some conclusions and future work.

2   Model Checking LLVM Bitcode

The LLVM system [13] is based on an assembly-level, single-static-assignment language. The main goal of the LLVM framework is to provide support for code transformations using the LLVM assembly language as both the input and output languages (most notably optimisation passes). Moreover, thanks to code generation and just-in-time compilation tools available for the LLVM assembly, LLVM can be transformed into a native program on a number of platforms, either in an ahead-of-time or a just-in-time manner. The combination of these traits makes LLVM a very attractive choice for compiler back ends: a number of optimiser passes are available on the LLVM level, and many code generation choices come basically for free. This fact is illustrated by the widespread support of LLVM-based code generation in various programming language compilers.[1]

The general work-flow for model checking of programs directly from their source code is following: the program source code is compiled into LLVM bitcode, the bitcode is verified with a model checker, and finally the target binary is generated from the verified bitcode. A major advantage of this approach is the ability to apply compiler optimisations before running the model checker. This enables discovery of subtle effects of unspecified behaviour in the presence of various levels of compiler optimisation. Of course, the code generation step still constitutes a possibly unfaithful translation which can introduce property violations unseen by the model checker at the LLVM bitcode level. Nevertheless, this fidelity gap is much narrower here than it is in high-level model checkers and static analysis tools operating on the "statement level" of the input language.

To explore the state space of the program (as described by a LLVM bitcode file), we employ a modified version of the upstream LLVM interpreter. We serialise and store states of the interpreter as byte vectors (one vector corresponding to one state) and each such vector fully describes the configuration of the virtual machine executing the program under consideration. Within a single state, each executing thread is represented by an execution stack, consisting of execution

---

[1] Including GCC (C, C++, Objective C, Java, Ada and Fortran), Clang (C and C++), GHC (Haskell) and others.

contexts. Each context in turn contains a set of LLVM registers[2], including a program counter. Additionally, the state contains a shared heap for both non-register local variables and for dynamically allocated memory. A thread to be executed is picked non-deterministically in each step, producing all the possible thread interleavings.

Thanks to symbolic debugging interfaces and metadata stored at the bitcode level, it is possible to reliably map assembly locations to original source locations and provide meaningful counterexample traces.

2.1 Preparing Programs

While the approach allows for model checking of unmodified programs, there is nevertheless a certain minimal amount of preparation that needs to be done to apply the DiVinE model checker. On the level of C/C++ source code, we provide a header file, `divine-llvm.h` which transparently maps native (pthreads and other supported API) calls to their equivalents required for model checking. In the current version, the header file defines or alters the definition of at least the following functions:

- `malloc(int)` – For the purpose of verification a non-deterministic choice is made whenever `malloc` is called by the program, with a `NULL` return in one of the branches and successful allocation in the other one. This behaviour can be suppressed by a CPP macro `NO_MALLOC_FAILURE`.
- `malloc_guaranteed(int)` – A variant of `malloc` that never fails, even when `NO_MALLOC_FAILURE` is not defined.
- `free(void *)` – Release heap-allocated memory. Deterministic.
- `assert(int)` – Semantically equivalent to `assert` provided in system's `assert.h`; however, the program is not terminated. Any transition that calls a failing assert will raise an "assertion failed" flag in the resulting state. Reachability analysis can be used to search for system states with the flag.
- `trace(...)` – A debugging API, printing a message to the console whenever the corresponding transition is executed by the verification core; implemented as standard error output in non-verification runs.
- `pthread_create(...)` – Create a new thread. Part of the pthread library.
- `arbitrary(n)` – An interface to data-based non-determinism, produces a value between 0 and n. In non-verification builds, a single random value is produced.

The pthread mutex API will be provided in a future revision, most likely in the form of spinlocks. Condition variables can be implemented in terms of a busy wait as well, although this needs to be done carefully to preserve the specified semantics.

The use of `divine-llvm.h` header file is, however, fully transparent. The header does not modify behaviour of the program in any way, unless a CPP

---

[2] The registers in LLVM are dynamic, and only created upon assignment. Their existence and values are lexically scoped. Allocation of processor registers is part of the native code generation step.

macro DIVINE is defined. Therefore, the only difference between a "normal"
compilation (producing a natively executable binary) and a "verification" build
(for model checking with DiVinE) is that in the latter case, -DDIVINE needs
to be passed on the command line to the compiler, and native code generation
needs to be suppressed. For C and C++ programs, there are two options to
generate the requisite LLVM bitcode:

- with **CLang** (which is the preferred option), passing -emit-llvm at the
  command line,
- using **GCC** with **dragonegg**, which will use the GCC compiler front- and
  middle- ends; the options -flto -fplugin=/path/to/dragonegg.so will
  cause GCC to emit LLVM bitcode.

The library of APIs available to programs under verification is of course subject
to future extensions. As of now, the arbitrary calls can be used to implement an
exhaustive search based on "arbitrary" inputs from the environment, and in turn
to provide verification-friendly implementations of system APIs that interact
with the environment. In its present form, though, this option is only viable for
the smallest of programs: a form of hybrid data-symbolic model checking will
be needed to at least partially lift this restriction. Therefore, support for data-
intensive forms of system interactions, like reading files or sockets, is not planned
until the model checking back end can usefully deal with the consequences.

3   Atomic Propositions and LTL in C/C++ Programs

Model checking of general LTL properties is often neglected in favour of safety
checking in the form of detection of deadlocks and assertion violations detection.

Of course, LTL model checking of programs per se is not devoid of difficulties: a language of atomic propositions needs to be devised. While it is tempting
to simply provide an expression language on top of global variables, such an
arrangement is likely to interfere with optimisers (and with store buffer simulation). In particular, the ordering of memory stores at the LLVM bitcode level is
not guaranteed to be the same as at the source-code level. Hence, augmenting
the source code with new global variables to represent properties of individual
states is not satisfactory. Moreover, there is further interference with "instantaneous" occurrences of atomic propositions: often, it is desirable to mark a single
state with a certain proposition – in context of programs, this is a very useful
ability – allowing the developer to mark, for an LTL formula, that program has
passed through a certain location in the source code. Setting and immediate
resetting a global variable is not quite a faithful simulation of the instantaneous
moment as the two succeeding modifications of the variable may interleave non-
deterministically with execution of a parallel thread.

Therefore, we propose (and implement) a system where assert-like statements
can be added to the program to encode the atomic propositions: there are two
statements, first of which is called simply ap, and its only parameter is the
identifier of the atomic proposition drawn from an enumerated type in the source
language. The other is called ap_set, and apart from the atomic proposition

identifier, it takes a value indicating the value to set the atomic proposition to (0/1 in the usual false/true interpretation). The `divine-llvm.h` header provides a no-op implementation for the `ap` procedure, so it can be used in real programs without the need to keep separate versions of the source.

Whenever a call to either of the `ap*` built-in procedures is found in the instruction stream, the atomic proposition values visible to LTL formulae may change. If we were to admit the `ap` mechanism as the only one, the valuation of atomic propositions could only ever change at these points. Of course, the `ap` calls can be, in the source language, guarded by arbitrary conditionals, making (virtually) the complete source language *the* language of atomic propositions. The `ap_set` calls comprise a very "imperative" approach to valuation of atomic propositions, while the `ap` calls are more declarative.

Nevertheless, even though as explained above, values of variables alone are not sufficient to satisfactorily formulate some of the desirable LTL properties, the `ap*` statements alone are not sufficient either: in some cases, the atomic propositions should be derived from variable content, and it is often inconvenient to encode such a relation in explicit `ap` calls. Unfortunately, a solution to this problem is only straightforward when we restrict our interest to global variables – in case of those, we can ask the user to supply a Boolean function which essentially becomes the "implementation" of an atomic proposition. However, not all variables of interest live in the global scope, and we need to deal with atomic propositions that depend on variables that may be currently out of scope. To allow flexible specification of atomic propositions by the users, we propose the following mechanism.

An atomic proposition relies on a function of the source language with Boolean return value and arbitrary parameters. To bind such a function to an actual atomic proposition, we need to match the input parameters of this Boolean function to specific variables in a given scope. This binding will work differently for global and differently for intermittent (lexically scoped) variables. On the global level, a macro `ap_global(proposition, function, N, p1, ..., pN)` is provided. This informs the model checker that whenever the value of the atomic proposition `proposition` is required, it can call the function `function` with values of variables p1 through pN as parameters. Since all these variables must be in the global scope, such a call can be made at any time.

On the other hand, scope-dependent propositions are somewhat more complicated. Whenever local variables may be declared in the program, we allow a macro `ap_local` to be used, similar to `ap_global`. The `ap_local` call may refer to the values of arbitrary in-scope variables, and their **current scope** values will be passed to the evaluation function. This means that when a different scope is entered, even if it shadows some of the parameters to `ap_local`, *the original, shadowed variable values* are used for evaluating the atomic proposition. For example, see Figure 1.

The main problem with these scope-dependent propositions arises in presence of recursive and parallel scopes (lexical scope combined with recursion or with

```
#include "divine-llvm.h"

enum AP { Progress };
LTL(GF(Progress));

bool progress( int x ) { return x == 2; }

void thread( int *x ) {
    int y = 0;
    ap_local( Progress, progress, y );
    while ( true ) {
        y = arbitrary( 3 ); // 0 - 2
        while ( *x == y ); // wait
        y = *x;
    }
}

int main() {
    int x = 0;
    pthread_t tid;

    pthread_create( &tid, NULL, thread, &x );
    while ( true )
        x = (x + 1) % 3;
    return 0;
}
```

Fig. 1. A simple example demonstrating the use of LTL and ap_local. Fair scheduling is assumed. The forked thread is expected to continue making progress (i.e. Progress is true infinitely often). This progress may fail to happen if the compiler moves the dereference of x out of the inner loop (which it may legally do).

multi-threading gives rise to this case): the same atomic proposition could be assigned different values in different contexts at the same time.

A similar problem arises with the "imperative" ap_set construct – in this case, however, it is the responsibility of the developer to manipulate atomic propositions carefully. The ap_set call is provided for completeness, and because there are corner cases that are difficult to address otherwise. Whenever possible, it should be avoided in favour of ap or of the "declarative" style.

Nevertheless, even the "declarative" approach has certain limits, related to the already mentioned "recursive" scopes. We define the value of a declarative atomic proposition as false whenever it has been bound to none of the currently active scopes. However, it can happen that multiple such scopes are active simultaneously (either "below" each other, in a recursion stack, or "beside" each other in multiple execution threads). In this case, we can no longer relegate this problem to the developer, since they have no explicit control over precedence in such cases. There are multiple candidates for a solution, but the simplest and most intuitive seems to be to set the atomic proposition to 1 (true) when *any*

```
#include "divine-llvm.h"

enum AP { Cap };
LTL(G(Cap));

bool cap( int x ) { return x < 3; }

int rec( int x ) {
    ap_local( Cap, cap, x );
    return x >= 3 ? x : rec( x + 1 );
}

int main() {
    int x = rec( 0 );
    trace( "%d", x );
    return 0;
}
```

rec(0)	rec(1)	rec(2)	rec(3)		result
–	–	–	–		¬Cap
Cap	–	–	–	→	Cap
Cap	Cap	Cap	–	→	Cap
Cap	Cap	Cap	–	→	Cap
Cap	Cap	Cap	¬Cap	→	Cap

Fig. 2. An example of recursive program where atomic proposition Cap is defined in multiple (shadowed) scopes with different value

of the active scopes proscribes this (in other words, we take the logical disjunction of the values given by all active scopes). For example, see Figure 2. This approach is also consistent with the behaviour of the ap built-in function.

As a future enhancement, we propose a scheme with (optional) parametric formulae that can use quantifiers over scopes, and a matching (and likewise optional) indexing scheme for atomic propositions. This will allow substantially more flexibility in specification of scope-dependent behaviours in multi-threaded and recursive programs.

4 τ-Reduction

The fine-grained nature of LLVM bitcode further aggravates the state-space explosion problem. To counter this, we have devised a very simple yet efficient state-space reduction technique that mimics the path reduction as presented in [21]. The reduction is based on the observation that some transitions in the state space are invisible for other active threads in the system. These, the so-called τ actions, of a thread or process, can be delayed over other τ actions of other processes. In fact, multiple subsequent τ actions of a single process/thread can be safely collapsed into a single transition without any effect on other threads or the observed property.

In traditional model-based model checking of asynchronous systems, this reduction would be quite ineffective, because τ chains are fairly rare in purpose-built models. On the other hand, they are extremely ubiquitous in the SSA bitcode produced by LLVM. While identifying all τ actions is very complex, there is a simple yet very efficient heuristic that will identify and collapse a majority of safe (i.e. not forming a loop) τ transitions. All transitions that:

- do not access memory (only registers), and
- are within a single basic block

can be safely treated as τ transitions. From experience, we know that assembly-level programs, especially in the RISC style with explicit loads and stores (as is the case of LLVM), contain a significant share of instructions (actions) that meet both these criteria.

To illustrate the extreme payoff for the otherwise very simple reduction, let's take our `peterson.c` running example (listing in Figure 3): the state space size (when compiled with `-O2`) before τ-reduction was 37482 states and 107533 transitions and safety verification took 135 seconds. The reduced state space has 5301 states, 14585 transitions and verifies in 18 seconds, which is about 7-fold decrease in state count and 7.5-fold decrease in verification runtime.

It should be noted that this reduction is closely related to "superstep POR" proposed in [22], although simpler. A good candidate for improving both these reductions may be a heuristic that would identify (some of the) memory writes that are (provably) invisible to any other threads. On the other hand, since the LLVM virtual machine has a possibly infinite register file, no register spilling happens (register spilling is normally a major source of invisible memory writes; in LLVM-based compilers, register allocation takes place in a later phase of code generation). This naturally limits the number of invisible writes, and easily explains why the reduction is so successfull despite its simplicity.

5   Related Work

According to [17], the MCP model checker uses a similar approach to model checking of C/C++ program. MCP uses the LLVM interpreter, while it is more closely following the Java PathFinder model, where safety checking is the primary goal. Later papers on MCP [18,19], however, present a different approach to model checking – through automatic code transformation (annotation). While MCP appears to provide extra flexibility by allowing user-level implementation of the thread scheduling algorithm, we focus on an out-of-the box experience. It is already possible to apply the DiVinE model checker to a subset of unmodified pthread-based programs, with semantics closely resembling those of actual (hardware and operating system) implementations. The major outstanding feature in this regard are store buffers [14], which will bring the model checking process to almost perfect fidelity with real executing code. On the other hand, we believe that the focus of MCP is more on dedicated, mission-critical systems, whereas our primary focus is on commodity hardware and software. However, while we do not rule out the possibility of allowing user override of the scheduler in future releases, we will very likely always provide a default behaviour that mimics real hardware and operating systems.

A different approach to LLVM-based explicit-state model checking is presented in [22] – instead of making it possible to directly model check programs, a hybrid approach is chosen. A program is supplemented with a driver written in the modelling language ProMeLa, which can then call into LLVM-compiled

Table 1. Verification results for `peterson.c`, as listed in Figure 3. The compiler used was *clang*, version 2.9. All figures are of the entire state space; no early termination was done in this experiment.

model variant	assertion	state count	transition count
-O0	safe	1992772	5323045
-O0, BUG	**unsafe**	1609112	4237118
-O1	safe	18631	49624
-O1, BUG	**unsafe**	23849	63804
-O2	safe	5842	16121
-O2, BUG	**unsafe**	12718	35439

functions, with proper interleaving. The approach employs SPIN as the model checking back end, with its respective advantages (maturity, speed) and disadvantages (lack of distributed memory support and limited parallelism). In many cases, this is more laborious, and requires knowledge of another programming language and of the special interface between SPIN and LLVM.

At the time of this writing, though, there was no support for LTL model checking in either of these tools, nor were the tools available publicly. The support for LTL model checking in the non-LLVM software model checkers (Java PathFinder, MoonWalker, and the like) is very limited as well.

6   Use Cases

A very simple, illustrative use case for the software model checking capabilities added to DiVinE is represented by the program listed in Figure 3. We can compile the listed program using the *clang* compiler into LLVM bitcode, either optimised or unoptimised. By running the model checker on the program (τ-reduction enabled), both with the bug indicated in the listing present and corrected, we obtain the results shown in Table 1. We can observe a significant decrease in state space size when compiler optimisations are enabled.

A second, much more elaborate use-case is the verification of a lock-free data structure for inter-thread communication. We do not include the listing of the source code in this paper due to space constraints, but it is available among the examples shipped as part of the development versions of DiVinE.[3] Since the implementation of the data structure as such is not a complete program, we have combined a *unit testing* approach with model checking, providing a small, standalone *unit* test-case (listing shown in Figure 5). Furthermore, since we know that the correctness of the implementation is independent of the actual data payload, we use the common approach of fixing the data values as part of the test-case, making the whole unit test into a closed system. This approach is ubiquitous whenever automated testing of software is performed. However, since

---

[3] In fact, this same data structure is employed by the parallel model-checking back end of DiVinE for fast shared-memory communication.

```
 1: #include "divine-llvm.h"
 2:
 3: struct state {
 4:     volatile int flag[2];
 5:     int turn;
 6:     volatile int in_critical[2];
 7: };
 8:
 9: struct p {
10:     int id;
11:     pthread_t ptid;
12:     struct state *s;
13: };
14:
15: void thread( struct p *p ) __attribute__((noinline));
16: void thread( struct p *p ) {
17:     p->s->flag[p->id] = 0; // BUG. Should assign 1 here.
18:     p->s->turn = 1 - p->id;
19:     while ( p->s->flag[1 - p->id] == 1 && p->s->turn == 1 - p->id ) ;
20:     p->s->in_critical[p->id] = 1;
21:     trace("Thread %d in critical.", p->id);
22:     assert( !p->s->in_critical[1 - p->id] );
23:     p->s->in_critical[p->id] = 0;
24:     p->s->flag[p->id] = 0;
25: }
26:
27: int main() {
28:     struct state *s = malloc( sizeof( struct state ) );
29:     struct p *one = malloc( sizeof( struct p ) ),
20:             *two = malloc( sizeof( struct p ) );
31:     if (!s || !one || !two)
32:         return 1;
33:
34:     one->s = two->s = s;
35:     one->id = 0;
36:     two->id = 1;
37:
38:     s->flag[0]    = 0;
39:     s->flag[1]    = 0;
40:     s->in_critical[0] = 0;
41:     s->in_critical[1] = 0;
42:     pthread_create( &one->ptid, NULL, thread, one );
43:     pthread_create( &two->ptid, NULL, thread, two );
44:     pthread_join( one->ptid, NULL );
45:     pthread_join( two->ptid, NULL );
46:     return 0;
47: }
```

Fig. 3. An example program implemented in C, using pthreads and shared memory for communication. The program implements Peterson's mutual exclusion.

```
===== Trace from initial =====

[ peterson.c:29 ]

[ peterson.c:42 ]

[ divine-llvm.h:54 ]
[ divine-llvm.h:57 ]
[ divine-llvm.h:53 ]

[ divine-llvm.h:54 ]
[ peterson.c:19, p = *0x800 { 0, 0, *0x400 { [ 0, 0 ], 1, [ 0, 0 ] } } ]
[ divine-llvm.h:53 ]

[ divine-llvm.h:54 ]
[ peterson.c:20, p = *0x800 { 0, 0, *0x400 { [ 0, 0 ], 1, [ 1, 0 ] } } ]
[ divine-llvm.h:53 ]

[ divine-llvm.h:54 ]
[ peterson.c:22, p = *0x800 { 0, 0, *0x400 { [ 0, 0 ], 1, [ 1, 0 ] } } ]
[ peterson.c:17, p = *0x80c { 1, 1, *0x400 <...> } ]

[ divine-llvm.h:54 ]
[ peterson.c:22, p = *0x800 { 0, 0, *0x400 { [ 0, 0 ], 1, [ 1, 0 ] } } ]
[ peterson.c:18, p = *0x80c { 1, 1, *0x400 <...> } ]

[ divine-llvm.h:54 ]
[ peterson.c:22, p = *0x800 { 0, 0, *0x400 { [ 0, 0 ], 0, [ 1, 0 ] } } ]
[ peterson.c:19, p = *0x80c { 1, 1, *0x400 <...> } ]

[ divine-llvm.h:54 ]
[ peterson.c:22, p = *0x800 { 0, 0, *0x400 { [ 0, 0 ], 0, [ 1, 0 ] } } ]
[ peterson.c:20, p = *0x80c { 1, 1, *0x400 <...> } ]

===== The goal =====

[ divine-llvm.h:54 ]
[ peterson.c:22, p = *0x800 { 0, 0, *0x400 { [ 0, 0 ], 0, [ 1, 1 ] } } ]
[ peterson.c:20, p = *0x80c { 1, 1, *0x400 <...> } ]
! ASSERTION FAILED
```

Fig. 4. Counterexample trace produced by DiVinE. The trace has been manually short-ened to fit a single page, but is otherwise verbatim. Each block separated by a blank line represents a single configuration of the system, and in each block, each row represents a single thread of execution. Within a single row, the program counter is represented first (whenever possible, through a source file location) and a listing of in-scope variables follows. Type information is used to format values: pointers are automatically deref-erenced, structures are shown using braces and arrays using square brackets. Aliased values are elided (shown as <...>), unless their occurrences are of different types.

```
LTL( G(Sent -> F(Recv)) )
void threads( Fifo< int > *f ) {
    int id = thread_create();
    if ( id ) {
        for ( int i = 0; i < 3; ++i ) {
            ap( Sent );
            f->push( i );
        }
        while( true );
    } else {
        for ( int i = 0; i < 3; ++i ) {
            int j = f->front( true ); // wait for value
            ap( Recv );
            f->pop();
        }
        assert( f->empty() );
        while ( true );
    }
}
```

Fig. 5. A parallel unit test for a lock-free queue

our unit test employs multiple threads, the outcome of the test when simply executed is not deterministic. A possible bug in the implementation may go uncovered for many testing iterations, for certain bug classes even thousands or millions. However, by applying our model checker to this multi-threaded unit test, we can guarantee that if *any* thread interleaving violates the property, the problem is found reliably, even though the particular interleaving may be extremely improbable and never found through testing alone.

This hybrid approach is best applied in situations where the control aspect of a parallel program is most important and the input data can be fixed or taken from a small set of specific examples. Moreover, writing test-cases for automated software testing is already part of many software development methodologies and as such familiar to development staff. For these reasons, we believe that this particular combination of unit testing and model checking can become a new and rather useful part of the developer's toolkit when dealing with multi-threaded applications.

7   Future Work and Conclusions

We have presented how LLVM bitcode interpreter can be used in concert with a distributed-memory, explicit-state LTL model checker DiVinE, enabling direct verification of program source code.[4] By using the DiVinE model checker, we

---

[4] The final implementation will be released as part of DiVinE 3.0. Nevertheless, the work currently in progress is already available in development versions of DiVinE – see http://divine.fi.muni.cz/development.html.

are tapping a powerful tool designed for high-performance model checking on platforms ranging from commodity multi-core workstations to high-end compute clusters. The powerful model checking back end is further assisted by a path reduction in the LLVM-based front end.

Moreover, to facilitate LTL model checking of software, we have proposed a novel mechanism to specify atomic propositions, one that is both practical and unobtrusive, while at the same time sufficiently expressive. This makes it possible to realistically specify properties in terms of LTL formulae, as part of real-world programs.

One of the priorities in our future work is to improve support for dynamic memory. Distinguishing states that are identical up to a symmetry with respect to their heap layout is a significant waste of the tool's capacity. Unfortunately, not all languages that the tool supports can provide reliable pointer tagging, which means that the more traditional heap canonisation approaches are not directly applicable. A conservative approach will be required, at least in cases where exact pointer tagging is not available in the given input language.

Additionally, while sequential DFS-based software model checkers can use delta compression to improve their memory efficiency, this is not straightforward with a distributed-memory system. Nevertheless, especially for software, where a single system state is often large, delta compression is usually very efficient and an adequate alternative is needed for use in distributed-memory environment.

Finally, we would like to investigate heuristics for obtaining efficient *ample* sets for LLVM bitcode sources, which would then enable the use of Partial Order Reduction as implemented in DiViNE.

References

1. Andrews, T., Qadeer, S., Rajamani, S.K., Rehof, J., Xie, Y.: Zing: A Model Checker for Concurrent Software. In: Alur, R., Peled, D.A. (eds.) CAV 2004. LNCS, vol. 3114, pp. 484–487. Springer, Heidelberg (2004)
2. Ball, T., Cook, B., Levin, V., Rajamani, S.K.: SLAM and Static Driver Verifier: Technology Transfer of Formal Methods inside Microsoft. In: Boiten, E.A., Derrick, J., Smith, G.P. (eds.) IFM 2004. LNCS, vol. 2999, pp. 1–20. Springer, Heidelberg (2004)
3. Barnat, J., Brim, L., Češka, M., Ročkai, P.: DiVinE: Parallel Distributed Model Checker (Tool paper). In: Parallel and Distributed Methods in Verification and High Performance Computational Systems Biology (HiBi/PDMC), pp. 4–7. IEEE (2010)
4. Bingham, B., Bingham, J., de Paula, F.M., Erickson, J., Singh, G., Reitblatt, M.: Industrial Strength Distributed Explicit State Model Checking. In: Parallel and Distributed Methods in Verification and High Performance Computational Systems Biology (HiBi/PDMC), pp. 28–36. IEEE (2010)
5. Chaki, S., Clarke, E., Groce, A.: Modular verification of software components in C. IEEE Transactions on Software Engineering, 385–395 (2003)
6. Corbett, J.C., Dwyer, M.B., Hatcliff, J., Laubach, S., Pasareanu, C.S., Zheng, H.: Bandera: Extracting Finite-state Models from Java Source Code. In: International Conference on Software Engineering, p. 439 (2000)

7. Aan de Brugh, N.H.M., Nguyen, V.Y., Ruys, T.C.: MOONWALKER: Verification of.NET Programs. In: Kowalewski, S., Philippou, A. (eds.) TACAS 2009. LNCS, vol. 5505, pp. 170–173. Springer, Heidelberg (2009)

8. DeMartini, C., Iosif, R., Sisto, R.: A Deadlock Detection Tool for Concurrent Java Programs. Software Practice and Experience 29(7), 577–603 (1999)

9. Henzinger, T.A., Jhala, R., Majumdar, R., Sutre, G.: Software Verification with BLAST. In: Ball, T., Rajamani, S.K. (eds.) SPIN 2003. LNCS, vol. 2648, pp. 235–239. Springer, Heidelberg (2003)

10. Holzmann, G.J.: Logic Verification of ANSI-C Code with SPIN. In: Havelund, K., Penix, J., Visser, W. (eds.) SPIN 2000. LNCS, vol. 1885, pp. 131–147. Springer, Heidelberg (2000)

11. Holzmann, G.J.: The SPIN model checker: primer and reference manual, 1st edn. Addison-Wesley Professional (2003)

12. Holzmann, G.J., Smith, M.H.: Software Model Checking: Extracting Verification Models from Source Code. Software Testing, Verification and Reliability 11(2), 65–79 (2001)

13. Lattner, C., Adve, V.: LLVM: A Compilation Framework for Lifelong Program Analysis & Transformation. In: International Symposium on Code Generation and Optimization (CGO), Palo Alto, California (March 2004)

14. Linden, A., Wolper, P.: An Automata-Based Symbolic Approach for Verifying Programs on Relaxed Memory Models. In: van de Pol, J., Weber, M. (eds.) SPIN 2010. LNCS, vol. 6349, pp. 212–226. Springer, Heidelberg (2010)

15. Musuvathi, M.S., Park, D., Chou, A., Engler, D.R., Dill, D.L.: CMC: A Pragmatic Approach to Model Checking Real Code. In: The Fifth Symposium on Operating Systems Design and Implementation (2002)

16. Peled, D.: Ten Years of Partial Order Reduction. In: Vardi, M.Y. (ed.) CAV 1998. LNCS, vol. 1427, pp. 17–28. Springer, Heidelberg (1998)

17. Brat, G., Thompson, S., Schimpf, K.: The MCP Model Checker (2008), Submitted to PEPM 2008

18. Thompson, S., Brat, G.: Verification of C++ Flight Software with the MCP Model Checker. In: 2008 IEEE Aerospace Conference, pp. 1–9 (March 2008)

19. Thompson, S., Brat, G., Venet, A.: Software model checking of ARINC-653 flight code with MCP. In: Muñoz, C. (ed.) Proceedings of the Second NASA Formal Methods Symposium (NFM 2010), NASA/CP-2010-216215, Langley Research Center, Hampton VA 23681-2199, USA, pp. 171–181. NASA (April 2010)

20. Visser, W., Havelund, K., Brat, G.P., Park, S.: Model Checking Programs. In: ASE, pp. 3–12 (2000)

21. Yorav, K., Grumberg, O.: Static Analysis for State-Space Reductions Preserving Temporal Logics. Formal Methods in System Design 25(1), 67–96 (2004)

22. Zaks, A., Joshi, R.: Verifying Multi-threaded C Programs with SPIN. In: Havelund, K., Majumdar, R., Palsberg, J. (eds.) SPIN 2008. LNCS, vol. 5156, pp. 325–342. Springer, Heidelberg (2008)

Integrating Statechart Components in Polyglot

Daniel Balasubramanian[1], Corina S. Păsăreanu[2], Jason Biatek[3],
Thomas Pressburger[4], Gabor Karsai[1],
Michael Lowry[4], and Michael W. Whalen[3]

[1] Vanderbilt University/ISIS, 1025 16th Ave S, Nashville, TN 37212
[2] Carnegie Mellon Silicon Valley, NASA Ames, M/S 269-2, Moffett Field CA 94035
[3] University of Minnesota, Dept. of Comp. Sci. and Eng., Minneapolis, MN 55455
[4] NASA Ames Research Center, M/S 269-2, Moffett Field, CA 94035

Abstract. Statecharts is a model-based formalism for simulating and
analyzing reactive systems. In our previous work, we developed Polyglot,
a unified framework for analyzing different semantic variants of Statechart models. However, for systems containing communicating, asynchronous components deployed on a distributed platform, additional
features not inherent to the basic Statecharts paradigm are needed.
These include a connector mechanism for communication, a scheduling framework for sequencing the execution of individual components,
and a method for specifying verification properties spanning multiple
components. This paper describes the addition of these features to Polyglot, along with an example NASA case study using these new features.
Furthermore, the paper describes on-going work on modeling Plexil execution plans with Polyglot, which enables the study of interaction issues
for future manned and unmanned missions.

Keywords: Statecharts, analysis, modeling, testing.

1 Introduction and Motivation

This paper reports on an on-going project at NASA Ames, whose goal is to
develop early, *design-level* automated techniques for error detection in the flight
control software developed for the next generation of manned and unmanned
space missions. The Ares-Orion abort scenario for the Constellation program
was an original motivating example for this work and is also used in this paper to
illustrate the technical capabilities of integrating different Statechart components
in our modeling and analysis framework.

During the Constellation Program, NASA was determined to provide a last-
chance option for astronaut survival if the Ares launch vehicle exploded during
launch – as did the rocket booster for the Space Shuttle Challenger in 1986 –
and therefore spent significant resources on a launch abort system. The driving requirement was to provide the Orion crew capsule with a powerful abort
rocket capable of rapidly pulling the capsule away in case of an explosion. The
Ares launch vehicle on-board fault diagnostics would interact with the Orion
spacecraft'scontrol system to detect an emerging hazard and execute either a
crew-initiated or automated firing of the launch abort rocket. Achieving the

A. Goodloe and S. Person (Eds.): NFM 2012, LNCS 7226, pp. 267–272, 2012.

rapid control capability for a launch abort became a major design driver for the Orion software architecture, the Ares software architecture, and also the interface between Ares and Orion.

The interface requirements between Ares and Orion were defined in an English-language Interface Control Document that included communication and control specifications to be implemented by the Ares and Orion flight software. Both Ares and Orion had adopted model-based software design methods. However, due to cultural reasons and the technical capabilities of different tools, a multitude of modeling formalisms were adopted: Enterprise Architect (UML 2.0) for Ares, Mathworks Simulink/Stateflow for math-intensive functions on Orion, and Rhapsody for the overall software framework for Orion. The Statechart control component for these different modeling formalisms each has different execution semantics. This makes performing conventional formal methods analysis of interacting systems developed with these different modeling formalisms difficult.

In previous work [2] we developed Polyglot, a framework for modeling and analysis of software using different Statechart formalisms. Polyglot uses a common intermediate representation with customizable Statechart semantics and leverages existing verification and test case generation technologies developed at Ames [1,4]. However, to study integration issues between asynchronous components described using different modeling formalisms, as in the Ares-Orion case study, additional features need to be added to Polyglot. These include a connector mechanism for modeling communication, an execution scheduling framework and a method for specifying verification properties that span multiple components. This paper describes the addition of these features to Polyglot, along with an analysis of the Ares-Orion abort scenario using these new features. We also describe on-going work on modeling Plexil [3] execution plans with Polyglot, which enables the study of interaction issues for future manned and unmanned (robotic) missions. Although we make our presentation in the context of a particular NASA project, we believe that our work should be relevant to other complex, safety critical model-based software that is built from multiple components modeled with different Statechart formalisms.

2   Integrating Statechart Components in Polyglot

Due to space constraints, we present here only a brief review of the typical usage of Polyglot; for a detailed description, see [2]. The basic Polyglot framework is used in the following way. First, the structure of the Statechart model (expressed in Matlab Stateflow, or Rational Rhapsody) is translated into a common intermediate representation (IR). The IR is then translated into Java code that represents the structure of the model. Only the structure of a model is translated because the semantics are provided as "pluggable" modules. Currently, modules implementing the semantics of Matlab Stateflow, Rational Rhapsody, and UML Statemachines are provided. The Java code representing the structure of the model is combined with one of these semantic modules, resulting in an executable component. Analysis can be performed using Symbolic Pathfinder

(SPF), the symbolic execution module of Java Pathfinder (JPF), which provides test-case generation and reachability analysis.

Polyglot can be used as described above to execute and analyze both individual models and also systems with simple communication between multiple models where the communication semantics matches that of Statecharts (i.e. event broadcast). However, large systems often contain components that execute in parallel and communicate asynchronously, and the basic Statecharts formalism does not provide a way to model either asynchrony or non-trivial communication between components. The remainder of this section gives a high-level overview of the connector and scheduling frameworks that were added to Polyglot for modeling communicating, asynchronous components, and also describes how properties spanning such components can be specified and checked.

Connectors. The connector framework provides a generic way for components to communicate. From a component's point of view, a connector is simply a source (destination, resp.) of inputs (outputs). Instead of reading data from or sending data directly to another component, data is read from or written to a connector. The connector is responsible for determining both how data is queued when it arrives and the order in which messages are delivered when data is read.

Our basic implementation of connectors exposes two methods, *recvFrom* and *sendTo*, which components call to receive data from or write data to the connector. Sending data to a connector is non-blocking, but attempting to read from a connector that has no available data will block the calling component. This block happens on the level of the scheduling framework, so that upon being blocked, the component returns control to the scheduler. A component becomes unblocked, and thus eligible to be run by the scheduler, when another component sends data to it through a connector. The connector that we used in the experiment in Section 4 was lossless and messages were delivered in FIFO order. Another connector that we developed implements ARINC-653[1] ports. Our intention is to develop an extensive library of connectors, modeling different communication mechanisms, including lossy communication and non-FIFO message delivery.

Scheduler. The scheduling framework is responsible for determining the order of component execution and invoking the property checking. We have developed a generic scheduler that can be instantiated with different scheduling mechanisms, e.g. non-deterministic, priority-based, calendar-based, etc. The default non-deterministic scheduler implementation works in the following way. First, each Statechart component is registered with the scheduler and marked as "ready" for execution. The scheduler is then run, and upon each step of its execution, it non-deterministically runs a single step of a component that is either "ready", meaning it previously ran without blocking and is ready again, or "unblocked", meaning that the component was blocked during its previous execution step (when trying to read data from an empty connector, for instance), but has since become unblocked by the occurrence of some external event (such as having data sent to it through a connector). Unblocked components are invoked so

---

[1] Avionics Application Standard Software Interface, Aeronautical Radio, Inc.

that they can continue executing at the point at which they last became blocked, if desired. After the selected component finishes a step of execution, properties (described below) are checked.

Additionally, the scheduler is implemented such that if JPF or SPF are being used, all of the feasible paths with respect to which eligible (i.e., ready or unblocked) component is chosen to run are explored. This allows JPF to explore all possible valid orderings of component execution.

Properties. Checking properties that span multiple components (i.e., the property involves the state configuration of more than one Statechart model) involves two main tasks. The first is specifying the property. The second is deciding when to check for property satisfaction. We specify properties using observer automata defined as Statechart models because it allows us to leverage the existing framework for translating high-level automata descriptions into Java code that can be executed directly by Polyglot. If the individual components are modeled in different tools, then the property can still be modeled as a Statechart in any one of those tools and then translated into Java.

The relevant state variables and state configuration of the components being observed are modeled as inputs to the observer automata. However, in the generated Java code, the values of these inputs are set directly by the observer automata by using references to the individual components. The observers can look directly inside the components being monitored thus eliminating the need for the Statechart components to pass any messages to the observer automata.

All properties are checked by the scheduler after each step of execution by a component, i.e. after each step of the state machine that implements the component. Because the properties are defined as observer automata using Statechart models, they are translated into Java code and executed like normal Statechart components (with the only difference being that the observers set the values of their inputs at each step by looking directly inside the monitored components). Properties that are not satisfied trigger an exception, which can be caught by SPF. The sequence of input values leading to the property violation is also reported by SPF.

3   Integrating Plexil

To further extend the reach of Polyglot, we have recently added support for Plexil [3], a PLan EXecution Language that is being used in developing various mission software for e.g., the K10 Rover [5] and human habitat automation. Plexil is based on hierarchical state machines, but unlike the other notations in Polyglot, the state machines in Plexil are implicit in the definitions of *nodes*, which describe the computational activities for executing a plan. In addition, Plexil has several language features useful for planning that are not included in the other notations, such as an extended type system in which all variables can take on the value "unknown", and a variety of different node types that have template behaviors for several activities commonly required for plan execution.

As it is likely that Plexil plans will be integrated into complex mission software involving Rhapsody, Simulink, and UML Statecharts, we want Polyglot to have the capability of simultaneously analyzing models in all of these notations. To that end, we have added support for translating Plexil plans into Polyglot state machines whose execution model matches the Rhapsody semantics. The most significant aspect of the translation is to make explicit the implicit state machines in the Plexil plan, and to add support for the extended type system used in Plexil plans. We have added the type extensions through a Java class library that in turn is loaded into Polyglot for interpretation in JPF. There are several benefits of translating into Rhapsody besides the obvious integration into Polyglot. First, it is possible to visualize the state machines involved in Plexil nodes using the IBM Rhapsody tool suite. Second, it is possible to use the tool suite to generate code for Plexil plans.

The translation is schematic in the structure of the Plexil plan and is based on the operational semantics of Plexil [3]. However, it is currently not well-optimized, and the Rhapsody semantics impose a certain amount of inefficiency on top of the analysis due to some mismatches between the Rhapsody and Plexil conception of state machines. In the future, we are planning to perform two additional steps with respect to Plexil. First, unlike the other Statecharts notations, there is a single semantics for the Plexil Statecharts. Therefore, there is not the same utility to "swapping out" of multiple Statecharts semantics for Plexil plans. We plan first to create a better optimized translation into Polyglot in which we create a custom interpreter for Plexil plans to better match the Plexil state machine semantics. In addition, we are examining a direct-to-Java code generation option for Plexil plans as it allows still more efficient analysis.

4   Experience

The extensions to Polyglot presented in this paper were applied to models representing the interaction between the Ares launch vehicle and the Orion Crew Exploration Vehicle described in Section 1. An Ares engineer modeled both Ares and Orion in Stateflow. The Ares Stateflow model consists of six concurrent regions, each containing a state machine, while the Orion Stateflow model consists of five concurrent regions, each containing its own state machine. The inputs for this model consist of ten different boolean signals. We analyzed the component interactions using the non-deterministic scheduler described in Section 2.

We analyzed the Ares-Orion communication during abort by formulating a property derived from the official flight software design documents and the software requirements specification available for Ares I. The property states that:
"Ares aborts only if Orion initiates abort or crew commands automatic abort."

We formulated the property as an observer automaton (as described in Section 2) which is advanced whenever the Ares or Orion components execute one step through their associated state machines. Using Symbolic Pathfinder to check this property resulted in a property violation in a 3 step sequence leading to the error. The generated test sequence revealed that Ares could also abort when there is loss of communication. Based on this analysis, we formulated a new property that, when analyzed with SPF, holds on the system.

Our analysis confirmed problems suspected by the engineer who developed the model, who had already submitted a request for a change to the Ares I design document. Even though NASA's manned space flight program has moved beyond project Constellation, the same cultural and technical factors that led to multiple modeling formalisms used in interacting safety-critical systems will persist for future missions. Our framework provides automated formal methods tools for the analysis of interactive components modeled with multiple Statechart formalisms, not only Stateflow as we discussed for this case study, as well as robotic plan execution represented by Plexil plans. This will be a key capability for verification and validation of future manned and unmanned missions.

We have implemented the component framework presented here in Java, and based on our profiling results with the Ares-Orion scenario and also with other examples, we made improvements to the performance of Polyglot when used with SPF. Our original analysis using the non-optimized version of Polyglot took a total of 4m, 15s. The optimized version of Polyglot took 2m, 2s, over 50% less time compared to the original version.

5   Conclusions

We have presented a high-level overview of three extensions to Polyglot that allow systems with communicating, asynchronous components to be modeled and analyzed. These extensions are a connector framework for modeling communication, a scheduling framework for sequencing component execution and a method for specifying properties spanning multiple, asynchronous components. A NASA case study using these extensions was described, as well as our on-going work to support the analysis of Plexil plans in Polyglot.

We continue to work on the Plexil integration and to apply our framework to the analysis of interacting software components developed for human and robotic missions. We also plan to investigate program specialization via symbolic execution to increase the speed of our analysis. This involves using SPF to specialize the Polyglot semantic modules with respect to particular Statechart models.

The Polyglot framework is available in open source form, and we plan to make the scheduling and connector framework available as well.

References

1. Java Pathfinder tool-set (2011), http://babelfish.arc.nasa.gov/trac/jpf
2. Balasubramanian, D., Pasareanu, C.S., Whalen, M.W., Karsai, G., Lowry, M.R.: Polyglot: modeling and analysis for multiple statechart formalisms. In: ISSTA (2011)
3. Dowek, G., Muñoz, C., Păsăreanu, C.: A small-step semantics of PLEXIL. Technical Report 2008-11, National Institute of Aerospace, Hampton, VA (2008)
4. Pasareanu, C.S., Mehlitz, P.C., Bushnell, D.H., Gundy-Burlet, K., Lowry, M.R., Person, S., Pape, M.: Combining unit-level symbolic execution and system-level concrete execution for testing nasa software. In: ISSTA, pp. 15–26 (2008)
5. Verma, V., Baskaran, V., Utz, H., Harris, R., Fry, C.: Demonstration of Robust Execution on a NASA Lunar Rover Testbed. In: iSAIRAS (2008)

Using PVS to Investigate Incidents through the Lens of Distributed Cognition

Paolo Masci*, Huayi Huang, Paul Curzon, and Michael D. Harrison

Queen Mary University of London
Mile End, London, United Kingdom
{paolo.masci,huayih,paul.curzon,michael.harrison}@eecs.qmul.ac.uk

Abstract. A systematic tool-based method is outlined that raises questions about the circumstances surrounding an incident: why it happened and what went wrong. The approach offers a practical and systematic way to apply a distributed cognition perspective to incident investigations, focusing on how available information resources (or the lack of them) may shape user action, rather than just on causal chains. This perspective supports a deeper understanding of the more systemic causes of incidents. The analysis is based on a higher order-logic model describing how information resources may have influenced the actions of those involved in the incident. The PVS theorem proving system is used to identify situations where available resources may afford unsafe user actions. The method is illustrated using a healthcare case study.

Keywords: Theorem proving, incident analysis, socio-technical system.

1   Introduction and Motivation

We explore whether automated reasoning tools, like PVS [11], informed by a distributed cognition perspective, can help investigators improve their understanding of the circumstances surrounding an incident. Distributed cognition [6] explains how people within a socio-technical system use information resources to support their actions and achieve their goals. These information resources may be external (on paper, signs, computers) or internal (in the head). Understanding how they are deployed and transformed as people perform actions helps to understand the socio-technical system and what might have led to an incident. The proposed method is illustrated using a medical incident example described in a comprehensive investigation report [3]. The analysis demonstrates that additional, and potentially error-inducing conditions that were not envisaged in the original report can be identified.

Contribution. (i) A distributed cognition perspective is demonstrated that could help investigators understand the circumstances surrounding an incident. This perspective focuses on how the availability of internal and external information resources (or the lack of them) may shape user action. (ii) The use of a

---

* Corresponding author.

A. Goodloe and S. Person (Eds.): NFM 2012, LNCS 7226, pp. 273–278, 2012.

built-in PVS type-checking mechanism is illustrated that can challenge investigators about their reconstruction of facts.

2   The Proposed Approach for Incident Investigation

We propose that automated reasoning tools can be used systematically to help investigators understand the factors contributing to an incident by (i) making explicit conjectures about the availability and use of resources; (ii) supporting exploration of the validity of the logical argument about how resources are used; (iii) challenging the validity of possible recommendations aimed at avoiding the recurrence of such incidents. This proposal is illustrated through the example incident using PVS.

The proposed constructive method to incident investigation focuses on information resources and their transformation. It involves the following steps: (1) modeling information resources used by those involved in the incident (e.g., infusion rate printed on a medication order); (2) modeling how information resources propagate within the system (e.g., how a medication order is entered into the pharmacy information system); (3) formulating and verifying conjectures about how resources were used (e.g., were relevant resources available at critical moments to relevant actors) and facts about the prescribed use of information resources (e.g., according to procedures and regulations).

This approach is not intended to replace existing accident analysis methods. Rather the aim is to further support the investigators' awareness about the circumstances surrounding an incident, enhancing the final recommendations. A variety of techniques have been proposed for conducting incident analysis. The Australian Transport Safety Bureau has developed an investigation analysis framework [2] based on Reason's model of organizational accidents [12]. Johnson's substantial and systematic review of the topic covers many of the more mature techniques [7]. Analyzing descriptions of incidents using formal techniques is not a new idea. For example, Ladkin's Why-Because analysis [9] uses formal proofs to verify the correctness and completeness of the causal argument hypothesized by the investigator. Petri Nets have also been used to generate alternative paths towards an incident [13]. A comprehensive overview of formal methods for incident investigation can be found in [8]. Leveson [10], Hollnagel [5] and others critique the basis of these approaches because they are largely based on event chains and because inappropriate classifications can bias the analysis. Leveson's STAMP approach aims to overcome some of the perceived deficiencies enabling an exploration of how *constraints* are propagated systemically contributing to the circumstances of the incident.

3   Illustrative Example

The example is described in a thorough report relating to an accident involving an intravenous infusion pump [3]. Documented incidents with a range of infusion pumps show that the wrong drug, or the wrong volume at the wrong rate, may

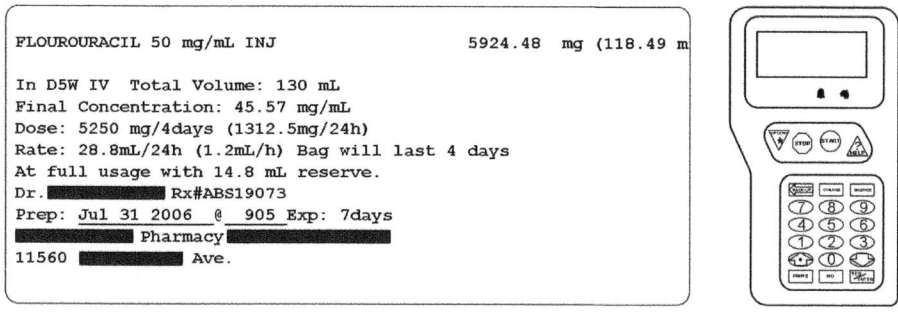

```
FLOUROURACIL 50 mg/mL INJ                    5924.48  mg (118.49 m

In D5W IV  Total Volume: 130 mL
Final Concentration: 45.57 mg/mL
Dose: 5250 mg/4days (1312.5mg/24h)
Rate: 28.8mL/24h (1.2mL/h) Bag will last 4 days
At full usage with 14.8 mL reserve.
Dr.▇▇▇▇▇▇▇▇ Rx#ABS19073
Prep: Jul 31 2006 @  905 Exp: 7days
▇▇▇▇▇▇▇▇ Pharmacy▇▇▇▇▇▇▇
11560 ▇▇▇▇▇▇▇ Ave.
```

Fig. 1. Reproduction of the label and of the pump used in the incident [3]

have disastrous consequences for the patient to whom the infusion was being administered [4]. In this case a pump that delivered a drug dose over a period of time to treat a patient in an oncology out-patients unit was programmed incorrectly. The events relevant to this incident included the prescription of the medication at the pharmacy; transferring the prescription to the out-patients unit; one nurse using the label attached to the drug bag to program the infusion pump; another nurse cross-checking, and commencement of the infusion process.

The circumstances surrounding the incident were explored by producing a PVS higher-order logic model, then using PVS to explore the facts and events. A number of questions are raised by the analysis that cannot be answered through the report. They highlight issues that may have warranted further investigation. We focus here for the purposes of illustration on one part of the incident. The complete PVS specification is available at [1].

3.1   Modeling Information Resources

Resource identification allows the analyst to externalize facts about the information that is available to the actors. Each resource is modeled using a different PVS datatype. The PVS *predicate subtyping* language mechanism which restricts the domain of already defined data-types is used extensively in the specifications of these models. When using expressions with subtypes, PVS automatically generates proof obligations. They identify type correctness conditions to ensure the valid use of the type. By this means issues in the incident are highlighted.

Information resources are first identified through the "initial understanding" of the incident described in the report: a nurse mistakenly programmed the infusion pump with the wrong rate (28.8 mL/h instead of 1.2 mL/h). The report notes that a label attached to the drug bag was used to program the pump. This printed label specified: unit of delivery, concentration, rate, and volume to be infused (see Figure 1). Further details are in the incident report [3]. The label provides a number of information resources. It is modeled using a record type [# a: A, b: B, ... #]. Each field represents a distinct information resource.

```
label_th: THEORY BEGIN
  drug_name_type: TYPE = { fluorouracil, cisplatin, %... }
```

```
rate_type: DATATYPE BEGIN mL_Xh(val: real, unit: nat): mL_Xh? END rate_type
% ...
bag_label_type: TYPE =
 [# drug_name    : drug_name_type,
     % ...
    rate_mL_24h : rate_type,
    rate_mL_h   : rate_type, % ... #]
END label_th
```

The label specification models the multiple fields contained in the bag label. If different resources can be specified with the same type, then such resources are potentially either replicated or have compatible content (e.g., in terms of values and/or units) but different meaning. Either case could lead to confusion. Checking these type matches can therefore reveal potential issues that may warrant further investigation. In this case the bag label contained information resources (e.g., rate, dose) specified multiple times in different formats. According to the report, this seemed to be the direct cause of the incident. One field on the label was used incorrectly in preference to another: *"The calculated rate (28.8 mL/h) was observed to match a number on the pharmacy label."* ([3], page 13).

The pump also contains information resources including the displays, labels that may have been attached to the pump, and audible alarms. Similarly to the bag label, the pump can be modeled as a record type, pump_type (not shown here, available from [1]). The predicate subtype used in each field reflects the constraints imposed by the pump on each information resource.

3.2   Modeling Transformations of Information Resources

Transformations are modeled as functions over resources. PVS generates proof obligations to ensure correct use of types. Discharging a proof obligation challenges the investigator's reconstruction of events and facts. Modeling the transformations helps the investigators to be clear about relations that hold among resources. Building a specification that correctly type-checks in the presence of these transformations can therefore help identify when and in what form resources are needed. An example transformation is the use of the information resources printed on the bag label by the nurse to enter the rate into the pump. Consider the information resource "rate". The constraints imposed by the bag label can be modeled easily as a PVS datatype (rate_type, defined in theory label_th) with constructor mL_Xh(val: real, unit: nat). The pump rate, on the other hand, is simply a non-negative real number below a maximum value (rate_type, defined as { x: nonneg_real | x <= max_rate } in theory pump_th). The transformation function is:

```
enter_rate(rate: label_th.rate_type): pump_th.rate_type = val(rate)
```

PVS generates a proof obligation to ensure the correct use of types:

```
enter_rate_TCC: OBLIGATION
    FORALL (rate: label_th.rate_type): val(rate) >= 0 AND val(rate) <= max_rate;
```

To discharge this proof obligation, it is necessary to show that the label rate ranges over values that can be entered in the pump — the pump rate is a bounded real number. This proof obligation, with the available information, cannot be discharged — the rate specified on the bag label is unbounded. Although mathematically trivial these results highlight implications for incident investigation that are potentially significant. They raise the question: What are the constraints on the rate value printed on the label? If answers are not available then this may suggest a weakness in the system and a potential for unsafe workarounds. The proof obligation also stimulates further investigation about rate value bounds: What is the procedure in practice when a nurse has to program a pump and the label indicates values that cannot be entered? These issues were not covered in the incident report [3].

3.3   Conjectures about the Use of Information Resources

Conjectures about the actual or prescribed use of information resources can be formulated as predicates over resources. They can be embedded in the specification of information resources – PVS then systematically generates proof obligations that ensure the conjectures hold.

One significant aspect of the incident was the safe limit of administration for the drug. A reasonable conjecture is that the resources available to the nurse provided appropriate information about safe infusion rates. The predicate subtype for the infusion rate in the label is {r: rate_type | safe_rate?(r, drug_name)}, where drug_name is another information resource provided by the label (PVS allows the specification of dependent subtypes). Instantiating the label (see Figure 1) automatically generates the proof obligation:

```
fluorouracil_bag_label_TCC: OBLIGATION safe_rate?(mL_Xh(28.8, 24), fluorouracil);
```

Given available information resources, this proof obligation cannot be discharged. Neither the label nor the pump provides information about safe limits. A bag label reporting safe limits could have helped the nurses or the patient catch the mistake, e.g., while reviewing the therapy parameters — recognition and pattern matching over recall from memory. Similarly, a pump with safeguards would have prompted a warning and, thus, could have helped catch the mistake. This seems to be a real problem in the incident: *"The calculation was not validated with a mental approximation"* ([3], page 18). A similar issue due to the propagation of information resources from the medication order to the computerized physician order entry can be highlighted with this approach. This issue is not explicitly covered in the incident report [3], though the report points out that *"a miscalculation occurred when the pharmacist initially reviewed the order in the clinic"* ([3], page 33).

4   Conclusions

This brief illustration indicates that applying a distributed cognition perspective to incident analysis can lead to insight that would help guide an incident investigator. Missing insight could of course just mean that this particular report was weak rather than our method useful. However, we argue that the method found issues beyond that related to direct causes of the particular incident. Insight can also relate to other issues that could lead to future mishaps. A traditional causal analysis method as used does not aim to highlight such issues. They would only be found through craft skill not the method. The proposed technique shares with STAMP [10] the notion that incident analysis is about discovering systemic failures rather than focusing on causal chains. It is necessary however to carry out more case studies to further explore the benefits of our approach.

A relatively simple use of a theorem prover can support this analysis. In the illustration sub-typing alone was used to raise issues and questions and it was not necessary for the analyst to formulate theorems. The analyst just models how information resources are transformed and propagated in the system. PVS automatically produces the obligations and proof attempts, demonstrating the satisfaction or otherwise of the investigators' understanding of the circumstances surrounding the incident. As more information is uncovered and modeled further proof obligations raise issues that may warrant further investigation or lead to further recommendations.

Acknowledgments. This work is supported by EPSRC (EP/G059063/1). Michael Harrison had support from Newcastle University. Dr. Astrid Mayer of the UCL Research Department of Cancer Biology, checked some details of the incident report.

References

1. Fluorouracil incident in PVS (December 2011),
   http://tinyurl.com/PVS-fluorouracil
2. Australian Transport Safety Bureau. Analysis, causality and proof in safety investigations, ATSB transport safety research report, AR-2007-053 (2007)
3. ISMP Canada. Fluorouracil incident root cause analysis report,
   http://www.ismp-canada.org/download/reports/
   FluorouracilIncidentMay2007.pdf
4. Zhang, J., et al.: Using usability heuristics to evaluate patient safety of medical devices. Journal of Biomedical Informatics, 36 (2003)
5. Hollnagel, E.: Barriers and accident prevention. Ashgate, Aldershot (2004)
6. Hutchins, E.: Cognition in the Wild. new edn. The MIT Press (1995)
7. Johnson, C.W.: Failure in Safety-Critical Systems: A Handbook of Accident and Incident Reporting. University of Glasgow Press, Glasgow (2003)
8. Johnson, C.W., Holloway, C.M.: A survey of logic formalisms to support mishap analysis. Reliability Engineering & System Safety 80(3), 271–291 (2003)
9. Ladkin, P., Sieker, B., Sanders, J.: Safety of Computer-Based Systems. Springer, Heidelberg (draft version from July 27, 2011)
10. Leveson, N.: A new accident model for engineering safer systems. Safety Science, 237–270 (2004)
11. Owre, S., Rushby, J.M., Shankar, N.: PVS: A Prototype Verification System. In: Kapur, D. (ed.) CADE 1992. LNCS (LNAI), vol. 607, pp. 748–752. Springer, Heidelberg (1992)
12. Reason, J.T.: Human error. Cambridge University Press (1991)
13. Vernez, D., Buchs, D., Pierrehumbert, G.: Perspectives in the use of coloured Petri Nets for risk analysis and accident modelling. Safety Science, 41(5) (2003)

Automated Analysis of Parametric Timing-Based Mutual Exclusion Algorithms

Roberto Bruttomesso[1], Alessandro Carioni[1],
Silvio Ghilardi[1], and Silvio Ranise[2]

[1] Università degli Studi di Milano, Milan, Italy
[2] FBK (Fondazione Bruno Kessler), Trento, Italy

Abstract. Deadlock-free algorithms that ensure mutual exclusion crucially depend on timing assumptions. In this paper, we describe our experience in automatically verifying mutual-exclusion and deadlock-freedom of the Fischer and Lynch-Shavit algorithms, using the model checker modulo theories MCMT. First, we explain how to specify timing-based algorithms in the MCMT input language as symbolic transition systems. Then, we show how the tool can verify all the safety properties used by Lynch and Shavit to establish mutual-exclusion, regardless of the number of processes in the system. Finally, we verify deadlock-freedom by following a reduction to "safety problems with lemmata synthesis" and using acceleration to avoid divergence. We also show how to automatically synthesize the bounds on the waiting time of a process to enter the critical section.

1 Introduction

In distributed systems, deadlock-free algorithms that ensure mutual exclusion crucially depend on timing assumptions. For example, the one proposed by Fischer cannot guarantee mutual exclusion when all the steps of a process do not take time in a fixed interval, while that proposed by Lynch and Shavit [16] guarantees that mutual exclusion is never violated even when the timing constraints are not satisfied. As witnessed by the pen-and-paper proofs in [16], the verification of such a class of algorithms is a subtle and time-consuming activity. This is so because of the following two main difficulties. First, the verification should be done regardless of the number n of processes in the systems, i.e., it must be *parametric* in n. Second, the waiting time of a process to enter the critical section is usually specified by means of a linear polynomial that is *parametric* in c_1 and c_2, where $[c_1, c_2]$ is the interval time in which any other step can be executed. Hence, for such a class of timing-based systems, there are two meanings of the word "parametric". This, in turn, implies that these systems have (at least) two dimensions along which they are infinite state. To overcome these difficulties, we first introduce a class of symbolic transition systems, called *parameterized timed systems*, that support the declarative specification of timing-based systems that are parametric in both the number of processes and the timing-constraints (Section 2) by using certain classes of formulae. We also sketch how the three algorithms for mutual exclusion in [16] can be formally

A. Goodloe and S. Person (Eds.): NFM 2012, LNCS 7226, pp. 279–294, 2012.

described as parameterized timed systems (Section 4). Then, we explain how to automatically solve reachability problems for parametric timed systems by using the Model Checker Modulo Theories (MCMT) [12] (Section 3). The tool uses Satisfiability Modulo Theories (SMT) techniques that cope with both kinds of parameters uniformly. Although the reachability problem for parameterized timed systems is undecidable, our experiments show that MCMT terminates when analyzing mutual exclusion and all the other safety properties considered in [16] for all the three algorithms (Section 5). Interestingly, safety properties can also be used to automatically verify deadlock-freedom by reducing the analysis of the liveness property to reachability problems as outlined below. The key observation is that the bound on the waiting time to enter the critical section is independent of the number n of processes in the system. Thus, deadlock-freedom reduces to show that it is impossible, *starting from a reachable state of the system*, to reach the states where an interval of time has passed which is longer than the bound without recording the event that a process has entered the critical section. In order to make the tool converge on these new problems, we use acceleration techniques. The role of lemmata is crucial to specify invariants overapproximating the notion of a "reachable state": first (Section 6.1), we show how MCMT is able to check the invariants identified in [16] and use them as lemmata to prove deadlock-freedom. Then (Section 6.2), we explain a technique to automatically synthesize such lemmata again by using MCMT and we report about our findings in its application for the fully automated verification of deadlock-freedom.

2   Parameterized Timed Systems

The notion of parameterized time system is an extension of that of parametrised timed network in [3] with shared variables and universal conditions in the time elapsing transitions. Informally, a parameterized timed system is formed by a collection of finitely many identical processes. Each process is a finite state automaton extended with data and clock variables, that may be local or shared. There are two kinds of transitions: one modelling the passing of time (specified by incrementing the clocks of the same amount of time) and another one in which data variables are updated and a given number of processes (usually 1 or 2) synchronize and change their states simultaneously. Transitions of the first kind (called *time elapsing*) may be guarded by "universal" conditions on the values of the clocks, i.e. by predicates involving the values of a finite but unknown number of clocks. If the guard is satisfied, all the clock variables are added of the same amount of time while the values stored in the data variables are left unchanged. Universal conditions in time elapsing transitions allow us to model the so-called *location invariants*, i.e. guards forcing a process to leave a certain location before a fixed amount of time has passed. Transitions of the second kind (called *location*) are guarded by "existential" conditions on the data and clock variables, i.e. by predicates involving a finite and known number of processes. If the guard is satisfied, both the data and clock variables of the involved processes are updated; for example, the value of some clocks may be reset. Initially, all the processes are in a distinguished initial state and their clock variables are set to zero. The value of the clocks is always positive and ranges over $\mathbb{R}$, thereby modeling a continuous flow of the time.

In the rest of this section, we explain how parameterized timed systems can be specified in the formal framework of [11] underlying the infinite state model checker MCMT [12]. The idea is to use guarded assignment transition systems whereby state variables are functions mapping a subset of the integers (used as identifiers of the processes) to either a finite subset of the integers representing the locations of the automaton or an infinite set of time points, representing the values of the clocks. For simplicity, we provide only an abstract characterization of the fragment of the MCMT input language that will be used to specify the class of parameterized timed systems; the concrete syntax can be found in the on-line user manual available at [22].

Formalization. We use multi-sorted first-order logic extended with the ternary expression constructor "if-then-else." We consider a sort INDEX for indexes of arrays, the sorts INT and REAL for elements of arrays, $\mathtt{ARRAY_{INT}}$ and $\mathtt{ARRAY_{REAL}}$ for arrays indexed over INDEX and storing elements of sort INT and REAL, respectively. We assume the availability of the arithmetic symbols of Linear Arithmetic (e.g., $+$ and $\leq$) and of the binary symbols $\_[\_]_{\mathtt{INT}} : \mathtt{ARRAY_{INT}}, \mathtt{INDEX} \to \mathtt{INT}$ and $\_[\_]_{\mathtt{REAL}} : \mathtt{ARRAY_{REAL}}, \mathtt{INDEX} \to \mathtt{REAL}$ to denote the array dereferencing operations (by abuse of notation, we omit the subscript INT or REAL when this is clear from the context). Semantically, we shall consider the class of structures where (i) INDEX is interpreted as a finite subset of the integers; (ii) INT is interpreted as $\mathbb{Z}$, REAL as $\mathbb{R}$, and the usual arithmetic symbols have their standard meanings; and (iii) $\mathtt{ARRAY_{INT}}$ and $\mathtt{ARRAY_{REAL}}$ are interpreted as the set of functions from a finite subset of the integers to $\mathbb{Z}$ and $\mathbb{R}$, respectively, and $\_[\_]$ is interpreted as function application. According to the SMT-LIB standard [19], a pair comprising a set of symbols and a class of structures (also called *models*) identifies a theory: the theory described above will be called PTS in the rest of the paper.

If $\underline{i}$ is a tuple of variables of sort INDEX and $\mathbf{a}$ a tuple of array variables, $\mathbf{a}[\underline{i}]$ is a tuple comprising all terms of the kind $a[i]$ for $a \in \mathbf{a}, i \in \underline{i}$; when writing $\phi(\underline{i}, \mathbf{a}[\underline{i}])$, we mean that ϕ is a quantifier-free formula, that the $\underline{i}$'s are the only variables of sort INDEX occurring in ϕ and that all the variables of sort INT or REAL occurring in ϕ have been replaced by the terms $\mathbf{a}[\underline{i}]$ of the corresponding sorts. A $\forall^I$-*formula* is a formula of the kind $\forall \underline{i} \phi(\underline{i}, \mathbf{a}[\underline{i}])$ and an $\exists^I$-formula is a formula of the kind $\exists \underline{i} \phi(\underline{i}, \mathbf{a}[\underline{i}])$.

A *parameterized timed system pts* is a tuple

$$\langle \mathbf{p}, \mathbf{a}, Ax, I, \{L_i(\mathbf{a}, \mathbf{a}')\}_i, E(\mathbf{a}, \mathbf{a}') \rangle$$

where $\mathbf{p}$ is a tuple of *parameters*, $\mathbf{a}$ is a tuple of *state variables*, Ax is a finite set of *system axioms*, I is the *initial state* formula, L_i is a finite set of *location* transitions, and E is a *time elapsing* transition. (Intuitively, $\mathbf{a}$ and $\mathbf{a}'$ denote the values of the state variables immediately before and after, respectively, of the execution of a transition.) We also assume the following proviso on the components of the *pts*.

Parameters. The tuple **p** is composed of an array constant id of sort $\texttt{ARRAY}_{\texttt{INT}}$ and a tuple $\mathbf{p}_r$ of constants of sort $\texttt{REAL}$. The constant id maps indexes to a finite (unknown) set of integers to allow for indirect dereference of arrays by integers. We assume id to be injective—i.e., it satisfies $\forall i,j.(id[i] = id[j] \rightarrow i = j)$—and its co-domain to be the set of positive integers—i.e., it also satisfies $\forall i.(id[i] > 0)$. In other words, id is a "casting" function from integers to indexes; for more details on the role of id, the reader is pointed to [4]. In the rest of the paper, for the sake of simplicity, we will simply write i in place of $id[i]$ (this syntactic sugar is also allowed by MCMT input language) and omit to list id among the parameters in **p**. The fact that 0 and negative integers cannot be considered as identifiers will turn out to be useful in the specification of the algorithms considered in this paper. The constants in the tuple $\mathbf{p}_r$ are called *real-valued parameters* and will be used to represent time bounds of a parameterized timed system which can be subject to some constraints, such as being strictly positive or one being larger than another. All the elements in **p** do not change their values over any run of the parameterized timed system.

State Variables. The tuple **a** is partitioned into two disjoint tuples **b** and **c** of sort $\texttt{ARRAY}_{\texttt{INT}}$ and $\texttt{ARRAY}_{\texttt{REAL}}$, respectively. The variables in **b** are the data variables and those in **c** are the clock variables. Concerning data variables, we assume that there exists a distinguished variable pc, short for *program counter*, mapping indexes to a finite (known) set of integers that represent the control locations of an automaton. Without loss of generality, we assume pc to be constrained by $\forall i.(1 \leq pc[i] \wedge pc[i] \leq \ell)$ (abbreviated as $pc \in [1,\ell]$) for some given value $\ell \geq 1$ (corresponding to the number of control locations). The updates to the clock variables in **c** model the flow of time. We assume that the tuple **c** contains a distinguished variable pc_{clock} that measures the time a process is staying in a given location. Thus, pc_{clock} is initialized to zero and reset every time the corresponding location is changed. In our framework, a shared (data or clock) variable a is modeled as a "constant" array, i.e. a is initialized and updated so that the invariant $\forall i,j.(a[i] = a[j])$ (abbreviated as $\texttt{global}(a)$) is maintained. In the rest of the paper, abusing notation, we shall write a instead of $a[i]$ or $a[j]$, etc. to emphasize that the exact value of the index used to dereference a constant array is immaterial.

System Axioms. Constraints on parameters **p** (linear inequalities and the like) are included in the set Ax of system axioms: these axioms are added to the theory PTS and used in the satisfiability tests modulo PTS mentioned in next Section. Obvious invariants known to the user (e.g., the fact that the values of the clocks are always nonnegative, the above assertions $pc \in [1,\ell]$, $\texttt{global}(a)$, etc.) can be introduced as further system axioms in MCMT specification files so that the tool can make use of them too.

Initial State Formula. We assume $I(\mathbf{a})$ to be a $\forall^I$-formula.

Location Transition Formulae. We assume $L_i(\mathbf{a}, \mathbf{a}')$ to be of the form

$$\exists i\, (\phi_L(i, \mathbf{a}[i]) \ \wedge \ \bigwedge_{a \in \mathbf{a}} a' = \lambda j.\ Upd_a(j, i, \mathbf{a}[i], \mathbf{a}[j])), \tag{1}$$

where i is a variable of sort INDEX, ϕ_L is a conjunction of literals, and the Upd_a are functions defined by cases, i.e., by suitably nested if-then-else expressions whose conditionals are again conjunctions of literals. To keep the technicalities to a minimum and since this is sufficient for the systems considered in this paper, we consider only one existentially quantified variable i in (1). However, the discussion can be generalized to location transitions with two quantified variables, which are supported by MCMT and allow one to model a wide class of systems, as observed in [11].

Time Elapsing Transition. We assume $E(\mathbf{a}, \mathbf{a}')$ to be of the form

$$\exists \varepsilon \geq 0 \ \left(\forall j \, \phi_G(j, \mathbf{a}[j], \varepsilon) \wedge \mathbf{b}' = \mathbf{b} \wedge \mathbf{c}' = \lambda j.(\mathbf{c}[j] + \epsilon) \right), \tag{2}$$

where ϕ_G is a quantifier free formula, ε is a variable of sort REAL, and equality of tuples of variables is interpreted as the conjunction of componentwise equalities. The universal guard $\forall j \, \phi_G(j, \mathbf{a}[j], \varepsilon)$ is typically used to model a location invariant.

3   Reachability for Parameterized Timed Systems

Let $\pi := \langle \mathbf{p}, \mathbf{a}, Ax, I, \{L_h(\mathbf{a}, \mathbf{a}')\}_h, E(\mathbf{a}, \mathbf{a}') \rangle$ be a parameterized timed system and $U(\mathbf{a})$ be an $\exists^I$-formula, i.e., a formula of the form $\exists \underline{i}.\phi(\underline{i}, \mathbf{a}[\underline{i}])$. Assuming that the unsafe formula is an $\exists^I$-formula allows us to express the complement of a large class of safety properties as these can usually be encoded as $\forall^I$-formulae. For example, if location 4 is the critical section location, the set of unsafe states violating the mutual exclusion property can be expressed by the $\exists^I$-formula $\exists i_1, i_2.(i_1 \neq i_2 \wedge pc[i_1] = 4 \wedge pc[i_2] = 4)$, saying that two distinct processes are in the critical section at the same time.

Given π and $U(\mathbf{a})$, the symbolic backward reachability procedure iteratively computes the set of backward reachable states $BR(\mathbf{a})$ as follows. Preliminarily, let us put $\tau(\mathbf{a}, \mathbf{a}') := \bigvee_h L_h(\mathbf{a}, \mathbf{a}') \vee E(\mathbf{a}, \mathbf{a}')$; define also (for $n \geq 0$) the n-pre-image of a formula $K(\mathbf{a})$ as

$$Pre^0(\tau, K) := K \quad \text{and} \quad Pre^{n+1}(\tau, K) := Pre(\tau, Pre^n(\tau, K)),$$

where $Pre(\tau, K) := \exists \mathbf{a}'.(\tau(\mathbf{a}, \mathbf{a}') \wedge K(\mathbf{a}'))$. Intuitively, $Pre^n(\tau, U)$ describes the set of backward reachable states in $n \geq 0$ steps. At the n-th iteration, the *backward reachability procedure* computes the formula $BR^n(\tau, U) := \bigvee_{i=0}^{n} Pre^i(\tau, U)$ representing the set of states which are backward reachable from the states in U with at most n steps. While computing $BR^n(\tau, U)$, the procedure also checks whether the system is unsafe by establishing if the formula $I \wedge Pre^n(\tau, U)$ is satisfiable modulo PTS (*safety test*) or whether a fix-point has been reached by checking if $(BR^n(\tau, U) \rightarrow BR^{n-1}(\tau, U))$ is PTS-valid or, equivalently, if the formula $BR^n(\tau, U) \wedge \neg BR^{n-1}(\tau, U)$ is PTS-unsatisfiable (*fix-point test*). If a safety test is positive, the procedure returns UNSAFE; if this does not happen and a fixed point is reached, the procedure returns SAFE.

The essential requirement in order to mechanize the procedure (which might be non-terminating for various known general reasons) is the closure of $\exists^I$-formulae under preimage computation. In this way, in fact, a formula in the sequence $BR_0, BR_1...$, is an $\exists^I$-formula and we need to check the satisfiability of conjunctions of $\exists^I$- and $\forall^I$-formulae, which is decidable by using a general result in [11]. Let K be an $\exists^I$ formula; while it is easy to show that $Pre(L, K)$ is equivalent to an $\exists^I$-formula for any location transition L, it is unfortunately impossible to prove it for $Pre(E, K)$. Although the existential variable ε can be eliminated by using a standard quantifier-elimination procedure for Linear Real arithmetic, the main difficulty is posed by the universal guard in (2), namely $\forall j.\phi_G(j, \mathbf{a}[j], \epsilon)$. In fact, it is known (see, e.g., [2]) that universal conditions are difficult to analyze automatically and require approximation techniques. In MCMT, the system is approximated by using the *stopping failures* model [17] (similar to the "approximate model" of [1, 2]). According to this model, processes can crash at any time and crashed processes remain so. In this way, the approximated system admits more runs than the original one and thus satisfies fewer safety properties. As a consequence, if the approximated system enjoys a safety property, then we are entitled to conclude that also the original system does so. In fact, establishing a safety property for the approximate system means that the system enjoys that property in a "fault-tolerant way", i.e., even in presence of failures. This will be the case for all safety properties considered in this paper and also for the deadlock freedom properties (modulo some provisoes discussed in Section 6 below). For a detailed description of how MCMT implements the stopping failures model and for information on how to check whether an unsafety trace applies to the original system, the reader is pointed to [4] (again, all unsafety traces found in the experiments of this paper can be proved to apply to the original version of the system without failures). We just point out that after moving to the stopping failures model the desired closure property of $\exists^I$-formulae under preimages holds.

4   The Lynch-Shavit Algorithm

Lynch and Shavit [16] develop a time-based algorithm for mutual exclusion by combining two other algorithms for mutual exclusion: a Lamport style asynchronous algorithm (see, e.g., [17]) and the well-known Fischer's timed mutual exclusion algorithm. The three algorithms presented in [16] consist of a finite (but unknown) number n of identical processes running concurrently. Each process is composed of four *regions* of code:

- *remainder*: the region of code not concerned with the access to critical resources;
- *trying*: the region of code where the process tries to acquire access to the critical region;
- *critical*: the region of code with exclusive access;
- *exit*: the region of code where the process exits from the critical region.

Process i:

repeat forever
  remainder region
  trying region
  critical region
  exit region
end repeat

Algorithm 1	Algorithm 2	Algorithm 3
x, y: shared registers initially $y = 0$	x: shared register, initially 0 delay: positive integer constant	x, y: shared registers initially 0 delay: positive integer constant
repeat forever	**repeat** forever	**repeat** forever
0b: *remainder exit$_i$*	0b: *remainder exit$_i$*	0b: *remainder exit$_i$*
L: $x := i$;	L: **if** $x \neq 0$ **then goto** L;	L: **if** $x \neq 0$ **then goto** L;
1: **if** $y \neq 0$ **then goto** L;	1: $x := i$;	1: $x := i$;
2: $y := 1$;	2: *pause(delay)*;	2: *pause(delay)*;
3: **if** $x \neq i$ **then goto** L;	3: **if** $x \neq i$ **then goto** L;	3: **if** $x \neq i$ **then goto** L;
4a: *critical entry$_i$*	4a: *critical entry$_i$*	% Start of Critical Region
4b: *critical exit$_i$*	4b: *critical exit$_i$*	4: **if** $y \neq 0$ **then goto** L;
5: $y := 0$;	5: $x := 0$;	5: $y := 1$;
0a: *remainder entry$_i$*	0a: *remainder entry$_i$*	6: **if** $x \neq i$ **then goto** L;
end repeat	**end repeat**	7a: *critical entry$_i$*
		7b: *critical exit$_i$*
		8: $y := 0$;
		% End of Critical Region
		9: $x := 0$;
		0a: *remainder entry$_i$*
		end repeat

(1)	(2)	(3)

Fig. 1. The three Algorithms from [16] (code for process i): (1) Lamport's Style Mutual Exclusion; (2) Fisher's Timed Mutual Exclusion; (3) Lynch-Shavit's Combined Mutual Exclusion

The pseudo-code of a process i belonging to the three algorithms (taken verbatim from [16]) is shown in Figure 1. Algorithm 1 is asynchronous while Algorithms 2 and 3 are timing-based: the time interval between successive steps of a process i is assumed to range in some interval of time when i is in its trying or exit region. The instruction $pause(k)$ causes the process to delay by a number $k - 1$ of steps. Intuitively, $pause(k)$ is equivalent to a sequence of $k - 1$ no-operations. The idea is to choose values for time parameters in Algorithms 2 and 3 so as to guarantee the two key properties:

- Mutual Exclusion (MEX): in any reachable state, at most one process is in its *critical* region;
- Deadlock Freedom (DF): in any execution, if some process is in the *trying* region, and no process is in the *critical* region, then eventually some process enters the *critical* region.

Algorithm 1 enjoys property MEX but not DF. Two timing constraints are crucial for Algorithms 2 and 3 [16]: (**TC1**) the time interval between successive steps of a process i should be contained in $[c_1, c_2]$ (for $0 < c_1 \leq c_2 < \infty$) when i is in its trying or exit region and (**TC2**) $delay \geq C = c_2/c_1$ where C is called the *time uncertainty*. If (**TC1**)-(**TC2**) are satisfied, then both Algorithms 2, 3 satisfy MEX and DF, otherwise Algorithm 2 satisfies only property DF and Algorithm 3 satisfies only property MEX. Since, ideally, timing-based algorithms should guarantee mutual exclusion regardless of the timing constraints, in this sense, Algorithm 3 is better designed than Algorithms 1 and 2.

Algorithm 1 can be formalized by a parameterized timed system

$$\pi_1 := \langle \emptyset, \langle pc, x, y \rangle, \{\texttt{global}(x), \texttt{global}(y), pc \in [1,9]\}, I, LT_1, \emptyset \rangle$$

where $I := \forall i.pc[i] = 1 \land y = 0$ and the integers $1, ..., 9$ stands for the labels $0b, ..., 0a$ in the pseudo-code, LT_1 contains the location transition corresponding to the various instructions in the pseudo-code. The tuple of parameters and the set of time elapsing formulae of π_1 are empty since time plays no role for an asynchronous algorithm like the Algorithm 1.

Algorithm $h \in \{2,3\}$ is formalized by a parameterized timed system of the form $\pi_h := \langle \mathbf{p}, \mathbf{a}_h, Ax_h, I_h, LT_h, TE_h \rangle$, where

$$\mathbf{p} := \langle C, F, G \rangle, \quad \mathbf{a}_2 := \langle pc, pc_{clock}, x \rangle, \quad \mathbf{a}_3 := \langle pc, pc_{clock}, x, y \rangle,$$
$$Ax_2 := Ax \cup \{pc \in [1,9]\}, \quad Ax_3 := Ax \cup \{pc \in [1,13], \texttt{global}(y)\},$$
$$\text{with } Ax := \{G \geq F, F \geq C, C \geq 1, \texttt{global}(x), \forall i.pc_{clock}[i] \geq 0\},$$
$$I_2 := \forall i.pc[i] = 1 \land x = 0, \quad I_3 := \forall i.pc[i] = 1 \land x = 0 \land y = 0,$$

the location transition formulae in LT_h are derived from the pseudo-code as for Algorithm 1, the time elapsing formulae in TE_h is of the form (2), and the matrix ϕ_G of the universal guard is a conjunction of formulae of the form

$$pc[j] = q \; \rightarrow \; pc_{clock}[j] + \varepsilon \leq B_q \qquad (3)$$

saying that the location q has a bound B_q that cannot be violated if $pc_{clock}[j]$ is updated to $pc_{clock}[j] + \varepsilon$. (Recall that transitions of the form (1) should have set the special clock variable $pc_{clock}[j]$ to 0 as soon as the process j enters location q.) Two clarifications about the role of the parameters C, F, and G are mandatory (the full formalization of Algorithm 2 is reported in the Appendix of [6]).

First observe that, without loss of generality, it is possible to assume $c_1 = 1$: in this way we will be able to use only Linear Arithmetic constraints, as prescribed by the definition of parameterized timed system of Section 2. Thus we have $C = c_2/c_1 = c_2$. Because of the timing constraint (**TC1**), a process is forced to remain in a location belonging to the *trying* or *exit* regions for at least 1 and at most C time units. This is encoded in π_h (for $h \in \{2,3\}$) with the two following conditions (i)-(ii). Condition (i) adds $pc_{clock}[i] \geq 1$ to the guards of those location transition formulae that modify a control location inside the *trying* or the *exit* regions. Condition (ii) adds a conjuct of the form (3) to the universal guard of the time elapsing formulae with B_q set to C, for each location inside the *trying* or *exit* regions; the only exception is for the location corresponding to the *pause* instruction, i.e., line 2 in the pseudo-code of Algorithms 2 and 3.

The second clarification is about the absence of the parameter *delay* and the presence of the parameters F and G that do not occur in the pseudo-code of Algorithms 2 and 3. The idea is to replace the obvious Non-Linear Arithmetic constraint in the formulae of π_h (for $h \in \{2,3\}$) modelling *pause(delay)* with a linear one involving F and G. In fact, the naive encoding of *pause(delay)* would require the use of the non-linear term $delay * C$ to count the the number of nullary operations that the process should wait before continuing its computations. Fortunately, as observed in [16], the key property of $pause(delay)$ is that its duration is greater than the time uncertainty C. Thus, the two additional parameters F and G are used to model the minimum and maximum time span that a process can spend inside $pause(delay)$. In this way, the time constraint (**TC2**) is encoded in π_h (for $h \in \{2, 3\}$) by adding (i) the condition $pc_{clock}[i] \geq F$ to the guard of the transition location in LT_h modifying the control location q and (ii) a conjunct of the form (3) in the universal guard of the time elapsing formulae in TE_h with B_q set to G, where q is the control location associated to the pause instruction (i.e., line 2 in the pseudo-code of Algorithms 2 and 3).

5   Automated Verification of Mutual Exclusion

We begin by reporting the results of our experiments on verifying the mutual exclusion and other safety properties of the three algorithms. All the specification files and scripts used in our experiments can be downloaded from the web page http://www.oprover.org/mcmt_lynch_shavit.html.

Table 1. Mutual exclusion experiments. Experiments were run on an Intel i7 2.70 GHz running Ubuntu Linux 11.10 32-bits.

Protocol	Property	Result	Time (s)	Notes
Lamport	MEX	safe	0.04	
	MEX	safe	2.64	T. c. specified
Fischer	MEX	unsafe	3.73	T. c. not specified
	MEX + I1	safe	(0.02 + 0.17) 0.19	Invariant added
	MEX	safe	24.39	T. c. specified
Lynch-Shavit	MEX	safe	353.91	T. c. not specified
	MEX abstr.	safe	8.56	Uses MCMT's abstraction

Algorithm 1. As it is clear from Table 1, MCMT verifies instantaneously the mutual exclusion (MEX) property of Algorithm 1. Although the related results are not shown in the Table, the same applies to the three properties of Lemma 3.2 of [16], which are used as helper properties to derive theorems in the original paper. We briefly discuss Property I3 because it is not a safety property. It is formulated as follows:

– I3: If $y = 1$ then some process i is not inside the *remainder* region.

Since MCMT proceeds by refutation, in order to be proved, I3 should be negated and formalized as the unsafety condition

$$y = 1 \ \wedge \ \forall i. \, (pc[i] \in \{0a, 0b\}) \tag{4}$$

which is not an existential formula, i.e., it cannot be handled directly by MCMT. However it is not difficult to build an existential formula whose negation implies the safety property represented by the negation of (4). The idea is to add a historical variable H that records the id of the process that set y to 1 last (initially $H = 0$); we use H to replace (4) with the weaker statement

$$y = 1 \ \wedge \ \exists i. \, (i = H \wedge pc[i] \in \{0a, 0b\}) \tag{5}$$

which corresponds to the invariant

- I3': If $y = 1$ then the process H that set $y = 1$ last is not inside the *remainder* region.

We shall implicitly use similar tricks to transform some other safety lemma statements arising in our experiments.

Algorithm 2. As discussed in Section 4, mutual exclusion for Algorithm 2 depends on timing constraints. This is confirmed by MCMT, as reported in the rows 2-3 of Table 1. Also, it appears that checking mutual exclusion with the help of Lemma 4.1 (I1), as suggested by [16], yields a substantial performance improvement, see row 4, MEX + I1 (in order to use a Lemma, MCMT first verifies it and then adds it to the set of system axioms).

Algorithm 3. Algorithm 3 combines the previous two and guarantees both mutual exclusion (even without timing constraints) and deadlock-freedom (with timing constraints). In Table 1 we check with MCMT that Algorithm 3 has the mutual exclusion property, even without timing constraints (rows 5-6). MCMT implements only a rudimentary form of abstraction which might be used during invariant search. Since mutual exclusion for Algorithm 3 does not depend on timing information at all, one can try to ask the tool to abstract away any timing information: with this proof strategy, mutual exclusion without timing constraints is established much quicker (compare lines 7 and 6 from Table 1). We just mention that it is possible to quickly check with MCMT also other lemmata from [16], e.g., those that are used as ingredients for the proof of the deadlock-freedom property.

6   Automated Verification of Deadlock Freedom

Algorithms 2 and 3 have the deadlock freedom property: interestingly, time bounds for waiting time are *independent* on the size of the network and can be expressed as linear polynomials $p(C, G)$ involving the parameters G and C. This raises the possibility of verifying deadlock-freedom using MCMT, even if MCMT can only accept safety problems. We first show how to do it with manual intervention and then we fully automatize the whole procedure by synthesizing both invariants and polynomial bounds.

6.1   Verification

We first suppose that we already know the linear polynomials giving the time bounds ($p(C, G) = 2 * G + 5 * C$ for Algorithm 2 and $p(C, G) = 2 * G + 9 * C$

for Algorithm 3); we just want to check that such bounds are correct by using MCMT. Thus we want to check that "if some process is in the trying region, and no process is in the critical region, then *before $p(C, G)$ time units have passed some process enters the critical region*". The first idea is the following:

(i) we add an absolute clock abs_{clock} and a boolean flag k to the specification (the Boolean flag k is permanently turned to true as soon as one process reaches the critical region);

(ii) we initialize the system by putting $abs_{clock} := 0$, $k :=$ false, and by saying that no process is in critical region and that the process having N as an id is in the trying region (N is a new parameter of type INT subject to the constraint $N > 0$);

(iii) we consider unsafe the states in which $abs_{clock} > p(C, G)$ and $k =$ false.

For various reasons, the above idea is not correct (indeed MCMT returns UNSAFE if you implement it, even if the chosen bound $p(C, G)$ is correct). We need to identify these reasons and make the suitable adjustments to our plan.

The reason for a *first adjustment* is clear: MCMT adopts the stopping failures model (due to the presence of universal quantifiers in transitions guards) and in the stopping failures model deadlock freedom does not hold (as a trivial counterexample, consider the run in which a process i sets the shared register x to i and then crashes, thus preventing any other process to access the critical region forever). However, crashes can be tolerated without losing deadlock freedom, provided *some key actors* do not crash: there is a limited (albeit sufficient) possibility to tell this to MCMT. Notice that, whenever MCMT adopts the stopping failures models, it automatically relativizes quantifiers to non-crashed processes (see [4] for details). Recall also that a process that crashes is crashed forever: as a consequence, *processes that are existentially quantified in the unsafety formula cannot be crashed*. Thus, the proposal is to use as unsafety formula the disjunction of the following three existential sentences:

$$\exists i_1 \exists i_2 \, (i_1 = N \wedge i_1 \neq i_2 \wedge x = i_2 \wedge k = \text{false} \wedge abs_{clock} > p(C, G)) \quad (6)$$

$$\exists i_1 \, (i_1 = N \wedge x = i_1 \wedge k = \text{false} \wedge abs_{clock} > p(C, G)) \quad (7)$$

$$\exists i_1 \, (i_1 = N \wedge x = 0 \wedge k = \text{false} \wedge abs_{clock} > p(C, G)) \quad (8)$$

In this way we are guaranteed that process N (i.e., the one who was trying to access the critical region from the very beginning) does not get crashed and that, in case an undesired state is reached, it will be reached either with an uninitialized shared register or with the shared register set to the id of a non crashed process. This is much weaker than saying that there are no crash failures at all, but it is sufficient for our problems.

Still, MCMT gives UNSAFE and now comes the reason for our *second adjustment*: we need to *constrain the initial states to be "reachable" states of our Algorithms 2 and 3*. The notion of "reachable state" needs not to be definable, however we can overapproximate it by using suitable lemmata. This is in a sense the strategy of [16]: suitable lemmata describing seemingly interesting properties of the reachable states are invented, then they are formally proved and finally

they are used when proving the correctness of time bounds for deadlock freedom. In our experiments, we proposed two lemmata for Algorithm 2 and three
lemmata for Algorithm 3; such lemmata are checked by using MCMT itself (in a
total amount of time of 8.95 sec. for Algorithm 2 and 236.51 sec. for Algorithm
3) and then they are added as system axioms to the specification file of the time
bounds for deadlock-freedom (we shall see below how to automatically synthesize the lemmata). In other words, we try to prove that the deadlock-freedom
property and the related time bound for the access to the critical region *apply to
all the states that satisfy the lemmata we found*, independently on whether such
states are really reachable or not.

But now another problem arises: MCMT diverges. In fact, termination is not
guaranteed at all, because we are outside the scope of decidability results known
from the literature. However, the divergence source is limited and we can fruit-
fully apply a well-know model checking technique, namely *acceleration* (this will
be our *third and last adjustment*). The point is that the sequence of the two
transitions formed by line code L (for a fixed process i) and time elapsing can be
indefinitely applied: we need to insert a further transition modeling n executions
of this sequence for an arbitrary n. This is definable in the format accepted by
MCMT (details are shown in the Appendix of [6]). After this last adjustment,
MCMT *is able to check the time bounds* in 80.97 sec. and in 1374.38 sec. for
Algorithms 2 and 3, respectively.

6.2   Synthesis

The insertion of the accelerated transition is the only manual intervention that
is actually needed. In fact, both the lemmata used to overapproximate the set
of reachable states and the polynomial $p(C, G)$ can be synthesized.

Invariant Synthesis. Suppose first the polynomial $p(C, G)$ is fixed; let us run
MCMT on our Algorithm 2 (or 3), with the unsafety formula given by the disjunction of (6)-(8) and with the initial formula $I(\mathbf{a}, abs_{clock}, k)$ given by the
statement suggested in 6.1(ii), namely

$$abs_{clock} = 0 \wedge k = \texttt{false} \wedge \forall i\, (i = N \rightarrow pc[i] \in Try) \qquad (9)$$

(here $pc[i] \in Try$) abbreviates a disjunction saying that $pc[i]$ is equal to one
of the locations of the trying region). The tool returns UNSAFE by producing
an $\exists^I$-formula $P := \exists \underline{i} \phi(\underline{i}, \mathbf{a}[\underline{i}], k, abs_{clock}, N)$, which means that that during the
safety check the formula

$$\exists \underline{i} \phi(\underline{i}, \mathbf{a}[\underline{i}], k, abs_{clock}, N) \wedge I(\mathbf{a}, abs_{clock}, k) \qquad (10)$$

is reported to be PTS-satisfiable. Now notice that (6)-(8) all contain the conjunct $i_1 = N$, which is not modified during the calculus of preimages, thus
$\phi(\underline{i}, \mathbf{a}[\underline{i}], k, abs_{clock}, N)$ is of the kind $i_1 = N \wedge \psi(i_1, \underline{j}, \mathbf{a}[i_1], \mathbf{a}[\underline{j}], k, abs_{clock}, N)$.
Taking into consideration (9) and the instantiation algorithm for PTS-satisfia-
bility given in [11], the PTS-unsatisfiability of (10) means that the formula

$$\psi(i_1, \underline{j}, \mathbf{a}[i_1], \mathbf{a}[\underline{j}], \texttt{false}, 0, i_1) \wedge \bigwedge_{i \in i_1, \underline{j}} (i = i_1 \rightarrow pc[i] \in Try) \qquad (11)$$

is not PTS-satisfiable. The idea is to *check whether the negation of this formula can be used as a lemma*, i.e., if it is an overapproximation of the set of reachable states. To check this, it is sufficient to run MCMT on the problem having the standard initialization of Algorithm 2 (resp. 3) and having precisely (11) as an unsafe formula. If the tool returns UNSAFE, then the bound $p(C, G)$ is not correct, because composing the two traces leading to the unsafe sets of states, we have a counterexample showing that the time bound can be violated. If the tool returns SAFE, then we can repeat our attempt of verifying the time bound, but in the new run we add the negation of (11) as a system axiom. As a consequence, the formula (10) is not satisfiable anymore and the tool won't exit if $\exists \underline{i} \phi(\underline{i}, \mathbf{a}[\underline{i}], k, abs_{clock}, N)$ is produced. Of course, the tool may still produce an UNSAFE outcome, in which case the procedure must be repeated. In the end, provided divergence does not arise, the tool either synthesizes enough lemmata and certifies that the time bound is correct or it finds a counterexample for it.

Time Bounds Synthesis. The above procedure works independently on the fact whether the time bound we suggest to the tool is correct or not, thus it is possible to use it in order to get the optimal polynomial $p(C, G)$. In fact, what we are looking for is a linear polynomial $\alpha * C + \beta * G$ with positive integers coefficients: we can just begin with $\alpha = 1, \beta = 1$ and then increment the values with a dichotomic search as soon as we get a counterexample. The statistics of our experiments, for values close to the optimum, are reported in Table 2.

Table 2. Time bounds synthesis. The table reports the attempts of checking a polynome with given α and β coefficients. The optimal values found (2 5 for Fischer and 2 9 for Lynch-Shavit) coincide with the known theoretical optimal bounds. Experiments were run on an Intel i7 2.70 GHz running Ubuntu Linux 11.10 32-bits.

Protocol	α, β	Bound Holds	Iterations	Time (s)
Fischer	2, 2	NO	1	62.06
	2, 4	NO	5	110.68
	2, 5	YES	8	155.56
	2, 6	YES	6	130.69
	2, 10	YES	3	51.25
Lynch-Shavit	2, 2	NO	1	224.74
	2, 8	NO	11	5764.42
	2, 9	YES	16	27995.78
	2, 10	YES	10	6935.91
	2, 14	YES	3	974.06

7   Discussion

We have described how mutual exclusion and deadlock-freedom of a class of timing-based algorithms can be specified and automatically verified by the model checker MCMT. We have highlighted how two kinds of being parametric are supported by our framework, namely with respect to the number of processes in the system and the symbolic constants in the timing constraints. We have illustrated our approach on the Lynch-Shavit algorithm.

To the best of our knowledge, it is the first time that a formal and automatic analysis of this algorithms is performed. Key to the automated verification of deadlock-freedom is the use of acceleration (to avoid non-termination) combined with the automated synthesis of invariants to be used as lemmata in the main proof (to realize a fully automated analysis procedure).

Related Work. To the best of our knowledge, analysis techniques for the verification of parameterized systems seldom consider the two dimensions of the parameters as we do here. For example, [10, 18] consider only finite-state processes while [3] presents a method for the verification of a parametric number of timed automata with real-valued clocks. Our notion of parameterized timed systems is strictly more general than that in [10, 18] by allowing for arithmetic variables and that of [3] by allowing for location invariants (see Section 2) in timed transitions.

There is also a substantial body of work on the analysis of safety properties for parameterized systems with an arbitrary number of processes operating on bounded and unbounded variables, see, e.g., [14, 15, 21]. These methods are not targeted to the verification of timing-based algorithms and consider only safety properties whereas we also tackle the problem of verifying a restricted class of liveness properties. The approach in [9] uses SMT techniques to verify systems with several dimensions in the parameters but it only supports invariant checking or bounded model checking.

In [5, 8, 20], SAL is used to model check several timed systems. In contrast to our approach that is fully automatic, these approaches require some amount of user interaction, which is reasonable given the large size of some of the systems (especially those in [5]). The model checker Uppaal [23] is capable of automatically checking both safety and liveness of timed automata without timing parameters. An extension of Uppaal described in [13] is capable of synthesizing linear parameter constraints for the correctness of the timed automata. Both of these approaches are not parametric in the number of processes. Our approach is parametric both in the number of processes and in the time constraints but does not attempt to perform the synthesis of linear arithmetic constraints although, in principle, this would possible and we leave it to future work. Here, we focus on the automated synthesis of invariants to be used as lemmata in proving deadlock-freedom.

In our previous work on MCMT [7, 11], we have only considered safety properties of parametric systems while here we verify also a restricted class of liveness properties. Furthermore, the analysis presented here is much more fine-grained than that in [7], because, for instance, specific time interval bounds are considered for each step of the protocols and not just for few relevant locations. This additional precision in the formalization significantly increases the difficulty of the verification tasks.

References

1. Abdulla, P.A., Delzanno, G., Ben Henda, N., Rezine, A.: Regular Model Checking Without Transducers (On Efficient Verification of Parameterized Systems). In: Grumberg, O., Huth, M. (eds.) TACAS 2007. LNCS, vol. 4424, pp. 721–736. Springer, Heidelberg (2007)
2. Abdulla, P.A., Delzanno, G., Rezine, A.: Parameterized Verification of Infinite-State Processes with Global Conditions. In: Damm, W., Hermanns, H. (eds.) CAV 2007. LNCS, vol. 4590, pp. 145–157. Springer, Heidelberg (2007)
3. Abdulla, P.A., Jonsson, B.: Model checking of systems with many identical timed processes. Theoretical Computer Science, pp. 241–264 (2003)
4. Alberti, F., Ghilardi, S., Pagani, E., Ranise, S., Rossi, G.P.: Universal Guards, Relativization of Quantifiers, and Failure Models in Model Checking Modulo Theories. JSAT 8, 29–61 (2012),
   http://jsat.ewi.tudelft.nl/content/volume8/
   JSAT8_2_Alberti.pdf
5. Brown, G.M., Pike, L.: Easy Parameterized Verification of Biphase Mark and 8N1 Protocols. In: Hermanns, H. (ed.) TACAS 2006. LNCS, vol. 3920, pp. 58–72. Springer, Heidelberg (2006)
6. Carioni, A., Bruttomesso, R., Ghilardi, S., Ranise, S.: Automated Analysis of Parametric Timing-Based Mutual Exclusion Algorithms (Extended Version) (2012),
   http://www.oprover.org/mcmt_lynch_shavit.html
7. Carioni, A., Ghilardi, S., Ranise, S.: MCMT in the Land of Parametrized Timed Automata. In: Proc. of VERIFY 2010 (2010)
8. Dutertre, B., Sorea, M.: Timed systems in sal. Technical Report SRI-SDL-04-03, SRI International, Menlo Park, CA (2004)
9. Faber, J., Ihlemann, C., Jacobs, S., Sofronie-Stokkermans, V.: Automatic Verification of Parametric Specifications with Complex Topologies. In: Méry, D., Merz, S. (eds.) IFM 2010. LNCS, vol. 6396, pp. 152–167. Springer, Heidelberg (2010)
10. Fang, Y., Piterman, N., Pnueli, A., Zuck, L.D.: Liveness with invisible ranking. Software Tools for Technology 8(3), 261–279 (2006)
11. Ghilardi, S., Ranise, S.: Backward reachability of array-based systems by SMT-solving: termination and invariant synthesis. LMCS 6(4) (2010),
    http://www.lmcs-online.org/ojs/
    viewarticle.php?id=694&layout=abstract
12. Ghilardi, S., Ranise, S.: MCMT: A Model Checker Modulo Theories. In: Giesl, J., Hähnle, R. (eds.) IJCAR 2010. LNCS, vol. 6173, pp. 22–29. Springer, Heidelberg (2010)
13. Hune, T., Romijn, J., Stoelinga, M., Vaandrager, F.W.: Linear Parametric Model Checking of Timed Automata. In: Margaria, T., Yi, W. (eds.) TACAS 2001. LNCS, vol. 2031, pp. 189–203. Springer, Heidelberg (2001)
14. Krstic, S.: Parameterized system verification with guard strengthening and parameter abstraction. In: AVIS (2005)
15. Lahiri, S.K., Bryant, R.E.: Predicate abstraction with indexed predicates. ACM Transactions on Computational Logic (TOCL) 9(1) (2007)
16. Lynch, N.A., Shavit, N.: Timing-based mutual exclusion. In: Proc. of IEEE Real-Time Systems Symposium, pp. 2–11 (1992)
17. Lynch, N.A.: Distributed Algorithms. Morgan Kaufmann (1996)

18. Pnueli, A., Ruah, S., Zuck, L.D.: Automatic Deductive Verification with Invisible Invariants. In: Margaria, T., Yi, W. (eds.) TACAS 2001. LNCS, vol. 2031, pp. 82–97. Springer, Heidelberg (2001)
19. Ranise, S., Tinelli, C.: The SMT-LIB Standard: Version 1.2. Technical report (2006), http://www.SMT-LIB.org/papers
20. Steiner, W., Dutertre, B.: Automated Formal Verification of the TTEthernet Synchronization Quality. In: Proc. of the NASA Formal Methods Symposium (2011)
21. Talupur, M., Tuttle, M.: Going with the flow: Parameterized verification using message flows. In: Proc. of FMCAD 2008, pp. 1–8 (2008)
22. MCMT web site, http://www.dsi.unimi.it/~ghilardi/mcmt/
23. Uppaal, http://www.uppaal.com

Efficient Symbolic Execution
of Value-Based Data Structures for Critical Systems*

Jason Belt[1], Robby[1], Patrice Chalin[1], John Hatcliff[1], and Xianghua Deng[2]

[1] Kansas State University
{belt,robby,chalin,hatcliff}@ksu.edu
[2] Google Inc.
wdeng@google.com

Abstract. Symbolic execution shows promise for increasing the automation of verification tasks in certified safety/security-critical systems, where use of statically allocated value-based data structures is mandated. In fact SPARK/Ada, a subset of Ada designed for verification and used for building critical systems, only permits data structures that are statically allocated. This paper describes a novel and efficient *graph-based* representation for programs making use of *value-based* data structures and procedure contracts. We show that our graph-based representation offers performance superior to a logic-based representation that is used in many approaches that delegate array reasoning to a decision procedure.

1 Introduction

The development of effective techniques for handling complex heap-based data structures has been a key enabler in the recent revival of symbolic execution (SymExe) [10, 16, 23]. While the application of SymExe has focused on the detection of common program faults and test case generation [8, 13, 16, 21, 22], we are exploring its effectiveness in highly-critical systems development. We aim to support the checking of formal contracts written in languages such as SPARK/Ada (SPARK for short) that are capable of capturing rich functional correctness properties.

SPARK [2] is a subset of Ada designed for programming and verifying high assurance applications. SPARK includes a notation for procedure contracts and deliberately omits constructs that are difficult to reason about such as dynamically allocated data, pointers, and exceptions. Even though SPARK and its static analysis tools are beneficial and easy to use for "lightweight" specifications, its full contract language is rarely used since the burden imposed on developers attempting to prove full functional correctness is too high. When the contract language is employed, its use is almost always limited to adding only enough pre/post-conditions necessary to establish absence of run-time errors such as array and numeric bounds violations. We are unaware of any industrial development effort that makes significant use of the SPARK contract notation for specifying full functional correctness properties.

* Work supported in part by the US National Science Foundation (NSF) CAREER award 0644288, the US Air Force Office of Scientific Research (AFOSR), Rockwell Collins, and the Natural Sciences and Engineering Research Council (NSERC) of Canada grant 261573.

A. Goodloe and S. Person (Eds.): NFM 2012, LNCS 7226, pp. 295–309, 2012.

One of our overall goals is to alleviate this burden by adapting SymExe to check interface specification languages for highly-critical systems such as those captured in SPARK, so that functional correctness can be checked fully automatically and with a high degree of efficiency—sufficiently so that SymExe could be run during normal compile-and-test cycles. In this paper, we address the important task of evaluating algorithms for the SymExe of *statically-allocated, value-based data structures* (SAVB DS). This is important because in most highly-critical systems, *dynamically allocated heap-based data structures are disallowed*; instead, data structures are most often implemented using SAVB arrays and records.

The technique generally employed for performing SymExe on SAVB arrays and records is to encode operations over these data structures in a **logical form** using the vocabulary of corresponding logical theories (of arrays and records) of an external decision procedure[1] (DP). This is the approach used by tools such as PEX [23] and KeY [1]. We have applied this logical approach to SPARK and it yields a simple and clean SymExe algorithm since the management of the state of composite objects is entirely handled by the decision procedure.

However, our prior experience in developing a Lightweight Decision Procedure (LDP) for SymExe [5] has shown that significant performance gains can be realized by optimizing the interface between an analysis engine and a DP. Unfortunately, the logic-based approach does not lend itself to such optimizations, so we decided to investigate whether or not the **graph-based** SymExe approaches used for analyzing dynamically-allocated reference-based data structures [10] could be adapted for SAVB DS as used in languages like SPARK. We conjectured that a graph-based representation would be beneficial since it would allow greater flexibility in optimizing not only our SymExe engine, with regards to how it handles composite objects, but also how constraints over composite objects are presented to the DP. As will be discussed in Section 2, SPARK's support of unconstrained arrays makes this adaptation non-trivial.

The main contributions of the research reported in this paper are as follows:

- We present a novel and efficient graph-based representation of SAVB DS, and its associated algorithms, as it might be used in the SymExe of a SPARK-like language (Section 3). This new approach addresses issues that arise when attempting to apply SymExe strategies originally developed for Java (*e.g.*, algorithms used in the Java PathFinder [16] and optimized approaches we subsequently developed [5, 10]) to SAVB DS.
- We describe aspects of the implementation (realized in *Bakar Kiasan*, our SymExe tool for SPARK [3]) of both a conventional logic-based and our new graph-based scheme for SAVB DS. This tool leverages various optimizations (Section 3) such as incremental solving (with native bindings to external decision procedures that we previously developed), as well as fast linear-solving of certain forms of arithmetic constraints, employing symbolic value representatives, constant propagation, term rewriting, and caching [5].
- Using various configurations of state-of-the-art decision procedures, we measure differences in SymExe times on a collection of SPARK examples, including modules

---

[1] Our use of the term "decision procedure" includes automated theorem provers.

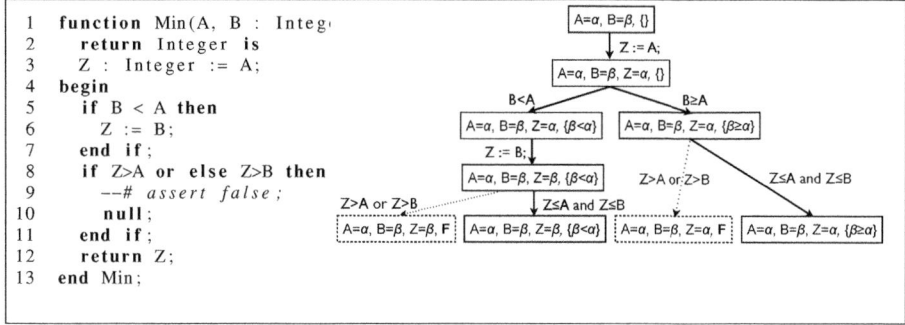

```
1   function Min(A, B : Integ
2     return Integer is
3     Z : Integer := A;
4   begin
5     if B < A then
6       Z := B;
7     end if;
8     if Z>A or else Z>B then
9       --# assert false;
10      null;
11    end if;
12    return Z;
13  end Min;
```

Fig. 1. Illustration of symbolic execution on a simple example

from our industrial partners. Results demonstrate that for composite structures, our graph-based approach is more efficient than the logic-based approach despite having optimized support for the latter by modern high-performance SMT solvers such as Yices [12] and Z3 [9]. Speed-ups range from 2×–6× when checking data structures with small bounds, and up to 65× when scaling up bounds (Section 4). The practical importance of this improved efficiency is that it enables checking with higher bounds on more complex examples within the normal compile-and-test cycle, as desired by our industrial collaborators at Rockwell Collins.

A novel representation, even if more efficient, is useless if incorrect. Thus, we have formalized both the logic- and our graph-based approaches. Space constraints prevent us from presenting the formalization here, but it is available in the following report [4]. The report also contains, or gives references to: full experimental data, tool output reports, source code for examples used, and a proof sketch of correctness. To our knowledge, this is the first time that two popular approaches to SymExe (logic- and graph-based) are realized within the same framework (*i.e.*, allowing controlled experiments), and empirically evaluated and compared.

2   Background and Motivation

In this section, we present a brief overview of SymExe and show how a logic-based representation of composite objects can be used. We then motivate the need for a new graph-based formalism for SAVB DS.

Overview of SymExe: SymExe characterizes values flowing through a program using logical constraints. This allows one to potentially explore all the possible paths through a program[2] without the need to construct concrete test cases, which in most situations would be infeasible. Consider the Min example of Figure 1 that computes the minimum of two values; in this case, we are interested in proving that the assertion at line 9 is never executed (*i.e.*, the true-branch of line 8 is infeasible) without knowing specific concrete values. Thus, we introduce special symbolic values α and β to act as placeholders for concrete values of A and B, respectively. The computation tree on the right

---

[2] In general, some form of bounding is required in order to ensure termination.

```
1    type AType is array (Integer range <>) of Integer;
        ...
2    A, B : AType;
        ...
3    A[1] := 5;  -- Instead of Ada parentheses, we use the more
4    B := A;     -- familiar square brackets for array access.
5    B[2] := 7;
6    --# assert A[3] = B[3];
7    --# assert A[2] = B[2];
```

Fig. 2. Running Example: SPARK/Ada-like Command Sequence Over Arrays

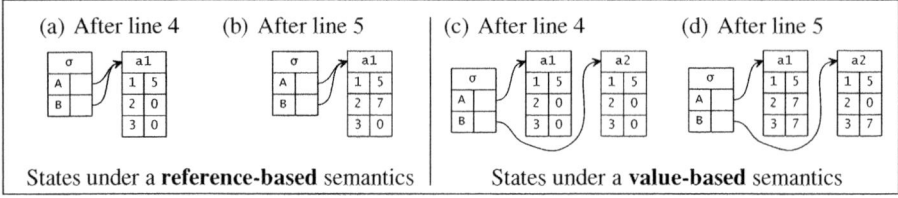

| (a) After line 4 | (b) After line 5 | (c) After line 4 | (d) After line 5 |

States under a **reference-based** semantics   States under a **value-based** semantics

Fig. 3. Concrete execution states contrasting a reference- and value-based semantic interpretation of part of our running example in Figure 2

side of Figure 1 illustrates SymExe on the procedure by keeping track of the symbolic values bound to each variable as well as logical constraints through the so-called *path condition* given in curly brackets {...}.

Initially, the constraint set is empty because we know nothing about the values of A and B. After executing line 3, we know that $Z=\alpha$, thus, (A=α and B=β and Z=α). At line 5, both the condition ($\beta < \alpha$) and its negation ($\beta \geq \alpha$) are satisfiable under the path condition, thus, we have to consider both program executions following the conditional's true-branch and its false-branch; hence, the initial path on the right side of the figure splits to cover the two possible cases. At line 8, the program state is characterized by either (A=α, B=β, Z=β, {$\beta < \alpha$}) or (A=α, B=β, Z=α, {$\beta \geq \alpha$}).

The constraints imply that the if-condition at line 8 is false in either situation (as indicated by the F for the path condition for the "true" cases) – there is no *feasible path* (no possible assignment of concrete values to inputs) that leads to line 9—and thus exploration along these paths is ignored. Decision procedures are used to determine if constraints in path conditions are satisfiable in order to make decisions at branching points (*e.g.*, lines 5 and 8). The symbolic value manipulation above is typically incorporated in a depth-first exploration.

Reference- vs. Value-based Semantics: The path condition constraints of Figure 1 are relatively simple since Min only contains integer based variables. Using the code excerpt of Figure 2 as a running example, in the next sections we detail techniques that can be used to reason about programs manipulating composite structures.

Before proceeding, we illustrate (Figure 3) the main difference between the reference-based semantics of arrays (of say, Java) vs. the value-based semantics of SPARK/Ada, for the purpose of making it obvious that a direct application of graph-based representations to SAVB DS will not work since it is incompatible with a value-based semantics. The SPARK program fragment shown in Figure 2 performs lookups and updates on two unconstrained (*i.e.*, whose lengths are unspecified) integer arrays. To simplify our illustration, assume that the arrays have an index range of 1..3 and that all the initial array element values are 0. The sub-figures 3(b) and 3(d) show the program states after executing line 5 under a reference-based and valued-based interpretation of the program, respectively. In a reference-based semantics, assignment of A to B at line 4 will cause A's reference value to be stored in B. Clearly then, both of the assertions at lines 6 and 7 will hold since any modification of B will necessarily impact A since both are pointing to the same object. In contrast, in a value-based semantics, the assignment at line 4 is handled by *copying* each element of A into B. Therefore the array update at line 5 will only impact B, and thus, the assertion at line 7 will fail since $0 \neq 7$.

Logic-Based Representation: Static analysis engines that generate constraints over composite structures can often defer the task of deciding satisfiability to external decision procedures such as SMT solvers. These solvers often contain highly optimized decision procedures for reasoning about composite objects, such as arrays and records, via theories over these structures. We illustrate this now on the example[3] of Figure 2.

We first start by assigning fresh symbolic values to A and B to obtain the initial state, s_0 : A=α_0, B=α_1, { }. To encode the effect of the assignment of line 3 in a logical-form, we assume the decision procedure provides function symbols such as SELECT and STORE that allow a client to perform *lookups* and functional *updates* on composite structures, respectively. Thus, execution of A[1] := 5 in line 3 will result in the state:

$$s_1 : \text{A}=\alpha_2, \text{B}=\alpha_1, \{\alpha_2 = \text{STORE}(\alpha_0, 1, 5)\}$$

The store operation returns a new composite value, leaving the original value unchanged, so a new symbolic value α_2 is introduced to be equal to the resultant array. The current value of A is thus set to α_2. For the assignment B := A at line 4 we simply assign A's value to B, s_2 : A=α_2, B=α_2, {α_2 = STORE(α_0, 1, 5)}. For the final update B[2] := 7 on line 5 we obtain the state:

$$s_3 : \text{A}=\alpha_2, \text{B}=\alpha_3, \{\alpha_2 = \text{STORE}(\alpha_0, 1, 5), \alpha_3 = \text{STORE}(\alpha_2, 2, 7)\}$$

As before, a new symbolic value α_3 is obtained and assigned to B. The DP should indicate that the false-branch of the assertion at line 6 is infeasible (*i.e.*, the assertion will never fail) since B was derived from A and no update was performed on the third index position in either array. However, as was seen in the concrete execution of the valued-based interpretation of our running example (Figures 3(c) and 3(d)), there exists paths for which the assertion at line 7 can fail. This is evident from the path condition in state s_3 where nothing is known about the second index position for the array stored in A (*i.e.*, α_2). Therefore, the DP will allow SymExe to enter the false-branch of the assertion and thus detect that it can be violated.

Assessment: This approach yields a very clean SymExe algorithm when applied to SPARK since the management of the state of composite objects is entirely handled by the decision procedure. However, it does not lend itself easily to the types of optimizations that we believe are necessary to scale to industrial applications for exhaustive

---

[3] For the remainder of the paper, we assume that array bounds checking is done by embedding assertions in programs for each array access; this simplifies subsequent discussions.

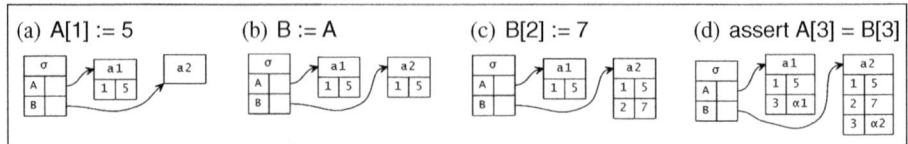

Fig. 4. States of our running example using a conventional graph-based representation

checking of strong behavioral contracts. Specifically, we have shown in previous work that because general purpose SMT solvers are not targeted to the specific patterns of constraints that are typically generated in SymExe and inherent properties of SymExe state-space exploration, they miss many opportunities for optimization when used in SymExe algorithms. To overcome this, we introduced a lightweight decision procedure LDP [5] that sits in front of a collection of conventional SMT solvers such as Yices and Z3. LDP performs SymExe-targeted optimizations including constant propagation in constraints and caching that significantly reduces the number of calls that need to be made to external decision procedures. Our experimental studies showed that significant performance gains can be realized by this approach. Unfortunately, there is little room for such optimizations when a logical form is used since so much of the symbolic representations of data must be pushed directly down into the SMT solvers.

Graph-Based Representation: In the graph-based approach for reference-based data structures introduced in [16], the SymExe engine internally maintains the state of objects and only employs the DP to reason about the values stored within them, not their actual structures. As was demonstrated earlier, one cannot naïvely apply these techniques on SAVB DS as the semantics are not compatible.

Although the concrete manipulation of value-based data structure copies values rather than sharing references, performing copies (*i.e.*, deep-clones) (upon executing an assignment) on the state that SymExe is maintaining for composite objects would not solve this issue since the techniques in [16] employ an optimization strategy known as *lazy-initialization* that allows SymExe to delay materializing values stored in composite objects until they are actually referenced. So for our running example, Figure 4 illustrates that the connection between non-materialized indices of the two arrays will be lost after the assignment of A to B of line 4 (Figure 4(b)). The assignment at line 5 would only affect B as expected (Figure 4(c)), however the assertion at line 6 will cause new symbolic values α_1 and α_2 to be materialized for A[3] and B[3], respectively (Figure 4(d)). As there is no relationship between these two values, SymExe would incorrectly conclude that the assertion at line 6 could be violated as the constraint $\alpha_1 \neq \alpha_2$ would be satisfiable. Moreover, deep-cloning is inefficient for large nested structures.

Instead of lazy-initialization, one could choose to fully materialize composite objects by assigning symbolic values to any unknown element. While this might be practical for records, it would not be for arrays since SPARK supports the notion of unconstrained arrays (at program development time) for which it may be impossible to statically determine the length and therefore the number of elements in the array (until the code is ready to be deployed in a specific configuration). In addition, always eagerly materializing structures are needlessly inefficient as the materialized elements may not be used in the context of unit-level or modular program analysis.

In the next section, we present a novel graph-based representation which allows for the efficient application of SymExe to languages with a value-based semantics for composite structures, while still realizing the performance benefits of lazy-initialization.

3 Our Approach

To recap the main points of the previous section, a key characteristic of the logic-based approach that allows it to faithfully represent the value-based semantics is that constraints in the path condition capture the necessary relationship between array copies. For example, consider the evaluation of B[2] := 7: this yields the new array α_3 for B such that $\alpha_3 = \text{STORE}(\alpha_2, 2, 7)$ where α_2 is the current array value assigned to A (cf. s_3 on page 299). The connection between A and B is implicitly captured by the equation $\alpha_3 = \text{STORE}(\alpha_2, \_, \_)$ within the path condition. In contrast, in a naïve application of the standard graph-based approach with deep cloning, the connection between the arrays is effectively lost after the statement B := A is evaluated.

To address these issues, we propose optimized graph-based data structures to represent symbolic values (abstractions) of arrays or records directly in the SymExe engine. Techniques for handling arrays are described next. Due to space constraints, and since records are handled similarly to arrays, readers interested in the technical details for record handling are referred to [4]. A key sub-structure used in our approach for arrays is termed the *base-array*; it mimics that aspect of the logical approach that keeps track of the origin of arrays and hence the relationship between array copies. Intuitively, arrays that share a base-array are considered to have a common base value. A shared base is necessary to ensure that index positions that have not been materialized will be the same when (if) they are eventually materialized. Using this adaptation we are able to correctly capture the value-based semantics for SymExe. Similar in spirit to previous graph-based representation of symbolic objects [10], analysis begins with an initial abstraction that reflects no specific knowledge about array content, it is then incrementally refined by materializing array entries as execution proceeds.

SymExe Trace of the Running Example in Figure 2: To provide a better understanding on how base arrays are used, we present here a simplified SymExe trace for our running example, but before doing so, we explain the notation that we use to graphically depict a SymExe state: for this purpose, refer to Figure 5(b). The node labeled σ_3 represents the store of the program which, for this example state, contains a mapping of variables to their values such as arrays A and B. The nodes a_0 and a_1 denote array values. Below each node label, a_i, is drawn (vertically) the list of materialized index-value pairs. Thus, the array value a_0 has a single materialized pair $1 \mapsto \alpha_1$. If an array has no materialized index-value pairs then only the array value label is shown (as is the case for a_1). The nodes labeled ρ_0 and ρ_1 denote base arrays and in state s_3, we see that the array values a_0 and a_1 have distinct base arrays, whereas s_{11} (Figure 5(f)) shows two array values sharing a base array. We now return to our running example.

Starting from the initial state s_1 (Figure 5(a)), the assignment A[1] := 5 is evaluated in two steps. The first step consists in evaluating the A[1] (as if to determine an L-value). Since nothing is known about a_0, we first materialize a new symbolic value α_1 at 1 and store it in a_0 as shown in the intermediate state s_3 (Figure 5(b)). Intuitively, we expect

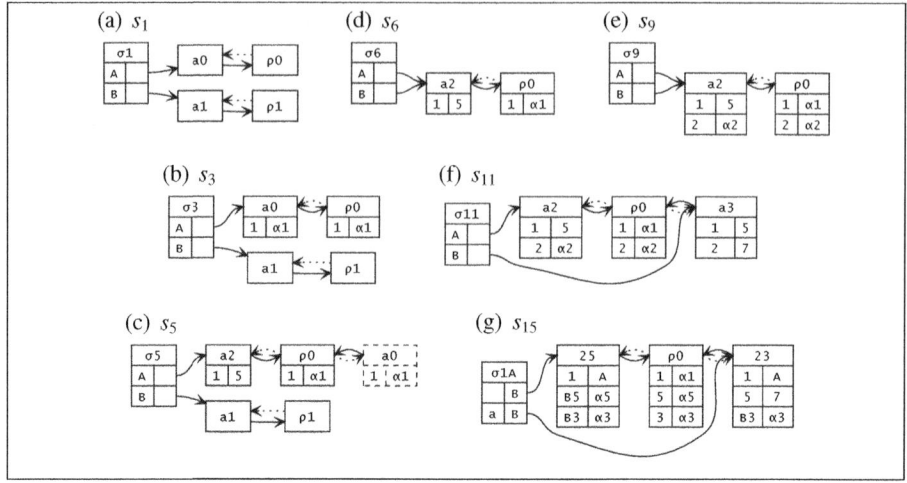

Fig. 5. States of our running example using our graph-based representation

the new mapping $1 \mapsto \alpha_1$ to also appear in any array whose origin was derived from A. In general, this implies that any index that has not been materialized in arrays which share a common origin should have a common value. The base array structure allows us to correctly propagate these materializations. This representation has an invariant that all arrays that share a common base array will have exactly the same set of indices. For this particular case, the base-array ρ_0 is used to propagate the mapping to all the arrays that depend on it, which in this case is just a_0. For reasons of efficiency, we use a copy-on-write scheme for modeling composite structures, thus, for the actual assignment, we need to build a new array a_2 with the same base-array and the mapping $1 \mapsto 5$. The resultant state s_5 is shown in Figure 5(c)—the array node a_0 is dashed because it can be safely ignored (along with reference to/from it) because no variable refers to it.

Next, evaluating A := B in s_5 simply changes the value of B to a_2 because we use copy-on-write; this results in s_6 (Figure 5(d)). Continuing with the execution of B[2]:=7, we get the intermediate state s_9 after evaluating (and thus materializing) B[2], and s_{11} (Figure 5(f)) after the assignment. Note that although B refers to a new array a_3, it still shares the base array ρ_0 since A was copied to B earlier.

Evaluating A[3] from assert A[3]=B[3] yields the state s_{15} with a new symbolic value α_3. Note that since A and B share a common base-array, the materialization $3 \mapsto \alpha_3$ is propagated to B when evaluating the sub-expression A[3]. Therefore, evaluating B[3] yields this propagated value which is clearly equal to A[3], thus the assertion always holds. For the final assert statement A[2] = B[2], evaluation causes a path split since both the true-branch and the false-branch of $\alpha_2 = 7$ are feasible because the value of α_2 is unconstrained in the path condition, thus, it simulates what may happen at runtime.

Array-Lookup Case Splitting: Previously, we only considered array lookup expressions in which the index was a constant. More generally, when performing a lookup on an array using a possibly symbolic value ι, we have three variants that case-split on whether the array has any information in its index mapping for ι. The three lookup cases represent the three possible situations for ι: (1) ι is already present in the index map, (2) ι is not already present in the mapping and is not equal to any of the existing materialized indices, and (3) ι is not already present in the mapping but is equal to one of the existing indices. Note that this is similar to the array approaches used for reference-based structures [10, 16], although it has been modified to take into account materialization propagation using base arrays. In case (1), where an element at ι already exists, the value v at that index in the array is simply returned (e.g., looking up B[3] in s_{15}). The other two cases will cause the path to split.

In case (2), we assume that ι is not equal to any of the existing indices in the array and its element value will be assigned to a fresh symbolic value. Since we assume that ι is distinct, the path condition needs to be updated to constrain ι to be different from all the other existing indices for the array. For the evaluation of A[1] in s_1, the path condition did not need to be updated since the index set for a_0 was empty. For the evaluation B[2] in s_6, a_2's index set was not empty yet the constraint $1 \neq 2$ does not add new information to the path condition and so can be ignored. To further illustrate this case, consider evaluating A[A[1]] in s_3. Since a_0 contains a mapping for 1, the expression reduces to A[α_1]. Under the assumption that α_1 is distinct, we add $\alpha_1 \neq 1$ to the path condition and then materialize a fresh symbolic value β_1 and use the base array ρ_0 to propagate the mapping $\alpha_1 \mapsto \beta_1$ to all the arrays that came from it. We obtain the following state, s_3', with the path condition $\{\alpha_1 \neq 1\}$:

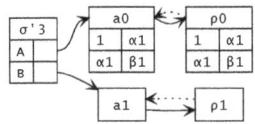

In case (3), we assume that ι is equal to some index ι' that is already mapped in the array's set of materialized indices and simply return the value at ι'; note that this case-splits (non-deterministic choice) for each index in the index set. In the resulting state, we constrain ι and ι' to be equal by adding an appropriate formula to the path condition. To illustrate this, again consider evaluating A[A[1]] in s_3: in this case we assume that α_1 is equal a materialized index of a_0. Since there is only one such index, namely $1 \mapsto \alpha_1$, we only have one path to explore for which we add the constraint $1 = \alpha_1$.

Array Element Update and Nested Structures: Care needs to be taken when implementing array element updates for value-based structures using a graph-based approach since an update does not actually cause an array to be updated but it rather constructs a new array value. This has special repercussions for nested structures. For example, consider a variable A which is an array of arrays of integers, and the instruction A[1] := B where B is an array of integers. Under a reference-based semantics, we would simply assign the reference value of B to A[1]. For a value-based interpretation, this is actually realized as a copy of the elements in B to the array stored at A[1]. To efficiently implement this for SymExe, we effectively treat the copy as a series of individual updates: e.g., A[1][1] := B[1]; A[1][2] := B[2]; etc. The end result is a new array value which is stored in A. This evaluation may result in case splitting similar to the previous example

```
1  proc compare(a₁, a₂) is begin
2  if a₁ and a₂ refer to the same structure then return TRUE
3  if types/bounds of a₁ and a₂ are not equal then return FALSE
4  foreach materialized ι₁ ↦ v₁ in a₁ do
5  │   v₂ ← a₂[ι₁]  // materializing a₂[ι₁] if needed
6  │   if v₁ ≠ v₂ then return FALSE
7  foreach materialized ι₂ ↦ v₂ in a₂ where ι₂ not in a₁ do
8  │   v₁ ← a₁[ι₂]  // materializing a₁[ι₂] if needed
9  │   if v₁ ≠ v₂ then return FALSE
10 if base-arrays of a₁ and a₂ are equal then return TRUE
11 if chooseBoolean() then return TRUE
12 // Generate a witness for inequality
13 Assume a₁, a₂ can be further materialized, and materialize fresh index α and values v₁, v₂
14 Add α ↦ v₁ to a₁ and α ↦ v₂ to a₂
15 Assume (add) v₁ ≠ v₂ and α is distinct to other indices in a₁ and a₂ to path condition
16 return FALSE
```

Fig. 6. Pseudocode for Graph-based Array Comparison

since the values stored in A[1] may not be known and hence must be materialized and propagated to any array sharing the same base as A.

Optimizations: As will be seen in our experimental results from Section 4, in comparison to the logic-based approach this graph-based representation can provide significant speed-ups by: (a) reducing calls to the decisions procedures, (b) exposing the opportunity to leverage an LDP's symbolic value representatives and constant propagation, as well as linear-solving (time and space) of some arithmetic operations and caching capabilities (which further reduce calls to the decisions procedures), and (c) providing an optimized and incrementally maintained representation of data structures that would otherwise need to be captured by solving (often deeply nested) STORE and SELECT terms.

For example, consider again the evaluation of A[A[1]] but this time using the logical approach. Because array reasoning is delegated to a decision procedure, the inner lookup operation would yield a fresh symbolic value and so we would never be able to infer that $1 = \alpha_1$. In contrast, the graph-based approach allowed us to discover that α_1 is in fact a constant value for any path emanating from this point. Aggressively applying optimization techniques such as those implemented in an LDP, we can now perform a fast constant propagation so that any reference to α_1 is replaced with 1. This can lead to more concrete executions being performed down the SymExe path and can significantly reduce the state-space (*i.e.*, due to the decrease in the number of nondeterministic choices for array lookup described previously), and thus analysis time.

Array Equality: Further, consider the case of deciding array equalities. This can be accomplished by simply iterating over the indices of two arrays and materializing index/value mappings as needed. However, this would cause large portions of arrays to be needlessly materialized and hence greatly increase the state-space SymExe needs to explore. Figure 6 presents pseudocode for an optimized test of array equality. The *compare* procedure takes two arrays, a_1 and a_2, and first checks whether the arrays point to the same structure at line 2. For example, *compare* would have returned TRUE had it been called on state s_6 in Figure 5(d) since A and B have the same value a_2. It next ensures the bounds of the two arrays are compatible at line 3. The for-loops at lines 4-9 perform the actual comparison of the elements materialized in the arrays by iterating over their index-value mappings. The array lookups performed at lines 5 and 8 may materialize new elements since the two arrays might not have the same materialized indices. It is important to note that this can introduce (desired) non-determinism and hence program splitting. For example, when evaluating the equality test A = B in state s_5 of Figure 5(c) in which A has the mapping $1 \mapsto \alpha_1$ and B's map is empty. The lookup at line 5 in Figure 6 will materialize a new mapping $1 \mapsto v_2$ in B's array which would cause the element comparison at line 6 to non-deterministically yield TRUE or FALSE.

If the for-loops do not return FALSE and the two arrays share a common base array (meaning they must be equal even on non-materialized indices) then *compare* returns TRUE at line 10. If the arrays share the same materialized indices but do not share a common base array, then the arrays may or may not be equal. Therefore we perform a case split by either assuming that they are equal and return TRUE (line 11), or by creating an index witness (α) and materializing index/value mappings for the two arrays, and then assuming the two array elements differ (lines 12-18).

When *compare* returns TRUE, the two arrays a_1 and a_2 are then unified as a means to optimize later queries further down the SymExe path. That is, when a_1 and a_2 are equal, an operation on a_1 can be replaced by the same operation on a_2 and vice versa. Thus, one can select a representative, for example, a_1, and always use a_1 whenever a_2 or a_1 is used. This unification can be done by simply replacing references to a_2 with a_1 in the state, or by efficiently leveraging LDP's symbolic value representative engine.

4   Empirical Evaluation

Table 1 presents experiment data for each procedure from our example applications (available online [25]). Each procedure has an associated contract that is checked, which specifies its full functional behavior[4]. In addition to standard sorting algorithms used for benchmarking, **IntegerSet** and **LinkedIntegerSet** are representative of data structures used to maintain data packet filtering and transformation in embedded security applications. **IntegerSet** set provides an array-based implementation of an integer set data structure that adds an element by inserting it at the end of the occupied slots in the array and deletes an element by sliding the contents of occupied slots at higher index positions down one slot to reclaim the slot at which the element was deleted. **LinkedIntegerSet** comes directly from a Rockwell Collins code base and uses arrays to provide a "linked list" set implementation (where links are represented as indices in an auxiliary array) with more efficient additions/deletions. The MMR (MILS Message Router) is an idealized version of a MILS infrastructure component (first proposed by researchers at the University of Idaho [18]) designed to mediate communication between partitions in a *separation kernel* [19]—the foundation of specialized real-time platforms used in security contexts to provide strong data and temporal separation. For each of these examples, **C-LoC** and **I-LoC** gives the number of lines of code in the

---

[4] Checking SPARK contracts involves translating them to an executable form as described in [3].

method contract and implementation respectively broken down as X/Y where X is the LoC appearing directly in the contract/implementation and Y is the LoC appearing in helper functions. Note that the examples are relatively small in terms of lines of code/contract as we focus on exhaustive modular checking of strong behavioral contracts instead of (selective-search) whole program bug-finding. Even with these small examples, many of which come from the embedded system domain, we can already show the performance benefits of our approach.

To rigorously evaluate both the logical and the graph-based state representation approaches, we used two highly-optimized decision procedures: Yices 1.0.29 [12] (Y) and Z3 2.15 [9] (Z)—we are in the process of adapting our optimized SMT solver bindings to Z3 3.x. In our use of both Yices and Z3, path conditions are incrementally pushed to a single instance (process) of the prover. We also included configurations that use LDP in combination with the underlying decision procedure package for the graph-based representation (YG and ZG); data for the logical representation with LDP only offers negligible benefits, thus we do not use it (YL and ZL). The k column indicates the bound on the number of array elements; beside this and a 2 hour timeout per test (overtime indicated by $O.T.$), no other bounding is used. Due to short running time with k up to 5, we averaged the timing data from 10 runs. The YL/YG and ZL/ZG columns give the speed-up rate of graph-based over logical-based.

Table 1. Experiment Data Excerpts (YL, YG, ZL, ZG are in seconds)

Package (P).Method (M)	C-LoC	I-LoC	Total-LoC
Sort.Bubble	1/23	14/4	42
Sort.Insertion	1/21	11/0	33
Sort.Selection	1/21	15/0	37
Sort.Shell	1/21	15/0	37
MMR.Fill_Mem_Row	3/1	6/1	11
MMR.Zero_Mem_Row	5/1	3/1	10
MMR.Zero_Flags	4/0	3/0	7
MMR.Read_Msgs	15/63	5/13	96
Total	31/151	72/19	273

Package (P).Method (M)	C-LoC	I-LoC	Total-LoC
IntegerSet.Get_Element_Index	7/0	8/0	15
IntegerSet.Add	8/29	4/2	43
IntegerSet.Remove	8/27	6/0	41
IntegerSet.Empty	1/0	2/0	3
LinkedIntegerSet.Get_Value	6/45	12/0	63
LinkedIntegerSet.Add	15/51	23/12	101
LinkedIntegerSet.Delete	14/45	22/0	81
LinkedIntegerSet.Init	1/37	10/0	48
Total	60/234	87/14	395

P	M	k	YL	YG	YL/YG	ZL	ZG	ZL/ZG
IntegerSet	Add	5	1.28	0.74	1.7×	2.14	0.79	2.7×
		6	1.90	0.91	2.1×	3.73	0.98	3.8×
		7	3.03	1.00	3.0×	6.35	1.08	5.9×
	Empty	5	0.03	0.02	1.8×	0.03	0.02	1.6×
		6	0.03	0.02	1.7×	0.03	0.02	1.6×
		7	0.03	0.02	1.7×	0.03	0.02	1.6×
	GetIdx	5	0.12	0.06	2.1×	0.15	0.06	2.3×
		6	0.16	0.07	2.2×	0.20	0.08	2.5×
		7	0.20	0.09	2.2×	0.26	0.09	2.8×
	Remv	5	1.15	0.61	1.9×	1.35	0.62	2.2×
		6	1.50	1.02	1.5×	2.39	1.04	2.3×
		7	2.15	1.26	1.7×	4.14	1.28	3.2×
Sort	Bubble	5	24.65	2.55	9.7×	33.61	3.13	10.7×
		6	317.00	14.62	21.7×	595.85	18.78	31.7×
		7	5468.32	153.62	35.6×	O.T.	198.21	—
	Insert	5	19.29	2.63	7.3×	28.58	3.20	8.9×
		6	222.48	14.71	15.1×	473.54	18.96	25.0×
		7	3131.62	156.42	20.0×	O.T.	210.46	—

P	M	k	YL	YG	YL/YG	ZL	ZG	ZL/ZG
LinkedIntegerSet	Add	5	16.47	2.14	7.7×	13.33	2.58	5.2×
		6	43.32	9.71	4.5×	35.37	12.95	2.7×
		7	201.39	72.91	2.8×	234.99	101.60	2.3×
	Delete	5	13.53	1.31	10.3×	10.10	1.34	7.6×
		6	32.76	2.00	16.4×	25.21	2.12	11.9×
		7	170.30	2.83	60.2×	193.41	3.30	58.7×
	GetVal	5	11.12	1.39	8.0×	9.11	1.43	6.4×
		6	27.85	2.02	13.8×	23.95	2.24	10.7×
		7	150.62	2.92	51.7×	190.05	3.28	58.0×
	Init	5	0.27	0.04	6.6×	0.12	0.05	2.7×
		6	0.41	0.05	9.1×	0.15	0.05	3.0×
		7	0.61	0.05	12.3×	0.19	0.05	3.4×
Sort	Select	5	22.52	3.13	7.2×	35.22	3.65	9.6×
		6	263.68	19.39	13.6×	686.75	24.58	27.9×
		7	3352.92	228.16	14.7×	O.T.	283.11	—
	Shell	5	22.34	2.78	8.0×	30.43	3.37	9.0×
		6	241.99	15.19	15.9×	502.20	18.95	26.5×
		7	3352.66	165.83	20.2×	O.T.	216.39	—

Package.Method	k	YL	YG	YL/YG	ZL	ZG	ZL/ZG
MMR.Fill_Mem_Row	3	0.62	0.21	3.0×	0.31	0.25	1.2×
MMR.Zero_Flags	3	0.15	0.05	3.2×	0.19	0.05	3.8×
MMR.Read_Msgs	3	1.05	0.24	4.4×	1.39	0.23	5.9×
MMR.Zero_Mem_Row	3	47.29	1.88	25.1×	115.18	1.78	64.8×

Test Machine	
CPU	2.4 GHz Intel Xeon X3430
RAM	8 GB
JVM	Oracle's 64-bit Server 1.6.0_22-b04
OS	Ubuntu Server 10.4 LTS 64-bit

The empirical data indicates that graph-based representation is strictly better than its logical counterpart. As the k-bound increases, the speed-up rate increases significantly (up to 65 times faster). This is because as SymExe proceeds, we observe that our approach gains more knowledge that enables a much faster concrete execution to be used, which is consistent with our experience with LDP [5]. It is interesting to note that for almost all of our benchmark examples, both approaches explore the same number of feasible paths. However, for **LinkedIntegerSet.Add**, the logical approach explores fewer paths because case-splits happen differently. In the graph-based approach, case-splits on an array index during array lookup happen explicitly as part of the SymExe state-space exploration; this contributes to the number of paths SymExe explores. In the logical approach, such case-splits happen internally inside the decision procedure, thus it does not contribute to different paths for SymExe to explore. Despite exploring less paths, the data suggests that this phenomenon does not necessarily translate to faster analysis time, which we found counter-intuitive at first. However, this illustrates the value of empirical evaluation in conjunction to formal treatment of our approaches.

5 Related Work

There has been much work on SymExe for programs that manipulate dynamically-allocated structures, *e.g.*, [8, 13, 14, 16, 20, 23]. With the exception of work by Khurshid *et al.* [16], most of these rely on decision procedures to reason about complex structures; *i.e.*, they use a logic-based representation, encoded using SELECT and STORE, as discussed in Section 2.

ACL2 supports value-based array structures similar to the logic-based representation with explicit compression operations [24]. That is, nested array update expressions can be compressed by removing overridden indices (e.g., STORE(STORE(α, β, 10), γ, 5) can be "simplified" to STORE(α, β, 10) when $\beta = \gamma$). In essence, our graph-based approach always eagerly compresses index-to-value mappings. In contrast, however, we have further leveraged previous work on LDP [5], allowing us to design an efficient graph-based representation whose use in SymExe can compete with specialized array decision procedure support offered by modern high-performance SMT solvers.

As was mentioned earlier, our graph-based representation is built on the lazy initialization algorithms that were originally designed for heap-allocated objects in Java [10, 16]. In this paper, we show how such an approach can be adapted to efficiently support value-based array structures—and records too, as they are simpler to handle than arrays. To our knowledge, this is the first time logic-based and graph-based approaches to handling composite structures in SymExe have been compared and evaluated in a controlled experiment.

In contrast to other work which uses SymExe primarily for bug-finding and test generation, our use of SymExe is targeted at bounded *verification* of behavioral contracts. Work on separation-logic-based techniques such as SmallFoot [6], jStar [11], and Veri-Fast [15] also aim to support contract checking via algorithms that use SymExe. A key difference relative to our work is that these focus on heap-based data structures instead of statically-allocated value-based structures.

Compared to theorem proving and other unbounded methods (*e.g.*, [17]), our approach not only provides precise analysis of code and contracts, but it excels at giving feedback/evidence, such as counter-examples when verification fails, and test-cases when the verification process succeeds. Furthermore, our approach does not require loop invariants (in contrast to, *e.g.*, [7]), which aligns with our goal of providing highly automated techniques to industrial developers. Of course, the trade-off here is that our approach is bounded, and thus it may potentially miss bugs only exposed by behaviors outside of the bounds. Thus, **it is complementary to other techniques**—we envision it being used earlier in system development and then followed by use of unbounded techniques (*e.g.*, semi-manual theorem proving) in later development phases.

6   Conclusion

We have presented part of our work in enhancing *Bakar Kiasan* to support both conventional logical and an efficient graph-based SymExe approaches for reasoning about value-based data structures used in critical systems programming. The graph-based representation uses an explicit-state approach, and decision procedure support is only used to handle scalar values that occur as leaf elements. Since the graph-based representation is implemented directly in the SymExe engine, it reduces the size of formulas as well as the number of calls to external decision procedures. The representation is tailored to the pattern of constraints generated by SymExe, and it enables a number of optimizations such as improved constant propagation, incrementally constructed data structures that more directly relate array indices to values (avoiding repeated rewritings in the logical representation in decision procedures), and an optimized form of copy-on-write state structures. An advantage of having both the logical and graph-based representations is redundancy: we were able to test the results of the two approaches against each other. This was helpful while developing and experimenting with our implementations.

The improvements in efficiency offered by the approach enable checking more complex examples with higher bounds to the extent that it can be used within the normal compile-and-test cycle of industrial software development. Due to the bounded nature of SPARK (*e.g.*, it forbids dynamic data structurs and recursion), we believe our bounded approach fits well with how developers use SPARK. That is, we believe our approach complements existing SPARK tools, and that it offers a different and worthwhile trade-off along the specification and verification effort/benefits space with respect to the ones offered by tools based on verification condition generation and automatic/semi-manual theorem proving.

References

1. Ahrendt, W., Baar, T., Beckert, B., Bubel, R., Giese, M., Hähnle, R., Menzel, W., Mostowski, W., Roth, A., Schlager, S., Schmitt, P.H.: The KeY tool. Software and Systems Modeling 4, 32–54 (2005)
2. Barnes, J.: High Integrity Software—the SPARK Approach to Safety and Security. AW (2003)
3. Belt, J., Hatcliff, J., Robby, Chalin, P., Hardin, D., Deng, X.: Bakar Kiasan: Flexible Contract Checking for Critical Systems Using Symbolic Execution. In: Bobaru, M., Havelund, K., Holzmann, G.J., Joshi, R. (eds.) NFM 2011. LNCS, vol. 6617, pp. 58–72. Springer, Heidelberg (2011)
4. Belt, J., Robby, Chalin, P., Hatcliff, J., Deng, X.: Efficient symbolic execution of programs for critical systems. Technical Report SAnToS-TR2011-01-10, Kansas State University (2011), http://people.cis.ksu.edu/~belt/SAnToS-TR2011-01-10.pdf

5. Belt, J., Robby, Deng, X.: Sireum/Topi LDP: A lightweight semi-decision procedure for optimizing symbolic execution-based analyses. In: Symposium on the Foundations of Software Engineering (ESEC/FSE), pp. 355–364 (2009)
6. Berdine, J., Calcagno, C., O'Hearn, P.W.: Symbolic Execution with Separation Logic. In: Yi, K. (ed.) APLAS 2005. LNCS, vol. 3780, pp. 52–68. Springer, Heidelberg (2005)
7. Bradley, A.R., Manna, Z., Sipma, H.B.: What's Decidable About Arrays? In: Emerson, E.A., Namjoshi, K.S. (eds.) VMCAI 2006. LNCS, vol. 3855, pp. 427–442. Springer, Heidelberg (2005)
8. Cadar, C., Dunbar, D., Engler, D.R.: Klee: Unassisted and automatic generation of high-coverage tests for complex systems programs. In: USENIX Symposium on Operating Systems Design and Implementation (OSDI), pp. 209–224. USENIX Association (2008)
9. de Moura, L., Bjørner, N.: Z3: An Efficient SMT Solver. In: Ramakrishnan, C.R., Rehof, J. (eds.) TACAS 2008. LNCS, vol. 4963, pp. 337–340. Springer, Heidelberg (2008)
10. Deng, X., Lee, J., Robby: Efficient and formal generalized symbolic execution. Automated Software Engineering, 1–69 Online First: 10.1007/s10515-011-0089-9 (to appear, 2012)
11. Distefano, D., Parkinson, M.J.: Jstar: towards practical verification for Java. In: Proceedings of the 23rd ACM SIGPLAN Conference on Object-Oriented Programming Systems Languages and Applications (OOPSLA 2008), pp. 213–226 (2008)
12. Dutertre, B., de Moura, L.: The Yices SMT solver. Tool paper (August 2006),
    http://yices.csl.sri.com/tool-paper.pdf
13. Godefroid, P., Klarlund, N., Sen, K.: DART: Directed automated random testing. In: ACM SIGPLAN 2005 Conference on Programming Language Design and Implementation (PLDI), pp. 213–223. ACM Press (2005)
14. Grieskamp, W., Tillmann, N., Schulte, W.: XRT - exploring runtime for.NET - architecture and applications. In: Workshop on Software Model Checking (2005)
15. Jacobs, B., Smans, J., Piessens, F.: A Quick Tour of the VeriFast Program Verifier. In: Ueda, K. (ed.) APLAS 2010. LNCS, vol. 6461, pp. 304–311. Springer, Heidelberg (2010)
16. Khurshid, S., Păsăreanu, C.S., Visser, W.: Generalized Symbolic Execution for Model Checking and Testing. In: Garavel, H., Hatcliff, J. (eds.) TACAS 2003. LNCS, vol. 2619, pp. 553–568. Springer, Heidelberg (2003)
17. Lev-Ami, T., Sagiv, M.: TVLA: A System for Implementing Static Analyses. In: SAS 2000. LNCS, vol. 1824, pp. 280–302. Springer, Heidelberg (2000)
18. Rossebo, B., Oman, P., Alves-Foss, J., Blue, R., Jaszkowiak, P.: Using SPARK-Ada to model and verify a MILS message router. In: Proceedings of the International Symposium on Secure Software Engineering (2006)
19. Rushby, J.: The design and verification of secure systems. In: 8th ACM Symposium on Operating Systems Principles, vol. 15(5), pp. 12–21 (1981)
20. Sen, K., Agha, G.: CUTE: A concolic unit testing engine for C. In: ACM SIGSOFT Symposium on the Foundations of Software Engineering (FSE), pp. 263–272 (2005)
21. Sen, K., Agha, G.: CUTE and jCUTE: Concolic Unit Testing and Explicit Path Model-Checking Tools. In: Ball, T., Jones, R.B. (eds.) CAV 2006. LNCS, vol. 4144, pp. 419–423. Springer, Heidelberg (2006)
22. Staats, M., Pasareanu, C.S.: Parallel symbolic execution for structural test generation. In: ISSTA, pp. 183–194 (2010)
23. Tillmann, N., de Halleux, J.: Pex–White Box Test Generation for.NET. In: Beckert, B., Hähnle, R. (eds.) TAP 2008. LNCS, vol. 4966, pp. 134–153. Springer, Heidelberg (2008)
24. ACL2 arrays,
    http://www.cs.utexas.edu/~moore/acl2/current/ARRAYS.html.
25. Benchmark example source, http://people.cis.ksu.edu/ belt/reports/ SAnToS-TR2011-01-10_src/

Generating Verifiable Java Code
from Verified PVS Specifications

Leonard Lensink[1], Sjaak Smetsers[1], and Marko van Eekelen[1,2]

[1] Institute for Computing and Information Sciences, Radboud University Nijmegen
[2] School of Computer Science, Open University of the Netherlands

Abstract. The use of verification tools to produce formal specifications of digital systems is commonly recommended, especially when dealing with safety-critical systems. These formal specifications often consist of segments which can automatically be translated into executable code.

We propose to generate both code and assertions in order to support verification at the generated code level. This is essential (and possible) when making modifications to the implemented code without revering to the verification tool, as the formal verification can be performed directly at the level of the adjusted code.

As a result of a feasibility study on this approach, we present a prototype of a code generator for the Prototype Verification System (PVS) that translates a subset of PVS functional specifications into Java annotated with JML assertions. To illustrate the tool's functionality a verified communication protocol from the NASA AirStar project is taken and a reference implementation in Java is generated. Subsequently, we experiment with verification on the Java level in order to show the feasibility of proving the generated JML annotations. In this paper we report on our experiences in this feasibility study.

1   Introduction

Safety critical systems [27] such as fault-tolerant avionics and air traffic management systems pose particular challenges due to the potential loss of life that could incur from a failure. Debugging techniques such as testing and model checking are often insufficient to ensure that the required safety guarantees hold. Heavyweight formal methods have been applied to such problems for many years, e.g. using the Prototype Verification System (PVS) [22] to model and mechanically prove that these models satisfy safety, correctness and completeness properties such as validity and agreement [18,24]. After verification of these properties, the models are implemented using traditional imperative programming languages. A huge improvement to this scenario would be to automatically derive the code and formal assertions from these proven models. As such, eliminating the potential introduction of errors during the coding phase. In addition, this opens possibilities for the use of verification condition generators or other software verification tools to check the correct implementation of the specified algorithm in the generated code, further improving code quality. Our approach follows the so-called Proof-Carrying Code principle: programs are accompanied by proofs that can be checked prior to execution.

Verifiable generated code is desirable for more reasons than that alone. When code is modified, for example, for maintenance issues or in an effort to improve efficiency,

A. Goodloe and S. Person (Eds.): NFM 2012, LNCS 7226, pp. 310–325, 2012.

the corresponding link to original PVS specification inevitably gets lost. In such a case, the generated formal assertions can be used to prove that the required safety properties still hold. In practice, one will make the step from formal specification to a concrete implementation only once. Any necessary adjustments in the generated code will then be made directly, instead of going back to the model, modifying it, and generating new code. As such, possibly introducing errors which would remain undetected, and thereby annihilating previous investments into quality.

Generally, two different techniques are employed for generating code from formal specifications. The first technique exploits the Curry-Howard isomorphism in order to extract programs from constructive proofs [17, 23]. The second technique translates the original specification into code assuming that the specification has been sufficiently refined such that it has been written in a pseudo-executable subset of the specification language [1, 12, 29, 32]. The latter technique is particularly appealing when generating code from specifications written in declarative languages, such as PVS. These languages encourage writing specifications in a style that is in large part functional and therefore, executable. Furthermore, in the absence of (constructive) proofs, the second technique is usually the only viable option.

In this paper, we present in Section 2 a prototype generator of annotated code for declarative specifications written in PVS. We currently derive Java code annotated with JML [6] assertions. Although PVS contains a Lisp code generator, we believe that in order to integrate with the traditional software engineering process, a widely used imperative language like Java is a logical choice. We will show the tool's functionality and usability. Firstly by specifying in PVS in Section 3 a communication protocol for a remotely operated aircraft. Subsequently, in Section 4 we illustrate code generation by examining the automatically derived Java code from the PVS specification. Finally, we reflect in Section 5 on our experiment using the state-of-the-art verification tool KeY, especially designed for the formal verification of Java programs with JML specifications. We use this tool to reconstruct the correctness proof, originally presented in PVS, by supplying KeY with our generated Java program annotated with JML assertions. Finally, we discuss related work and conclude (Sections 6 and 7).

2   Overview of the Approach

The input to our code generator is a specification written in PVS, a theorem prover with specification language based higher-order logic. Since we aim at a wide range of applications, we do not fix the target language a priori. Indeed, the tool first generates code in Why, an intermediate language for program verification [9]. Our current prototype translates Why code with proof obligations into Java with JML annotations.

In addition to enabling multi-target generation of code, another benefit of an intermediate language is that transformations and analysis that are independent from the target language can be applied to the intermediate code directly. Besides, our code generator supports exporting Why code in XML format. This relieves the developer of translation and/or analysis tools from delving into the internals of the generator or from having to write a custom parser.

In order to increase the confidence on the generated code, the generator annotates the code with logical assertions such as pre-, postconditions, and invariants. These assertions are extracted from the declarations, definitions, and lemmas in the formal PVS model. Therefore, the generated code can be the input of a verification condition generator (*VCG*), e.g. Krakatoa [8] or KeY [2]. The annotated code is also amenable to static analysis, software model checking, and automated test generation. The Why tool is used as the back-end of verification condition generators. Indeed, the same team that develops Why, develops the tools Krakatoa and FramaC, which are front-ends for Java and C VCG's, respectively.

The figure to the right illustrates the approach with multiple target languages. The feasibility case study in this paper concerns the framed part of the figure. Via the intermediate Why language different programming languages can be targeted. In order to ease the translation from PVS to Why, we have extended the Why language with several features. For instance, we have added records, and a simple notion of modules. Tuples are treated the same as records. Modules only provide a naming scope for a set of Why

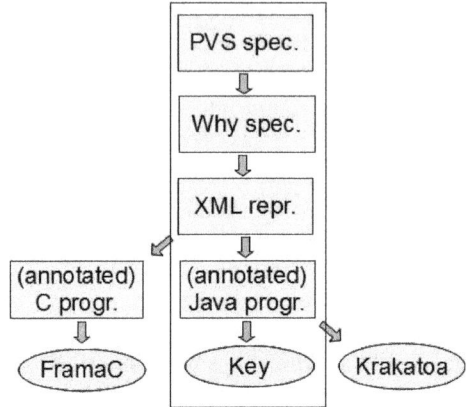

declarations. We note that a more general notion of module that includes interfaces is currently being added to the Why core language [31].

The key aspect of our approach is that not only code is transformed but also proven properties. Why proof obligations derived from PVS properties are transformed to annotations on the target language level (Java JML annotations in the case study). In this way proofs can be reconstructed on the target level even when the generated program has been changed for maintenance reasons. Consequently, maintenance of the program and its proof can be fully dealt with on the target language level.

In order to make this possible, interfaces to target level libraries are constructed on the PVS level. This is achieved via the definition of *enriched abstract interfaces* using dependently typed uninterpreted PVS functions: functions without a body for which the properties are given in a dependent type. These functions can be instantiated on the target language level with language specific libraries. Of course, the translated properties should hold for the library functions.

In the next two sections, we will perform a feasibility study on a specific case (AirStar, described in Section 3) generating Java with JML and verifying it with KeY. Further technical details of the approach are given while applying it on the case study. This will include how to deal with non-executable PVS specifications, destructive updates and higher-order functions.

3   The AirStar Model

AirSTAR [3] is a dynamically scaled experimental aircraft designed and built by NASA's Langley Research Center (LaRC) for use as a testbed for research on software health management and flight control. This Remotely Operated Aircraft (ROA) is a distributed system where its critical components are dispersed between the airborne vehicle and the ground station. Commands from the ground based pilot are broadcast to the aircraft and telemetry data from the aircraft are sent to the ground station. Communication between the air and ground components are critical for the safe operation of the vehicle.

The AirSTAR team was instructed to study [20] a small protocol that provides a guarantee of eventual message delivery, but would be simpler, and more verifiable than say User Datagram Protocol (UDP)/Transmission Control Protocol (TCP), which are considered to be too complex to be used in AirSTAR. Flight commands and telemetry data are treated differently. Flight commands are time sensitive in the sense that if a message is lost or corrupted in transit, it should not be resent, because the information would be stale by the time the new copy arrives. This requirement of the protocol is called the *weak delivery requirement*. On the other hand, engineers and researchers on the ground need to receive all data produced by the aircraft in order to analyze aircraft performance as well as to plan future aircraft flights. Hence, the protocol should guarantee that all telemetry data produced during the flight is eventually delivered. This requirement is called the *guaranteed delivery requirement*. Due to these differences, the protocol has been structured as two separate protocols: the weak delivery protocol (WDP) and the guaranteed delivery protocol (GDP).

The complete protocol is a simplified version of the standard OSI-model. It is structured in a protocol stack, where each layer handles a different aspect of message processing. As a message moves down the stack, each layer performs some processing and possibly adds packet headers. As a message moves up the stack, the corresponding packet headers are removed. The proposed protocol stack consist of four layers depicted in the figure on the right.

In this section we will describe only the layers/parts of the protocol that are used in the remainder of our paper.

The Ether layer is actually an abstraction of the concrete physical layer: an unreliable medium where messages can sometimes be dropped or duplicated, or corrupted by noise.

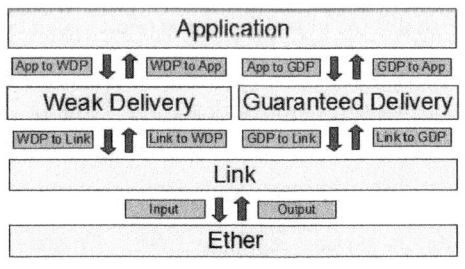

The behavior of this layer is described using *bags*, more concretely, the layer consists of an input and an output channel, both represented by a bag of messages. In PVS these bags will be modeled as functions that when applied to a message return the number of copies that were made of this message. E.g. if a bag returns 0 for a given message this might signify that the message was not yet sent or dropped.

The link layer is the interface between the WDP and GDP layers and the physical layer. It provides common services, such as error detection. Additionally, it multiplexes messages sent from the WDP and GDP layers, wrapping them in a common header and demultiplexes them on the receiving end. A link layer frame is composed of a checksum and either a WDP or a GDP frame. The link interface consists of four queues: gdp_to_ll, wdp_to_ll, ll_to_gdp and ll_tp_wdp, which are used to store the messages that were sent to or received from the the upper layers. For passing messages to the communication medium, the link state contains a reference to the Ether layer.

The weak delivery protocol is composed out of two sequences to_wdp and from_wdp, two queues app_to_wdp and wdp_to_app and a shared link interface. Sending a message is modeled as removing a message from the app_to_wdp queue and adding it to the wdp_to_11 queue.

3.1   AirStar PVS Specification

All layers of the communication protocol are modeled in PVS. The Ether and Link layer are both represented by a process taking care of the receiving and sending in that layer. For the protocol layer (WDP and GDP) as well as for the application layer, separate sending and receiving processes are assumed. This resulted in a number of PVS theorems, each corresponding to one of these processes. Due to the nature of the problem, the theories could be structured uniformly. More specifically, each process P_i consists of:

- A *process state* PS_i consisting of connections to the layer above and below (if exists), and possibly some local information, i.e: $PS_i = PI_i \times PS_{i-1} \times PL_i$, where PI_i contains the information to connect P_i with process P_{i+1}, and PL_i consists of local data.
- A set of *local actions* PA_i describing the kind of actions that can occur in P_i.
- A local *step function* $Pstep_i$ of type: $PA_i \times PS_i \rightarrow PS_i$
- A global *transition relation* $Pnext_i$ specifying that a transition is either a local step or a transition at a lower level.

Usually, the global transition relation can be defined in the following way:

$$Pnext_i(s, n : PS_i) : \texttt{bool} = \begin{cases} \exists(a : PA_i) : n = Pstep_i(a, s) & or \\ Pnext_{i-1}(PS_{i-1}(s), PS_{i-1}(n)) \ and \\ \quad PI_i(s) = PI_i(n) \ and \ PL_i(s) = PL_i(n) \end{cases}$$

For example, for the link layer, this resulted in the following PVS theory.

```
Link[GDPFrame, WDPFrame:TYPE] : THEORY BEGIN
  LinkInterface : TYPE = [#
    gdp_to_11 : fifo[GDPFrame],
    11_to_gdp : fifo[GDPFrame],
    wdp_to_11 : fifo[WDPFrame],
    11_to_wdp : fifo[WDPFrame]  #]

  LinkFrame : DATATYPE BEGIN
    GDP(gdp:GDPFrame,cs: CheckSum): GDPFrame?
    WDP(wdp:WDPFrame,cs: CheckSum): WDPFrame?
  END LinkFrame

  LinkState : TYPE = [# link : LinkInterface, ether: EtherState #]

  LinkAction : DATATYPE BEGIN
```

```
    SendWDP : SendWDP?
    SendGDP : SendGDP?
    Receive(linkframe:LinkFrame) : Receive?
  END LinkAction

  step(a:LinkAction, s:LinkState) : LinkState = CASES a OF
    SendWDP :
      IF ¬empty_fifo?(s'link'wdp_to_ll)
      THEN LET wdpf  = topof(s'link'wdp_to_ll) IN
        s WITH [ 'link'wdp_to_ll := dequeue(s'link'wdp_to_lll),
                 'ether'input := add(WDP(wdpf,checksum(wdpf)),s'ether'input)
      ELSE s ENDIF,
    SendGDP          : ...
    Receive(linkframe): ...
  ENDCASES

  next(s,n: LinkState): boolean = ∃(a:LinkAction):
    n = step(a,s) ∨ next(s'ether, n'ether) ∧  s'link = n'link
END Link
```

In the step function, only one action is specified; the other actions are defined similarly. Since this layer does not require any local administration, the local part of LinkState could be omitted. Observe that, the next predicate is not recursive: the call to next refers to the predicate in the lower ether layer.

Our aim is to prove correctness of WDP. To formulate this property, we introduce an auxiliary theory that combines the sender and receiver processes of both the WDP and the Application layer. The states that are involved in the communication are collected in a single record WDPState. To activate one of the underlying processes, the complete state is divided into two parts: the part that is needed by the active process and the part that remains unaffected. For this reason we introduce for each of the communication processes P a function $PSplit$ that takes the WDPState record and returns a pair consisting of the PState component of this state and the unaffected part.

```
WDPState : TYPE = [# ... a record containing all the states required by
    WDPSender, WDPReceiver, AppWDPSender and AppWDPReceiver ... #]

WDPSenderSplit  (s: WDPState) : ( WDPSenderState, ... ) = ...
WDPReceiverSplit(s: WDPState) : ( WDPReceiverState, ... ) = ...

WDP (s, n: WDPState) : bool =
    LET (ss,sr) = WDPSenderSplit(s), (ns,nr) = WDPSenderSplit(s)
    IN WDPSender(ss,ns) ∧  sr = nr
    OR
    LET (ss,sr) = WDPReceiverSplit(s), (ns,nr) = WDPReceiverSplit(s)
    IN WDPReceiver(ss,ns) ∧  sr = nr
    OR
    ...
    OR
    s = n
```

The invariant that expresses correctness of WDP is defined as follows:

wdp_sound?: pred[WDPState] = λ(s:WDPState): s′from_wdp ⊆ s′to_wdp

Here to_wdp and from_wdp are fields of WDPState representing collections of data. to_wdp contains the frames that were sent by the application layer, and from_wdp that were sent back to the application layer. In essence, the invariant says that only genuine data will actually arrive at its destination.

4   Generating Code

Generating Why from PVS is straightforward for the most part: each language construct in the executable subset of PVS has an almost identical counterpart in Why. Indeed, like PVS, Why can be used as a purely functional programming language.

An important difference between the PVS and Why is that Why separates the logical from computational expressions: program functions cannot be used in logical expressions. In PVS, this is not the case: one can mix both expressions freely. In principle, this problem can be circumvented easily, by generating for each PVS construct both a logical and computational version.

For brevity, the intermediate translation into Why is left out in this paper in favor of the translation to Java.

Subtype Annotations. The predicate subtyping capability of PVS enables the creation of a new subtype corresponding to an arbitrary predicate. One can use this feature to specify functions more accurately, akin to pre- and postconditions in traditional Hoare logic-based specification languages [26]. For instance, the square root function in PVS can be typed as follows: sqrt(x:real | x ≥ 0) : y:real | y ≥ 0∧ x = y*y.

This typing states that sqrt is a function that takes a non-negative real x and returns a non-negative real y such that x=y*y.

The predicate subtypes in function arguments and result are translated into JML annotations as *requires* and *ensures* clauses, respectively. More concretely, if a function uses a subtype of a type σ (specified by the predicate P on σ), the translation into Java will employ σ, and lifts the additional subtyping requirement P to the *requires* and/or *ensures* clause.

If the predicates are executable, the resulting functions will be regular pure Java functions. However, if a predicate contains non-executable constructs, the result will be an abstract function, where the non-executable fragments are transformed into *requires/ensures* clauses. This holds for most of the quantified expressions. In this transformation, the case in which a subtype is used for a quantifier variable is treated special. For instance, a universally quantified expression $\forall(x : (P))^1 : Q(x)$ becomes $\forall(x : \sigma) : P(x) \implies Q(x)$, whereas the existential quantifier $\exists(x : (P)) : Q(x)$ is translated to $\exists(x : \sigma) : P(x) \land Q(x)$. Here P, Q are predicates on types σ and $P(\sigma)$, respectively.

As an example, consider the predefined PVS functions nat and below , both representing the subsets of int, and heavily used in the AirStar model. These functions are

---

[1] PVS allows the notation (P) as a shorthand for $\{x : \sigma | P(x)\}$.

executable, so they will be translated into pure Java methods. A pure Java method is generated whenever a function is used in a JML annotation.

```
boolean /*@ pure */ nat (int s) { return 0 <= s; }
boolean /*@ pure */ below (int s, int i) { return nat(s) && s < i; }
```

Destructive Updates. Another difference between PVS and Why is that the latter supports typically imperative features like references and side effects. The efficiency of code generated from Why could be significantly improved if some of the PVS constructs are translated into an imperative Why version. For instance, PVS supports record and array overriding. Translating these updates into a destructive (in-situ) updates, when it is safe to do so, would significantly improve the efficiency of the resulting code.

PVS includes a code generator that translates PVS expressions into Lisp [28]. In the generated Lisp code, a PVS overriding expression is translated into two variants: one that destructively updates the data structure and one that constructs a new copy. The decision about which version to use is based on a conservative approximation of the runtime behaviour of the program. This static analysis is fairly coarse: nested applications and higher-order operations mostly lead to inefficient but safe copying.

Our translation from PVS to Why and subsequently into Java uses a different approach. For every function we generate a destructive variant only. However, if the alias analysis determines that a particular variable is referenced more than once, and it cannot be destructively updated, we create of (deep) copy of the corresponding object *before* performing the function which destructively updates the copy. In this way possibly large structures are copied only once instead of multiple times leading to increased efficiency. A more detailed description can be found in [16].

4.1 PVS to Java

Although the translation of PVS to Java occurs via Why, we will explain it as if it was performed in a single step, i.e. we will describe the translation by showing how basic PVS constructs are represented in Java.

Each PVS theory results in a generic Java class in which theory parameters are represented by generic variables.[2]. Record definitions are directly represented as class. Abstract data types (containing constructors and recognizers) are translated into a collection of classes extending an abstract base class representing the type itself whereas each derived class corresponds to a constructor. Recognizers are Boolean methods defined in the base class returning false by default, and which are overridden in the derived class.

For the Link theory this leads to the following class definition:

```
public class Link<GDPFrame, WDPFrame> {

    /* The LinkInterface record */
    public class LinkInterface {
```

---

[2] Actually, this only holds for TYPE parameters: value or other more complex kinds of parameters a treated differently.

```
    public FiFo<GDPFrame> gdp_to_ll, ll_to_gdp;
    public FiFo<WDPFrame> wdp_to_ll, ll_to_wdp;
    public LinkInterface update (
        FiFo<GDPFrame> gdp_to_ll, FiFo<GDPFrame> ll_to_gdp,
        FiFo<WDPFrame> wdp_to_ll, FiFo<WDPFrame> ll_to_wdp) {
      this.gdp_to_ll = gdp_to_ll; this.ll_to_gdp = ll_to_gdp;
      this.wdp_to_ll = wdp_to_ll; this.ll_to_wdp = ll_to_wdp;
      return this; } }

  public class LinkState {
    public LinkInterface link;
    public Ether.EtherState ether;
    /* ... other methods as above .../  }

  /* The abstract data type LinkFrame */
  public abstract class LinkFrame {
    public boolean isGDPFrame () { return false; }
    public boolean isWDPFrame () { return false; } }

  public  class WDP  extends LinkFrame {
    public WDPFrame wdp;
    public int cs;
    public WDP (WDPFrame wdp, int cs) {
        this.wdp = wdp; this.cs = cs;     }

    @Override
    public boolean isWDPFrame () { return true; } }

  public class GDP  extends LinkFrame { /* ... as above ... */ }

  /* The abstract data type LinkAction is similar to LinkFrame */

  /* The local step function */
  public LinkState step ( LinkAction a, LinkState s ) {
   if (a.isSendWDP()) {
     if (! s.link.wdp_to_ll.isempty()) {
      WDPFrame wdpf = s.link.wdp_to_ll.topof ();
      LinkInterface link_update = s.link.update (s.link.gdp_to_ll,
         s.link.ll_to_gdp, s.link.wdp_to_ll.dequeue(), s.link.ll_to_wdp);
      Ether<LinkFrame>.EtherState ether_update = s.ether.update(
         Bag.add (new WDP(wdpf, checksum(wdpf)),s.ether.input),
                   s.ether.output);
      return s.update(link_update, ether_update);
     } else { return s; }
   } else { /* sendWDP and receive */ } } }
```

The next predicate in the Link theory contains a non executable part. It is translated into
a function in a separate abstract interface class. The required behavior is guaranteed by
the *ensures* annotation.

```
public interface class LinkAbstract {
  /*@ ensures \result == (\exists LinkAction a; n = step(a,s)
         || next(s.ether,n.ether) && s.link == n.link);
  public /*@ pure */ next(LinkState s, LinkState n);
}
```

Higher-order Functions/Closures. The current version of Java does not (yet) support higher-order functions. For this reason, we use a common technique to implement closures, namely by introducing the following interface:

```
public interface Lambda <ARG,RES>{ RES apply (ARG arg); }
```

The higher-order functions mainly arise due to the way bags are modeled in PVS, namely as a function from LinkFrame to int. In Java, we have modeled these bags by the following helper class providing elementary operations as *services* (i.e. public static methods):

```
public class Bag<E> {
  public static <E> Lambda<E,Integer> emptyBag () {
    return new Lambda<E,Integer> () {
        public Integer apply(E arg) { return 0; }}; }

  public static <E> Lambda<E,Integer> add (
      final E elem, final Lambda<E,Integer> bag) {
    return new Lambda<E,Integer> () {
        public Integer apply (final E arg) {
            if (arg.equals(elem)) {
                return bag.apply(arg) + 1;
            } else { return bag.apply(arg); } }}; }

  public static <E> Lambda<E,Integer> remove (
      final E elem, final Lambda<E,Integer> bag) {
    /* similar to add */  } }
```

Inheritance and abstract classes are also used to enable the integration of the generated code with existing code. This integration is particularly useful when a given function is uninterpreted in the original specification. Take, for example, the case of the square root function in PVS. Since a constructive version of this function is not available, this function (and the class in which it is defined) is declared as abstract in Java. Since the pre- and postconditions of sqrt are still generated, any VCG should be able to generate proof obligations guaranteeing that the provided function satisfies the specification of the uninterpreted one.

5   Verification of Weak Delivery Protocol

There are several theorem prover tools available that can be used to prove Java code correct. The most notable are *KeY* and *Krakatoa*. The Krakatoa tool could not handle inline classes at time of the experiment, so we decided to use KeY.

Soundness of the protocol is expressed by assuming that there are two nodes running the WDP protocol, consisting of a sender and a receiver. The invariant states that all WDP messages delivered by the receiver to the application layer originated from the sender's application layer.

5.1   Invariants

One of the challenges is deducing which part of the specification describes the invariants on the model that should be turned into pre- and postconditions. The translator recognizes these functions by matching them with a template.

The theories will be scanned for special predicates[3]. There should be an initialization predicate F_{init} of type pred[S][4], a transition relation R of type pred[S, S]] and a predicate that defines the invariant: P of type pred[S]. The type S can be any tuple or record that holds state variables. For matching functions there should exist a theorem or lemma that states $\forall r, n : P(r(n))$, where r is defined as mapping from natural numbers to S and n is a natural number. For the mapping r, $F_{init}(r(0))$ and $R(r(n), r(n+1))$ should hold. A second template looks for functions that match a simple transition schema: $\forall P, S : F_{init}(S) \implies P(S)$ and $\forall P, S_1, S_2 : R(S_1, S_2) \wedge P(S_1) \implies P(S_2)$.

For functions that match the above structures pre- and postconditions are generated. F_{init}, will have to *ensure* P, while R will *require* P and will have to *ensure* P.

In the WDP theory, functions that match the first template are used, defined as a separate PVS theory, with the state, the initial state predicate and the transition relation as theory parameters.

The proofs require that the invariants are maintained by the transition relation. For the weak delivery protocol, the transition relation is WDP. The soundness invariant is_subset[5] depends on the invariant wdp_in_app_to_wdp in its proof. All other invariants that are required for proofs of the invariant are added to the precondition of the transition relation WDP.

The no_null_pointers predicate is generated by the translator under the assumption that all the generated code is properly initialized. The predicate simply states that all fields of objects are properly initialized.

```
/*@ requires  no_null_pointers(s)
  @           && WDPAbstract.wdp_in_app_to_wdp(s)
  @           && WDPAbstract.is_subset(s);
  @ ensures \result ==>  WDPAbstract.is_subset(n); */
  public boolean WDP(final WDPState s, final WDPState n)
```

5.2   Proof Construction in KeY

The KeY theorem prover uses dynamic logic, which includes an operator $< p >$, where p is a sequence of Java statements. The formula $< p > \phi$ expresses that the program

---

[3] Instead of predicates, the initial and transition functions could also be functional in nature: f_{init} of type S and f_{trans} as a function from [S → S]. The template match is adjusted accordingly.

[4] In PVS pred[t] is a shorthand for t -> bool.

[5] Java can not handle special characters in identifiers, therefore identifiers with them, like subset? are translated into an meaningful identifier excluding character.

p terminates in a state where ϕ holds. For the soundness invariant the program p is a function call to WDP(s,n) and it should terminate in a state where the postcondition holds. To prove that, the operator needs to be eliminated. The logic rules of the operator are constructed in such a way that they perform a symbolic execution of the Java statements.

A prominent feature of the AirStar model are transitions between states where one aspect of the model is active and the rest remains unchanged. For instance, in the WDP transition relation defined at the end of Section 3, will generate Java code like the following:

```
(   WDPSender(ss,ns)
&& sr.equals(nr)
) ||
(   WDPReceiver(ss,ns)
&& .. etc..
)
```

This will be symbolically evaluated using the <> operator. While symbolically evaluating the code it will lazily translate the && and || into if statements. The result of the evaluation of WDPSender(ss,ns) is assigned to a variable and if it is false, it will skip further evaluation and directly assign the value to the variable representing the first argument of the or statement.

```
b0 = { b1 = WDPSender(ss,ns);
        if (!b1) b2 = false;
        else b2 = sr.equals(nr);
        return b2; };
if (b0) b3 = true;
else b3 = { b4 = WDPReceiver(ss,ns);
             if (b4) ...
```

The rules defined for the symbolic execution of the if statements will force the creation of two separate goals, one where WDPSender(ss,ns) holds and one where it does not. Within each of these cases it will have to evaluate WDPReceiver(ss,ns). The number of goals increases with each branching point added to the transition relation.

When running fully automatically, KeY should be able to dispatch most of these goals. However, in the proofs we sometimes need to choose between regular method expansion and the use of an operation contract. On its own, KeY prefers the regular method expansion. For the proofs to succeed, sometimes the operation contract is needed. The KeY theorem prover can be instructed to halt at points where it needs to expand methods in order to let the user choose. Combined with branching factor, the user is quickly overwhelmed by the amount of manual labor needed to complete the proof goals. By hiding the branching points behind *pure* function definitions, it is possible to delay the branching until a more opportune moment.

Not all of the invariants of the original PVS specification have been completely verified in Key. However, the experiment gave us sufficient confidence in our proposed approach. That is, with some more sophisticated support from the KeY prover environment a semi-automatic proof of all the invariants for the WDP as well as for the GDP protocol should be feasible.

5.3   Feasibility Case Study Evaluation

The case study shows that generating Java code from the PVS models for small to medium sized models is definitely feasible. The translated model roughly doubled in size by translating it into Java. Generating annotations made it possible to prove parts of the properties of the original model. However, generating annotated Java for the current set of Java code verifiers is still a cumbersome process. There were some issues we ran into when using KeY to prove the generated Java code correct.

- KeY only supports Java 1.4 language constructs. Specifically the use of generics is not allowed. Although it has a built-in procedure that can remove generics, this did not seem to work properly for our model. Instead we removed it using *Declawer*, a tool used to strip the generics from Java source code.
- Proving properties of complex Java code statements is cumbersome when there are multiple branching points in the code. These branching points multiply goal generation in the KeY theorem prover. However, using pure Java methods it is possible to postpone evaluation of these branching points until evaluation is opportune.
- KeY has no problem with defining static abstract functions, while Java does. This is not directly an issue, but might lead to proving programs correct that do not compile.
- KeY properly demands that references to fields within an object can only happen when the object itself is not null. These checks make up a great deal of the prover activity. This condition can be relaxed, however, only for regular method calls. When used in conjunction with operation contracts, the user still has to prove the existence of the object.
- KeY sometimes refuses to load saved proofs due to parser issues.

Although the translator supports all executable language constructs that PVS provides, some minor changes had to be made to the original models in order to be successfully translated. The changes all have to do with clashing name spaces, non-translatable characters in identifiers and the fact that the translation requires all fields of a record to be updated. All these issues can be easily resolved. Furthermore, the generation of pre- and postconditions is still work in progress.

The case study presented in the previous section should be viewed in terms of "proof of concept." Although several features are still missing, the case study demonstrates the potential for integrating heavy-weight formal methods tools into the software development cycle.

6   Related Work

Two major fields of computer science come together in generating code from formal specifications: theorem proving and compiler construction.

Within the theorem proving community all the major theorem provers have some form of code generation to a functional language from their specification language. The theorem prover Isabelle/HOL even provides two code generators. There is the original generation from higher-order logic to ML, described by Berghofer and Nipkow [4].

A second translator, developed by Haftmann [10], targets multiple languages. Unlike our generator, however, these languages are all functional programming languages like Haskell, OCaml and SML. ACL2's [13] specification language *is* a subset of Common Lisp. The theorem prover Coq [5] has its generator [17] that extracts lambda terms and translates them in either Haskell or OCaml. As mentioned before, PVS [22] provides a code generator for Lisp. A PVS translation into the functional programming language Clean is in its prototype stage [11]. Using semantic attachments or analog mechanisms to tie executable code and logical statements together has been studied by Ray in ACL [25], and by Rushby et al [7] and Muñoz [19] in PVS.

Integrating formal methods into the software engineering process has been the main goal of the B-method [1], a collection of mathematically based techniques for the specification, design and implementation of software components. The main difference with our method is that PVS as a specification language allows for higher-order functional specification and is a more powerful theorem prover than those that come with the B-tool suite. The added expressiveness of the specification language allows for code generation to functional languages, unlike the B-method where only C or ADA code can be generated. A similar approach is taken with the Vienna Development Method (VDM) [12]. This also is a collection of formal methods and tools that aim at using mathematical techniques in the software development process. It does support higher-order functions and can generate Java as well as C++ code. However, their code generator uses a standard library of VDM concepts, instead of translating more directly into the target language. Both VDM and the B-method do not annotate their generated code, which makes it harder to check whether the generated code is indeed correct.

From within the compiler construction community, work has been done on source to source translators from functional languages to imperative ones: A source code translator between Lisp and Java has been constructed by Leitao [15]. However, not all language constructs of Lisp are supported. Another translator from ML to Java was proposed by Koser et al in [14]. Instead of Java, Ada has also been used as a target language by Tolmach [30].

7 Conclusion and Future Work

Integrating formal methods into the software engineering process requires tools that provide support without unnecessarily constraining the design and implementation choices. We present an approach designed to generate annotated code from declarative PVS specifications for multiple functional and imperative target languages. We reported on a feasibility study using a prototype tool. The key advantages:

- Independently **verifiable** code: The generated code is accompanied by annotations that allow for proof obligation generation. The generated code can be verified, changed and verified again.
- The generated code is **readable** and it allows for **integration** with existing code.
- The generated code is reasonably **efficient**, due to the nature of the translation from an executable subset, as well as by using destructive update optimization techniques. Since we are using an intermediate language, further optimizations such as tail recursion elimination can be easily added.

The attractiveness of our approach is that we have tied together existing techniques into a complete package targeting both functional and imperative languages in such a way that maintenance and verification can be done on the target language level.

Future Work. The code generator presented in this paper is still a proof of concept. Many features have to be improved to be really useful in a large scale software engineering process. For example, currently, only a subset of the specification language of PVS can be translated. Many models are only partially executable. In particular, formal models of protocols typically use a relational specification style to describe functional behaviors. These models cannot directly be translated into an executable program. Being able to generate code for these models, by providing syntactic restrictions on their specification, is one of our next goals. For this we need to add support for guarded non-determinism.

In the spirit of proof carrying code [21], another venue of progress would be to extend the Why logic and the extraction mechanism so that annotated programs carry with them a reference to the correctness lemmas in the original specification and enough information for discharging the proof obligations from these lemmas. Thus, eliminating most of the burden of mechanically proving the correctness of the generated code.

Another interesting issue that can be addressed is whether it is possible to maintain a correspondence between the proofs in the original PVS model and the generated Java code, JML specifications and KeY proofs.

Acknowledgements. The first author's visit to NASA was partially supported by the National Aeronautics and Space Administration under Cooperative Agreement NNX08AE37A. We want to thank Alwyn Goodloe and César A. Muñoz for their guidance, their support and their hospitality.

References

1. Abrial, J.-R.: The B-Book: Assigning Programs to Meanings. Cambridge University Press (1996)
2. Ahrendt, W., Baar, T., Beckert, B., Bubel, R., Giese, M., Hähnle, R., Menzel, W., Mostowski, W., Roth, A., Schlager, S., Schmitt, P.H.: The KeY tool. Software and System Modeling (2005)
3. Bailey, R., Hostetler, R., Barnes, K., Belcastro, C., Belcastro, C.: Experimental validation subscale aircraft ground facilities and integrated test capability. In: Proceedings of the AIAA Guidance Navigation, and Control Conference and Exhibit 2005, San Francisco, California (2005)
4. Berghofer, S., Nipkow, T.: Executing Higher Order Logic. In: Callaghan, P., Luo, Z., McKinna, J., Pollack, R. (eds.) TYPES 2000. LNCS, vol. 2277, pp. 24–40. Springer, Heidelberg (2002)
5. Bertot, Y., Castéran, P.: Interactive Theorem Proving and Program Development. Coq'Art: The Calculus of Inductive Constructions. Texts in Theoretical Computer Science (2004)
6. Burdy, L., Cheon, Y., Cok, D.R., Ernst, M.D., Kiniry, J.R., Leavens, G.T., Leino, K.R.M., Poll, E.: An overview of JML tools and applications. Int. J. Softw. Tools Technol. Transf. 7(3), 212–232 (2005)
7. Crow, J., Owre, S., Rushby, J., Shankar, N., Stringer-Calvert, D.: Evaluating, testing, and animating PVS specifications. Technical report, Computer Science Laboratory. SRI International, Menlo Park, CA (March 2001)

8. Filliâtre, J.-C., Marché, C.: The Why/Krakatoa/Caduceus Platform for Deductive Program Verification. In: Damm, W., Hermanns, H. (eds.) CAV 2007. LNCS, vol. 4590, pp. 173–177. Springer, Heidelberg (2007)
9. Filliâtre, J.-C.: Why: a multi-language multi-prover verification tool. Research Report 1366, LRI, Universit Paris Sud (March 2003)
10. Haftmann, F., Nipkow, T.: A code generator framework for Isabelle/HOL. In: Schneider, K., Brandt, J. (eds.) Theorem Proving in Higher Order Logics: Emerging Trends Proceedings, number 364/07 (August 2007)
11. Jacobs, B., Smetsers, S., Wichers Schreur, R.: Code-carrying theories. Formal Asp. Comput. 19(2), 191–203 (2007)
12. Jones, C.B.: Systematic Software Development using VDM, 2nd edn. Prentice Hall (1990)
13. Kaufmann, M., Moore, J.S., Manolios, P.: Computer-Aided Reasoning: An Approach. Kluwer Academic Publishers, Norwell (2000)
14. Koser, J., Larsen, H., Vaughan, J.: SML2Java: a source to source translator. In: Proceedings of DP-Cool, PLI 2003, Uppsala,Sweden (2003)
15. Leitao, A.M.: Migration of Common Lisp programs to the Java platform -the Linj approach. In: CSMR 2007: Proceedings of the 11th European Conference on Software Maintenance and Reengineering, pp. 243–251. IEEE Computer Society, Washington, DC (2007)
16. Lensink, L., Muñoz, C.A., Goodloe, A.E.: From verified models to verifiable code. Technical Report NASA/TM2009-215943, NASA Langley Research Center (2009)
17. Letouzey, P.: A New Extraction for Coq. In: Geuvers, H., Wiedijk, F. (eds.) TYPES 2002. LNCS, vol. 2646, pp. 200–219. Springer, Heidelberg (2003)
18. Miner, P., Geser, A., Pike, L., Maddalon, J.: A Unified Fault-Tolerance Protocol. In: Lakhnech, Y., Yovine, S. (eds.) FORMATS/FTRTFT 2004. LNCS, vol. 3253, pp. 167–182. Springer, Heidelberg (2004)
19. Muñoz, C.: Rapid prototyping in PVS. Report NIA Report No. 2003-03, NASA/CR-2003-212418, NIA-NASA Langley, National Institute of Aerospace, Hampton, VA (May 2003)
20. Muñoz, C., Goodloe, A.E.: Design and verification of a distributed communication protocol. Technical Report NASA/CR-2009-215703 (2008)
21. Necula, G.: Proof-Carrying Code. In: Proc. of POPL 1997, pp. 106–119. ACM Press (1997)
22. Owre, S., Rushby, J.M., Shankar, N.: PVS: A Prototype Verification System. In: Kapur, D. (ed.) CADE 1992. LNCS, vol. 607, pp. 748–752. Springer, Heidelberg (1992)
23. Paulin-Mohring, C., Werner, B.: Synthesis of ML programs in the system Coq. J. Symb. Comput. 15(5/6), 607–640 (1993)
24. Pike, L., Maddalon, J., Miner, P., Geser, A.: Abstractions for Fault-Tolerant Distributed System Verification. In: Slind, K., Bunker, A., Gopalakrishnan, G.C. (eds.) TPHOLs 2004. LNCS, vol. 3223, pp. 257–270. Springer, Heidelberg (2004)
25. Ray, S.: Attaching Efficient Executability to Partial Functions in ACL2. In: Kaufmann, M., Moore, J.S. (eds.) Fifth International Workshop on the ACL2 Theorem Prover and Its Applications (ACL2 2004), Austin, TX (November 2004)
26. Rushby, J., Owre, S., Shankar, N.: Subtypes for specifications: Predicate subtyping in PVS. IEEE Transactions on Software Engineering 24(9), 709–720 (1998)
27. Rushby, J., von Henke, F.: Formal verification of algorithms for critical systems. IEEE Transactions on Software Engineering 19(1), 13–23 (1993)
28. Shankar, N.: Efficiently executing PVS. Technical report, Menlo Park, CA (1999)
29. Shankar, N.: Static analysis for safe destructive updates in a functional language (2002)
30. Tolmach, A.P., Oliva, D.: From ML to Ada: Strongly-typed language interoperability via source translation. Journal of Functional Programming 8(4), 367–412 (1998)
31. Urribarrí, W.: A module system for Why. Personal Communication (2008) manuscript
32. Wordsworth, J.: Software Development with Z. Addison-Wesley (1992)

Belief Bisimulation for Hidden Markov Models
Logical Characterisation and Decision Algorithm

David N. Jansen[1], Flemming Nielson[2], and Lijun Zhang[2]

[1] Radboud Universiteit Nijmegen, The Netherlands
[2] Technical University of Denmark

Abstract. This paper establishes connections between logical equivalences and bisimulation relations for hidden Markov models (HMM). Both standard and belief state bisimulations are considered.

We also present decision algorithms for the bisimilarities. For standard bisimilarity, an extension of the usual partition refinement algorithm is enough. Belief bisimilarity, being a relation on the continuous space of belief states, cannot be described directly. Instead, we show how to generate a linear equation system in time cubic in the number of states.

1 Introduction

Probabilistic models like Markov chains allow to describe processes whose behaviour is governed by probabilistic distributions. Together with extensions with nondeterministic choices, reward structures and continuous time, they are widely used in networked and distributed systems. During the last twenty years, efficient model-checking algorithms of Markov chains and their extensions have been extensively studied, allowing for performance evaluation and formal reasoning.

Markov chains are fully observable, in the sense that at any time, an observer can determine the exact state and infer the probability to be in a specific state at later times. This may be too restrictive in many applications: Intuitively, the underlying state space of a Markov chain may contain fine-grained information, which is not always visible from the outside. For instance, a meteorologist might use a Markov chain with states for several kinds of snow [19] to model the weather behaviour. Non-expert observers only see whether it is snowing or not, implying that the states of the Markov chain are not fully visible to them.

Hidden Markov models (HMM) [15] enhance Markov chains with *observations*. These reveal partial information about the state, while the actual state remains unknown. Given the sequence of produced observations, we may infer a probability distribution over the states, a so-called *belief state*.

HMMs have received much attention in the area of speech recognition [10], communication channel modelling [16], and biological systems [5]. Recently, they have also been used to analyse stochastic dynamic systems [17]. A typical problem is to find the most probable state after a given observation sequence and perhaps other constraints. For example, in speech recognition, a sequence of sound recordings is given, and the sentence that has probably been pronounced is sought.

A. Goodloe and S. Person (Eds.): NFM 2012, LNCS 7226, pp. 326–340, 2012.

As for Markov chains, model checking and other algorithms depend on the size of the HMM, which is usually very large. Bisimulation equivalences have been shown to be an effective way to amend the state space problem for Markov chains [11]. In contrast, behavioral equivalences for HMMs have only been introduced recently by Castro et al. [4]. It is, however, not clear whether such equivalences agree with the logical properties in HMMs. To pave the way for efficient algorithms using reduction techniques, we study various bisimulation equivalences and characterise them logically with variants of the logic POCTL* (probabilistic observation-CTL*) introduced in [20].

Contributions. Our main contribution is the *logical characterisation* for three variants of bisimulation for HMMs, and their corresponding *decision algorithms.* For standard state-based bisimulation, we show that the logic POCTL* is sound and complete. Since Markov chains are special instances of HMMs, this result conservatively extends the logical characterisation for Markov chains [2]. More interesting are the strong and weak belief bisimulations defined by [4]. (We shall follow [4] and call the two equivalence relations strong and weak belief bisimulation, although this differs from the usual distinction between strong and weak bisimulation.) We show that these relations are too coarse for POCTL*: the nested probabilistic operator, conjunction and some forms of the until operator can distinguish belief bisimilar states. We introduce two sublogics SBBL* and WBBL*, which correspond to strong and weak belief bisimilarity, respectively. The key difference between SBBL* and WBBL* is that the latter cannot describe requirements on the most probable state after a certain sequence of observations.

We also present decision algorithms for the bisimilarities. For standard bisimilarity, an extension of the usual partition refinement algorithm [14] is enough. Belief bisimilarity is a relation over distributions and cannot be computed with partition refinement. Instead, we extend the approach in [7]: we generate a linear equation system that is satisfied by two belief states iff they are bisimilar. The time to construct the system is in $\mathcal{O}(|S|^3)$ for weak and in $\mathcal{O}(|S|^3 \cdot |\Omega|)$ for strong belief bisimilarity, where $|S|$ is the number of states and $|\Omega|$ the number of observations. Since the bisimulation for labelled Markov chains considered in [7] can be regarded as a special case of strong belief bisimulation, our results apply also in that setting. This produces another logical characterisation. More interestingly, our algorithm improves their complexity $\mathcal{O}(|S|^4)$.

We believe that our results are of practical relevance. We have identified the properties corresponding to the bisimulation relations considered, so the model checker can choose the appropriate relation and reduce the size of the HMM under consideration, using our efficient decision algorithm. Such characterisation and decision algorithms will make it possible to analyse HMMs of larger size.

Organisation of the paper. In Section 2 we recall the definition of HMM, belief states, probabilistic measures on them and the logic POCTL*. Section 3 discusses the three different notions of bisimulation for HMMs. The corresponding logical characterisations are presented in Section 4. The decision algorithm is presented in Section 5. We discuss related work in Section 6.

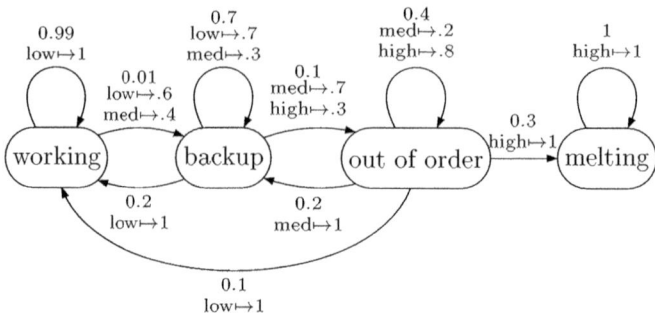

Fig. 1. A hidden Markov model for a cooling system

2   Hidden Markov Models and the Logic POCTL*

In this section we recall the definition of hidden Markov models (HMM) [4] and some related notions. On that basis, we can define the logic POCTL*.

2.1   Hidden Markov Models

Definition 1. *A* hidden Markov model *is a sextuple* $M = (S, P, L, \Omega, O, \alpha)$, *where S is a finite set of states; $P : S \times S \to [0,1]$ is a probabilistic transition relation satisfying $\sum_{s' \in S} P(s, s') = 1$ for every $s \in S$; $L : S \times AP \to \{0,1\}$ describes the truth values of atomic propositions; Ω is a finite set of observations; the partial function $O : S \times S \to Dist(\Omega)$ assigns a probability distribution over the observations to each transition in $P^{-1}((0,1])$; $\alpha : S \to [0,1]$ is the initial distribution.*

Note that we assign observations to transitions. Many other definitions assign observations to states, but in that case, the observations would be almost the same as atomic propositions. Our choice is inspired by [4].

Example 1. In Fig. 1, a simple HMM that describes a small part of a nuclear power plant is depicted. It describes the state of the cooling system and how much information about this state can be obtained based on the incomplete information provided by temperature sensors, a situation that may occur in a partially broken power plant. For example, if the temperature sensor produces a "high" reading, it is not completely clear whether the power plant is melting down, renouncing all hope of repair, or it is only in state "out of order", so a repair should be attempted.

2.2   Belief States

In this and the following sections, we assume we are given a fixed set of atomic propositions AP and a hidden Markov model $M = (S, P, L, \Omega, O, \alpha)$.

In a hidden Markov model, only the observation can be seen, and a standard problem is to guess the real state of the HMM based on the observations. We can summarize the history of observations in a *belief state* (or *information state*) [15].

Definition 2. *A belief state is a probability distribution over S. Moreover, we let $\mathbf{1}_s$ be the characteristic belief state for $s \in S$ defined by: $\mathbf{1}_s(s) = 1$.*

A belief state is not really a state of the HMM. Rather, it is a way to describe what we know about the state. The set of all belief states is called the *belief space* and is denoted by $\mathcal{B}$. The labelling function can easily be extended to belief states by: $L(b,a) := \sum_{s \in S} b(s) \cdot L(s,a)$. Intuitively, $L(b,a)$ gives the probability of satisfying a in belief state b.

The belief state b_n at time $n \geq 0$, i.e. the distribution over S at time n given the observation history $\omega_0, \ldots, \omega_{n-1}$, captures all information about the past. We can inductively calculate the next belief state b_{n+1} based on the previous belief state b_n and the current observation ω_n. More details will be given after introducing probability spaces for HMMs.

2.3   Paths in HMM and Probability Spaces over Paths

Given $M = (S, P, L, \Omega, O, \alpha)$ (as fixed above), we first introduce some notation. A path σ of M is a sequence $s_0, \omega_0, s_1, \omega_1 \ldots \in (S \times \Omega)^\omega$ where $P(s_i, s_{i+1}) > 0$ and $O(s_i, s_{i+1})(\omega_i) > 0$ for all $i \in \mathbb{N}$. For $i \in \mathbb{N}$, let $\sigma(i)_s = s_i$ denote the $(i+1)$th state of σ, and $\sigma(i)_o = \omega_i$ denote the $(i+1)$st observation of σ. Let $\sigma(i \ldots)$ denote the suffix path of σ starting with $\sigma(i)_s$, i.e., $s_i, \omega_i, s_{i+1}, \omega_{i+1}, \ldots$

Let $Path^M$ denote the set of all paths in M, and $Path^M(s)$ denote the set of paths in M that start in s. The superscript M is omitted whenever it is clear from the context. We define a probability space on paths of M using the standard cylinder construction. For a finite state–observation sequence $s_0, \omega_0, s_1, \omega_1, \ldots, s_n$, its induced *basic cylinder set* is $\mathcal{C}(s_0, \omega_0, s_1, \omega_1, \ldots, s_n) := \{\sigma \in Path \mid \forall i \leq n : \sigma(i)_s = s_i \wedge \forall j < n : \sigma(j)_o = \omega_j\}$. This set consists of all paths σ starting with $s_0, \omega_0, s_1, \omega_1, \ldots, s_n$. Let Cyl contain all basic cylinder sets for all finite state–observation sequences. Given a finite sequence $C_0, \Upsilon_0, C_1, \Upsilon_1, \ldots, C_n$ of state sets and observation sets, we define the *cylinder set* to be the (disjoint) union of the basic cylinder sets with state–observation sequences picked from the sequence of sets:

$$\mathcal{C}(C_0, \Upsilon_0, \ldots, C_n) := \bigcup_{s_0 \in C_0} \bigcup_{\omega_0 \in \Upsilon_0} \cdots \bigcup_{s_n \in C_n} \mathcal{C}(s_0, \omega_0, \ldots, s_n)$$

Given a belief state b, we define the premeasure $Prob_b$ on Cyl by induction on n as: $Prob_b(\mathcal{C}(s_0)) = b(s_0)$ and, for $n > 0$, $Prob_b(\mathcal{C}(s_0, \omega_0, \ldots, s_n))$ equals: $P(s_{n-1}, s_n)O(s_{n-1}, s_n)(\omega_{i-1}) \cdot Prob_b(\mathcal{C}(s_0, \omega_0, \ldots, s_{n-1}))$. By induction, we get:

$$Prob_b(\mathcal{C}(s_0, \omega_0, \ldots, s_n)) = b(s_0) \prod_{i=1}^{n} O(s_{i-1}, s_i)(\omega_{i-1})P(s_{i-1}, s_i)$$

The above premeasure can be extended uniquely (Carathéodory's theorem, see e.g. [18, page 272]) to a measure on the σ-algebra generated by Cyl. We introduce a few shorthand notations, which will be used frequently later on:

- $Prob_b(\omega, s') := \sum_{s \in S} O(s, s')(\omega) P(s, s') b(s)$ is the probability to get observation ω and end in some state s'. So, $Prob_b(\omega, s') = Prob_b(\bigcup_{s \in S} \mathcal{C}(s, \omega, s'))$. For a set of states $A \subseteq S$, let $Prob_b(\omega, A) := \sum_{s' \in A} Prob_b(\omega, s')$.
- $Prob_b(\omega) := Prob_b(\omega, S)$ is the probability to get observation ω in belief state b. For a set of observations Υ, let $Prob_b(\Upsilon) := \sum_{\omega \in \Upsilon} Prob_b(\omega)$.
- $\tau(b, \omega)(s') := \frac{Prob_b(\omega, s')}{Prob_b(\omega)}$. Then, $\tau(b, \omega)$ is the resulting belief state under the condition that we take a transition from belief state b and that we get observation ω.
- $Prob_b(b') := \sum_{\omega \in \Omega} Prob_b(\omega) \cdot \mathbf{1}_{b' = \tau(b, \omega)}$ is the probability of getting to b' in the next step, starting from b. Here $\mathbf{1}_{b' = \tau(b, \omega)}$ equals 1 if $b' = \tau(b, \omega)$, and 0 otherwise. For a set of belief states B, we define $Prob_b(B) := \sum_{b' \in B} Prob_b(b')$.

For belief state $b = \mathbf{1}_s$, we sometimes write s when clear from the context. The updating of belief state described above can now be written by: $b_{n+1} = \tau(b_n, \omega_n)$.

2.4 Syntax of POCTL*

In our article [20], we defined a logic POCTL* to describe properties of HMMs. In POCTL*, we distinguish state formulas (denoted Φ), path formulas (denoted φ), and belief state formulas (denoted ε). Its syntax is:

$$\Phi ::= \mathbf{true} \mid a \mid \neg\Phi \mid \Phi \wedge \Phi \mid \varepsilon$$

$$\varphi ::= \Phi \mid \neg\varphi \mid \varphi \wedge \varphi \mid X_\Upsilon \varphi \mid \varphi \, \mathcal{U}^{\leq n} \varphi$$

$$\varepsilon ::= \neg\varepsilon \mid \varepsilon \wedge \varepsilon \mid \mathcal{P}_{\bowtie p}(\varphi)$$

where a is an atomic proposition, Υ is a set of observations, n is a natural number or ∞, $\bowtie$ is a comparison operator $\in \{<, \leq, \geq, >\}$, and p is a probability bound $\in [0, 1]$.[1]

The disjunction $\vee$ is defined as usual as an abbreviation. If $\Upsilon = \Omega$, we will sometimes suppress the index of a next-state operator: $X \varphi := X_\Omega \varphi$. The future-operator $\Diamond^{\leq n}\varphi$ abbreviates $\mathbf{true} \, \mathcal{U}^{\leq n} \varphi$.

The semantics of Φ and φ is mostly defined in the same way as for CTL over states and paths, respectively [20,1]. A few examples: $s \models \varepsilon$ iff $\mathbf{1}_s \models \varepsilon$, $\sigma \models X_\Upsilon \varphi$ iff $\sigma(0)_o \in \Upsilon$ and $\sigma(1\ldots) \models \varphi$, and $b \models \mathcal{P}_{\bowtie p}(\varphi)$ iff $Prob_b\{\sigma | \sigma \models \varphi\} \bowtie p$. POCTL* can be applied to the typical problem (based on the sequence of observations and perhaps other constraints, find a probable state) by verifying a formula like $\mathcal{P}_{\geq 0.25}(X_{\omega_1} X_{\omega_2} X_{\omega_3} \mathbf{true})$.

---

[1] Some formulas, e.g. state formula $\neg\varepsilon$, have two derivations: either use the negation of state formulas or the negation of belief state formulas. However, this will not pose problems because the two are semantically equivalent.

3   Bisimulation Notions for HMMs

In this section we define various bisimulations for HMMs. First, one can simply extend standard bisimulation of Markov chains [13] to the HMM setting:

Definition 3. *Let $R \subseteq S \times S$ be an equivalence relation on the states of M. R is a strong bisimulation if it respects the following conditions for every $(s,t) \in R$:*

1. *For all atomic propositions $a \in AP$, we have $s \models a$ iff $t \models a$.*
2. *For all observations $\omega \in \Omega$, we have $Prob_{1_s}(\omega) = Prob_{1_t}(\omega)$.*
3. *For all observations $\omega \in \Omega$, and all R-equivalence classes $C \in S/R$, we have $\tau(s,\omega)(C) = \tau(t,\omega)(C)$.*

Two states $s,t \in S$ are strongly bisimilar if there exists a strong bisimulation R with $s\ R\ t$. We denote this as $s \sim t$. Bisimilarity can be extended to paths: two paths $\sigma, \rho \in Path$ are strongly bisimilar if $\sigma(i)_o = \rho(i)_o$ and there exists a strong bisimulation R such that $\sigma(i)_s\ R\ \rho(i)_s$ for all $i \in \mathbb{N}$.

Note that Conditions 2 and 3 can be subsumed to: For all observations $\omega \in \Omega$ and all R-equivalence classes $C \in S/R$, we have $Prob_s(\omega, C) = Prob_t(\omega, C)$. Since probabilities agree on bisimilar states, we sometimes denote $Prob_s$ by $Prob_{[s]_R}$. The definition conservatively extends bisimilarity on Markov chains: if $|\Omega| = 1$, HMM bisimilarity reduces to standard bisimilarity for Markov chains [13].

The state-based bisimulation defined above does not take into account that states in HMMs are hidden, i.e., only indirectly observable. Recently, Castro et al. [4] introduced two new notions of bisimulation relations, not on the states of the HMM, but on the belief states, i.e., on distributions over states. We recall their definitions and adapt them to our fully probabilistic setting.

Definition 4. *Let $R \subseteq \mathcal{B} \times \mathcal{B}$ be an equivalence relation on the belief states. R is a strong belief bisimulation if it respects the following conditions for every $(b,c) \in R$:*

1. *For all atomic propositions $a \in AP$, we have $L(b,a) = L(c,a)$.*
2. *For all observations $\omega \in \Omega$, we have $Prob_b(\omega) = Prob_c(\omega)$.*
3. *For all observations $\omega \in \Omega$, we have $\tau(b,\omega)\ R\ \tau(c,\omega)$.*

Two belief states $b,c \in \mathcal{B}$ are strongly belief bisimilar if there exists a strong belief bisimulation R with $b\ R\ c$. This is denoted $b \sim_{sb} c$.

The first condition requires that b and c have the same labelling. The second condition states that the probability of observing ω is the same from b or c. The new condition is the third one, stating that the updated belief states with respect to ω must also be in the relation R. It is weaker than the third condition of state-based bisimulation: The following example illustrates the difference.

Example 2. Consider the HMM depicted in Fig. 2. Assume $L(s_3) \neq L(s_4)$, and other states have the same labelling. First, $s_1 \nsim t_1$, independent of the observations. The reason is that s_2 cannot be bisimilar with either t_2 or t_3. Now let $b = 1_{s_1}$ and $c = 1_{t_1}$. It is easy to verify that $b \sim_{sb} c$.

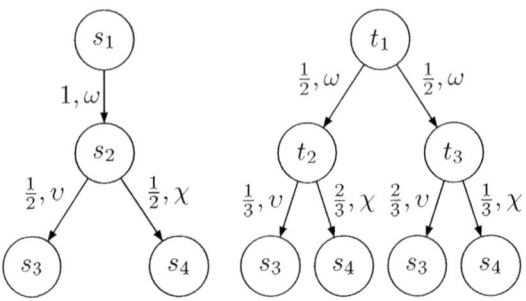

Fig. 2. State-based strong bisimulation and strong belief bisimulation differ

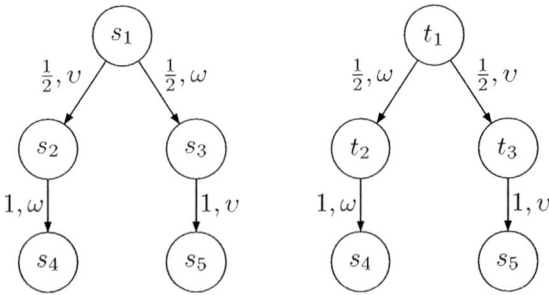

Fig. 3. Strong and weak belief bisimilarity differ

Now we recall weak belief bisimulation for HMMs, based on [4]:

Definition 5. *Let $R \subseteq \mathcal{B} \times \mathcal{B}$ be an equivalence relation on the belief states. R is a weak belief bisimulation if it respects the following conditions for every $(b, c) \in R$:*

1. *For all atomic propositions $a \in AP$, we have $L(b, a) = L(c, a)$.*
2. *For all observations $\omega \in \Omega$, we have $Prob_b(\omega) = Prob_c(\omega)$.*
3. *For all R-equivalence classes $B \in \mathcal{B}/R$, we have $Prob_b(B) = Prob_c(B)$.*

Two belief states $b, c \in \mathcal{B}$ are weakly belief bisimilar if there exists a weak belief bisimulation R with $b \; R \; c$. This is denoted $b \sim_{wb} c$.

Indeed, it holds that $\sim_{sb} \subset \sim_{wb}$, where the inclusion is strict [4]. Intuitively, while strong belief bisimulation requires that the updated belief states must be in the relation, in weak belief bisimulation we require only that the updated belief states evolve with the same probability to each $B \in \mathcal{B}/R$.

The example in Fig. 3, taken from [4], illustrates the difference: $\mathbf{1}_{s_1}$ and $\mathbf{1}_{t_1}$ are not strongly belief bisimilar, but they are weakly belief bisimilar.

4   Characterising Bisimilarity

This section presents the logical characterisation results for the three bisimilarities for HMMs. We first show that state-based bisimilarity agrees with the logical equivalence induced by POCTL*. Then, we shall identify two sublogics of POCTL* to characterise strong and weak belief bisimilarities, respectively.

4.1   Strong Bisimilarity

We show that the equivalence induced by POCTL* agrees with state-based bisimilarity. As a preparation, we introduce bisimulation-closed sets of paths.

Definition 6. *A set of paths is* bisimulation-closed *if it is a (disjoint) union of equivalence classes induced by strong bisimilarity on paths.*

Lemma 1. *Assume that s is strongly bisimilar to t. Then, for all bisimulation-closed sets of paths Π, we have that $Prob_s(\Pi) = Prob_t(\Pi)$.*

Proof. It is enough to show equality for a $\cap$-closed generator of the σ-algebra of all bisimulation-closed sets of paths. Therefore, assume w.l.o.g. that Π is a cylinder set $\mathcal{C}(C_0, \omega_0, C_1, \omega_1, \ldots, C_n)$, where the C_i are bisimulation equivalence classes, and assume that $s \in C_0$. Bisimilarity implies $t \in C_0$. Clearly,

$$Prob_s(\Pi) = Prob_{C_0}(\omega_0, C_1) \cdot Prob_{C_1}(\omega_1, C_2) \cdots Prob_{C_{n-1}}(\omega_{n-1}, C_n) = Prob_t(\Pi)$$

where $Prob_{C_i} = Prob_{s_i}$ for some $s_i \in C_i$; as C_i is a bisimulation equivalence class, $Prob_{C_i}$ is well-defined. The intersection of two such cylinder sets is either the smaller of the two or empty.

The following theorem shows that the equivalence induced by the logic POCTL* agrees with strong bisimulation:

Theorem 1. *The logic POCTL* characterises strong bisimilarity, i.e., two states are strongly bisimilar iff they satisfy the same POCTL* state formulas, and two paths are (statewise) strongly bisimilar iff they satisfy the same POCTL* path formulas.*

The proof is mostly based on the proof of Theorem 10.67 of [1], adapted to the setting of HMMs – details appear in [9]. The completeness proof does not rely on the until operator being part of the logic; therefore, the sublogic of POCTL* without until formulas is sufficient to characterise state-based strong bisimilarity. Thus, it conservatively extends the result for Markov chains [1].

4.2   Strong Belief Bisimilarity

In this section we will present a logical characterisation of strong belief bisimilarity. First, in Subsection 4.2, we will discuss that several operators of POCTL* are too discriminative with respect to belief bisimilarity. Then, we define the logic SBBL*, which characterises strong belief bisimilarity.

POCTL* Is too Discriminative. In the example of Fig. 4, we illustrate why we shall have to remove a few operators to characterise strong belief bisimilarity. Every transition in the HMM produces the same observation.

- The nested probabilistic operator $\varepsilon_1 := \mathcal{P}_{\geq 0.5}\,(\mathcal{P}_{\geq 1}\,(X\,a_3))$. Consider belief state b_1 defined by $b_1(s_2) = b_1(s_4) = 0.5$, and b_2 defined by $b_2(s_1) = b_2(s_3) = 0.5$. It follows that $b_1 \sim_{\mathrm{sb}} b_2$, but $b_1 \models \varepsilon_1$, while $b_2 \not\models \varepsilon_1$. The distinguishing power of ε_1 comes from the fact that s_2 (in the support of b_1) satisfies the inner probabilistic formula, whereas no state in the support of b_1 does so.
- The conjunction $\varepsilon_2 := \mathcal{P}_{\geq 0.5}\,(a_1 \wedge a_2)$. For the belief states b_1 and b_2 defined above, it holds then $b_2 \models \varepsilon_2$ but $b_1 \not\models \varepsilon_2$.
- The conjunction after the path operator $\varepsilon_3 := \mathcal{P}_{\geq 0.5}\,(X\,(a_1 \wedge a_2))$, and belief states $\mathbf{1}_{s_7} \sim_{\mathrm{sb}} \mathbf{1}_{s_8}$. We again have $\mathbf{1}_{s_7} \not\models \varepsilon_3$ but $\mathbf{1}_{s_8} \models \varepsilon_3$.
- The until formula $(X\,a_1)\,\mathcal{U}^{\leq \infty}\,a_2$ is satisfied by paths in $\mathcal{C}(s_8, \omega, s_3)$, but not by any path starting in s_7. Therefore, $\mathbf{1}_{s_7} \models \mathcal{P}_{=0}\,((X\,a_1)\,\mathcal{U}^{\leq \infty}\,a_2)$, but $\mathbf{1}_{s_8}$ does not satisfy this formula.
- The nested until formula $\neg a_1\,\mathcal{U}^{\leq \infty}\,(a_2\,\mathcal{U}^{\leq \infty}\,a_3)$ holds on paths in $\mathcal{C}(s_8, \omega, s_3, \omega, s_6)$, so similarly $\mathbf{1}_{s_7} \models \mathcal{P}_{=0}\,(\neg a_1\,\mathcal{U}^{\leq \infty}\,(a_2\,\mathcal{U}^{\leq \infty}\,a_3))$.

The Logic SBBL*. Based on the discussion above, we present a sublogic of POCTL* to characterise strong belief bisimilarity. We call this logic SBBL*:

$$\Phi ::= \textbf{true} \mid a \mid \neg \Phi$$
$$\varphi ::= \Phi \mid X_{\Upsilon}\,\varphi$$
$$\varepsilon ::= \neg \varepsilon \mid \varepsilon \wedge \varepsilon \mid \mathcal{P}_{\bowtie p}\,(\varphi)$$
$$\mid \mathcal{P}_{\bowtie p}\,(\Phi\,\mathcal{U}^{\leq n}\,\Phi)$$

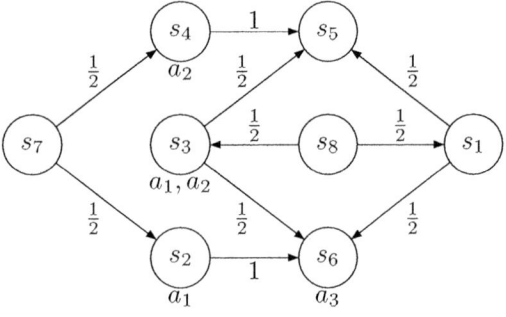

Fig. 4. A hidden Markov model

Theorem 2. *The logic SBBL* characterises strong belief bisimilarity, i.e., two belief states are strongly belief bisimilar iff they satisfy the same SBBL* belief state formulas.*

Proof. We prove *soundness* by induction over the structure of the formulas; in contrast to Theorem 1, the induction runs only over the belief state formulas.

We assume given two belief states $b \sim_{\mathrm{sb}} c$ and a belief state SBBL*-formula ε; we prove that $b \models \varepsilon$ iff $c \models \varepsilon$. For symmetry reasons, it is enough to prove one direction, so assume that $b \models \varepsilon$; then it remains to be proven that $c \models \varepsilon$.

- $\varepsilon = \neg \varepsilon'$ and $\varepsilon = \varepsilon_1 \wedge \varepsilon_2$. These two cases are simple consequences of the induction hypothesis.
- $\varepsilon = \mathcal{P}_{\bowtie p}\,(\textbf{true})$ or $\mathcal{P}_{\bowtie p}\,(\neg\textbf{true})$. Trivial.
- $\varepsilon = \mathcal{P}_{\bowtie p}\,(a)$ or $\mathcal{P}_{\bowtie p}\,(\neg a)$. A simple consequence of Condition 1 of Def. 4.
- $\varepsilon = \mathcal{P}_{\bowtie p}\,(\varphi)$, where $\varphi = X_{\Upsilon_1}\,X_{\Upsilon_2}\,\cdots X_{\Upsilon_k}\,\Phi$. Let Π be the set of paths satisfying φ. So, $\Pi = \{\sigma \mid \sigma(0)_o \in \Upsilon_1 \wedge \sigma(1)_o \in \Upsilon_2 \ldots \sigma(k-1)_o \in \Upsilon_k \wedge \sigma(k\ldots) \models \varphi'\}$.

Note that $Prob_b(\Pi)$ is a product of factors of the form $Prob_b(\omega_1)$ for $\omega_1 \in \Upsilon_1$, $Prob_{\tau(b,\omega_1)}(\omega_2)$ for $\omega_2 \in \Upsilon_2$, all constructed using $Prob_{(\cdot)}$ and $\tau(\cdot, \cdot)$. Similarly, $Prob_c(\Pi)$ can be described using $Prob_c(\omega_1)$, $Prob_{\tau(c,\omega_1)}(\omega_2)$ etc. All these terms for b and c are equal, because $\tau(b, \omega_1) \sim_{sb} \tau(c, \omega_1)$ for all ω_1 (Condition 3 of Def. 4), $Prob_b(\omega_1) = Prob_c(\omega_1)$ (Condition 2 of Def. 4), etc.

- $\varepsilon = \mathcal{P}_{\bowtie p} \left(\Phi_1 \, \mathcal{U}^{\leq n} \, \Phi_2 \right)$. First assume that $n < \infty$. We evaluate this property on a modified HMM M'. It has the same states, labels and observations as M, but $(\Phi_2 \vee \neg \Phi_1)$-states are made absorbing. This does not change the truth values of Φ_1 or Φ_2. Further, once a path has reached a $(\Phi_2 \vee \neg \Phi_1)$-state, it has become clear whether it satisfies φ. So modifying transitions out of these states does not change the truth value of ε. On M', the formula ε is equivalent to $\mathcal{P}_{\bowtie p} (X X \cdots X \Phi_2)$ (n next-operators); then, the argumentation for the next-operator can be used to complete the proof.
  Now, if $n = \infty$, note that the sequence $(Prob_b(\Phi_1 \, \mathcal{U}^{\leq i} \, \Phi_2))_{i \in \mathbb{N}}$ is a nondecreasing sequence in a compact interval, so it does have a limit, which is $Prob_b(\Phi_1 \, \mathcal{U}^{\leq \infty} \, \Phi_2)$. The corresponding sequence for $Prob_c$ consists of the same elements, so it must have the same (unique) limit.

This finishes the proof of soundness. To show *completeness*, we define the equivalence relation on belief states $R := \{(b, c) \mid \forall \text{ SBBL*-belief state formulas } \varepsilon : b \models \varepsilon \text{ iff } c \models \varepsilon\}$. We have to show that this relation is a strong belief bisimulation. Assume given two belief states b and c such that $b \, R \, c$.

- Condition 1. One sees easily that $L(b, a) = \sup \{r \mid b \models \mathcal{P}_{\geq r}(a)\}$. Obviously, $\{r \mid b \models \mathcal{P}_{\geq r}(a)\} = \{r \mid c \models \mathcal{P}_{\geq r}(a)\}$; therefore $L(b, a) = L(c, a)$.
- Condition 2. The same reasoning with $\sup \{r \mid b \models \mathcal{P}_{\geq r}(X_\omega \, \mathbf{true})\} = Prob_b(\omega)$.
- Condition 3. Assume given any $\omega \in \Omega$. We prove that $b' := \tau(b, \omega) \, R \, \tau(c, \omega) =: c'$. Assume given a belief state formula ε such that $b' \models \varepsilon$; if we can prove that $c' \models \varepsilon$, then we get the desired result.
  First assume that ε has the special form $\mathcal{P}_{\bowtie p}(\varphi)$. Then, $b \models \mathcal{P}_{\bowtie p \cdot Prob_b(\omega)} (X_\omega \, \varphi)$, as $Prob_b(X_\omega \, \varphi) = Prob_b(\omega) \cdot Prob_{b'}(\varphi)$. From the definition of R, we know that $c \models$ the same formula, and therefore $c' \models \mathcal{P}_{\bowtie p}(\varphi)$.
  Now assume that ε is constructed from the special form above using negation and conjunction, then a trivial induction over the structure of ε shows $c' \models \varepsilon$.

Again, from the completeness proof we see that the sublogic of SBBL* without until formulas is sufficient to characterise strong belief bisimilarity.

4.3   Weak Belief Bisimilarity

In this section we present logical characterisation results for weak belief bisimilarity. We restrict SBBL* further to the following logic, named WBBL*:

$$\Phi ::= \mathbf{true} \mid a \mid \neg \Phi$$

$$\varphi ::= \Phi \mid X_\Upsilon \, \mathbf{true} \mid X \varphi$$

$$\varepsilon ::= \neg \varepsilon \mid \varepsilon \wedge \varepsilon \mid \mathcal{P}_{\bowtie p}(\varphi) \mid \mathcal{P}_{\bowtie p} \left(\Phi \, \mathcal{U}^{\leq n} \, \Phi \right)$$

Essentially, the operator $X_\Upsilon \varphi$ in SBBL* is replaced by two subformulas X_Υ **true** and $X \varphi$. Note that properties like $\mathcal{P}_{\geq 0.25} (X_{\omega_1} X_{\omega_2} X_{\omega_3}$ **true**$)$ are not in WBBL*, so it cannot be used to describe to solve the corresponding standard problem. The following theorem shows the main result:

Theorem 3. *The logic WBBL* characterises weak belief bisimilarity,* *i. e., two belief states are weakly belief bisimilar iff they satisfy the same WBBL*-belief state formulas.*

Proof. We proceed as in the previous two cases. To prove *soundness*, assume given two belief states b and c that are weakly belief bisimilar and a belief state WBBL*-formula ε such that $b \models \varepsilon$. We have to prove that $c \models \varepsilon$.

- $\varepsilon = \neg\varepsilon'$, $\varepsilon = \varepsilon_1 \wedge \varepsilon_2$, $\varepsilon = \mathcal{P}_{\bowtie p} (\textbf{true})$, $\mathcal{P}_{\bowtie p} (\neg\textbf{true})$, $\mathcal{P}_{\bowtie p} (a)$, $\mathcal{P}_{\bowtie p} (\neg a)$, or $\mathcal{P}_{\bowtie p} (\Phi \, \mathcal{U}^{\leq n} \, \Phi)$. These cases are handled as in Theorem 2.
- $\varepsilon = \mathcal{P}_{\bowtie p} (X_\Upsilon \textbf{true})$: The set Π of paths that satisfy X_Υ **true** has probability $Prob_b(\Pi) = Prob_b(\Upsilon)$. From Condition 2 of Def. 5, it follows that this is equal to $Prob_c(\Upsilon) = Prob_c(\Pi)$.
- $\varepsilon = \mathcal{P}_{\bowtie p} (X \varphi)$. From the induction hypothesis, we can conclude that $b' \sim_{\text{wb}} c'$ implies $b' \models \mathcal{P}_{>p} (\varphi)$ iff $c' \models \mathcal{P}_{>p} (\varphi)$, so for every weak belief bisimilarity class $B \in \mathcal{B}/\sim_{\text{wb}}$, $Prob_B(\varphi)$ is well-defined. Therefore, $Prob_b(X \varphi) = \sum_{B \in \mathcal{B}} Prob_B(\varphi) \cdot Prob_b(B)$, and $Prob_c(X \varphi) = \sum_{B \in \mathcal{B}} Prob_B(\varphi) \cdot Prob_c(B)$. The right-hand sides are equal because of Condition 3 of Def. 5.

To show *completeness,* we define the equivalence relation on belief states

$$R := \{(b, c) \mid \forall \text{ WBBL*-belief state formulas } \varepsilon : b \models \varepsilon \text{ iff } c \models \varepsilon\}$$

We have to show that this relation is a weak belief bisimulation. Assume given two belief states b and c such that $b \, R \, c$.

- Conditions 1 and 2 are handled as in Theorem 2.
- Condition 3. Assume given any R-equivalence class B. We prove $Prob_b(B) = Prob_c(B)$ by regarding the satisfaction sets $Sat(\varepsilon)$ for all WBBL*-belief state formulas with rational probability bounds, i. e., every subformula $\mathcal{P}_{\bowtie p} (\cdot)$ has $p \in \mathbb{Q}$. Let $Sat_\mathbb{Q}$ contain all such satisfaction sets, and let $\mathcal{F}$ be the σ-algebra generated from $Sat_\mathbb{Q}$. Then, $B \in \mathcal{F}$, because B is a countable intersection of elements of $Sat_\mathbb{Q}$. Note that $Sat_\mathbb{Q}$ is $\cap$-closed. Therefore, if two premeasures agree on $Sat_\mathbb{Q}$, then their extensions to measures on $\mathcal{F}$ also agree. From the definition of R, it follows easily that $Prob_b(\varphi) = Prob_c(\varphi)$ for any belief states $b \, R \, c$, because $Prob_b(\varphi) = \sup\{q \in \mathbb{Q} | b \models \mathcal{P}_{>q} (\varphi)\}$. So, $Prob_b$ and $Prob_c$ agree on $Sat_\mathbb{Q}$.

5 Decision Algorithms

In this section we present decision algorithms for the three different bisimilarities. The state-based strong bisimilarity is the easiest one, as it can be computed by a simple extension of the usual partition refinement algorithm [14,6,11]. The complexity is linear in the number of transitions and observations and logarithmic in the number of states. We do not go further into that matter, as details can be found in [3].

As the belief states are probability distributions, the set of belief states is uncountable. Therefore, one cannot describe the belief state bisimulation quotient as a partition of the state space as for standard bisimilarity or ordinary lumping of Markov chains. Another approach has been proposed by [7]: two belief states b and c are belief bisimilar if they are a solution to a specific equation system over $b(s)$ and $c(s)$, for all $s \in S$. We adapt their algorithm to our setting and show an improved time bound. The equation system is constructed as follows.

Let $\{s_1, s_2, \ldots, s_n\}$ be an order of the states. We denote $b(s_j)$ as b_j and $c(s_j)$ as c_j; these variables will be the unknowns in the system. We construct the equation system iteratively. We start with the system

$$\bigwedge_{a \in AP} \sum_{s_i \models a} b_i - c_i = 0 \wedge \bigwedge_{\omega \in \Omega} \sum_{s_i \in S} Prob_{s_i}(\omega) \cdot (b_i - c_i) = 0$$

The base case is the same for strong and weak belief bisimilarity: the first conjunction corresponds to the condition on the labelling, and the second one to the condition that the probability of observing $\omega \in \Omega$ agrees with b and c. Considering b and c and row vectors, this equation system can be written as

$$\begin{pmatrix} A_1 & -A_1 \end{pmatrix} \cdot (b, c)^T = \mathbf{0}$$

where A_1 is an $(|AP| + |\Omega|) \times n$-matrix. We assume that A_1 is brought to upper triangular form (i. e., a matrix with zeroes below the main diagonal) immediately, and the equations that turn out to be linearly dependent are removed. Let k_1 be the number of rows in A_1 (after the triangular transformation), i.e., $k_1 \leq |AP| + |\Omega|$. There can be at most n linearly independent equations of this form (since A_1 has n columns); this property will be used to ensure termination. If $k_1 = n$, we stop immediately.

5.1 Deciding Weak Belief Bisimilarity

Now we describe the iteration step for weak belief bisimilarity – corresponding to the third condition of weak belief bisimulation in Def. 5. In the ith iteration step, we assume given an equation system of the form

$$\begin{pmatrix} A_1 & -A_1 \\ \vdots & \vdots \\ A_i & -A_i \end{pmatrix} \cdot (b, c)^T = \mathbf{0} \tag{1}$$

with at most $n - 1$ equations (n equations cannot occur because the algorithm would have terminated earlier in that case), all of them linearly independent, in upper triangular form. From this, we construct an extended equation system of the same form, but possibly with more equations:

$$\begin{pmatrix} A_1 & -A_1 \\ \vdots & \vdots \\ A_i & -A_i \\ A_{i+1} & -A_{i+1} \end{pmatrix} \cdot (b,c)^T = \mathbf{0} \tag{2}$$

If it does not have more equations, we have reached a fixpoint. In that case, or if the new system has n equations, we can stop after the ith iteration step.

To find A_{i+1}, we first add new equations to the system: the new equations are produced from equations in (1) by replacing the variable b_j with $\sum_{\omega \in \Omega} Prob_b(\omega, s_j)$ and replacing the c_j with $\sum_{\omega \in \Omega} Prob_c(\omega, s_j)$. It is enough to add the new equations for the rows of A_i, as equations for $A_1, A_2, \ldots, A_{i-1}$ have been added earlier. This adds at most k_i equations, where k_i is the number of rows in A_i. Then, we bring the matrix in equation 2 with all these new equations into upper triangular form, to find out which ones are linearly dependent. As $A_1, \ldots, A_i$ are already in upper triangular form, we only have to do calculations with A_{i+1}. Finally, we drop the linearly dependent equations, giving us $k_{i+1} \le k_i$ additional equations.

Time complexity. The algorithm generates an equation system in upper triangular form. It basically interleaves (i) steps where k_i new equations are generated, corresponding to A_{i+1}, with (ii) steps where these new equations are brought into upper triangular form, and the linear dependent ones are removed. Some equations then turn out to be linearly dependent; this will happen exactly $|AP| + |\Omega|$ times in total, because we started with this number of equations. To see this, remember that every single row in A_1 (i.e., a single equation) is transferred to A_2, A_3, ... by the variable substitution described above. In one of those transfers, the generated equation turns out to be linearly dependent, and from that iteration on, it is dropped completely. Therefore, at most $n + |AP| + |\Omega|$ equations over $2n$ variables are generated. The most costly step is to turn the equation system into upper triangular form. For all equations together, this takes time $\in \mathcal{O}(n^2(n + |AP| + |\Omega|))$. Notably, this improves the time bound $\mathcal{O}(n^4)$ in [7] by a factor of n, as in most cases $|AP| + |\Omega| \ll n$.

5.2   Deciding Strong Belief Bisimilarity

The argumentation for strong belief bisimilarity is almost the same as for weak belief bisimilarity; only in the iteration step, one set of equations for each observation $\omega \in \Omega$ is generated – corresponding to the third condition of strong belief bisimulation in Def. 4. In the ith iteration step, assume that we start with an equation system of the form in (1). We similarly add new equations to it, but now, for every $\omega \in \Omega$, we add a set of equations where we replace b_j by $Prob_b(\omega, s_j)$ and c_j by $Prob_c(\omega, s_j)$. So, A_{i+1} consists of rows of the form $A_i \cdot P \cdot O(\ldots, \ldots)(\omega)$. It adds at most $k_i|\Omega|$ equations to the system.

We similarly bring all these equations into upper triangular form and eliminate the linearly dependent ones.

Time complexity. The final equation systems contains at most n equations. In the worst case, from every of these equations, we generated $|\Omega|$ new equations in some iteration, brought them into upper triangular form and found them (almost) all linearly dependent. So, at most $n|\Omega|$ equations over $2n$ variables have been generated. Turning them into upper triangular form takes time $\in \mathcal{O}(n^3|\Omega|)$.

6   Related Work

The three bisimulation relations we have considered here are based on existing definitions in the literature for Markov chains and their extensions. State-based strong bisimulation was considered earlier [3] and is a simple extension of the bisimulation for Markov chains [13], by incorporating the notion of observations. Our logic POCTL* [20] is an extension of the logic PCTL* [8]. Moreover, our logical characterisation for state-based bisimulation also conservatively extends the logical characterisation of Markov chains presented e. g. in [2].

The strong and weak belief bisimulations we have used were taken from [4], where they are defined for a general model with nondeterministic choices. The new concept here is to match distributions with distributions, instead of states with states as in the classical setting. This notion of equivalence has also been studied in [7], where bisimulation between distributions is defined for labelled Markov chains: strong belief bisimulation can be considered as an extension of the definition in [7] with the observation function attached to the transitions. In HMMs where all transitions generate the same trivial observation, it agrees with the definition in [7]. Thus, inspired by the work in [7], we have presented an algorithm for deciding strong belief bisimulation. As we have noted, our time bound improves theirs. Because of the mentioned connection to [7], our logical characterisation also carries over to the setting of labelled Markov chains.

Finally, we want to mention the recent related paper [12] in which the algorithm in [7] was – independently – improved to cubic as well: they have a similar observation as our paper by keeping the basis in a canonical orthogonal set. Moreover, they have proposed a randomized algorithm with quadratic complexity which could be applied in our setting as well.

Acknowledgements. This work was partly done while David N. Jansen was visiting MT-LAB at the Technical University of Denmark; he was partially supported by the NWO/DFG Bilateral Research Programme ROCKS, the EU FP7 under grant number ICT-214755 (Quasimodo) IDEA4CPS and by MT-LAB, a VKR Centre of Excellence.

References

1. Baier, C., Katoen, J.-P.: Principles of model checking. MIT Press, Cambridge (2008)
2. Baier, C., Katoen, J.-P., Hermanns, H., Wolf, V.: Comparative branching-time semantics for Markov chains. Inf. Comput. 200(2), 149–214 (2005)

3. Bicego, M., Dovier, A., Murino, V.: Designing the Minimal Structure of Hidden Markov Model by Bisimulation. In: Figueiredo, M., Zerubia, J., Jain, A.K. (eds.) EMMCVPR 2001. LNCS, vol. 2134, pp. 75–90. Springer, Heidelberg (2001)
4. Castro, P.S., Panangaden, P., Precup, D.: Equivalence relations in fully and partially observable Markov decision processes. In: Boutilier, C. (ed.) Proc. of the Twenty-First Intl. Joint Conference on Artificial Intelligence, IJCAI-2009, pp. 1653–1658. AAAI Press, Menlo Park (2009)
5. Christiansen, H., Have, C.T., Lassen, O.T., Petit, M.: Inference with constrained hidden Markov models in PRISM. TPLP 10(4-6), 449–464 (2010)
6. Derisavi, S., Hermanns, H., Sanders, W.H.: Optimal state-space lumping in Markov chains. Inf. Proc. Lett. 87(6), 309–315 (2003)
7. Doyen, L., Henzinger, T.A., Raskin, J.F.: Equivalence of labeled Markov chains. Int. J. Foundations of Computer Science 19(3), 549–563 (2008)
8. Hansson, H., Jonsson, B.: A logic for reasoning about time and reliability. Formal Asp. Comput. 6(5), 512–535 (1994)
9. Jansen, D.N., Nielson, F., Zhang, L.: Belief bisimulation for hidden Markov models: logical characterisation and decision algorithm. Tech. Rep. ICIS-R12002, Radboud Universiteit: ICIS, Nijmegen (2012)
10. Jurafsky, D., Martin, J.H.: Speech and language processing: an introduction to natural language processing, computational linguistics, and speech recognition. Prentice Hall, Upper Saddle River (2000)
11. Katoen, J.-P., Kemna, T., Zapreev, I., Jansen, D.N.: Bisimulation Minimisation Mostly Speeds Up Probabilistic Model Checking. In: Grumberg, O., Huth, M. (eds.) TACAS 2007. LNCS, vol. 4424, pp. 87–101. Springer, Heidelberg (2007)
12. Kiefer, S., Murawski, A.S., Ouaknine, J., Wachter, B., Worrell, J.: Language Equivalence for Probabilistic Automata. In: Gopalakrishnan, G., Qadeer, S. (eds.) CAV 2011. LNCS, vol. 6806, pp. 526–540. Springer, Heidelberg (2011)
13. Larsen, K.G., Skou, A.: Bisimulation through probabilistic testing. Inf. Comput. 94(1), 1–28 (1991)
14. Paige, R., Tarjan, R.E.: Three partition refinement algorithms. SIAM J. Comput. 16(6), 973–989 (1987)
15. Rabiner, L.R.: A tutorial on hidden Markov models and selected applications in speech recognition. Proc. IEEE 77(2), 257–286 (1989)
16. Salamatian, K., Vaton, S.: Hidden Markov modeling for network communication channels. In: Proc. 2001 ACM SIGMETRICS Intl. Conf. on Measurement and Modeling of Computer Systems, vol. 29, pp. 92–101. ACM, New York (2001)
17. Sistla, A.P., Žefran, M., Feng, Y.: Monitorability of Stochastic Dynamical Systems. In: Gopalakrishnan, G., Qadeer, S. (eds.) CAV 2011. LNCS, vol. 6806, pp. 720–736. Springer, Heidelberg (2011)
18. Stein, E.M., Shakarchi, R.: Real analysis: measure theory, integration, and Hilbert spaces, Princeton lectures in analysis, vol. III. Princeton Univ. Pr., Princeton (2005)
19. Woodward, A., Penn, R.: The wrong kind of snow. Hodder & Stoughton, London (2007)
20. Zhang, L., Hermanns, H., Jansen, D.N.: Logic and Model Checking for Hidden Markov Models. In: Wang, F. (ed.) FORTE 2005. LNCS, vol. 3731, pp. 98–112. Springer, Heidelberg (2005)

Abstract Model Repair

George Chatzieleftheriou[1], Borzoo Bonakdarpour[2],
Scott A. Smolka[3], and Panagiotis Katsaros[1]

[1] Department of Informatics, Aristotle University of Thessaloniki
54124 Thessaloniki, Greece
[2] School of Computer Science, University of Waterloo
200 University Avenue West Waterloo N2L3G1, Canada
[3] Department of Computer Science, Stony Brook University
Stony Brook, NY 11794-4400, USA

Abstract. Given a Kripke structure M and CTL formula φ, where $M \not\models \varphi$, the problem of *Model Repair* is to obtain a new model M' such that $M' \models \varphi$. Moreover, the changes made to M to derive M' should be minimal with respect to all such M'. As in model checking, *state explosion* can make it virtually impossible to carry out model repair on models with infinite or even large state spaces. In this paper, we present a framework for model repair that uses *abstraction refinement* to tackle state explosion. Our model-repair framework is based on Kripke Structures, a 3-valued semantics for CTL, and Kripke Modal Transition Systems (KMTSs), and features an abstract-model-repair algorithm for KMTSs. Application to an Automatic Door Opener system is used to illustrate the practical utility of abstract model repair.

Keywords: Model Repair, Model Checking, Abstraction Refinement.

1 Introduction

Given a model M and temporal-logic formula φ, *model checking* is the problem of determining if $M \models \varphi$. When this is not the case, a model checker will typically provide a *counterexample* in the form of an execution path along which ϕ is violated. The user should then process the counterexample manually to correct the model.

An extended version of the model-checking problem is that of *model repair*: given a model M and temporal-logic formula φ, where $M \not\models \varphi$, obtain a new model M' such that $M' \models \phi$. The problem of Model Repair was introduced for the first time in the context of Kripke structures and the CTL temporal logic in [4].

State explosion is a well known problem in automated formal methods, such as model checking and model repair, which limits their applicability to systems having large or even infinite state spaces. Different techniques have been developed to cope with this problem. In the case of model checking, *abstraction* is used to create a smaller, more abstract version $\hat{M}$ of the initial concrete model

A. Goodloe and S. Person (Eds.): NFM 2012, LNCS 7226, pp. 341–355, 2012.

M, and model checking is performed on this smaller model. For this technique to work as advertised, it should be the case that $\hat{M} \models \varphi$ iff $M \models \varphi$.

Motivated by the success of abstraction-based model checking, we present in this paper a new framework for Model Repair that uses *abstraction refinement* to tackle state explosion. The resulting *Abstract Model Repair* (AMR) methodology makes it possible to repair models with large state spaces, and to speed-up the repair process through the use of smaller abstract models. The major contributions of our work are as follows:

- We provide an AMR framework that uses Kripke structures (KSs) for the concrete model, Kripke Modal Transition Systems (KMTSs) for the abstract model, and a 3-valued semantics for interpreting CTL over KMTSs. An abstract KMTS model is refined whenever the 3-valued CTL model-checking problem returns a value of undefined. Repair is initiated on the KMTS when a value of false is returned.
- We strengthen the Model Repair problem by additionally taking into account the following *minimality* criterion (refer to the definition of Model Repair above): the changes made to M to derive M' should be minimal with respect to all M' satisfying φ. To handle the minimality constraint, we define a metric space over KSs that quantifies the structural differences between KSs.
- A key feature of our Abstract Model Repair framework is a repair algorithm for KMTSs, which takes into account the minimality criterion.
- We illustrate the utility of our approach by applying it to the repair of an Automatic Door Opener system [1].

The rest of this paper is organized as follows. Sections 2 and 3 introduce KS, KMTSs, and the concepts of abstraction and refinement for a 3-valued semantics for CTL. Section 4 defines a metric space for KSs and gives the problem statement for Model Repair. Section 5 presents our framework for Abstract Model Repair, while Section 6 highlights our model-repair algorithm for KMTSs. Section 7 considers related work, while Section 8 offers our concluding remarks.

2   Kripke Modal Transition Systems

Let AP be a set of *atomic propositions*. Also, the set Lit of *literals* is given by:

$$Lit = AP \cup \{\neg p : p \in AP\}$$

Definition 1. *A Kripke Structure (KS) is a quadruple* $M = (S, S_0, R, L)$, *where:*

1. S *is a finite set of* states.
2. $S_0 \subseteq S$ *is the set of* initial states.
3. $R \subseteq S \times S$ *is a transition relation that must be total; i.e.,* $\forall s \in S, \exists s' \in S$ *such that* $R(s, s')$.
4. $L : S \rightarrow 2^{Lit}$ *is a state labeling function such that* $\forall s \in S, \forall p \in AP,$ $p \in L(s) \Leftrightarrow \neg p \notin L(s)$.

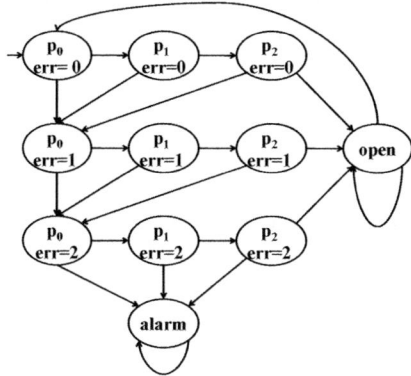

Fig. 1. The Automatic Door Opener (ADO) System

The fourth condition in Def. 1 ensures that an atomic proposition $p \in AP$ has one and only one truth value at any state.

Example. We use the Automatic Door Opener system (ADO) of [1] as a running example throughout the paper. The system, given as a KS in Fig 1, requires a three-digit code (p_0, p_1, p_2) to open a door, allowing for a wrong digit to be entered at most twice. Variable *err* counts the number of errors, and an alarm is rung if its value exceeds two. For the purposes of our paper, we use a simpler version of the ADO system, given as the KS M in Fig. 2a, where the set of atomic propositions $AP = \{q\}$, $q \equiv (open = true)$.

Definition 2. *A* Kripke Modal Transition System *(KMTS) is a 5-tuple* $\hat{M} = (\hat{S}, \hat{S}_0, R_{must}, R_{may}, \hat{L})$, *where:*

1. $\hat{S}$ *is a finite set of* states.
2. $\hat{S}_0 \subseteq \hat{S}$ *is the set of* initial states.
3. $R_{must} \subseteq \hat{S} \times \hat{S}$ *and* $R_{may} \subseteq \hat{S} \times \hat{S}$ *are transition relations such that* $R_{must} \subseteq R_{may}$.
4. $\hat{L} : \hat{S} \to 2^{Lit}$ *is a state-labeling such that* $\forall \hat{s} \in \hat{S}$, $\forall p \in AP$, $\hat{s}$ *is labeled by at most one of* p *and* $\neg p$.

A KMTS has two types of transitions: *must-transitions*, which exhibit *necessary* behavior, and *may-transitions*, which exhibit *possible* behavior. The "at most one" condition in the fourth part of Def. 2 makes it possible for the truth value of an atomic proposition at a given state to be *unknown*. This relaxation of truth values in conjunction with the existence of may-transitions in a KMTS constitutes a *partial modeling* formalism.

Verifying a CTL formula ϕ over a KMTS may result in an undefined answer ($\bot$). We use the *3-valued semantics* [13] of a CTL formula ϕ at a state $\hat{s}$ of KMTS $\hat{M}$ (denoted $[(\hat{M}, \hat{s}) \models^3 \phi]$). From the 3-valued semantics, it follows that must-transitions (under-approximation) are used to check the truth of existential

CTL properties, while may-transitions (over-approximation) are used to check the truth of universal CTL properties. This works inversely for checking the refutation of CTL properties. When we get $\perp$ from the 3-valued model checking of a CTL formula ϕ on a KMTS, the result of model checking property ϕ on the corresponding KS can be either true or false. In the rest of the paper, we use $\models$ instead of $\models^3$ in order to refer to 3-valued satisfaction relation.

3   Abstraction and Refinement for 3-Valued CTL

3.1   Abstraction

Abstraction is a state-space reduction technique that produces a smaller abstract model from an initial *concrete* model, so that the models behave similarly. In order for the result of verifying an abstract model to hold for its concrete model, the abstract model should be produced with certain requirements [7,10].

Definition 3. *Let* $M = (S, S_0, R, L)$ *be a KS. For any pair of total functions* $\Re = (\alpha : S \to \hat{S}, \gamma : \hat{S} \to 2^S)$, *where* $\forall s \in S$, $\hat{s} \in \hat{S}$, $\alpha(s) = \hat{s}$ *if and only if* $s \in \gamma(\hat{s})$, *a KMTS* $\hat{M} = (\hat{S}, \hat{S}_0, R_{must}, R_{may}, \hat{L})$ *is defined as follows:*

1. $\hat{s} \in \hat{S}_0$ *iff* $\exists s \in \gamma(\hat{s})$ *such that* $s \in S_0$
2. $lit \in \hat{L}(\hat{s})$ *only if* $\forall s \in \gamma(\hat{s})$ *it holds that* $lit \in L(s)$
3. $R_{must} = \{(\hat{s_1}, \hat{s_2}) \mid \forall s_1 \in \gamma(\hat{s_1}) \exists s_2 \in \gamma(\hat{s_2})$ *such that* $R(s_1, s_2)\}$
4. $R_{may} = \{(\hat{s_1}, \hat{s_2}) \mid \exists s_1 \in \gamma(\hat{s_1}) \exists s_2 \in \gamma(\hat{s_2})$ *such that* $R(s_1, s_2)\}$

For a given KS and pair of abstraction and concretization functions, Def. 3 introduces a KMTS with a set $\hat{S}$ of *abstract states*. In our AMR framework, we view the given KS as the *concrete model* and the derived KMTS as the *abstract model*. A state of the abstract KMTS is initial *if and only if* at least one of its concrete states is initial. An atomic proposition is true (or false) in an abstract state, *only if* this atomic proposition is true (or false) in all of its concrete states. *Only if* allows for the value of an atomic proposition to be unknown at a KMTS state. Between two abstract states $\hat{s_1}, \hat{s_2}$, there exists a must-transition if there are transitions from all the concrete states of $\hat{s_1}$ to at least one concrete state of $\hat{s_2}$ ($\forall\exists - condition$), while on the other side, there exists a may-transition if there is a transition from at least one concrete state of $\hat{s_1}$ to at least one concrete state of $\hat{s_2}$ ($\exists\exists - condition$).

Definition 4. *[8,11] Let* $M = (S, S_0, R, L)$ *be a concrete KS, and let* $\hat{M} = (\hat{S}, \hat{S}_0, R_{must}, R_{may}, \hat{L})$ *be an abstract KMTS. A relation* $H \subseteq S \times \hat{S}$ *for* M *and* $\hat{M}$ *is called a* mixed simulation, *when* $H(s, \hat{s})$ *implies:*

- $\hat{L}(\hat{s}) \subseteq L(s)$
- *if* $r = (s, s') \in R$, *then there exists some* $\hat{s}' \in \hat{S}$ *such that* $r_{may} = (\hat{s}, \hat{s}') \in R_{may}$ *and* $(s', \hat{s}') \in H$.
- *if* $r_{must} = (\hat{s}, \hat{s}') \in R_{must}$, *then there exists some* $s' \in S$ *such that* $r = (s, s') \in R$ *and* $(s', \hat{s}') \in H$.

Abstraction function α in Def. 3 is a mixed simulation for KS M and KMTS $\hat{M}$.

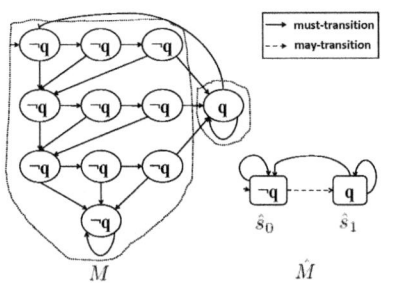
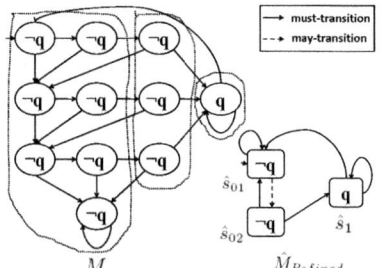

(a) The KS and initial KMTS   (b) The KS and refined KMTS

Fig. 2. The KS and KMTSs for the ADO system

Theorem 1. [11] *Let $H \subseteq S \times \hat{S}$ be a mixed simulation from a KS $M = (S, S_0, R, L)$ to a KMTS $\hat{M} = (\hat{S}, \hat{S}_0, R_{must}, R_{may}, \hat{L})$. Then, for every CTL formula φ and every $(s, \hat{s}) \in H$ it holds that*

$$[(\hat{M}, \hat{s}) \models \varphi] \neq \bot \Rightarrow [(M, s) \models \varphi] = [(\hat{M}, \hat{s}) \models \phi]$$

Theorem 1 ensures that if a CTL formula ϕ has a definite truth value (true or false) in the abstract KMTS then it has the same truth value in the concrete KS.

Example. An abstract KMTS $\hat{M}$ is presented in Fig. 2a, where all the states labeled by q are grouped together, as are all states labeled by $\neg q$.

3.2   Refinement

When the answer to verifying a CTL formula φ on an abstract model using the 3-valued semantics is $\bot$, then a *refinement* step is needed to acquire a more precise abstract model. A number of refinement frameworks specialized for 3-valued model checking have been proposed [10,16]. The refinement technique that we use in our framework is a two-step process: (1) identify a *failure state* in the KMTS, and (2) produce a new abstract KMTS such that this failure state is refined into several states. The cause of failure for a state s stems from an atomic proposition having an undefined value in s, or from an outgoing may-transition from s. In both cases, s is refined in a way that the cause of failure is eliminated.

Example. Consider the case where the ADO system requires a mechanism for opening the door from any state with a direct action. This could be an action done by an expert if an immediate opening of a door is required. This property can be expressed in CTL as the formula $\varphi = AGEXq$. Observe that in $\hat{M}$ of Fig. 2a, the absence of a must-transition from $\hat{s}_0$ to $\hat{s}_1$, where $[(\hat{M}, \hat{s}_1) \models q] = true$, in conjunction with the existence of a may-transition from $\hat{s}_0$ to $\hat{s}_1$, thus to a state where $[(\hat{M}, \hat{s}_1) \models q] = true$, results in an undefined answer to the model-checking question for $\hat{M}$ and φ. State $\hat{s}_0$ is identified as the failure state, and the may-transition from $\hat{s}_0$ to $\hat{s}_1$ as the cause of the failure. Consequently, $\hat{s}_0$ is refined into two states, $\hat{s}_{01}$ and $\hat{s}_{02}$, such that the former has no transition to $\hat{s}_1$ and the latter has an outgoing must-transition to $\hat{s}_1$. As such, we eliminate the may-transition which led to the undefined answer of model checking $var\phi$ over $\hat{M}$. The refined KMTS $\hat{M}_{Refined}$ together with the initial KS is shown in Fig. 2b.

4   The Model Repair Problem

In this section, we give the problem statement for Model Repair and define a metric space over Kripke structures to quantify their structural differences such that the *minimality of changes* can be taken into account as a criterion for Model Repair.

Let G be a function on the set of all functions $F : X \to Y$ such that:

$$G(F : X \to Y) = \{(x, F(x)) : x \in X\}$$

Let $F : X \to Y$ be a function defined over a set X. A *restricting operator* ($\upharpoonright$) for the domain of function F can be defined such that

$$F \upharpoonright_{X_1} = \{(x, F(x)) : x \in X_1\}$$

where $X_1 \subseteq X$. Finally, we let S^C denote the complement of a set S.

Definition 5. *Let K_M be the set of all KSs $M' = (S', S'_0, R', L')$ derived from the KS $M = (S, S_0, R, L)$, where $S' = (S \cup S_{IN}) - S_{OUT}$ for some $S_{IN} \subseteq S^C$, $S_{OUT} \subseteq S$, $R' = (R \cup R_{IN}) - R_{OUT}$ for some $R_{IN} \subseteq R^C$, $R_{OUT} \subseteq R$, $L' = S' \to 2^{LIT}$. A distance function d can be defined over K_M such that*

$$d(M, M') = |S \Delta S'| + |R \Delta R'| + \frac{|G(L \upharpoonright_{S \cap S'}) \Delta G(L' \upharpoonright_{S \cap S'})|}{2}$$

where $A \Delta B$ represents the symmetric difference $(A - B) \cup (B - A)$.

For any two KSs defined over the same set of atomic propositions AP, function d counts the number of differences $|S \Delta S'|$ in the state space of M, the number of differences $|R \Delta R'|$ in their transition relation and the number of common states with altered labeling.

Proposition 1. *The ordered pair (K_M, d) is a metric space.*

Definition 6. *Given a KS M and a CTL formula φ where $M \not\models \varphi$, the Model Repair problem is to find a KS M', such that $M' \models \varphi$ and $d(M, M')$ is minimal with respect to all such M'.*

The Model Repair problem aims at modifying a KS such that the KS satisfies a CTL formula that it originally does not. We focus on repair with minimal changes to the original KS.

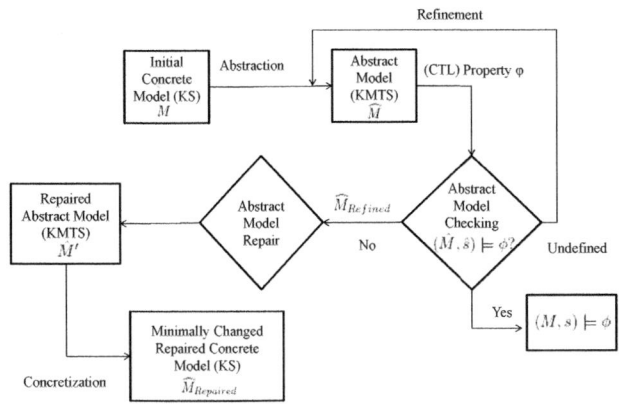

Fig. 3. Abstract Model Repair Framework

5 Abstract Model Repair Framework and Algorithm

Our AMR framework integrates 3-valued model checking, model refinement, and a new algorithm for ordering the basic repair operations to be performed on the abstract model. The goal of this algorithm is to order the repair operations in such a way that the number of corresponding structural changes applied to the concrete model is minimized. The basis for this algorithm is a partial order over the basic repair operations. This section describes the steps involved in our AMR framework, the basic repair operations, and the operations-ordering algorithm.

5.1 The Abstract Model Repair Process

The process steps shown in Fig. 3 rely on the KMTS abstraction of Def. 3. These are the following:

Step 1. Given a KS M, a state s of M, and a CTL property φ, let us call $\hat{M}$ the KMTS obtained as in Def. 3.

Step 2. For state $\hat{s} = \alpha(s)$ of $\hat{M}$, we check whether $(\hat{M}, \hat{s}) \models \varphi$ by 3-valued model checking.

  Case 1. If the result is *true*, then, according to Theorem 1, $(M, s) \models \varphi$ and there is no need for repair.

  Case 2. If the result is *undefined*, $\hat{M}$ is refined to an $\hat{M}_{Refined}$ and control is transferred to Step 2.

  Case 3. If the result is *false*, then, from Theorem 1, $(M, s) \not\models \varphi$ and the repair process follows.

Step 3. The *AbstractRepair* algorithm is called for the KMTS $\hat{M}$ (or $\hat{M}_{Refined}$ if refinement occurred), the state $\hat{s}$ and the property φ.

  Case 1. *AbstractRepair* returns an $\hat{M}'$ for which $(\hat{M}', \hat{s}) \models \varphi$.

  Case 2. *AbstractRepair* fails to find an $\hat{M}'$ for which the property holds.

Step 4. If *AbstractRepair* returns an $\hat{M}'$, then the process ends with a set of KSs, resulting from the concretization of $\hat{M}'$, whose structural distance d from the original KS M is minimized.

5.2   Basic Repair Operations

We decompose the repair process of the KMTS into seven basic repair operations:

AddMust. Adding a must-transition
AddMay. Adding a may-transition
RemoveMust. Removing an existing must-transition
RemoveMay. Removing an existing may-transition
ChangeLabel. Changing the labeling of a KMTS state
AddState. Adding a new KMTS state
RemoveState. Removing a disconnected KMTS state

Definition 7 (AddMust). *For a given KMTS $\hat{M} = (\hat{S}, \hat{S}_0, R_{must}, R_{may}, \hat{L})$ and $\hat{r}_n = (\hat{s}_1, \hat{s}_2) \notin R_{must}$ with $\hat{s}_1, \hat{s}_2 \in \hat{S}$, $AddMust(\hat{M}, \hat{r}_n)$ is a KMTS $\hat{M}' = (\hat{S}', \hat{S}_0', R_{must}', R_{may}', \hat{L}')$ such that $\hat{S}' = \hat{S}$, $\hat{S}_0' = \hat{S}_0$, $R_{must}' = R_{must} \cup \{\hat{r}_n\}$, $R_{may}' = R_{may} \cup \{\hat{r}_n\}$ and $\hat{L}' = \hat{L}$.*

Fig. 4 shows how the basic repair operation *AddMust* modifies a given KMTS.

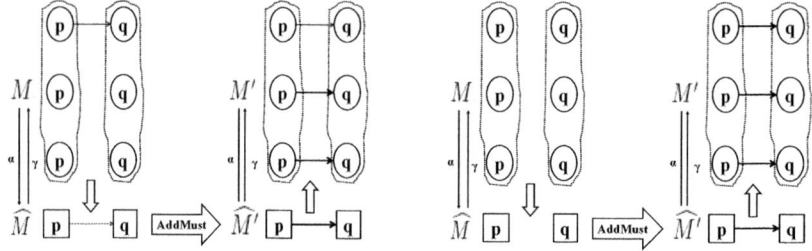

(a) May-transition exists   (b) May-transition doesn't exist

Fig. 4. *AddMust*: Adding a new must-transition

Definition 8. *Let M be a KS, $\hat{M}$ be a KMTS derived as in Def. 3, and $\hat{M}' = AddMust(\hat{M}, \hat{r}_n)$ for some $\hat{r}_n = (\hat{s}_1, \hat{s}_2) \notin R_{must}$ with $\hat{s}_1, \hat{s}_2 \in \hat{S}$. The set of KSs, derived from the concretization of $\hat{M}'$, whose structural distance d from M is minimized is given by:*

$$K_{min} = \{M' = (S', S_0', R', L') \mid S' = S, S_0' = S_0, R' = R \cup R_n, L' = L\} \quad (1)$$

where
$R_n = \{r_n = (s_1, s_2) \mid$ *for every $s_1 \in \gamma(\hat{s}_1)$ such that $\nexists s \in \gamma(\hat{s}_2)$ with $(s_1, s) \in R$, and only one $s_2 \in \gamma(\hat{s}_2)\}$.*

Def. 8 implies that when the *AbstractRepair* algorithm applies *AddMust* on the abstract KMTS $\hat{M}$, then a set of KSs are retrieved from the concretization of $\hat{M}'$.

The same holds for the other basic repair operations for which their definition is omitted for the sake of brevity. Consequently when *AbstractRepair* finds a repaired KMTS, one or more KSs can be obtained for which property φ holds.

Proposition 2. *For all $M' \in K_{min}$, it holds that $1 \leq d(M, M') \leq |S|$.*

From Prop. 2, we conclude that a lower and upper bound exists for the distance between M and any $M' \in K_{min}$.

Minimality of Changes Ordering for Basic Repair Operations. Based on the upper bound given by Prop. 2 and the corresponding results for the other basic repair operations, we introduce the ordering shown in Fig. 5. We use this ordering in the *AbstractRepair* algorithm to *heuristically* select at each step the basic repair operation that generates the KSs with the least changes. The alternative to check at each step all possible repaired KSs in order to identify the proper basic repair operation, would cancel the benefits of using abstraction. The reason is that such a check inevitably depends on the size of the KS.

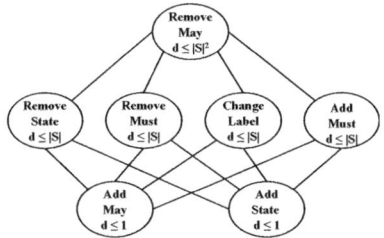

Fig. 5. Minimality of changes ordering of the set of basic operations

6   The Abstract Model Repair Algorithm

The *AbstractRepair* algorithm used in Step 3 of our repair process is a recursive, CTL syntax-directed algorithm. The repair of an abstract KMTS is accomplished by successive calls of primitive repair functions that handle atomic formulas, logical connectives and CTL operators.

The main routine of *AbstractRepair* is presented in Algorithm 1. A set of constraints $C = \{(\hat{s}_{c1}, \phi_{c1}), (\hat{s}_{c2}, \phi_{c2}), ..., (\hat{s}_{cn}, \phi_{cn})\}$ which is initially empty is passed as an argument in the successive recursive calls of *AbstractRepair*. If C is not empty, then for the KMTS $\hat{M}'$ returned from *AbstractRepair*, it holds that $(\hat{M}', \hat{s}_{ci}) \models \phi_{ci}$ for all $(\hat{s}_{ci}, \phi_{ci}) \in C$. C is used for handling conjunctive formulas of the form $\phi = \phi_1 \wedge \phi_2$ for some state $\hat{s}$. In this case, *AbstractRepair* is called for the KMTS $\hat{M}$ and property ϕ_1 with $C = \{(\hat{s}, \phi_2)\}$. The same is repeated for property ϕ_2 with $C = \{(\hat{s}, \phi_1)\}$ and the two results are combined appropriately.

For any CTL formula ϕ and KMTS state $\hat{s}$, *AbstractRepair* either outputs a KMTS $\hat{M}'$ for which $(\hat{M}', \hat{s}) \models \phi$ or else returns FAILURE if such a model cannot be found. This is the case when the algorithm handles conjunctive formulas and a KMTS that simultaneously satisfies all conjuncts cannot be found.

Algorithm 1. AbstractRepair

Input: $\hat{M} = (\hat{S}, \hat{S}_0, R_{must}, R_{may}, \hat{L})$, $\hat{s} \in \hat{S}$, a CTL property ϕ for which $(\hat{M}, \hat{s}) \not\models \phi$,
   and a set of constraints $C = \{(\hat{s}_{c1}, \phi_{c1}), (\hat{s}_{c2}, \phi_{c2}), ..., (\hat{s}_{cn}, \phi_{cn})\}$ where $\hat{s}_{ci} \in \hat{S}$ and
   ϕ_{ci} is a CTL formula.
Output: $\hat{M}' = (\hat{S}', \hat{S}_0', R_{must}', R_{may}', \hat{L}')$ and $(\hat{M}', \hat{s}) \models \phi$ or FAILURE.
1: $\phi_{pos} := PositiveNormalForm(\phi)$
2: **if** ϕ_{pos} is $\bot$ **then**
3:     **return** FAILURE
4: **else if** $\phi_{pos} \in LIT$ **then**
5:     **return** $AbstractRepair_{ATOMIC}(\hat{M}, \hat{s}, \phi_{pos}, C)$
6: **else if** ϕ_{pos} is $\phi_1 \wedge \phi_2$ **then**
7:     **return** $AbstractRepair_{AND}(\hat{M}, \hat{s}, \phi_{pos}, C)$
8: **else if** ϕ_{pos} is $\phi_1 \vee \phi_2$ **then**
9:     **return** $AbstractRepair_{OR}(\hat{M}, \hat{s}, \phi_{pos}, C)$
10: **else if** ϕ_{pos} is $OPER\phi_1$ **then**
11:     **return** $AbstractRepair_{OPER}(\hat{M}, \hat{s}, \phi_{pos}, C)$
12:     where $OPER \in \{AX, EX, AU, EU, AF, EF, AG, EG\}$

6.1 Primitive Functions

For a simple atomic formula, $AbstractRepair_{ATOMIC}$ updates the label of the input state with the given atomic proposition. While conjunctive formulas are handled by the algorithm with the use of constraints, disjunctive formulas are handled by repairing any of the disjuncts.

Algorithm 2 describes the primitive function $AbstractRepair_{AG}$ which is called when $\phi = AG\phi_1$. When $AbstractRepair_{AG}$ is called with state $\hat{s}$ as argument, it recursively calls $AbstractRepair$ for all states that are reachable from $\hat{s}$ through successive may-transitions and do not satisfy ϕ_1. If the found KMTS $\hat{M}'$ does not violate any constraint in C, then $(\hat{M}', \hat{s}) \models \phi$ and $AbstractRepair_{AG}$ returns the found solution. If a KMTS does not satisfy all the constraints in C, then $AbstractRepair_{AG}$ tries to repair the input KMTS by removing all may-transitions through which the state violating ϕ_1 is reached.

$AbstractRepair_{EX}$ presented in Algorithm 3 is the primitive function for handling properties of the form $EX\phi_1$ for some state $\hat{s}$. Initially, this function tries to repair the KMTS by adding a must-transition from $\hat{s}$ to a state that satisfies property ϕ_1. If the obtained KMTS does not satisfy all constraints in C, then $AbstractRepair$ is recursively called for an immediate successor of $\hat{s}$ through a must-transition, such that ϕ_1 is not satisfied. If a constraint in C is still violated, then (i) a new state is added, (ii) $AbstractRepair$ is called for the new state and (iii) a must-transition from $\hat{s}$ to the new state is added.

6.2 Well-definedness and Soundness

$AbstractRepair$ is *well-defined*, in the sense that all possible cases are handled and each algorithm step is deterministically defined. This feature distinguishes our approach from related concrete model repair solutions which entail nondeterministic behavior [19,5].

Algorithm 2. *AbstractRepair$_{AG}$*

Input: $\hat{M} = (\hat{S}, \hat{S}_0, R_{must}, R_{may}, \hat{L})$, $\hat{s} \in \hat{S}$, a CTL property $\phi = AG\phi_1$ for which $(\hat{M}, \hat{s}) \not\models \phi$, and a set of constraints $C = \{(\hat{s}_{c1}, \phi_{c1}), (\hat{s}_{c2}, \phi_{c2}), ..., (\hat{s}_{cn}, \phi_{cn})\}$ where $\hat{s}_{ci} \in \hat{S}$ and ϕ_{ci} is a CTL formula.

Output: $\hat{M}' = (\hat{S}', \hat{S}_0', R_{must}', R_{may}', \hat{L}')$ and $(\hat{M}, \hat{s}) \models \phi$ or FAILURE.

1: **if** $(\hat{M}, \hat{s}) \not\models \phi_1$ **then**
2:    $RET := AbstractRepair(\hat{M}, \hat{s}, \phi_1, C)$
3:    **if** $RET == FAILURE$ **then**
4:       **return** FAILURE
5:    **else**
6:       $\hat{M}' := RET$
7: **else**
8:    $\hat{M}' := \hat{M}$
9: $\hat{M}'' := \hat{M}'$
10: **for all** reachable states $\hat{s}_k$ through may-transitions from $\hat{s}$ such that $(\hat{M}', \hat{s}_k) \not\models \phi_1$ **do**
11:    $RET := AbstractRepair(\hat{M}', \hat{s}_k, \phi_1, C)$
12:    **if** $RET == FAILURE$ **then**
13:       BREAK
14:    **else**
15:       $\hat{M}' := RET$
16: **if** $\hat{M}' \models \phi$ && $\hat{M}' \models C$ **then**
17:    **return** $\hat{M}'$
18: **else**
19:    $\hat{M}' := \hat{M}''$
20:    **for all** $\hat{\pi}_{may} := [\hat{s}, \hat{s}_1, ..., \hat{s}_i, \hat{s}_k]$ for which $(\hat{M}', \hat{s}_k) \not\models \phi_1$, $(\hat{M}', \hat{s}_i) \models \phi_1$ and $\not\exists \hat{s}_j \in \hat{\pi}_{may}$ such that $(\hat{M}', \hat{s}_j) \not\models \phi_1$ and $\hat{s}_j \in Pre_{may}(\hat{s}_i)$ **do**
21:       $\hat{r}_m := (\hat{s}_i, \hat{s}_k)$, $\hat{M}' := RemoveMay(\hat{M}', \hat{r}_m)$
22:       **if** $\hat{s}_i$ is a dead-end state **then**
23:          $\hat{r}_n := (\hat{s}_i, \hat{s}_i)$, $\hat{M}' := AddMay(\hat{M}', \hat{r}_n)$
24:    **if** $\hat{M}' \models C$ **then**
25:       **return** $\hat{M}'$
26:    **else**
27:       **return** FAILURE

Theorem 2 (Soundness). *Let $\hat{M}$ be a KMTS and ϕ a CTL formula for which $(\hat{M}, \hat{s}) \not\models \phi$ for some state $\hat{s}$ of $\hat{M}$. If AbstractRepair$(\hat{M}, \hat{s}, \phi)$ returns a KMTS $\hat{M}'$, then $(\hat{M}', \hat{s}) \models \phi$.*

Proof. The proof is done by structural induction over ϕ.

Theorem 2 shows that *AbstractRepair* is *sound* in the sense that if it returns a KMTS $\hat{M}'$, then $\hat{M}'$ satisfies property ϕ. In that case, from Def. 8 it follows that one or more KSs are obtained for which property ϕ holds true.

6.3 Application

We present the application of *AbstractRepair* to the ADO system from Section 2. After the first two steps of our repair process, *AbstractRepair* is called for the

---

Algorithm 3. *AbstractRepair$_{EX}$*

Input: $\hat{M} = (\hat{S}, \hat{S}_0, R_{must}, R_{may}, \hat{L})$, $\hat{s} \in \hat{S}$, a CTL property $\phi = EX\phi_1$ for which $(\hat{M}, \hat{s}) \not\models \phi$, and a set of constraints $C = \{(\hat{s}_{c1}, \phi_{c1}), (\hat{s}_{c2}, \phi_{c2}), ..., (\hat{s}_{cn}, \phi_{cn})\}$ where $\hat{s}_{ci} \in \hat{M}$ and ϕ_{ci} is a CTL formula.

Output: $\hat{M}' = (\hat{S}', \hat{S}_0', R'_{must}, R'_{may}, \hat{L}')$ and $(\hat{M}', \hat{s}) \models \phi$ or FAILURE.

```
 1: if there exists ŝ₁ ∈ Ŝ such that (M̂, ŝ₁) ⊨ φ₁ then
 2:     for all ŝᵢ ∈ Ŝ such that (M̂, ŝᵢ) ⊨ φ₁ do
 3:         r̂ₙ := (ŝ, ŝᵢ), M̂' := AddMust(M̂, r̂ₙ)
 4:         if M̂' ⊨ C then
 5:             return M̂'
 6: else
 7:     for all ŝᵢ ∈ Postₘᵤₛₜ(ŝ) do
 8:         RET := AbstractRepair(M̂, ŝᵢ, φ₁, C)
 9:         if RET ≠ FAILURE then
10:             M̂' := RET
11:             return M̂'
12:     M̂' := AddState(M̂, ŝ₁'), r̂ₙ := (ŝ, ŝ₁'), M̂' := AddMust(M̂', r̂ₙ)
13:     if ŝ₁' is a dead-end state then
14:         r̂ₙ := (ŝ₁', ŝ₁'), M̂' := AddMay(M̂', r̂ₙ)
15:     RET := AbstractRepair(M̂', ŝ₁', φ₁, C)
16:     if RET ≠ FAILURE then
17:         M̂' := RET
18:         return M̂'
19:     else
20:         return FAILURE
21: return FAILURE
```

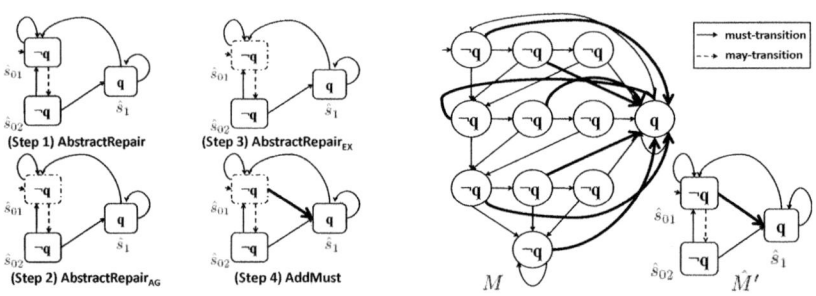

(a) Application of *AbstractRepair*   (b) The repaired KMTS and KS

Fig. 6. Repair of ADO system using abstraction

KMTS $\hat{M}_{Refined}$ that is shown in Fig. 2b, the state $\hat{s}_{01}$ and the CTL property $\phi = AGEXq$.

AbstractRepair calls *AbstractRepair$_{AG}$* with arguments $\hat{M}_{Refined}$, $\hat{s}_{01}$ and $AGEXq$. The *AbstractRepair$_{AG}$* algorithm at line 2 triggers a recursive call of *AbstractRepair* with the same arguments. Eventually, *AbstractRepair$_{EX}$* is called with arguments $\hat{M}_{Refined}$, $\hat{s}_{01}$ and EXq, that in turn calls *AddMust* at line 3, thus adding a must-transition from $\hat{s}_{01}$ to $\hat{s}_1$. *AbstractRepair* terminates by returning a KMTS $\hat{M}'$ that satisfies $\phi = AGEXq$. The repaired KS M' is the single element in the set of KSs derived by the concretization of $\hat{M}'$. The execution steps of *AbstractRepair* and the obtained repaired KMTS and KS are shown in Fig. 6a and Fig. 6b respectively.

Although the ADO is not a system with a large state space, it is shown that the repair process is accelerated by the proposed use of abstraction. If on the other hand model repair was applied directly to the concrete model, adding transitions to the state labeled with *open* would have to take place for all states with a different labeling. The number of these states is seven but in a system with a large state space this number can be significantly higher. Direct repair of such a model without using abstraction is impractical.

7   Related Work

To the best of our knowledge this is the first work that suggests the use of abstraction as a means to counter the state space explosion in the search for a solution to the Model Repair problem. In [18], abstract interpretation is used in *program synthesis*, a problem related to Model Repair but much different.

A first attempt for introducing the Model Repair problem in the context of CTL has been done in [4], where a repair algorithm with high computational cost is presented based on the AI techniques of abductive reasoning and theory revision. A formal algorithm for Model Repair in the context of KSs and CTL is presented in [19]. The authors acknowledge that the repair process strongly depends on the size of the model, while they do not implement explicitly in their algorithm how the constraints can be used to handle conjunctive formulas. An effort for making repair applicable to large KSs, is done by the authors of [6]. They use "table systems", a concise representation of KSs, implemented in the NuSMV model checker. A certain limitation for their approach is that table systems cannot represent any KS. In [20], tree-like local model updates are introduced with the aim of making repair process applicable to large scale domains, but their approach is limited to the universal fragment of CTL formulas. For better handling of the constraints in the repair process and thus, ensuring completeness of it, the use of constraint automata for ACTL formulas [14] and the use of protected models for an extension of CTL [5] have been proposed. Both methods are not directly applied to formulas of full CTL. An extension of the Model Repair problem in the context of Labeled Transition Systems has been examined in [9].

The Model Repair problem has been addressed in [2] in the context of probabilistic systems. A slightly different problem, that of Model Revision, has been studied for UNITY properties in [3] and for CTL in [12]. Finally, the *program repair* problem that does not consider KSs as the repair model, has been examined in prior work [17,15].

8   Conclusions

In this paper, we have shown how abstraction can be used to fight state explosion in Model Repair. Our model-repair framework is based on Kripke Structures, a 3-valued semantics for CTL, and Kripke Modal Transition Systems, and features an abstract-model-repair algorithm for KMTSs. To demonstrate its practical utility, we applied our framework to an Automatic Door Opener system.

As future work, we plan to apply our method to case studies with larger state spaces, and investigate how abstract model repair can be used in different contexts and domains.

Acknowledgments. We thank the anonymous reviewers for their valuable comments. Research supported in part by NSF Grants CCF-0926190 and CCF-1018459, and AFOSR Grant FA0550-09-1-0481 . Researchers from Greece were supported by the GSRT/COOPERATION/TRACER Project (09SYN-72-942). The second author's research is supported in part by Canada's NSERC DG 357121-2008, ORF RE03-045, ORE RE04-036, and ORF-RE04-039.

References

1. Baier, C., Katoen, J.-P.: Principles of Model Checking. Representation and Mind Series. The MIT Press (2008)
2. Bartocci, E., Grosu, R., Katsaros, P., Ramakrishnan, C.R., Smolka, S.A.: Model Repair for Probabilistic Systems. In: Abdulla, P.A., Leino, K.R.M. (eds.) TACAS 2011. LNCS, vol. 6605, pp. 326–340. Springer, Heidelberg (2011)
3. Bonakdarpour, B., Ebnenasir, A., Kulkarni, S.S.: Complexity results in revising UNITY programs. ACM Trans. Auton. Adapt. Syst. 4, 5:1–5:28 (2009)
4. Buccafurri, F., Eiter, T., Gottlob, G., Leone, N.: Enhancing model checking in verification by AI techniques. Artif. Intell. 112, 57–104 (1999)
5. Carrillo, M., Rosenblueth, D.A.: Nondeterministic Update of CTL Models by Preserving Satisfaction through Protections. In: Bultan, T., Hsiung, P.-A. (eds.) ATVA 2011. LNCS, vol. 6996, pp. 60–74. Springer, Heidelberg (2011)
6. Carrillo, M., Rosenblueth, D.A.: A method for CTL model update, representing Kripke Structures as table systems. IJPAM 52, 401–431 (2009)
7. Clarke, E.M., Grumberg, O., Long, D.E.: Model checking and abstraction. ACM Trans. Program. Lang. Syst. 16, 1512–1542 (1994)
8. Dams, D., Gerth, R., Grumberg, O.: Abstract interpretation of reactive systems. ACM Trans. Program. Lang. Syst. 19, 253–291 (1997)
9. de Menezes, M.V., do Lago Pereira, S., de Barros, L.N.: System Design Modification with Actions. In: da Rocha Costa, A.C., Vicari, R.M., Tonidandel, F. (eds.) SBIA 2010. LNCS, vol. 6404, pp. 31–40. Springer, Heidelberg (2010)
10. Godefroid, P., Huth, M., Jagadeesan, R.: Abstraction-Based Model Checking Using Modal Transition Systems. In: Larsen, K.G., Nielsen, M. (eds.) CONCUR 2001. LNCS, vol. 2154, pp. 426–440. Springer, Heidelberg (2001)
11. Godefroid, P., Jagadeesan, R.: Automatic Abstraction Using Generalized Model Checking. In: Brinksma, E., Larsen, K.G. (eds.) CAV 2002. LNCS, vol. 2404, pp. 137–150. Springer, Heidelberg (2002)

12. Guerra, P.T., Wassermann, R.: Revision of CTL Models. In: Kuri-Morales, A., Simari, G.R. (eds.) IBERAMIA 2010. LNCS, vol. 6433, pp. 153–162. Springer, Heidelberg (2010)

13. Huth, M., Jagadeesan, R., Schmidt, D.A.: Modal Transition Systems: A Foundation for Three-Valued Program Analysis. In: Sands, D. (ed.) ESOP 2001. LNCS, vol. 2028, pp. 155–169. Springer, Heidelberg (2001)

14. Kelly, M., Pu, F., Zhang, Y., Zhou, Y.: ACTL Local Model Update with Constraints. In: Setchi, R., Jordanov, I., Howlett, R.J., Jain, L.C. (eds.) KES 2010, Part IV. LNCS, vol. 6279, pp. 135–144. Springer, Heidelberg (2010)

15. Samanta, R., Deshmukh, J.V., Emerson, E.A.: Automatic generation of local repairs for boolean programs. In: FMCAD 2008, pp. 27:1–27:10 IEEE Press, Piscataway (2008)

16. Shoham, S., Grumberg, O.: Monotonic Abstraction-Refinement for CTL. In: Jensen, K., Podelski, A. (eds.) TACAS 2004. LNCS, vol. 2988, pp. 546–560. Springer, Heidelberg (2004)

17. Staber, S., Jobstmann, B., Bloem, R.: Finding and Fixing Faults. In: Borrione, D., Paul, W. (eds.) CHARME 2005. LNCS, vol. 3725, pp. 35–49. Springer, Heidelberg (2005)

18. Vechev, M., Yahav, E., Yorsh, G.: Abstraction-guided synthesis of synchronization. In: POPL 2010, pp. 327–338. ACM, New York (2010)

19. Zhang, Y., Ding, Y.: CTL model update for system modifications. J. Artif. Int. Res. 31, 113–155 (2008)

20. Zhang, Y., Kelly, M., Zhou, Y.: Foundations of tree-like local model updates. In: ECAI 2010, pp. 615–620. IOS Press, Amsterdam (2010)

CLSE: Closed-Loop Symbolic Execution

Rupak Majumdar[1,2], Indranil Saha[1], K. C. Shashidhar[2], and Zilong Wang[2]

[1] University of California Los Angeles
[2] Max-Planck Institute for Software Systems
{rupak,indranil}@cs.ucla.edu, {shashi,zilong}@mpi-sws.org

Abstract. We present CLSE, a *closed-loop* symbolic execution engine for control system implementations. CLSE takes as input the description of a physical plant represented by a system of linear ordinary differential equations, the software implementation and execution frequency for a discrete-time controller that senses and actuates the plant, and a time horizon, and symbolically executes the closed-loop system —the combination of the plant and the controller— up to the time horizon. The execution helps capture the bounded-time dynamics of the system in terms of the finite sequences of the plant's sampled state-sets and symbolic control inputs. We show the use of CLSE in symbolic execution of a set of control systems benchmarks. Using the symbolic execution engine, we also build a robustness analysis tool which computes the maximum deviation of the states of the plant due to measurement uncertainties in the controller up to the time horizon.

1 Introduction

Software controllers for physical systems are at the core of many safety-critical systems. The combination of physical behavior, given by the dynamics of continuous state variables, and discrete behavior, implemented in software, makes the end-to-end behavior of these systems hard to design and to reason about.

The need for effective analysis techniques for cyber-physical systems combining physical plants and software controllers has been long recognized. Most current techniques, though, take one of two approaches. In the first approach, the system is modeled as a hybrid automaton [1,19,8,23] —a finite-state machine that is endowed with dynamics over continuous variables— and symbolic reachability analysis is performed on this model. This captures the intended semantics of the plant but usually the software-based controller is "abstracted away" to a mathematical function that is typically modeled as a second automaton running in parallel with the plant. While recent progress in reachability analysis for hybrid automata [14] have shown the potential for symbolic techniques to scale to systems with many continuous variables, the simplistic modeling of the controller, in particular, the omission of programming-language level features whose interaction with the physical system often leads to errors, makes it hard to provide any guarantees for the *implementation* of the feedback control system.

In the second approach, techniques used in analysis of programs, based on abstract interpretation, precisely model features of the controller program, but

A. Goodloe and S. Person (Eds.): NFM 2012, LNCS 7226, pp. 356–370, 2012.

usually "abstracts away" the plant's dynamics, assuming that the sensors can read arbitrary values from their range in each cycle. While tools like Astrée [4] and Fluctuat [17] have been successful in proving various safety properties of controller programs, such as the absence of arithmetic or buffer overflows, the absence of a plant model makes it hard to verify properties of the entire feedback control system that depend on the interaction between the plant and the controller.

In order to address the drawbacks in these two approaches, it is clear that analysis techniques need to model the full *closed-loop* control system —both the plant and the controller code— in analyzing embedded control software in cyber-physical systems. This need has been expressed before [7,16,3,11], and some tools to perform closed-loop *simulation* of feedback control systems have been developed recently (cf. [2,24,27]).

We present CLSE, a symbolic execution engine for feedback control systems. CLSE takes as input the description of a feedback control system in two parts: a plant model given as a set of linear ordinary differential equations, and a software implementation of a controller for the plant. For a given time horizon and a sampling rate for the controller, and a given set of initial states of interest for the plant, CLSE performs symbolic simulation of the plant and the controller up to the time horizon. The simulation is guaranteed to provide a complete coverage of the initial state set, that is, the bounded-time evolution of the system starting from *any* state in the initial state set is included in the simulation. The symbolic analyses of the controller uses concolic execution techniques (cf. [15,29]), together with decision procedures for non-linear arithmetic [20]. We have implemented CLSE for controller implementations in the C language and describe its application to examples of closed-loop control systems.

We also show how the symbolic execution engine of CLSE can be used to build additional analyzers for closed-loop control systems. We develop a symbolic robustness analyzer on top of CLSE. In addition to the closed-loop system and the time horizon, the robustness analyzer takes as input a bound on the disturbance on sensor measurements, and computes the maximum deviation between the plant state without disturbance and the plant state with disturbance up to the time horizon.

Outline of the paper. We first describe the class of closed-loop systems we address in our work in Section 2. CLSE's closed-loop analysis algorithm is then presented in Section 3. In Section 4 we show how such an analysis can be used to realize a robustness analyzer. Experimental evaluation of the analyses is provided on a few example closed-loop control systems in Section 5. We conclude the paper after a discussion of related work in Section 6.

2   Closed-Loop System Model

We consider the standard model of a closed-loop control system that is composed of a plant and a controller that are connected via sensors and actuators (see Fig. 1). The plant captures the continuous dynamics of the environment

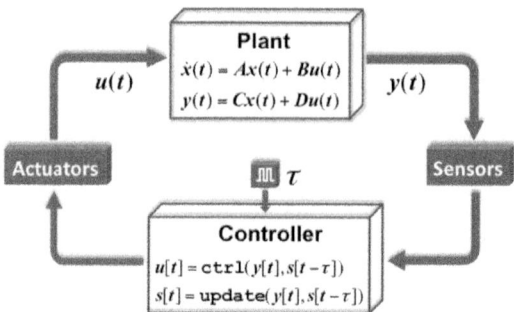

Fig. 1. Model of a closed-loop control system

that is being controlled and the controller represents the software program that implements the control algorithm.

In an execution of the closed-loop system, the state of the plant is sensed by the controller at discrete time instants, called sampling times, based on which the control inputs to the plant are computed. Our analysis performs a symbolic simulation of the system up to a bounded-time horizon, $T = N \times \tau$, where $N \in \mathbb{N}$ and τ is the sampling period of the system. In what follows, we characterize the dynamics of the two main components in detail.

2.1 Plant Dynamics

We consider linear dynamical systems

$$\dot{x}(t) = Ax(t) + Bu(t), \quad x(0) = x_0 \tag{1}$$

where $x(t) \in \mathbb{R}^n$ is the continuous state vector, A and B are $n \times n$ and $n \times m$ matrices, respectively, and the *control input* $u(t)$ is a piecewise continuous function from $\mathbb{R}^+$ to a convex, compact set $U \subseteq \mathbb{R}^m$ of control actions. A function $\xi : (0, \infty) \to \mathbb{R}^n$ is said to be a *trajectory* of the dynamical system if $\xi(0) = x_0$ and there exists a control input v such that $\dot{\xi}(t) = A\xi(t) + Bv(t)$ for almost all $t \in (0, \infty)$.

In a *sampled-data* control system, there is an a priori fixed sampling time $\tau \in \mathbb{R}^+$. The plant state is sensed at the end of each sampling period of τ time units. The control inputs are computed based on the plant state and applied to the plant at the beginning of the next sampling period. The control input remains constant throughout the next sampling period. Thus, control inputs in a sampled-data control system are piecewise constant curves of duration τ, i.e., a control input $v : \mathbb{R}^+ \to U$ satisfies $v(t) = v((k-1)\tau)$ for all $t \in [(k-1)\tau, k\tau)$ and $k \in \mathbb{N}$. For a sampled-data control system $\dot{x}(t) = Ax(t) + Bu(t)$ with sampling time τ and control actions U, and a given set of initial states X_0, we define $\mathcal{X}(X_0, U, Ax(t) + Bu(t), \tau)$ to be the set of all possible trajectories starting from some state in X_0.

2.2   Controller Dynamics

We assume that the controller of the closed-loop system is implemented as a software program. The controller program may have state variables that retain their values at the end of an execution to be used in the next execution. We model controller software in a simple imperative language. In our implementation we handle more general features such as pointers, arrays, and function calls.

We represent a controller program as a *control flow graph (CFG)* $P = (Y, \mathcal{L}, \ell_i, \ell_o, \mathsf{op}, E)$ consisting of (1) a set of variables Y, containing disjoint subsets Y_0 of *input* variables, Y_s of *state* variables, and Y_u of *output* variables, (2) a set of control locations (or program counters) $\mathcal{L}$ which includes a special start location $\ell_i \in \mathcal{L}$ and an output location $\ell_o \in \mathcal{L}$, (3) a function op labeling each location $\ell \in \mathcal{L}$ with one of the following basic operations:

- an assignment $y := e$, where $y \in Y$ and e is an arithmetic expression over Y, and
- a conditional **if** (e) **then** ℓ' **else** ℓ'', where e is a side-effect free expression and ℓ', ℓ'' are locations in $\mathcal{L}$

and (4) a set of directed edges $E \subseteq \mathcal{L} \times \mathcal{L}$ defined as follows. The set of edges E is the smallest set such that (i) every node ℓ where $\mathsf{op}(\ell)$ is an assignment statement has exactly one node ℓ' with $(\ell, \ell') \in E$, (ii) every node ℓ such that $\mathsf{op}(\ell)$ is **if** (e) **then** ℓ' **else** ℓ'' has two edges (ℓ, ℓ') and (ℓ, ℓ'') in E. For a location $\ell \in \mathcal{L}$ where $\mathsf{op}(\ell)$ is an assignment operation, we write $N(\ell)$ for its unique neighbor. Thus, the locations of a CFG correspond to program locations with associated commands, and edges correspond to control flow from one operation to the next. The program ends on reaching the location ℓ_o and outputs the values for all variables $u \in Y_u$. A *path* is a sequence of locations $\ell^1, \ell^2, \ldots, \ell^n$ in the CFG. A location $\ell \in \mathcal{L}$ is reachable from $\ell' \in \mathcal{L}$ if there is a path $\ell', \ldots, \ell$ in the CFG. We assume that every node in $\mathcal{L}$ is reachable from ℓ_i and ℓ_o is reachable from every node. Note that, even though we do not include a loop construct here, our implementation handles static control loops via unrolling.

The concrete semantics of the program is given using a *memory* that maps variables in Y to values. For a memory M, we write $M[y \mapsto v]$ for the memory mapping y to v and every other variable $z \in Y \setminus \{y\}$ to $M(z)$. For an expression e, we denote by $M(e)$ the value obtained by evaluating e where each variable y occurring in e is replaced by the value $M(y)$.

Execution starts from a memory M_0 containing initial values for input variables in Y_0, final values for the state variables in Y_s from the execution on the previously sampled plant output and constant default values for variables in Y_u, at the entry location ℓ_i. When the program runs for the first time, the values of the variables in Y_s are equal to their initial values. Each operation updates the memory and the control location. Suppose the current location is ℓ and the current memory is M. If $\mathsf{op}(\ell)$ is $y := e$, then the new location is $N(\ell)$ and the new memory is $M[y \mapsto M(e)]$. If $\mathsf{op}(\ell)$ is **if** (e) **then** ℓ' **else** ℓ'' then e is evaluated based on the current memory M. If the evaluated value is 0, then the new location is ℓ'', otherwise the new location is ℓ'. In either case, the memory remains unchanged. On reaching ℓ_o, the program terminates and outputs the values $M(v)$ for each $v \in Y_u$. Execution of the program starting from a memory M_0 defines a path in the CFG in a natural way. A path is executable if it is the path corresponding to program execution from some initial memory M_0.

3   Closed-Loop Symbolic Execution

In this section, we first discuss symbolic execution of the plant and concolic execution of the control program individually, and then discuss how their combination is handled by CLSE.

3.1   Symbolic Execution of the Plant

In sampled-data control systems, the evolution of the plant state over time can be viewed as a discrete-time sequence of sets $X_0, X_1, \ldots$, where X_0 is the set of initial states and X_i denotes the set of states reached in the ith sampling time (i.e., $t = i\tau$). Computation of the set X_i depends on the set of states X_{i-1} of the plant at the preceding sampling time, the continuous equations governing the dynamics of the plant, and the set of control inputs actuating the plant.

Suppose we are given a set X_k of states of a plant, corresponding to the sampling time $t = k\tau$, and suppose we let the plant evolve due to its continuous dynamics for one sampling period to reach the set X_{k+1}. For the given set of states X_k, the set of control inputs U_k, and the dynamics of the plant $\dot{x} = Ax(t) + Bu(t)$, we define $\texttt{Reach}(X_k, U_k, Ax(t) + Bu(t), \tau)$, the set of states X_{k+1} that the plant can be in after the elapse of a time period τ, as follows:

$$X_{k+1} = \texttt{Reach}(X_k, U_k, Ax(t) + Bu(t), \tau)$$
$$= \{\xi(\tau) \mid \xi \in \mathcal{X}(X_k, U_k, Ax(t) + Bu(t), \tau)\}.$$

The set X_{k+1} may be hard to compute and represent exactly. Thus, in practice, we approximate X_{k+1}.

We have tried two techniques in CLSE. First, using symbolic reachability techniques for linear systems for a bounded time interval (cf. [23]), we can get arbitrarily precise approximations but at high computation costs. Second, for relatively coarse, but computationally practical approximations, we used continuous-time to discrete-time model conversion based on the *zero-order hold* method [12]. Since the control inputs in our system model are piecewise constant over the sampling period, for a given continuous time dynamics of the plant $\dot{x} = Ax(t) + Bu(t)$, and a sampling period, the method provides the discrete-time dynamics of the plant defined by $x[k+1] = A_d x[k] + B_d u[k]$.

3.2   Concolic Execution of the Controller

For closed-loop analysis of a control system we need to have the symbolic outputs and symbolic path constraints for all possible executable paths in the controller program. We obtain them by analyzing the controller program via *concolic execution*, in which the program is executed on symbolic inputs in addition to concrete inputs [15,29]. The concolic execution algorithm executes the program while maintaining two additional artifacts: a *symbolic memory* μ which maps variables in Y to symbolic expressions over a set of symbolic constants, and a *path constraint* κ, which collects predicates over symbolic constants along the execution path. The symbolic memory map and the symbolic path constraint are updated during the course of execution.

Concolic execution proceeds as follows. Starting at location ℓ_i in the control flow graph, the symbolic memory μ maps each input variable $y \in (Y_0 \cup Y_s)$ to a fresh symbolic constant α_y and each variable $z \in Y \setminus (Y_0 \cup Y_s)$ to some default constant value. Initially, the path constraint is *true*.

For an assignment $y := e$ at location ℓ, the symbolic memory μ is updated to $\mu[y \mapsto \mu(e)]$, where $\mu(e)$ denotes the symbolic expression obtained by evaluating e using the map μ. The path constraint is unchanged. The control location is updated to $N(\ell)$. For a conditional **if** (e) **then** ℓ' **else** ℓ'' at location ℓ, if none of these branches has been explored with a path that agrees with the current path up to location ℓ, a branch is arbitrarily chosen, otherwise the branch which has not been explored is taken. Based on which branch is chosen, the control location is updated to either ℓ' or ℓ''. If the new control location is ℓ', the path constraint is updated to $\kappa \wedge \mu(e) \neq 0$, and if the new control location is ℓ'', the path constraint is updated to $\kappa \wedge \mu(e) = 0$. In each case, the new symbolic memory is still μ. Execution ends when the control location is ℓ_o. At this point, κ is the path constraint, and the restriction $\mu|_{Y_u \cup Y_s}$ maps each $y \in Y_u \cup Y_s$ to $\mu(y)$. We denote $\mu|_{Y_u \cup Y_s}$ by λ.

At the end of an execution, a new execution is created by selecting a conditional $\ell :$ **if** (e) **then** ℓ' **else** ℓ'' along the path that was executed such that (1) the current execution took the **then** (respectively, **else**) branch of the conditional, and (2) the path that agrees with the current execution up to ℓ but then takes the **else** (respectively, **then**) branch of this conditional has not been explored before. In this way, each control path in the program is explored. At the end of symbolic execution, we get the set *controllerPaths* of tuples $\langle \kappa, \lambda \rangle$ of path constraints and output maps for each explored path.

3.3   Combining the Two

Our closed-loop execution is based on the interaction of the symbolic execution of the plant and concolic execution of the controller implemented in software. A single iteration in our execution begins at the point the plant's state is sensed and ends after the plant evolves for one sampling period based on the controller's actions on the sensed data. We make the usual assumption that the time taken by the controller is negligible when compared to the chosen sampling period.

The algorithm for closed-loop execution is outlined in `closedLoopExecution` function in Algorithm 1. It takes the following as inputs: (1) t, the time instant at the current sampling (an integer multiple of τ), (2) X_t, the set of states of the plant at time instant t, (3) $S_{t-\tau}$, the set of states of the controller at time instant

---

Algorithm 1. Closed-loop execution of a system.

Input: A closed-loop system with the dynamics of the plant captured as *flow* and the controller captured in *controllerPaths*, sampling time τ, simulation time bound T, initial states of the plant X_0 and initial states of the controller $S_{-\tau}$.

Output: Reach set sequences and path sequences that cover the initial states up to time T.

function closedLoopExecution($t, X_t, S_{t-\tau}, reachSetSeq, pathSeq$)

begin

    $intersectingPaths \leftarrow \{\langle \kappa, \lambda \rangle | \langle \kappa, \lambda \rangle \in controllerPaths, (X_t \wedge S_{t-\tau} \wedge \kappa)$ is satisfiable$\}$

    foreach $\langle \kappa, \lambda \rangle \in intersectingPaths$ **do**

        $X_{\text{init}} \leftarrow \exists Y_s, X_t(Y_0) \wedge S_{t-\tau}(Y_0, Y_s) \wedge \kappa(Y_0, Y_s)$

        $U_t \leftarrow \{u | u = \lambda(y)(x, s), y \in Y_u, x \in X_t, s \in S_{t-\tau}\}$

        $S_t \leftarrow \{s' | s' = \lambda(y)(x, s), y \in Y_s, x \in X_t, s \in S_{t-\tau}\}$

        $X_{t+\tau} \leftarrow$ Reach$(X_{\text{init}}, U_t, flow, \tau)$

        $reachSetSeq' \leftarrow$ append$(reachSetSeq, X_{t+\tau})$

        $pathSeq' \leftarrow$ append$(pathSeq, \langle \kappa, \lambda \rangle)$

        if $t + \tau < T$ **then**

            closedLoopExecution$(t + \tau, X_{t+\tau}, S_t, reachSetSeq', pathSeq')$

        else

            $reachSetSequences \leftarrow reachSetSequences \cup \{reachSetSeq'\}$

            $pathSequences \leftarrow pathSequences \cup \{pathSeq'\}$

begin

    global $reachSetSequences \leftarrow \emptyset$

    global $pathSequences \leftarrow \emptyset$

    closedLoopExecution$(0, X_0, S_{-\tau}, [X_0], [])$

    return $(reachSetSequences, pathSequences)$

$t - \tau$, (4) *reachSetSeq*, the sequence of plant's state sets $X_0, X_\tau, \ldots, X_t$ reached in the current execution at successive sampling instants until time t, and (5) *pathSeq*, the sequence of controller paths traversed in the current execution. Initially, closedLoopExecution function is called with $t = 0$, $X_t = X_0$, where X_0 is the set of initial states of the plant, $S_{t-\tau} = S_{-\tau}$, where $S_{-\tau}$ is the set of initial states of the controller, *reachSetSeq* $= [X_0]$ and *pathSequences* $= []$.

At any time instant t, the closedLoopExecution function executes in the following manner. The algorithm first identifies *intersectingPaths*, a subset of paths in the controller program (*controllerPaths*), that can be executed starting from some state in X_t and $S_{t-\tau}$. This is computed by checking the satisfiability of $X_t \wedge S_{t-\tau} \wedge \kappa$, for each *path* $= \langle \kappa, \lambda \rangle \in controllerPaths$. Satisfiability implies that there exists a state with $x \in X_t$ and $s \in S_{t-\tau}$ that can execute the controller program along *path*. For each controller path, *path* $= \langle \kappa, \lambda \rangle$, we first compute three sets: X_{init}, U_t and S_t. The set X_{init} denotes the set of initial states of the plant for the evolution in the next sampling period when *path* is executed

---

Algorithm 2. Robustness analysis of a closed-loop system.

Input: A closed-loop system with the dynamics of the plant captured as *flow* and the controller captured in *controllerPaths*, sampling time τ, simulation time bound T, initial states of the plant X_0 and initial states of the controller $S_{-\tau}$ and an upper-bound ε on the sensor errors.

Output: An upper-bound Δ on the deviation in the plant's states after time T.

function computeDeviation($t, X_t, S_{t-\tau}, X'_t, S'_{t-\tau}, \delta_t$)

begin

$\quad X_t^\varepsilon \leftarrow X'_t \oplus \mathcal{E}$

$\quad intersectingPaths \leftarrow \{\langle \kappa, \lambda \rangle | \langle \kappa, \lambda \rangle \in controllerPaths, (X_t \wedge S_{t-\tau} \wedge \kappa) \text{ is satisfiable}\}$

$\quad intersectingPaths^\varepsilon \leftarrow \{\langle \kappa^\varepsilon, \lambda^\varepsilon \rangle |$
$\quad\quad\quad\quad \langle \kappa^\varepsilon, \lambda^\varepsilon \rangle \in controllerPaths, (X_t^\varepsilon \wedge S'_{t-\tau} \wedge \kappa^\varepsilon) \text{ is satisfiable}\}$

$\quad$ **foreach** $\langle \kappa, \lambda \rangle \in intersectingPaths$ **do**

$\quad\quad X_{\text{init}} \leftarrow \exists Y_s, X_t(Y_0) \wedge S_{t-\tau}(Y_0, Y_s) \wedge \kappa(Y_0, Y_s)$

$\quad\quad U_t \leftarrow \{u | u = \lambda(y)(x, s), y \in Y_u, x \in X_t, s \in S_{t-\tau}\}$

$\quad\quad S_t \leftarrow \{s' | s' = \lambda(y)(x, s), y \in Y_s, x \in X_t, s \in S_{t-\tau}\}$

$\quad\quad X_{t+\tau} \leftarrow \text{Reach}(X_{\text{init}}, U_t, flow, \tau)$

$\quad\quad$ **foreach** $\langle \kappa^\varepsilon, \lambda^\varepsilon \rangle \in intersectingPaths^\varepsilon$ **do**

$\quad\quad\quad X_{\text{init}}^\varepsilon \leftarrow \exists Y_s, X'_t(Y_0) \wedge S'_{t-\tau}(Y_0, Y_s) \wedge (\kappa(Y_0, Y_s) \oplus \mathcal{E})$

$\quad\quad\quad U_t^\varepsilon \leftarrow \{u' | u' = \lambda^\varepsilon(y)(x, s), y \in Y_u, x \in X'_t, s \in S'_{t-\tau}\}$

$\quad\quad\quad S_t' \leftarrow \{s'' | s'' = \lambda^\varepsilon(y)(x, s), y \in Y_s, x \in X'_t, s \in S'_{t-\tau}\}$

$\quad\quad\quad X'_{t+\tau} \leftarrow \text{Reach}(X_{\text{init}}^\varepsilon, U_t^\varepsilon, flow, \tau)$

$\quad\quad\quad \delta_{t+\tau} \leftarrow \text{findOutputDeviation}(X_t, X'_t, X_{t+\tau}, X'_{t+\tau}, \langle \kappa, \lambda \rangle, \langle \kappa^\varepsilon, \lambda^\varepsilon \rangle, \delta_t)$

$\quad\quad\quad$ **if** $t + \tau < T$ **then**

$\quad\quad\quad\quad$ computeDeviation($t + \tau, X_{t+\tau}, S_t, X'_{t+\tau}, S'_t, \delta_{t+\tau}$)

$\quad\quad\quad$ **else**

$\quad\quad\quad\quad \Delta \leftarrow \max(\Delta, \delta_{t+\tau})$

begin

$\quad$ **global** $\Delta \leftarrow 0$

$\quad$ computeDeviation($0, X_0, S_{-\tau}, X_0, S_{-\tau}, 0$)

$\quad$ **return** Δ

---

in the controller program due to the current sampled states of the plant. The sets U_t and S_t denote the next set of control inputs to the plant and the set of controller states at time t, respectively. Now we let time evolve for τ units to reach sampling time $t + \tau$ and obtain the set $X_{t+\tau}$ of the plant's states by applying Reach to X_{init} and U_t. The set $X_{t+\tau}$ thus computed is appended to the current sequence of reach sets *reachSetSeq'*, and the current controller path, *path*, is appended to the current sequence of paths *pathSeq'*.

If $t + \tau$ is less than the time horizon $T = N\tau$, we repeat the algorithm recursively, otherwise, the current sequence of reach sets *reachSetSeq'* is added to the set *reachSetSequences* of all state set sequences and the current sequence of paths *pathSeq* is added to the set *pathSequences* of all path sequences.

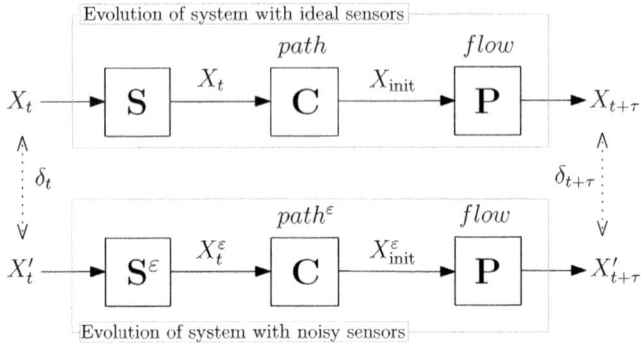

Fig. 2. Snapshot of an iteration instance in the computation of $\delta_{t+\tau}$

4   Closed-Loop Robustness Analysis

We now describe a closed-loop robustness analyzer based on CLSE. Given a closed-loop control system, the robustness analyzer carries out two symbolic executions over plant's state variables: (1) the *reference evolution* in which all sensors are ideal so that they do not introduce any numerical errors in the inputs to the controller; and (2) the *perturbed evolution* in which all sensors can possibly introduce errors that are bounded by ε. The objective is to compute the maximum deviation Δ in the state of the plant due to sensor errors for a given set of initial states up to a bounded time horizon T (which is a multiple of τ).

The computation of the deviation Δ is outlined in Algorithm 2. Figure 2 shows one step of the computation. On the top, we perform closed loop symbolic execution on the plant and controller as described in the previous section. At the bottom, we again perform closed loop symbolic execution, but assume that instead of reading X_t' for the set of states, there is some sensor error that adds additional "noise" to the measurement, which leads to execution of an erroneous path in the controller, $path^\varepsilon$, instead of $path$. Hence, instead of continuing the simulation with X_t', we continue with X_t^ε defined as follows.

We assume a simple sensor error model where each variable in the plant state x is sensed by a sensor with an error bound of ε. Then the bounds on the perturbations due to individual sensors define an *error box* $\mathcal{E} = [-\varepsilon, +\varepsilon]^n$. The perturbed set X_t^ε is defined by the Minkowski sum of X_t' and the error box, that is, $X_t^\varepsilon = X_t' \oplus \mathcal{E}$.

The closed-loop symbolic execution proceeds as previously described for the reference and perturbed evolutions by identifying the intersecting paths based on the sets X_t and X_t^ε, respectively. For each pair of intersecting paths from the two evolutions, the deviation in the next plant states, $\delta_{t+\tau}$, between $X_{t+\tau}$ and $X_{t+\tau}'$, is computed based on the deviation δ_t from the previous iteration. If $t + \tau < T$, we repeat the algorithm recursively. We now describe the different operations used in the above computation.

Updating the Initial Sets. In each iteration of plant's evolution, we need to provide the initial set that is a subset of the reach set of the plant in the previous

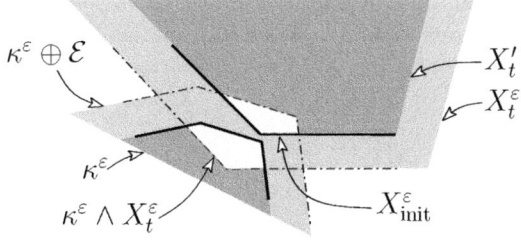

Fig. 3. Computation of the set $X_{\text{init}}^{\varepsilon}$

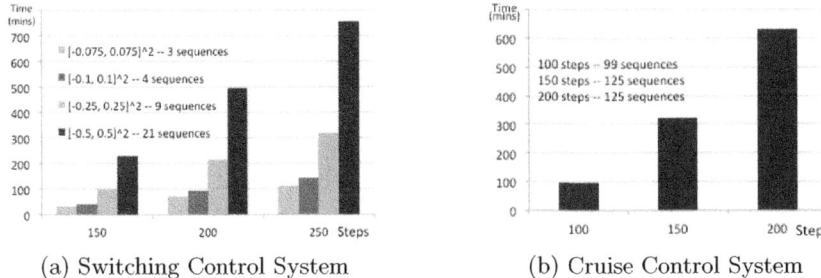

(a) Switching Control System   (b) Cruise Control System

Fig. 4. Performance of CLSE on increasing initial state set sizes and simulation lengths for example systems

iteration. For the reference evolution, the controller senses the set X_t and the initial set is computed as described previously in Section 3.3 for *path*, the current path of the controller. For the perturbed evolution, however, the controller senses the set X_t^{ε} that is obtained by adding the effect of sensor noise $\mathcal{E}$ to the plant's reach state X_t'. Computing the initial set of states for the perturbed evolution has to take into account the fact that, *path*$^{\varepsilon}$, the current path of the controller is due to erroneously sensed states of the plant. Therefore, we need to identify the subset of states in X_t^{ε}, which when perturbed lead to the selection of *path*$^{\varepsilon}$, as follows. We first bloat the set κ^{ε} with $\mathcal{E}$ to obtain the set $\kappa^{\varepsilon} \oplus \mathcal{E}$. The intersection of this set with the reach set of the plant X_t' gives the set $X_{\text{init}}^{\varepsilon}$, which is the initial set for the next execution of the plant for the perturbed evolution. The different sets described above are illustrated in Fig. 3.

Finding Deviation in the Plant's State. The function findOutputDeviation computes the deviation in plant's state for the reference and perturbed evolution. It takes as inputs: (1) two sets of states of plant variables, X_t and X_t', at time instant t for the reference and perturbed evolutions, respectively, (2) two sets of states of plant variables, $X_{t+\tau}$ and $X_{t+\tau}'$, at time instant $t+\tau$ for the reference and perturbed evolutions, respectively, (3) the pair of intersecting paths, *path* and *path*$^{\varepsilon}$, and (4) the current deviation δ_t, for a state $x_0 \in X_0$ between the two evolutions at time instant t. The computation of the deviation due to elapse of a time period τ is based on the relation between states in the sets X_t

and $X_{t+\tau}$ that is given by a discretization of the plant dynamics in Equation 1. Suppose that the discrete-time dynamics is given by: $x[k+1] = A_d x[k] + B_d u[k]$, where $k \in \mathbb{N}$, then the computation of the deviation is formalized as a constraint solving problem as below:

$$\delta_{t+\tau} := |x_{t+\tau} - x'_{t+\tau}|$$
$$\text{where, } x_{t+\tau} = A_d x_t + B_d u_t, \quad \text{and} \quad x'_{t+\tau} = A_d x'_t + B_d u'_t,$$
$$path = \langle \kappa, \lambda \rangle \in intersectingPaths,$$
$$path^\varepsilon = \langle \kappa^\varepsilon, \lambda^\varepsilon \rangle \in intersectingPaths^\varepsilon,$$
$$u_t = [\lambda(y_1), \ldots, \lambda(y_m)]^T,$$
$$u'_t = [\lambda^\varepsilon(y_1), \ldots, \lambda^\varepsilon(y_m)]^T, \text{where } Y_u = \{y_1, \ldots y_m\},$$
$$x_{t+\tau} \in X_{t+\tau} \quad \text{and} \quad x'_{t+\tau} \in X'_{t+\tau},$$
$$|x_t - x'_t| = \delta_t, \quad x_t \in X_{\text{init}} \quad \text{and} \quad x'_t \in X^\varepsilon_{\text{init}}.$$

The expression $|x_{t+\tau} - x'_{t+\tau}|$ denotes the vector with components obtained by taking the absolute values of the difference of the corresponding components in $x_{t+\tau}$ and $x'_{t+\tau}$.

Updating the Maximum Deviation. After the computation of $\delta_{t+\tau}$, if $t + \tau$ reaches the simulation time bound T, we compare $\delta_{t+\tau}$ with Δ that is the maximum deviation found so far. If $\delta_{t+\tau} > \Delta$, Δ is updated to $\delta_{t+\tau}$.

5   Experiments

We have implemented CLSE using c2d, a model discretization routine provided by MathWorks' Control System Toolbox, to capture plant dynamics, and Splat [31] for concolic execution of programs. In addition, CLSE calls upon iSAT [20] as the constraint solver of choice. We have run experiments using our implementation on a 64-bit Linux on a machine with Intel Xeon X5650 2.66GHz processor and 48GB memory. In what follows, we present performance of CLSE for two examples of feedback-control systems, followed by the results from the robustness analyzer for one of them.

Example 1: Switching Control System. We evaluate CLSE first on an example of a switching control system that is representative of the class of systems that we handle. In this system, obtained from the literature [10], depending on the current state of the plant, one of two controller gains is selected. The plant is represented as a transfer function and the controller uses an integrator. We have substituted the transfer function model of the plant by its equivalent state-space model, and the continuous integrator in the controller by a discrete-time integrator. The plant model has two continuous state variables and the controller (about 50 lines of C code) has two paths. The performance of CLSE on this example is as shown in Fig. 4(a) for increasing sizes of initial state sets. We observe that, in this example, the number of unique path sequences that covers each of the selected initial state sets does not increase with increase in the simulation

Table 1. The dimension-wise maximum deviation computed by our robustness analyzer and the time taken to compute them

No. of steps	Δ for $[-0.1, 0.1]^2$		time (mins)	Δ for $[-0.25, 0.25]^2$		time (mins)	Δ for $[-0.5, 0.5]^2$		time (mins)
	x_1	x_2		x_1	x_2		x_1	x_2	
50	0.041	0.059	20.28	0.053	0.064	40.12	0.071	0.084	65.12
75	0.068	0.081	60.13	0.076	0.093	74.07	0.134	0.151	91.27

length. This is due to the fact that the control system stabilizes quickly under the effect of the controller, and after that the controller can execute only one control path corresponding to the stable state of the plant.

Example 2: Cruise Control System. This automotive system is obtained from the demonstration suite of the Reactis test-generation tool [28]. The plant model has four inputs and one continuous state variable *speed*, whereas the controller has eight inputs and two state variables. The controller code is about 200 lines, with 24 unique control paths. Among the inputs to the controller, 6 are user inputs that are Boolean, which leads to 2^6 configurations in which the closed-loop system operates. We have simulated the system for all the configurations for varying simulation lengths for speed in the range $[0, 80]$. The worst case simulation time, which is observed for 2 of the 2^6 configurations, for each simulation length is as shown in Fig. 4(b).

Robustness of the Switching Control System. For robustness analysis of the switching control system, we choose an error bound of $[-0.01, 0.01]^2$ on the sensors for the two state of the plant. The maximum bounds obtained by our implementation for simulation lengths 50 and 75, for increasing sizes of initial sets, are as shown in Table 1.

6 Related Work

CLSE is similar to several recent projects in model-based testing and verification of hybrid systems. Alur *et al.* [2,21] have proposed a method for symbolic simulation of closed-loop Simulink/Stateflow (SL/SF) models for the purpose of test-case generation. Our method bears a resemblance to this work in that both adopt the same notion of equivalence of closed-loop trajectories and provide coverage over the space of initial states of the plant. However, CLSE implements a forward analysis that guarantees full coverage for the selected initial set of states unlike the backward analysis implemented in [2]; and CLSE directly handles the software implementation of the controller.

Lerda *et al.* [24] present a closed-loop analysis technique to find bugs in control software which is coupled with a continuous plant. They perform systematic simulation of the controller program using an explicit state software model checker and perform numerical simulations on the plant in the Simulink environment. In a similar vein, Păsăreanu *et al.* [27] propose a framework which supports translation of different modeling formalisms, including SL/SF, into a common intermediate representation and then use model checking and symbolic execution tools for property verification and test-case generation. Given a set of initial states of the system, our objective is not to model-check the controller software, but to compute the set of all sequences of paths executed in the controller software due to the closed-loop evolution up to a bounded time. Due to this, CLSE provides coverage over the input space of the system and not over the structure of the controller software.

HybridFluctuat [5] is a static analyzer for closed-loop systems that deals with software implementations directly just as CLSE does. It provides an assertion-based mechanism for specifying interaction of the software controller with the plant that can help build custom analyzers. It is based on Fluctuat [17] and leverages techniques from abstract interpretation in order to deal with path explosion.

Now we turn to related work on robustness analysis. Robustness of control system has been studied widely in the last thirty years (cf. [32]). There are software tools available to help design robust control systems [22]. However, the above-mentioned theory and software tools only help in analyzing a mathematical model of the control system. They do not consider the case when the controller is implemented as software. Analyzing software programs for robustness has been undertaken in a few recent works [25,6], where controller programs are analyzed independently without the presence of the plant.

Robustness of a trajectory of a system has been studied in the recent past. Fainekos and Pappas [9] introduce a notion of robust satisfaction of a Linear or Metric Temporal Logic formula which is interpreted over finite timed state sequences in some metric space. Fainekos et al. [10] present a framework for reporting points where a simulation of a Simulink model may not be robust in the presence of both uncertainties in the model and internal computation errors. Robustness of hybrid automaton models for control systems have been studied before [18,13]. However, unlike these works, our algorithm and tool provides a quantitative guarantee on the robustness of the system's output through the reachability analysis of the continuous plant, and program path exploration by symbolic execution of the controller program.

7   Conclusions

We believe CLSE is a step toward closed-loop static analyzers that incorporate both plant and software dynamics into the analysis. Our experiences suggest that symbolic execution-based techniques, while precise, suffer from path explosion. It will be interesting to design an algorithm combining symbolic execution (for precision) and abstract interpretation (for scalability). In particular, we would like to study how path merging techniques that have been developed in the abstract interpretation setting can help scale our method. While our tool accepts the description of the plant as a set of linear differential equations, it is possible (but not trivial) to "compile" a Simulink/Stateflow model of the system to such a description [30,26]. Adding support for non-linear systems is also an open direction.

References

1. Alur, R., Courcoubetis, C., Halbwachs, N., Henzinger, T., Ho, P.-H., Nicollin, X., Olivero, A., Sifakis, J., Yovine, S.: The algorithmic analysis of hybrid systems. Theoretical Computer Science 138, 3–34 (1995)
2. Alur, R., Kanade, A., Ramesh, S., Shashidhar, K.C.: Symbolic analysis for improving simulation coverage of Simulink/Stateflow models. In: de Alfaro, L., Palsberg, J. (eds.) EMSOFT, pp. 89–98. ACM (2008)
3. Anta, A., Majumdar, R., Saha, I., Tabuada, P.: Automatic verification of control system implementations. In: EMSOFT, pp. 9–18. ACM (2010)
4. Blanchet, B., Cousot, P., Cousot, R., Feret, J., Mauborgne, L., Miné, A., Monniaux, D., Rival, X.: A Static Analyzer for Large Safety-Critical Software. In: PLDI (2003)
5. Bouissou, O., Goubault, E., Putot, S., Tekkal, K., Vedrine, F.: HybridFluctuat: A Static Analyzer of Numerical Programs within a Continuous Environment. In: Bouajjani, A., Maler, O. (eds.) CAV 2009. LNCS, vol. 5643, pp. 620–626. Springer, Heidelberg (2009)
6. Chaudhuri, S., Gulwani, S., Lublinerman, R., Navidpour, S.: Proving programs robust. In: SIGSOFT FSE, pp. 102–112. ACM (2011)
7. Cousot, P.: Integrating Physical Systems in the Static Analysis of Embedded Control Software. In: Yi, K. (ed.) APLAS 2005. LNCS, vol. 3780, pp. 135–138. Springer, Heidelberg (2005)
8. Dang, T., Le Guernic, C., Maler, O.: Computing Reachable States for Nonlinear Biological Models. In: Degano, P., Gorrieri, R. (eds.) CMSB 2009. LNCS, vol. 5688, pp. 126–141. Springer, Heidelberg (2009)
9. Fainekos, G.E., Pappas, G.J.: Robustness of Temporal Logic Specifications. In: Havelund, K., Núñez, M., Roşu, G., Wolff, B. (eds.) FATES/RV 2006. LNCS, vol. 4262, pp. 178–192. Springer, Heidelberg (2006)
10. Fainekos, G.E., Sankaranarayanan, S., Ivančić, F., Gupta, A.: Robustness of model-based simulations. In: IEEE RTSS, pp. 345–354 (2009)
11. Feron, E.: From control systems to control software. IEEE Control Systems Magazine 30(6), 50–71 (2010)
12. Franklin, G.F., Powell, D.J., Workman, M.: Digital Control of Dynamic Systems. Prentice Hall (1997)
13. Frazzoli, E., Dahleh, M., Feron, E.: Robust hybrid control for autonomous vehicle motion planning. In: Proceedings of IEEE Conference on Decision and Control, vol. 1, pp. 821–826. IEEE (2000)
14. Frehse, G., Le Guernic, C., Donzé, A., Cotton, S., Ray, R., Lebeltel, O., Ripado, R., Girard, A., Dang, T., Maler, O.: SpaceEx: Scalable Verification of Hybrid Systems. In: Gopalakrishnan, G., Qadeer, S. (eds.) CAV 2011. LNCS, vol. 6806, pp. 379–395. Springer, Heidelberg (2011)
15. Godefroid, P., Klarlund, N., Sen, K.: Dart: directed automated random testing. In: Sarkar, V., Hall, M.W. (eds.) PLDI, pp. 213–223. ACM (2005)
16. Goubault, E., Martel, M., Putot, S.: Some future challenges in the validation of control systems. In: ERTS 2006 (2006)
17. Goubault, É., Putot, S., Baufreton, P., Gassino, J.: Static Analysis of the Accuracy in Control Systems: Principles and Experiments. In: Leue, S., Merino, P. (eds.) FMICS 2007. LNCS, vol. 4916, pp. 3–20. Springer, Heidelberg (2008)
18. Gupta, V., Henzinger, T., Jagadeesan, R.: Robust Timed Automata. In: Maler, O. (ed.) HART 1997. LNCS, vol. 1201, pp. 331–345. Springer, Heidelberg (1997)

19. Henzinger, T., Ho, P.-H., Wong-Toi, H.: HyTech: a model checker for hybrid systems. Software Tools for Technology Transfer 1, 110–122 (1997)
20. iSAT solver, AVACS project, http://isat.gforge.avacs.org
21. Kanade, A., Alur, R., Ivančić, F., Ramesh, S., Sankaranarayanan, S., Shashidhar, K.C.: Generating and Analyzing Symbolic Traces of Simulink/Stateflow Models. In: Bouajjani, A., Maler, O. (eds.) CAV 2009. LNCS, vol. 5643, pp. 430–445. Springer, Heidelberg (2009)
22. Kao, C.Y., Megretzki, A., Jonsson, U., Rantzer, A.: A MATLAB toolbox for robustness analysis. In: Computer-Aided Control Systems Design. IEEE (2004)
23. Le Guernic, C., Girard, A.: Reachability analysis of linear systems using support functions. Nonlinear Analysis: Hybrid Systems 4(2), 250–262 (2010)
24. Lerda, F., Kapinski, J., Maka, H., Clarke, E., Krogh, B.: Model checking in-the-loop: Finding counterexamples by systematic simulation. In: ACC (2008)
25. Majumdar, R., Saha, I.: Symbolic robustness analysis. In: IEEE RTSS (2009)
26. Manamcheri, K., Mitra, S., Bak, S., Caccamo, M.: A step towards verification and synthesis from Simulink/Stateflow models. In: HSCC (2011)
27. Păsăreanu, C.S., Schumann, J., Mehlitz, P., Lowry, M., Karsai, G., Nine, H., Neema, S.: Model based analysis and test generation for flight software. In: 3rd Intl. Conf. on Space Mission Challenges for IT, pp. 83–90. IEEE (2009)
28. Reactis, Reactive Systems, http://www.reactive-systems.com
29. Sen, K., Marinov, D., Agha, G.: Cute: a concolic unit testing engine for c. In: Wermelinger, M., Gall, H. (eds.) ESEC/SIGSOFT FSE, pp. 263–272. ACM (2005)
30. Tiwari, A.: Formal semantics and analysis methods for Simulink/Stateflow models. Technical report. SRI International (2002)
31. Xu, R.-G., Godefroid, P., Majumdar, R.: Testing for buffer overflows with length abstraction. In: Ryder, B.G., Zeller, A. (eds.) ISSTA, pp. 27–38. ACM (2008)
32. Zhou, K., Doyle, J.C.: Essentials of Robust Control. Prentice-Hall (1998)

On the Development and Formalization of an Extensible Code Generator for Real Life Security Protocols

Michael Backes[1,2], Alex Busenius[1], and Cătălin Hrițcu[1,3]

[1] Saarland University
[2] MPI-SWS
[3] University of Pennsylvania

Abstract. This paper introduces Expi2Java, a new code generator for crypto-graphic protocols that translates models written in an extensible variant of the Spi calculus into executable code in a substantial fragment of Java, featuring concurrency, synchronization between threads, exception handling and a sophisticated type system with generics and wildcards. Our code generator is highly extensible and customizable, which allows it to generate interoperable implementations of complex real life protocols from detailed verified specifications. As a case study, we have generated an interoperable implementation of TLS v1.0 client and server from a protocol model verified with ProVerif. Furthermore, we have formalized the translation algorithm of Expi2Java using the Coq proof assistant, and proved that the generated programs are well-typed if the original models are well-typed. This constitutes an important step towards the first machine-checked correctness proof of a code generator for cryptographic protocols.

1 Introduction

Implementing cryptographic protocols is a difficult and notoriously error-prone task, where even the smallest error can cause very serious security vulnerabilities[1]. One way to prevent many such vulnerabilities is to model the protocol in a process calculus [4], check the security of the model [3], and then automatically generate a secure implementation from the protocol model. Automatic tools exist that can generate protocol implementations starting from such verified models together with configuration information that allows them to produce code that is interoperable with other implementations of the same protocol.

This paper introducesExpi2Java[2], a new tool that brings code generators for security protocols even closer to reality. Expi2Java is highly extensible and customizable: The user can easily add new cryptographic primitives, configure all the relevant parameters, customize implementation classes, and even customize the code generation process by editing the provided templates. Commonly used cryptographic primitives and data types are supported out of the box, and the user has full control not only over the high-level

---

[1] For example, see the security advisories of OpenSSL
(http://www.openssl.org/news/), a well-known implementation of various SS-L/TLS versions.
[2] http://www.infsec.cs.uni-saarland.de/projects/expi2java

A. Goodloe and S. Person (Eds.): NFM 2012, LNCS 7226, pp. 371–387, 2012.

design of the protocol, but also over all the low-level details, including the precise bit-level representation of messages. To illustrate the flexibility of our tool we have generated an interoperable implementation of TLS v1.0 client and server from a protocol specification verified with ProVerif [3, 9]. TLS is orders of magnitude more complex and sophisticated than the protocols used by the previous code generation experiments.

The types in the generated Java code cannot be synthesized out of thin air, so we ask the user to provide type annotations while writing the protocol model. Our code generator uses the type information in the model to generate the typing annotations needed by the Java type-checker. Additionally, it is important to detect all the typing errors as soon as possible, before the code generation phase, so that we can guide the user with helpful error messages. Our tool uses a type-checker with variant parametric types and subtyping to prevent the incorrect usage of cryptographic primitives in the protocol model. This source-level type-checker rejects models that could lead to ill-typed Java code early on, and produces errors in terms of the model the user has actually written, not in terms of the automatically generated code that the user would have a hard time understanding. Moreover, our source-level type-checker performs type inference, which greatly decreases the annotation burden on the user.

All the features that make Expi2Java a usable and useful tool in practice come at a price though: the tool is rather complex; it currently has more than 16.000 lines of Java code, and this code needs to be trusted to generate correct implementations that preserve the security of the original protocol models. In this paper we take the first step towards formally bridging the gap between secure protocol models and automatically generated protocol implementations: We provide the first mechanized formalization of a code generator for cryptographic protocols. And we do this for an idealized code generator that is fairly close to the core of our Expi2Java tool, without "sweeping under the carpet" the interesting details. We formalize the semantics of both the source language of the code generator, an extensible and customizable variant of the Spi calculus, and the target language, which is a substantial fragment of the Java programming language featuring concurrency, synchronization between threads, exception handling and a sophisticated type system with generics and wildcards. We formally prove that our translation generates well-typed programs when starting from well-typed protocol models. This confirms the validity of our source-level type-checker, and constitutes an important step towards the longer-term goal of proving the correctness of the translation.

Outline. §2 discusses related work. §3 introduces the features and workflow of our tool while §4 reports on the TLS case study. §5 presents our source language, the Extensible Spi calculus. §6 describes the formalization of our target language, Variant Parametric Jinja. In §7 we give a high-level overview of our translation. In §8 we discuss our formalization, proofs, and some of the lessons we have learned. Finally, §9 discusses directions for future work and concludes. The implementation and documentation of Expi2Java are available online[3], together with the Coq formalization and proofs. The details that had to be omitted in this paper due to the lack of space are available in an online appendix.

---

[3] http://www.infsec.cs.uni-saarland.de/projects/expi2java

2 Related Work

The idea of a code generator for protocol implementations is not new, and several such tools were developed in the past. One of the early approaches is the AGVI toolkit [21] that uses the Athena protocol analyzer, and can generate Java implementations. The CIL2Java tool [20] can also generate Java code from the CAPSL intermediate language CIL. SPEAR II [18] is a tool aimed at rapid protocol engineering. It has a graphical user interface and generates Java code that uses the ASN.1 standard for data encoding and various cryptographic libraries. The Sprite tool [25] translates simple protocols written in "standard notation" to the Spi calculus from which it can generate Java code. The CPPL compiler [19] generates OCaml code that can be deployed on stock Web servers.

Spi2Java [22,23] is the first code generator that attempts to be flexible and configurable. It uses the original Spi calculus [4] as the input language together with a very simple type system, and provides a (quite primitive) way to specify the low-level information needed for generating interoperable protocol implementations. The generated implementation of a simple SSH client [22] demonstrates that interoperability with standard implementations is indeed achievable. Nevertheless, the practicality of Spi2Java is quite limited: its customization mechanism is very hard to use, and requires manually editing huge highly-redundant XML files for associating configuration information to individual terms. Also, Spi2Java cannot handle all the cryptographic primitives involved in really complex protocols like TLS, and has no extension mechanism to circumvent this problem — this lead us to develop Expi2Java, initially as an extension of Spi2Java, but rewritten by now. Moreover, none of the existing tools has a thorough formalization of their translation or a proof that their translation preserves any interesting property of the original model.

We have based our formalization of the target language on Jinja with Threads [16,17], since, to the best of our knowledge, this is the most complete and thorough formalization of Java that supports concurrency. In order to be able to use Jinja with our Coq formalization of the Extensible Spi calculus, we have manually translated the formalization of Jinja with Threads from Isabelle/HOL to Coq. We have extended Jinja with Threads with a type system based on the variant parametric types needed to express Java 5.0 generics and wildcards [26], which are pervasively used by our code generator. This very expressive type system was first introduced by Igarashi et al. [14] as an extension of Featherweight GJ, itself an extension of Featherweight Java [13].

3 Expi2Java: An Extensible Code Generator for Security Protocols

Expi2Java is a new code generator for security protocols designed to be highly extensible and customizable. It takes a protocol specification written in the Extensible Spi calculus (Expi calculus) together with a configuration that provides the low-level information needed for code generation and produces interoperable Java code.

The diagram in Figure 1 shows the workflow of Expi2Java. The Expi calculus model can be verified using ProVerif and type-checked with respect to the Expi2Java configuration(we defer the discussion about the type system to §5.3). The code generator takes the model and the configuration, and generates the symbolic library and the protocol implementation using code templates (special snippets of Java code).

The user can easily cus-
tomize and extend the in-
put language by adding new
types or new cryptographic
primitives (constructors and
destructors) to the default
configuration, and by instantiating them with specific cryptographic algorithms (DES encryption, RSA signature, etc.).
The Expi2Java configuration
also specifies which Java class
should be used to implement
each of the cryptographic algorithms and allows to pass

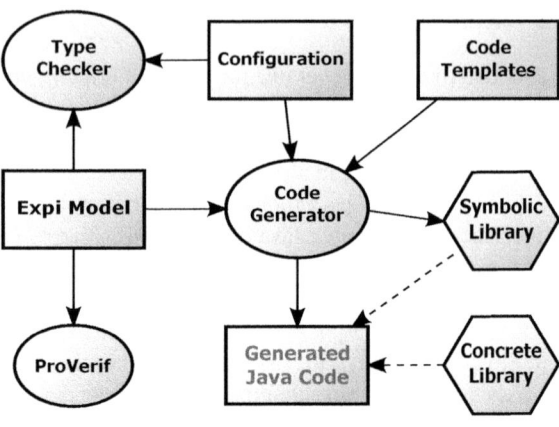

Fig. 1. Expi2Java Workflow

user-defined parameters to the implementation class. This can be used, for example,
to specify which encryption mode and padding should be used to encrypt data or to
specify the length and endianness of an integer nonce to match the protocol specification, basically giving the user full control over the low-level data format in the generated
protocol messages.

The syntax used by Expi2Java for writing models is very similar to the one used by
ProVerif [9] and includes support for ProVerif-specific constructs such as events and
queries. The main difference is that our calculus is typed and we therefore need to annotate some terms with their types. Expi2Java can pretty-print the protocol specifications in
ProVerif syntax, so that it can be formally verified before generating the implementation.

The translation used for code generation can also be customized, since Expi2Java
uses templates to generate the protocol implementation. The user can change the templates and the class stubs to simplify integration of the generated code into existing
applications. Our formalization targets the default translation used in the tool, with only
minor simplifications.

The generated code relies on a runtime library containing classes that implement
the cryptographic primitives used by the protocol specifications. We provide two interchangeable versions of the runtime library: a symbolic library that is used in our
formalization and proofs, and a concrete library that implements real networking and
cryptography. This separation allows us to abstract away from the complexity of network communication and cryptography when testing and debugging locally. The symbolic library is for the most part automatically generated by our tool and is described in
more detail in §6. The concrete library contains implementations for most of the common cryptographic primitives and data types out of the box, and can easily be extended
by the user. It uses the standard Java cryptographic providers to ensure interoperability
with other existing protocol implementations.

In the six major releases over the last three years we made a lot of progress in the form
of both practical features and usability improvements, turning Expi2Java from a prototype into a mature, useful and usable tool. We provide a detailed user manual and a tutorial that uses the Perrig-Song mutual authentication protocol as a running example [21].

More sample protocols such as Needham-Schroeder-Lowe, Andrew and Fiaba are provided in the Expi2Java distribution, together with the TLS implementation described in §4. Expi2Java is free software and is distributed under the terms of the GNU GPLv3.

4   TLS Case Study

In order to show the potential of Expi2Java we have generated a fully functional and interoperable implementation of the widely used Transport Layer Security (TLS) protocol [12] from a model verified with ProVerif. TLS provides a secure transport layer for many popular application protocols such as HTTP, FTP, SMTP, and XMPP.

TLS is a very complex protocol, it supports many different cryptographic schemes, optional parameters and extensions. Our TLS model implements TLS v1.0 [12] with the AES extension [10] and the Server Name Indication (SNI) extension [8]. The model includes both client and server sides, the Handshake, the Application Data Protocol and the Alert Protocol. We support dynamic cipher suite selection between 6 different cipher suites (including AES, RC4 and 3DES encryption with different key lengths, SHA1 or MD5 HMACs and RSA key exchange). One of these cipher suites is dynamically chosen during the handshake. Supporting multiple cipher suites in TLS in older tools such as Spi2Java would require duplicating the whole protocol model for each of the cipher suites. In order to prevent this, Expi2Java allows parameterizing processes with respect to the employed cryptographic algorithm. Using the parameterized processes, we could add support for 6 cipher suites with only a few lines of code. The only noteworthy TLS features we have not implemented are: session resumption (a performance tweak where a vulnerability [1] was discovered in 2009), the client-authenticated handshake (a rarely used feature) and record fragmentation (unsupported by some popular servers and clients, and therefore a very rarely used feature). The handshake messages that are used for key exchange algorithms (other than RSA) are also not supported.

Our model of TLS consists of an Expi calculus process (about 850 lines) and a configuration file (625 lines). It includes all the steps of the Handshake and the consequent message exchange through the Application Data Protocol. We check the validity of each received message, including MACs and certificate chain (when provided with the list of trusted CA certificates) and respond with an alert message on errors. The structure of all messages is modeled completely in the Extensible Spi calculus, while the data formats of encryptions and certificates are defined in the configuration and implemented in corresponding Java classes. For comparison, the most sophisticated generated protocol implementation we are aware of is a much simpler SSH client generated by Spi2Java[4]. Their model is 250 lines in size and needs 1390 lines of XML configuration (i.e., more than 5-to-1 ratio between configuration and model).

We have verified some security properties of our TLS model with ProVerif. In particular, we have used 3 secrecy queries showing that the secret key of the server, the "pre-master secret" nonce used to derive the session keys and initialization vectors, and the request that the client sends to the server using the Application Data Protocol are not leaked. We have used 3 correspondence queries adapted from FS2PV [7] to show message authentication for the certificate and pre-master secret. Additionally, we have

---

[4] The latest version (Apr. 19, 2011) is available at http://spi2java.polito.it/

used 9 reachability queries providing a consistency check to ensure that different parts of the handshake are reached. The verification process for all 15 queries took about 9 minutes on a laptop with an Intel® Core2™ Duo P7450 CPU.

The generation of Java code for the TLS model (on the same hardware) takes only about 12 seconds. In addition to the sample web server and web client we generate a verified implementation of a TLS channel for the concrete library. We use this channel to generate a simple web server offering a web page over HTTPS and a web client downloading it. We have tested that the resulting implementation is interoperable with common browsers and web servers.

5 The Extensible Spi calculus

The source language of our translation is a variant of the Spi calculus [4], a process calculus for cryptographic protocols. We start with the variant of the Spi calculus by Abadi and Blanchet [3] – a subset of the language accepted by ProVerif [9]. We extend it with the Expi2Java configuration in §5.2 and define a type system for it in §5.3.

5.1 Syntax

The syntax of the Extensible Spi calculus (Expi calculus) is defined in Table 1. In this calculus, *terms* are used to model cryptographic operations symbolically. Terms are obtained by applying *constructors* to other terms, *variables*, and *Expi names*. Constructor applications are parameterized by the name of the cryptographic algorithm A that should be used to implement the constructor in the generated code. Differently parameterized constructors are treated as different constructors, so for instance enc_{DES} is different than enc_{AES}. The global configuration (see §5.2) defines, amongst others, the set of constructor identifiers $\mathcal{F} = \{f_1, \ldots, f_n\}$ and their type, as well as the set of possible cryptographic algorithms that can be used to implement each constructor.

Table 1. The Syntax of the Extensible Spi calculus

$K, L, M, N ::=$	terms
$\quad a, b, m, n, k$	Expi names
$\quad x, y, z, v, w$	variables
$\quad f_A\langle T_1, \ldots, T_m\rangle(M_1, \ldots, M_n)$	constructor application
$G ::= g_A\langle T_1, \ldots, T_m\rangle(M_1, \ldots, M_n)$	destructor applications
$P, Q, R ::=$	processes
$\quad \mathrm{out}(M, N).P$	output
$\quad \mathrm{in}(M, x).P$	input
$\quad !\,\mathrm{in}(M, x).P$	replicated input
$\quad \mathrm{new}\ a : T.P$	restriction
$\quad P \mid Q$	parallel composition
$\quad \mathbf{0}$	null process
$\quad \mathrm{let}\ x = G\ \mathrm{in}\ P\ \mathrm{else}\ Q$	destructor evaluation

Notation: We use u to refer to both Expi names and variables.

In our calculus, *destructors* are partial functions that are applied to terms and can produce a term or fail. Similar to constructors, we parameterize destructors by a cryptographic algorithm name. The global configuration also defines the set of destructor identifiers $\mathcal{G}$, and the cryptographic algorithms supported for each destructor identifier. Constructors and destructors can have a parametric type, in which case we must provide a list of type annotations $\langle T_1, \ldots, T_m \rangle$ for instantiating the type variables. In our Expi2Java tool these annotations are automatically inferred in most cases.

Processes are used to model the behavior of protocol participants and the communication between them. A specific characteristic of our calculus is that replication (the bang symbol "!") can only appear before an input process [6]. This is the most common way to use replication in specifications, and at the same time it is reasonably easy to implement as a server that waits for requests on a channel and spawns a new thread to process each incoming message. The Expi name in the restriction process has a type annotation, otherwise the syntax of processes is standard [3,4,6].

In our Coq formalization, we use a *locally-nameless* representation [5] for Expi names (bound by the restriction process) and variables (bound by the let and input processes) to avoid the issues related to α-renaming. Nevertheless, for the sake of readability, throughout this paper we use the more familiar, named representation.

5.2   Configuration

The crucial features of the Expi calculus are its extensibility and customizability. The user can extend the sets of types, constructors and destructors, redefine the reduction relation for destructors and provide different implementations for the constructors and destructors. Our whole calculus is parameterized over what we call a *configuration* – a collection of several sets and functions that define the behavior of types, constructors and destructors (see Table 2). We have defined a default configuration that is sufficient to model most cryptographic protocols, please refer to the online appendix for more details.

Table 2. Configuration

$$\left(\mathcal{T}, \mathcal{F}, \mathcal{G}, alg(t), alg(f), alg(g), variance(t,i), Gen, f : \forall \widetilde{X}. (\widetilde{T}) \mapsto T, g : \forall \widetilde{X}. (\widetilde{T}) \mapsto T, \Downarrow \right)$$

$\mathcal{T} = \{t_1, \ldots, t_n\}$	a finite set of type identifiers
$\mathcal{F} = \{f_1, \ldots, f_m\}$	a finite set of constructor identifiers
$\mathcal{G} = \{g_1, \ldots, g_l\}$	a finite set of destructor identifiers
$alg(t_i) = \{A_i^1, \ldots, A_i^{n_i}\}$	cryptographic algorithms for each type
$alg(f_j) = \{A_j^1, \ldots, A_j^{n_j}\}$	cryptographic algorithms for each constructor
$alg(g_k) = \{A_k^1, \ldots, A_k^{n_k}\}$	cryptographic algorithms for each destructor
$variance(t,i) \in \{+, -, \circ\}$	the variance of each type identifier (see §5.3)
$Gen \subseteq \mathcal{T}$	a set of generative types
$f : \forall \widetilde{X}. (\widetilde{T}) \mapsto T$	the type of each constructor (see §5.3)
$g : \forall \widetilde{X}. (\widetilde{T}) \mapsto T$	the type of each destructor
$g(M_1, \ldots, M_n) \Downarrow N$	the destructor reduction relation

Notation: We use $\widetilde{T}$ to denote type sequences of form $T_1, \ldots, T_n$ for some n.

5.3   Type System

Since Java is an explicitly typed language, any code generator targeting Java needs to generate type annotations for variables, fields, method arguments and return values, etc. These type annotations cannot be generated out of thin air, and asking the user to manually annotate types in the automatically generated Java "spaghetti code" would be a usability disaster. We solve this problem by asking the user to provide type annotations while writing the protocol model. Our code generator uses the type information in the model to generate the typing annotations needed by the Java type-checker.

Additionally, we want to prevent that a user who makes mistakes in a protocol model finds out that the generated implementation does not even compile in Java because of typing errors in the automatically generated code, which the user does not understand. So we provide a type-checker for the Expi calculus that prevents generating ill-typed Java code. Our type-checker immediately reports inconsistent usage of terms in an understandable way – in terms of the specified protocol model the user has written. In §8.1 we show that if the original protocol model is well-typed with respect to our Expi calculus type system then our translation is guaranteed to generate a well-typed Java program.

Our type system for the Expi calculus features subtyping and parametric polymorphism [11]. This makes the type system very expressive and has allowed us to devise a very precise unification-based type inference algorithm. This decreases the type annotation burden on the user and improves the readability of the protocol models. Parametric polymorphism also allows us to have only a small number of "generically" typed constructors and destructors and still be able to specialize them. Parametric types can be nested, which naturally corresponds to the types of the nested terms and allows us to keep more information about the inner terms even after several destructor or constructor applications. The nested types allow us, for instance, to express the type of messages sent and received over a channel, or to model the fact that an encryption and the corresponding key work on messages of the same type.

In order to ensure the correct use of cryptographic primitives we also parameterize types, constructors and destructors with respect to cryptographic algorithms. We use this feature to statically prevent mixing up different algorithms, e.g., decrypting an AES encrypted message with a DES key. As the result, our type system can express many complex types, such as specializations of channels, encryptions etc. using a reasonably small set of core types, which is better suited for formalization.

Table 3. Syntax of Expi Types

$T, U ::=$	X	Top	$\mathsf{Channel}_A\langle T \rangle$	$t_A\langle T_1, \ldots, T_m \rangle$

Our type system has only two fixed types, which are required for the correct typing of processes: type Top and type $\mathsf{Channel}_A\langle T \rangle$. The type variables X are used in the parametric types of constructors and destructors. The types $t_A\langle T_1, \ldots, T_m \rangle$ represent user-defined parametric types defined using the set $\mathcal{T}$ from the configuration (see §5.2). Additionally, types $\mathsf{Channel}_A\langle X \rangle$ and $t_A\langle T_1, \ldots, T_m \rangle$ are parameterized by a cryptographic algorithm name A, which allows us to define subsets of related types such as TCP or TLS channels, AES keys and AES-encrypted messages. The configuration also defines a set of *generative* types, the only types that can be used in well-typed restriction processes.

The subtyping relation $<:$ is defined in Table 4. Type Top is a supertype of all other types. Subtyping for channel types $\mathsf{Channel}\langle X \rangle$ is invariant. The custom types $t_A\langle T_1, \ldots, T_m \rangle$ are subtyped according to the variance of their arguments, as defined in the configuration using function $variance(t, i)$. Subtyping is reversed for contravariant $(-)$ arguments, runs in the same direction for covariant $(+)$ arguments and requires the invariant $(\circ)$ arguments to be the same.

Table 4. Subtyping: $T <: U$

(SUB-TOP)	(SUB-CHANNEL)
$\dfrac{T \vdash \diamond}{T <: \mathsf{Top}}$	$\dfrac{T <: U \qquad U <: T}{\mathsf{Channel}_A\langle T \rangle <: \mathsf{Channel}_A\langle U \rangle}$

$$\dfrac{t \in \mathcal{T} \qquad \forall i \in [1, n]. \quad \begin{array}{l} (variance(t, i) = + \;\Rightarrow\; T_i <: U_i) \\ (variance(t, i) = - \;\Rightarrow\; U_i <: T_i) \\ (variance(t, i) = \circ \;\Rightarrow\; U_i <: T_i \;\wedge\; T_i <: U_i) \end{array}}{t_A\langle T_1, \ldots, T_n \rangle <: t_A\langle U_1, \ldots, U_n \rangle} \quad \text{(SUB-NESTED)}$$

The type of terms is defined by relation $\Gamma \vdash M : T$ from Table 5. This relation is very simple and general. The type of constructor applications is defined in the configuration and instantiated by the given type annotations. Finally, using the subsumption rule TERM-SUB a term can be used in any context that expects a supertype of its type.

Table 5. Term Typing: $\Gamma \vdash M : T$

(TERM-ENV)	(TERM-SUB)
$\dfrac{\Gamma \vdash \diamond \qquad T \vdash \diamond \qquad u : T \in \Gamma}{\Gamma \vdash u : T}$	$\dfrac{\Gamma \vdash M : T \qquad T <: U}{\Gamma \vdash M : U}$

$$\dfrac{f_A : \forall \widetilde{X}. (\widetilde{T}) \mapsto T \qquad \forall i \in [1, m]. \, U_i \vdash \diamond \qquad \forall i \in [1, n]. \, \Gamma \vdash M_i : T_i[\widetilde{X} := \widetilde{U}]}{\Gamma \vdash f_A\langle U_1, \ldots, U_m \rangle(M_1, \ldots, M_n) : T[\widetilde{X} := \widetilde{U}]} \quad \text{(TERM-CONSTR)}$$

6 Variant Parametric Jinja

Our target language, Variant Parametric Jinja (VPJ) is based on Jinja with Threads [16, 17] and the type system with variant parametric types by Igarashi et al. [14]. In this section, we briefly describe VPJ with an emphasis on the modifications we have made; please refer to the original papers for more details [14–17]. We chose Jinja with Threads over other formalized Java fragments because of its comprehensiveness. Expi2Java needs concurrency with shared memory and synchronization to model channels and message passing, a class hierarchy with inheritance and casting to model data types, a type system that supports Java generics and wildcards and, optionally, exceptions to simplify modeling destructor applications. Jinja with Threads supports all these features except for generics and wildcards, which we added in the form of variant parametric types [14]. We have also extended the values and types with support for a new base type, **string**, and removed support for arrays in order to simplify the semantics.

Syntax. The syntax of VPJ is very similar to Jinja with Threads, and therefore differs from Java in several ways. The most visible simplification is the absence of distinction between expressions and statements. Other differences inherited from Jinja are simplifications in the form of a few artificial expressions like **insync** (@n) {e} and **unit**, slightly different typing and reduction rules, and the lack of syntactic sugar [16]. A difference unique to VPJ is that the object instantiation expression **new** takes an additional list of expressions $e_1, \ldots, e_k$ that are used to initialize all fields declared in the corresponding class.

Table 6. Syntax of Variant Parametric Jinja (VPJ)

$v ::=$	**unit** | **null** | **true** | **false** | i | "xyz" | @n	VPJ value
$\otimes ::=$	$= \| \neq \| < \| \leq \| > \| \geq \| + \| - \| \times \| \wedge \| \vee \| \oplus$	binary operation
$e ::=$		VPJ expression
	new $C\langle T_1^P, \ldots, T_n^P \rangle (e_1, \ldots, e_k)$	object instantiation
	$(T^J)e$	typecast
	v	literal value
	$e_1 \otimes e_2$	binary operation
	x, y, z	variable access
	$x := e$	variable assignment
	$e.f\{C\}$	field access
	$e_1.f\{C\} := e_2$	field assignment
	$e.\langle T_1^P, \ldots, T_n^P \rangle m(e_1, \ldots, e_k)$	parametric method invocation
	$\{T^J x := \lfloor v \rfloor ; e\}$	variable declaration block
	synchronized $(e_1) \{e_2\}$	**synchronized** statement
	insync (@n) {e}	locked **synchronized** statement
	$e_1 ; e_2$	sequential composition
	if $(e_1) \{e_2\}$ **else** $\{e_3\}$	conditional
	while $(e_1) \{e_2\}$	while loop
	throw e	exception throwing
	try $\{e_1\}$ **catch** $(C\langle \widetilde{T^P} \rangle x) \{e_2\}$	exception catching
$T^J ::=$	**void** | **boolean** | **int** | **string** | **null**$_T$ | T^P	VPJ type
$T^P ::=$	$X^J \| T^C$	parametric type
$T^C ::=$	$C\langle T_1^P \mathcal{V}, \ldots, T_n^P \mathcal{V} \rangle, D\langle T_1^P \mathcal{V}, \ldots, T_n^P \mathcal{V} \rangle$	variant parametric class type
$\mathcal{V} ::=$	$+ \| - \| \circ \| \star$	co-, contra-, in- and bivariant subtyping
$L ::=$	**class** $C\langle \widetilde{X^J} \lhd \widetilde{U^C} \rangle \lhd D\langle \widetilde{T^P} \rangle \{\widetilde{T^J} \widetilde{f} ; \widetilde{M}\}$	class definition
$M ::=$	$\langle \widetilde{X^J} \lhd \widetilde{U^C} \rangle T^J m(\widetilde{T^J} \widetilde{x}) \{\textbf{return } e\}$	method definition

Notation: We write $\widetilde{X^J} \lhd \widetilde{T^C}$ for the sequence $X_1^J \lhd T_1^C, \ldots, X_n^J \lhd T_n^C$

Just like Java, VPJ allows to define custom classes with methods and fields. The syntax is similar to the Java generics, with type variables and upper bounds (denoted by $\lhd$ in VPJ, and **extends** in Java). In VPJ variant parametric types require variance annotations on each type argument. In addition to the variance annotations from the Expi calculus the variance in VPJ can also be *bivariant*; any two instantiations of a bivariant type constructor are subtypes of each other. The type system of VPJ is a mixture of the Jinja type system [16] with variant parametric types [14]. We have extended the subtyping relation with two additional rules stating that the null type null_T is the subtype of **string** and all reference types T^P. Additionally, we require the subclass relation to be *well-founded* (i.e., that it does not have infinite descending chains) to ensure termination of subclass checks.

Memory Model. VPJ uses a realistic memory model with a shared heap and thread-local stacks. The stacks are maps from variable names to values. The heap maps memory locations $@n$ to a class type T^C and the field values of the corresponding instance. This heap model differs slightly from the one used in Java, since we store the exact parametric type of each object, while in Java the information about the type parameters is lost.[5] We have decided to store the parametric types to simplify the formalization of variant parametric types in Coq and avoid the problems arising from Java-like semantics such as the need for run-time type-checks to enforce type soundness.

Semantics. The semantics of VPJ can be split into two parts, the *single-threaded core* and the (very complex) *multi-threaded semantics* [16]. The single-threaded semantics is defined using the reduction relation $\rightarrow_J$, a *labeled* relation that takes a conversion function and reduces some expression together with some state (i.e., stack and heap) to another expression and a state, possibly producing a *thread action* (usually visualized as a label on the reduction arrow). The thread actions are used for inter-thread communication in the multi-threaded semantics. The multi-threaded semantics is exactly the same as in Jinja with Threads [16], we have only made changes that had to be done to adapt the formalization for Coq.

Symbolic Library. In addition to the standard classes like Object required by VPJ we have defined an additional set of classes that are used by the translated protocols. We call this the *symbolic library*, because the implementation of the cryptographic primitives and data types does not perform any "real" cryptography or networking. The symbolic library provides a symbolic abstraction of the cryptographic primitives and channel communication. It is designed to be simple enough to simplify proving the translation secure in the future. Finally, the symbolic library can also be used for debugging purposes. Our Expi2Java tool also has a concrete library that performs "real" cryptographic operations and uses the actual network, but this is not formalized in Coq. The two libraries (concrete and symbolic) can be used interchangeably by the generated code.

The fixed part of the symbolic library (i.e., the one that is not generated by our translation from the configuration) consists of the 7 classes shown in Figure 2. AbstractBase is the base class of the class hierarchy for translations of all Expi types.

---

[5] In the Java community, this process is usually called "type erasure", although not all type information is erased, just the generics, and that just for backward compatibility.

`AbstractGenerativeBase` is the superclass for generative types. The generated flat class hierarchy reflects the subtyping relation from the Expi calculus.

Semaphore implements a counting semaphore using the synchronization primitives of VPJ. This class is used in the implementation of inter-thread message passing. Expi channels are modeled using the class `AbstractChannel`$\langle X^J \rangle$. It implements the synchronous semantics of Expi channels, where different processes are implicitly synchronized when a message is sent from one process to another over a channel.

Finally, `ELibrary` is the base class for `ELetProcess` and `EDestructor`,

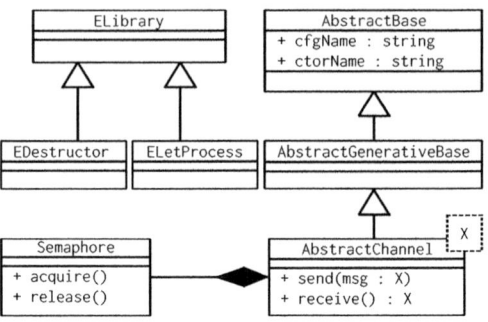

Fig. 2. Symbolic Library Classes

the two exception classes used to model failing destructor applications. The translation of the let $x = G$ in P else Q process uses the exception to decide when to take the else branch. We have proved that all symbolic library classes are well-typed with respect to our type system (see §8.1).

7   Translation Overview

The translation from the Expi calculus to VPJ is performed in two steps. The first step translates protocol models to the *Global Expi calculus*, a variant of the Expi calculus with a different semantics for name binders. The second and much more complex step translates Global Expi processes into VPJ programs.

7.1   First Step: Translating Expi Calculus Models to the Global Expi Calculus

The usual semantics of the Expi calculus heavily relies on α-renaming and scope extrusion to reduce processes. Scope extrusion allows to move name binders (i.e., restriction processes new $a : T$) out of parallel compositions ($P \mid Q$). The problem with this semantics is that it cannot be easily implemented in a mainstream programming language, because it would require changing the scope of variables and moving them across different threads. Instead of extruding restrictions around we define a variant of the Expi calculus in which restrictions generate globally fresh names. Such a global semantics was proposed for the pi calculus by Wischik [27], and we adapt this idea to the Expi calculus. The resulting Global Expi calculus uses the same terms and types as the Expi calculus and has only one different process, the generation process gen $x : T$ in P instead of the restriction process new $a : T.P$. The semantics of gen $x : T$ in P is to generate a globally fresh name a and substitute variable x with a in P, like for ML references. This first translation step brings our semantics closer to the one of the Java implementation, which should ease any future simulation proof.

7.2   Second Step: Translating Global Expi Models to VPJ Programs

In the second step of the translation we implement the behavior of Expi terms and processes in VPJ; this is much more complicated than the fist step.

Table 7. Translation Overview

Global Expi	VPJ
Configuration:	
- Type identifiers (t)	$\rightsquigarrow$ Class declarations
- Algorithm names (A)	$\rightsquigarrow$ String constants stored in fields
- Constructors (f)	$\rightsquigarrow$ Special methods in class **Fun**
- Destructors (g)	$\rightsquigarrow$ Special methods in class **Fun**
Expi Types	$\rightsquigarrow$ Variant parametric classes
Terms	$\rightsquigarrow$ Expressions (variables, method calls)
Processes	$\rightsquigarrow$ Expressions (variable declarations, control flow)
- Parallel composition $(P \mid Q)$	$\rightsquigarrow$ Threads that are spawned and joined
Free names	$\rightsquigarrow$ Variables in main (passed by reference to threads)

Table 7 gives an overview of how the different Expi calculus constructs are represented in VPJ. Expi calculus types are modeled as additional generated classes in our symbolic library, and variant parametric types are used to represent type parameters. The instances of these generated classes represent Expi terms of the corresponding type. The cryptographic algorithm names used in the destructor reduction relation are stored in fields of type **string** in the generated classes. Expi constructors and destructors are represented as special methods in a symbolic library class named Fun. We use a simple naming convention to distinguish constructor and destructor methods. Terms are translated to VPJ expressions that either access local variables or call constructor methods. Processes are translated into larger code blocks that create and modify the terms stored in local variables and use the symbolic library to interact with each other. We use threads to model parallel composition of processes and shared memory to pass data between them. Please refer to the online appendix for a more detailed description of the translation process.

8   Proofs

We have used the Coq proof assistant [2] to develop the mechanized formalization of our translation and machine-check our proofs. We believe that this is the only way to stay honest with ourselves when proving something about a formalization that is so complex. A proof assistant helps organizing the formal development as a software project, ensures that the definitions are well-formed and in sync with each other, provides automation for proving routine tasks, and checks that the proofs are correct, without missing any corner-case and without forgetting any assumption that was made. This gives us high confidence in the final result, which one cannot easily obtain by "handwaving". The formalization encompasses about 17K lines of Coq code in total.

8.1   Symbolic Library and Generated Code are Well-Typed

We have proved that the symbolic library and the VPJ code generated by the translation are well-typed. This is an important consistency check for our definitions (it helped us to find out, for instance, that the first version of our translation was using the field access in an inconsistent way) and justifies the usage of our Expi calculus type system to prevent generating ill-typed VPJ programs. In the longer-term perspective these proofs will be needed for proving the correctness of our translation.

We have shown that each class of the symbolic library is well-typed in a VPJ program containing a small number of Java standard library classes (NullPointerException, Object, etc.) and the symbolic library classes. We show that all types used in class declarations are well-formed and all declared methods are well-typed. The proof is by case analysis on the corresponding expression and using the right case of the expression typing relation. It is not complicated, but quite long and tedious, because we need to give the correct type of each subexpression and show all premises of the typing rules.

Showing that the code generated by the translation is well-typed was much harder. Most typing rules require providing the exact types of all subexpressions and only fail in the last moment if a wrong type was chosen. Another problem was finding the right invariants to type-check the code generated by recursive functions. An incorrect invariant (i.e., the type of the expression we are trying to type-check and the preconditions) usually becomes noticeable only when applying the induction hypothesis, after having constructed a big part of the failed proof attempt.

We prove that the expressions generated for the Global Expi processes, the constructor and destructor methods and the classes representing Expi types are well-typed assuming that our invariants hold. In the end, we use these results to show the following theorem:

Theorem 1 (Trans-WT). *If P is locally-closed, $\Gamma \vdash P$ and $P \rightsquigarrow (e, L)$, then L is well-typed and (e, L) is well-typed in a preallocated heap and an empty stack.*

This theorem shows that our translation generates a self-contained VPJ program (e, L) that is well-typed in the initial heap and stack. The assumptions about the heap and stack are inherited from Jinja. They are needed for the typing relation, because an arbitrary Jinja expression can contain local variables or throw system exceptions (e.g., NullPointerException). At the program start, we assume that the heap is "preallocated", i.e., it contains instances of system exception classes at some fixed addresses that are reserved for the system exceptions, and the stack contains no local variables.

The Coq proof of this theorem is done in great detail, and so are the proofs of most of the helper lemmas. Due to the lack of time we did not prove some rather obvious properties of the translation that are used in the proof of the theorem, e.g., that processes are translated only to expressions of type **void** or that all used types were generated and added to the program. Furthermore, we did not prove some list and substitution rewriting lemmas, and other similar helper lemmas in cases that looked trivially true but were tedious to prove in Coq. We have also assumed that the translation of terms and the declaration of free Expi names are well-typed.

8.2   Destructor Consistency Proof for Default Configuration

The flexible nature of the Expi calculus makes it impossible to prove the type system sound for an arbitrary configuration. The destructor reduction relation could be defined in a way that conflicts with the typing of the destructor or of some of the constructors used by the reduction rule. For instance, it would be possible to give an identity destructor (which returns its only argument) the type $(\mathsf{Int}) \mapsto \mathsf{Bool}$. Such a destructor would be inconsistent, because we cannot give any term two different types that are not even subtypes of each other. To prevent such inconsistency in our default configuration we have proved the following theorem in Coq:

Theorem 2 (Destructor Consistency). *If a destructor has type* $g_A : \forall[X_i]_n . ([T_j]_m)$ $\mapsto T$, $g_A \langle [U_i]_n \rangle ([M_j]_m) \Downarrow M$ *and* $\forall j \in [1, m].\ \Gamma \vdash M_j : T_j[\widetilde{X} := \widetilde{U}]$, *then* M *can be typed to the instantiated return type of the destructor:* $\Gamma \vdash M : T[\widetilde{X} := \widetilde{U}]$.

Destructor consistency is the crucial step in the subject-reduction proof of the Expi calculus. It is the only part of the soundness proof that depends on the configuration and should therefore be re-proven if the user wants to change the default configuration. Once one proves destructor consistency, soundness follows directly.

8.3   Lessons Learned

Formalizing our code generator in a proof assistant turned out to be harder than we expected. Like the large majority of similar tools before it [18, 20, 21, 23–25], Expi2Java targets the Java programming language. This has many pragmatic advantages for building a usable and secure tool: from type and memory safety, an extensive standard library and the cryptographic service provider architecture, to the ease of integrating the generated code into existing applications. However, Java is a very complex programming language, and merely adapting an existing mechanized formalization of a subset of Java [16, 17] to suit our particular needs turned out to be a quite daunting task. About 7k LOC out of the 17k LOC of our formalization are solely concerned with defining our target language, VPJ.

Moreover, Java is an imperative language, and lacks certain functional features that would have greatly simplified the symbolic representation of terms: immutable data structures, structural equality, and pattern matching. A programming language like Scala[6] or F#[7], which integrate features of both functional and imperative object-oriented programming, would have been a better match for implementing our symbolic library, while preserving most of the pragmatic advantages of Java. On the other hand, we are not aware of any mechanized formalizations of comprehensive subsets of these languages, so the upfront effort needed just to formalize the target language would have been even larger.

9   Conclusion

In this paper we have introduced Expi2Java, an extensible code generator for security protocols. We have illustrated the flexibility of Expi2Java by generating interoperable

---

[6] http://www.scala-lang.org/
[7] http://www.fsharp.net/

implementations of a client and a server for TLS v1.0 from a protocol model verified with ProVerif. We have formalized our source and target languages as well as the translation between them using the Coq proof assistant, and proved that the generated code is well-typed if the original protocol model is well-typed. This increases our confidence in the translation, and justifies the usage of our Expi calculus type system to catch all type errors as early as possible and present understandable error messages. Additionally, we have proved the consistency of the destructors in our default configuration.

In the future it would be very interesting to show that the translation presented here preserves the trace properties, and, more ambitiously, the security properties (e.g., the robust safety) of the original protocol model. The former could be achieved by using weak labeled simulation to relate VPJ programs to Expi processes, while for the later one would have to show that this simulation is contextual and that each VPJ attacker can be mapped back to an attacker in the Expi calculus. We believe that the current work builds a solid ground on which the preservation of security properties can be formally investigated.

References

1. CVE-2009-3555. Man-in-the-Middle Vulnerability in TLS via Session Renegotiation (2009)
2. The Coq proof assistant, Version 8.3, Home page (2010), http://coq.inria.fr/
3. Abadi, M., Blanchet, B.: Analyzing security protocols with secrecy types and logic programs. Journal of the ACM 52(1), 102–146 (2005)
4. Abadi, M., Gordon, A.D.: A calculus for cryptographic protocols: The Spi calculus. Information and Computation 148(1), 1–70 (1999)
5. Aydemir, B.E., Charguéraud, A., Pierce, B.C., Pollack, R., Weirich, S.: Engineering formal metatheory. In: POPL 2008, pp. 3–15 (2008)
6. Backes, M., Hriţcu, C., Maffei, M.: Type-checking zero-knowledge. In: CCS 2008, pp. 357–370. ACM Press (October 2008)
7. Bhargavan, K., Fournet, C., Gordon, A.D., Tse, S.: Verified interoperable implementations of security protocols. ACM Transactions on Programming Languages and Systems 31(1) (2008)
8. Blake-Wilson, S., Nystrom, M., Hopwood, D., Mikkelsen, J., Wright, T.: Transport Layer Security (TLS) Extensions RFC 4366 (April 2006)
9. Blanchet, B.: ProVerif v1.14pl4 (Automatic Cryptographic Protocol Verifier) User Manual (February 2008), http://www.proverif.ens.fr/
10. Chown, P.: Advanced Encryption Standard (AES) Ciphersuites for Transport Layer Security (TLS) RFC 3268 (Informational) (June 2002)
11. Curien, P.-L., Ghelli, G.: Coherence of subsumption, minimum typing and type-checking in $F_\leq$. Mathematical Structures in Computer Science 2(1), 55–91 (1992)
12. Dierks, T., Allen, C.: The TLS Protocol Version 1.0 RFC 2246 (January 1999)
13. Igarashi, A., Pierce, B.C., Wadler, P.: Featherweight Java: a minimal core calculus for Java and GJ. ACM TOPLAS 23(3), 396–450 (2001)
14. Igarashi, A., Viroli, M.: Variant parametric types: A flexible subtyping scheme for generics. ACM TOPLAS 28, 795–847 (2006)
15. Klein, G., Nipkow, T.: A machine-checked model for a Java-like language, virtual machine and compiler. ACM TOPLAS 28(4), 619–695 (2006)
16. Lochbihler, A.: Type safe nondeterminism - a formal semantics of Java threads. In: FOOL (2008)

17. Lochbihler, A.: Verifying a Compiler for Java Threads. In: Gordon, A.D. (ed.) ESOP 2010. LNCS, vol. 6012, pp. 427–447. Springer, Heidelberg (2010)
18. Lukell, S., Veldman, C., Hutchison, A.: Automated attack analysis and code generation in a multi-dimensional security protocol engineering framework. In: Southern African Telecommunications Networks and Applications Conference (2003)
19. McCarthy, J.A., Krishnamurthi, S., Guttman, J.D., Ramsdell, J.D.: Compiling cryptographic protocols for deployment on the web. In: WWW 2007, pp. 687–696 (2007)
20. Millen, J., Muller, F.: Cryptographic protocol generation from CAPSL. Technical Report SRI-CSL-01-07, SRI International (December 2001)
21. Song, D., Perrig, A., Phan, D.: AGVI - Automatic Generation, Verification, and Implementation of Security Protocols. In: Berry, G., Comon, H., Finkel, A. (eds.) CAV 2001. LNCS, vol. 2102, pp. 241–245. Springer, Heidelberg (2001)
22. Pironti, A., Sisto, R.: An experiment in interoperable cryptographic protocol implementation using automatic code generation. In: ISCC (2007)
23. Pozza, D., Sisto, R., Durante, L.: Spi2Java: Automatic cryptographic protocol Java code generation from spi calculus. In: AINA, pp. 400–405. IEEE Computer Society Press (2004)
24. Song, D.X., Berezin, S., Perrig, A.: Athena: A novel approach to efficient automatic security protocol analysis. Journal of Computer Security 9(1/2), 47–74 (2001)
25. Tobler, B., Hutchison, A.: Generating network security protocol implementations from formal specifications. In: CSES (2004)
26. Torgersen, M., Hansen, C.P., Ernst, E., von der Ahé, P., Bracha, G., Gafter, N.M.: Adding wildcards to the Java programming language. In: SAC 2004, pp. 1289–1296 (2004)
27. Wischik, L.: Old names for nu. Presented at Dagstuhl Seminar 04241 (2004)

Incremental Verification with Mode Variable Invariants in State Machines*

Temesghen Kahsai[1], Pierre-Loïc Garoche[1,2],
Cesare Tinelli[1], and Mike Whalen[3]

[1] The University of Iowa
[2] Onera, The French Aerospace Lab
[3] University of Minnesota

Abstract. We describe two complementary techniques to aid the automatic verification of safety properties of synchronous systems by model checking. A first technique allows the automatic generation of certain inductive invariants for mode variables. Such invariants are crucial in the verification of safety properties in systems with complex modal behavior. A second technique allows the simultaneous verification of multiple properties incrementally. Specifically, the outcome of a property—valid or invalid—is communicated to the user as soon as it is known. Moreover, each property proven valid is used immediately as an invariant in the model checking procedure to aid the verification of the remaining properties. We have implemented these techniques as new options in the KIND model checker. Experimental evidence shows that these two techniques combine synergistically to increase KIND's precision as well as its speed.

1 Introduction

Embedded systems often contain complex modal behavior that describes how the system will interact with its environment. In these systems, the *modes* of the software drive the behavior of the device. In a flight guidance system, these modes cause a particular control algorithm to be chosen; an *approach* mode enables a control algorithm that attempts to land the airplane, while a *go-around* mode enables a controller that attempts to climb the aircraft to a suitable safe altitude. These modes are often designed as state machines or mode transition tables. In addition, embedded systems typically have several parallel mode machines that communicate with one another to define the control state of the system. For instance, in flight guidance, there are separate lateral and vertical modes that manage the lateral and vertical aspects of flight.

Understanding which variables in a system's model represent system modes, and discovering relationships between such variables often determine whether or not a property can be proven about a system. However, such variable, which from now on we will refer informally to as *mode variables*, may not be easily identifiable among all of the system's variables. In addition, once identified, determining correct invariants between different mode variables is non-trivial. As an example of these challenges consider the

---

* This work was partially supported by AFOSR grant #AF9550-09-1-0517, NSF grant CNS-1035715, DGA/MRIS/ERE support and FNRAE Cavale project.

A. Goodloe and S. Person (Eds.): NFM 2012, LNCS 7226, pp. 388–402, 2012.

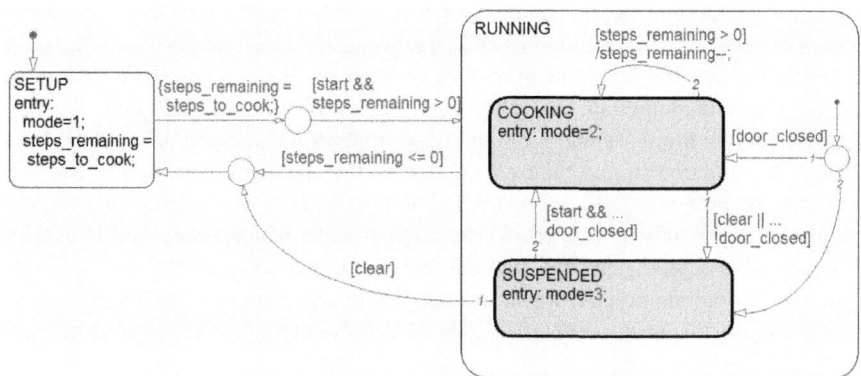

Fig. 1. State machine of a microwave model

hierarchical state machines (HSMs) described in Figure 1 that illustrates the modal behavior of a microwave.[1] HSMs are used by model-based development notations such
as Simulink and SCADE which are becoming widespread for software development in
avionics and other industries. In the example, mode information is encoded both explicitly, in the states of the HSM, and implicitly, through the integer variable **mode**. This
sort of hybrid encoding of mode information occurs regularly in industrial models that
we have analyzed. An additional complication is that when HSMs are compiled to a
lower level modeling language or to code, their state are usually encoded into integer
variables that are not immediately distinguishable from other integer variables.

The focus of this paper is on leveraging mode information for k-*induction*-based
model checking. In this approach, a prover attempts to prove a safety property of a
system inductively by showing for some $k \geq 0$ that (i) the property holds in all states
reachable in up to k steps, and (ii) for all sequences of $k + 1$ states along the system's transition relation, the last state satisfies the property whenever all the previous
ones do. As with mathematical induction, sometimes safety properties are not strong
enough to be provable by k-induction, for any k. In that case, it is helpful to strengthen
the induction hypotheses with known invariants properties of the system. We believe
that invariants involving mode variables are critical for the success of inductive methods when proving properties of control systems, and we provide initial experimental
evidence to support this conjecture.

This paper describes two complementary techniques to aid, more generally, the automatic verification of safety properties of synchronous systems with modal behavior.
The first technique, described in Section 3, allows the automatic identification of likely
mode variables, and the discovery of invariant relationships among them by adapting
an invariant generation method, described in Section 2, we developed in previous work.
We heuristically consider as a mode variables any system variable that (i) ranges over
a (small) finite set of values and (ii) whose next-state value is determined in part by its
current value. We generalize this idea slightly to *mode variable sets* in which strongly-
connected variables define a particular system mode. We develop a general invariant

---

[1]   We thank Steve Miller and Lucas Wagner at Rockwell Collins for the example.

generation method to identify implicative relationships between values of mode variables. The second technique, described in Section 4 and motivated by the industrial use of model checkers for the verification of large numbers of safety properties on the same system, allows the simultaneous and incremental verification of multiple properties. It is incremental in the sense that the status of a property—whether it holds in the system or not—is communicated to the user as soon as the checker determines it. Moreover, each property proven to hold is used immediately as an auxiliary invariant to aid the verification of the remaining properties. Our experimental results, described in Section 5 for selected benchmarks, indicate that our two techniques are quite effective in practice, especially in combination. As we show, using them together considerably increases the number of provable safety properties, as well as speeding up the verification process.

Related Work. Automatic invariant generation has been intensively investigated since the 1970s, producing a large body of literature. Manna and Pnueli [13] provide an early compendium of this research and an extensive set of references. In this paper, we focus on discovering invariants related to system modes. The idea of automatically discovering mode machines for hardware is (very briefly) referenced in [7]. The idea of generation of mode-specific invariants for the SCR notation was introduced in [8] and improved in [9] then generalized to LTSs by Damas [2,1]. This work supports discovery of invariants between a (known) state machine and variables used in guard expressions for transitions of the machine, using syntactic fixpoint algorithm that operates over the state machine graph. Our approach, based on our own previous work invariant discovery [10], is more general; it automatically identifies mode (state machine) variables and uses symbolic analysis to discover a superset of the implications in [9,2] to include variables not explicitly referenced in the definition of the state machine. On the other hand, the other approaches can quickly determine "local" mode invariants through simple graph traversal algorithms. It may be possible to combine both approaches to improve the scalability of invariant generation. In [4], query checking is used to discover mode invariants. That work uses symbolic methods and is in principle more general (a single query can discover *all* state invariants), but has serious scaling problems. The idea of simultaneous verification of multiple properties is not new [12, e.g.]. Our approach contrasts with previous work by using a parallel architecture that allows the incorporation of invariant generators to enhance the basic verification process.

1.1   Formal Preliminaries

Our work is built on logic-based model checking techniques that phrase reachability problems as entailment problems in a suitable logic for which efficient solvers exist. Relevant examples of such logics are propositional logic or any of the many logics used in SMT. For generality, we consider any of these logic $\mathcal{L}$ (with classical semantics) extending propositional logic, and rely on $\mathcal{L}$'s notion of variable, term, formula, free variable, model, satisfiability in a model, and entailment (which we denote as $\models_{\mathcal{L}}$). If F is a formula of $\mathcal{L}$ and $(x_1, \ldots, x_k)$ a tuple of distinct variables, we write $F[x_1, \ldots, x_k]$ to express that the free variables of F are in $(x_1, \ldots, x_k)$. If $t_1, \ldots, t_k$ are any terms, we write $F[t_1, \ldots, t_k]$ to denote the formula obtained from $F[x_1, \ldots, x_k]$ by simultaneously replacing each occurrence of x_i in F by t_i, for $i = 1, \ldots, k$. We denote finite tuples of elements by letters in bold font, and use comma (,) for tuple concatenation.

Let Q be a set of *states*, a *state space*. We assume some encoding of the state space Q in terms of n-tuples of ground terms in $\mathcal{L}$, for some fixed n.[2] Then, we say that (the encoding of) a state q *satisfies* a formula $F[x]$, where x is an n-tuple of distinct variables, if $F[x]$ is satisfied by every model of $\mathcal{L}$ interpreting x as q. This terminology extends to formulas over several n-tuples of free variables in the obvious way.

A *transition system* $\mathcal{S}$ over Q is a pair $(\mathcal{S}_I, \mathcal{S}_T)$ where $\mathcal{S}_I \subseteq Q$ is the set of $\mathcal{S}$'s *initial states*, and $\mathcal{S}_T \subseteq Q \times Q$ is $\mathcal{S}$'s *transition relation*. A state $q \in Q$ is *0-reachable* if $q \in \mathcal{S}_I$; it is *k-reachable* with $k > 0$ if it is $(k-1)$-reachable or $(s, q) \in \mathcal{S}_T$ for some $(k-1)$-reachable state s. A state is *($\mathcal{S}$-)reachable* if it is k-reachable for some $k \geq 0$. A *(state) property* is any formula $P[x]$ for some n-tuple x of variables. It is *invariant (for $\mathcal{S}$)* if it is satisfied by all $\mathcal{S}$-reachable states. For automated verification purposes one does not work directly with a transition system $\mathcal{S}$ itself, but with an *encoding* of it in some logic $\mathcal{L}$, namely, a pair $(I[x], T[x, x'])$ of formulas of $\mathcal{L}$, with x and x' both of size n, where

- $I[x]$ is a formula satisfied exactly by the initial states of $\mathcal{S}$;
- $T[x, x']$ is a formula satisfied by two reachable states q, q' iff $(q, q') \in \mathcal{S}_T$.

k-**Induction** Given an $\mathcal{L}$-encoding $(I[x], T[x, y])$ of some transition system $\mathcal{S}$, one can prove that a property P is invariant for $\mathcal{S}$ by showing that P is k-inductive.

Definition 1. *A state property $P[x]$ is k-inductive (wrt T) for some $k \geq 0$ if*

$$I[x_0] \wedge T[x_0, x_1] \wedge \cdots \wedge T[x_{k-1}, x_k] \models_{\mathcal{L}} P[x_0] \wedge \cdots \wedge P[x_k] \qquad (1)$$

$$T[x_0, x_1] \wedge \cdots \wedge T[x_k, x_{k+1}] \wedge P[x_0] \wedge \cdots \wedge P[x_k] \models_{\mathcal{L}} P[x_{k+1}] \qquad (2)$$

When entailment in $\mathcal{L}$ is decidable and an $\mathcal{L}$-solver is available for that, the k-inductiveness of a property P can be established by asking the $\mathcal{L}$-solver to prove both entailments in the definition above for some initial choice of k, retrying with an increasingly larger k until either the base case (1) is shown not to hold or both the base and the induction step (2) are shown to hold. In the second situation, P has been shown to hold for all reachable states, which means it is invariant. In the first situation, P is not invariant and a counterexample path can be generated from a counter-model of (1) above if the $\mathcal{L}$-solver is able to return models.

Since k-inductiveness is a sufficient condition for invariance, the k-induction procedure above is a sound verifier for invariance. The procedure, however, is not complete since there exist systems with invariant properties that are not k-inductive for any k. For those properties, the procedure will keep increasing k indefinitely. A number of improvements are possible to increase the procedure's *precision*, the set of invariant properties it can prove [15,3,5]. In particular, if Y is another state property already known to be invariant, one can strengthen the antecedent of the entailment in the induction step (2) by adding (conjunctively) the formula $Y[x_0] \wedge \cdots \wedge Y[x_{k+1}]$ to it. The strengthening is beneficial for eliminating *spurious counter-examples* to the induction step, i.e., counter-models involving unreachable states.

---

[2] Depending on $\mathcal{L}$, states may be encoded for instance as n-tuples of Boolean constants or as n-tuples of integer constants, and so on.

2   Template-Based Invariant Generation

In previous work [10] we described a general invariant discovery scheme that produces k-inductive invariants for a given transition system $\mathcal{S}$ from a *template* $R[x, y]$, a formula of $\mathcal{L}$ representing a decidable binary relation over one of the system's data types. The discovered invariants are instances $R[s, t]$ of the template generated with terms s, t from a set U of terms over the same n-tuple $\boldsymbol{x}$ of variables. The set U can be constructed heuristically in any number of ways from $\mathcal{S}$ and a given set of properties to be proven invariant for $\mathcal{S}$. In the experiments reported in [10], U included terms occurring in a given $\mathcal{L}$-encoding of $\mathcal{S}$, as well as a few distinguished constants.

The general scheme relies on the availability of efficient reasoning engines, such as SAT and SMT solvers, for the given logic $\mathcal{L}$, and capitalizes on their ability to quickly generate counter-models. It consists of a simple two-phase procedure, with an optional third phase not discussed here. Given the template R and the term set U, the first phase starts with the (very crude) conjecture that the state property $C[\boldsymbol{x}] = \bigwedge_{s,t \in U} R[s, t]$ is invariant. Then, it uses the $\mathcal{L}$-solver to weaken that conjecture by eliminating from it as many conjuncts $R[s, t]$ as possible that have a counterexample—specifically, all conjuncts falsified by a k-reachable state, for some heuristically determined k. The resulting formula C is passed to the second phase, which attempts to prove C k-inductive by checking that it satisfies the inductive step of k-induction. Any counter-examples there are used, conservatively, to weaken C further by eliminating additional conjuncts until no counter-examples exists. The final formula—the empty conjunction in the worst case—is by construction k-inductive, and so invariant.

The scheme above is impractical in its full generality because the number of instances of R over U can be very large. So we devised two specializations to relations R that are partial orders, one for general posets and one specific to binary posets. These specializations rely on the properties of partial orders to represent the conjunctive conjecture C compactly, and weaken it efficiently. In the following, we briefly illustrate the case of binary posets (see [10] for a more formal treatment and for the general case).

Invariant Generation for Binary Posets. For concreteness, and because it is relevant to our goal of learning invariants on mode variables, let us consider the poset $(\{\bot, \top\}, \rightarrow)$ of the Booleans, with logical implication $\rightarrow$ as the partial order, and with linear integer arithmetic as $\mathcal{L}$. In this case, the instances of R have the form $F \rightarrow G$, where F and G are any arithmetic predicates, i.e., quantifier-free arithmetic formulas.

The invariant generation procedure maintains a directed acyclic graph where each node contains a set of arithmetic predicates and stands for the conjecture that those predicates all imply each other (i.e., are all equivalent) in every reachable state of $\mathcal{S}$. An edge from a node A to a node B in the graph represents the weaker conjecture that the predicates in A imply the predicates in B, again in all reachable states.

The graph starts with a single node containing all the predicates in the candidate set U; it is then updated incrementally using a sequence $(M_1, M_2, \ldots)$ of models of $\mathcal{L}$, each containing a reachable state q that falsifies one of the conjectures in the current graph. Let G_0 be the initial graph and G_i the version of the graph updated after observing model M_i. The graph G_i is updated to G_{i+1} using model M_{i+1} as follows. If M_{i+1} falsifies (the conjecture expressed by) an edge of G_i, the edge is removed; if it falsifies

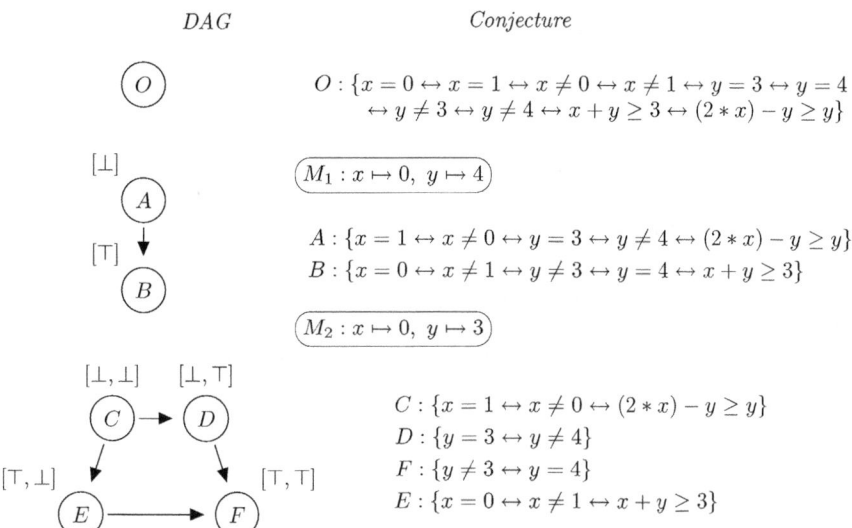

DAG   Conjecture

$O : \{x = 0 \leftrightarrow x = 1 \leftrightarrow x \neq 0 \leftrightarrow x \neq 1 \leftrightarrow y = 3 \leftrightarrow y = 4$
$\leftrightarrow y \neq 3 \leftrightarrow y \neq 4 \leftrightarrow x + y \geq 3 \leftrightarrow (2 * x) - y \geq y\}$

$M_1 : x \mapsto 0, \; y \mapsto 4$

$A : \{x = 1 \leftrightarrow x \neq 0 \leftrightarrow y = 3 \leftrightarrow y \neq 4 \leftrightarrow (2 * x) - y \geq y\}$
$B : \{x = 0 \leftrightarrow x \neq 1 \leftrightarrow y \neq 3 \leftrightarrow y = 4 \leftrightarrow x + y \geq 3\}$

$M_2 : x \mapsto 0, \; y \mapsto 3$

$C : \{x = 1 \leftrightarrow x \neq 0 \leftrightarrow (2 * x) - y \geq y\}$
$D : \{y = 3 \leftrightarrow y \neq 4\}$
$F : \{y \neq 3 \leftrightarrow y = 4\}$
$E : \{x = 0 \leftrightarrow x \neq 1 \leftrightarrow x + y \geq 3\}$

Fig. 2. In each graph, a node stands for a set of predicates that have evaluated to the same Boolean value ($\perp$ or $\top$) in each model considered until them. The predicates in a node are shown, as a double implication chain, in the *Conjecture* column. The list of observed values for the predicates in each node is shown on top of the node.

a node N, then (i) the node is split in two new nodes $N_\perp$ and $N_\top$ connected with an edge from $N_\perp$ and $N_\top$, (ii) N's predicates are assigned to $N_\perp$ and $N_\top$ depending on whether they are respectively falsified or satisfied by M_{i+1}, and (iii) all edges involving N are updated so that the set of conjectures represented by G_{i+1} is consistent with all the models observed so far and weakens the previous set only as little as needed to accommodate M_{i+1}. The procedure is perhaps best illustrated with an example.

Example 1. Consider a system whose $\mathcal{L}$-encoding contains exactly the predicates $x + y \geq 3$ and $(2*x) - y \geq y$, with $x \in [0..1]$ and $y \in [3..4]$, say. In the invariant generation procedure in [10], the set U would be just $\{x + y \geq 3, (2 * x) - y \geq y\}$. In the version we discuss here, if x and y are identified as mode variables of interest (see later) U is augmented with the predicates from the following set

$$V := \{x = 0, x = 1, y = 3, y = 4\} \cup \{x \neq 0, x \neq 1, y \neq 3, y \neq 4\} \,.$$

Figure 2 shows how the graph evolves with a sample sequence of two models. The procedure starts with the implication graph consisting of the node O, conjecturing that all predicates in U are equivalent. Nodes A and B are the result of splitting O. Nodes C and D are the result of splitting A, and node E and F of splitting B. ∎

The addition to U of predicates like those in the set V in Example 1 allows our procedure to discover, among others, invariants of the form $x = a \rightarrow y = b$ where x and y are mode variables and a and b are specific values in their range. Together with range constraints, negative predicates of the form $y \neq b$, allow the procedure to discover, in effect, also invariants of the form $x = a \rightarrow \bigvee_{i \in I} y = b_i$ where $[b_1..b_n]$ is y's range and $I \subseteq 1, \ldots, n$.[3] We will call these two kinds of invariants *mode invariants*.

3   Identifying Mode Variables

In this section we propose a technique to identify a relatively tight number of system variables as mode variables and a set of predicates on them to be used to produce mode invariants as described in the previous section. The overall goal is to capture with these invariants enough mode information about a software system under analysis—or, more accurately, about its encoding as a transition system in some logic $\mathcal{L}$.

The logic $\mathcal{L}$ used here will be the two-sorted logic consisting of the quantifier-free fragment of (mixed integer and real) linear arithmetic.

3.1   State Machines in Synchronous Models

Embedded systems, controllers for instances, are usually modeled as a set of synchronous dataflow computations governed by an overall mode logic. In aircraft control command, the mode logic could be a finite state machine iterating through the phases: taxi, take-off, flying, landing. For a car cruise controller, it could be a state machine describing how the controller engages and disengages depending on a number of parameters and actions. As mentioned in the introduction, when encoding these models as transition systems for verification purposes, the state machine expressing the original system's mode logic is often encoded with the introduction of *mode variables* to model the mode logic's finite state machine. These are variables over an enumeration type or, more often, Boolean variables or variables over a finite integer range.

While this approach is rather general, it has the disadvantage that the structure of state machine gets lost in the translation. This has important consequences for verification methods based on inductive arguments, such as k-induction, because the logical encoding ends up creating a state space with states that do not correspond to any state of the original state machine, and so are unreachable by the resulting transition system. These states are problematic because they typically lead to spurious counter-examples for the inductive step of the verification process.

To illustrate the problem with an example, consider again the microwave model of Figure 1, but without the variable mode. Consider then a layered encoding of the model into a transition system where a mode variable $top \in [1..2]$ represents the top states SETUP and RUNNING, with $top = 1$ for the first and $top = 2$ for the second, and a mode variable $running \in [0..2]$ represents the running state, with 0 meaning not running, 1 meaning SUSPENDED and 2 COOKING. The state space of this transition system contains the unreachable states $\{top \mapsto 1, run \mapsto 1\}$ and $\{top \mapsto 1, run \mapsto 2\}$ which may cause problems during induction. Those states can be ruled out during the verification process if $(top = 1) \leftrightarrow (running = 0)$ is discovered to be an invariant for the system.

---

[3] The reason is that such an invariant is equivalent to $\bigwedge_{j \in [b_1..b_n] \setminus I} (x = a \rightarrow y \neq b_j)$.

$$
\begin{aligned}
T_1 := \; & x' = z \\
& \wedge \; y' = \text{if } c_1' \text{ then 2 else} \\
& \qquad \text{if } c_2' \text{ then 1 else } x' \\
& \wedge \; z' = \text{if } c_3' \text{ then 0 else } y' \\
& \wedge \; x, y, z \in [0..2]
\end{aligned}
$$

$$
\begin{aligned}
T_2 := \; & a' = z \wedge x' = b \\
& \wedge \; b' = \text{if } c_4' \wedge a' = 2 \text{ then 1 else if } c_5' \text{ then } y' \text{ else 2} \\
& \wedge \; y' = \text{if } c_1' \text{ then 2 else if } c_2' \text{ then 1 else } x' \\
& \wedge \; z' = \text{if } c_3' \text{ then 0 else } y' \\
& \wedge \; a, b, x, y, z \in [0..2]
\end{aligned}
$$

Fig. 3. Transition relations over integer and Boolean variables. The latter are unconstrained just for simplicity.

3.2 Selecting Mode Variables

To generate mode invariants for a transition system S it is necessary to identify its mode variables in the first place. In the absence of explicit user-provided information, a possibility is to perform interval analysis on S to uncover variables that have a finite domain in all reachable states, and treat all such variables as mode variables. In general, examples of finite domain variables would be Boolean variables, enumeration type variables, and integer variables over a finite range. Then, one can strengthen S's transition relation as needed with the discovered finite domain constraints on those variables, and apply the invariant generation technique presented in Section 2 based on a set of predicates that contains all equations of the form $x = v$ and their negation, for each finite domain variable x and value v in its domain.

One problem with this approach is that it does not to scale well with respect to the number of finite domain variables or the size of their domains. Furthermore, many finite domain variables are uninteresting from a mode invariant generation perspective because they simply store intermediate values in the system's computation. For example, consider the two transition relations T_1 and T_2 in Figure 3, already strengthened with finite domain constraints for some of their variables. While artificial and somewhat contrived, they illustrate a common situation in which several of the finite domain variables can be ignored for depending functionally on other variables.

It is easy to see that in T_1 the values of x', y' and z'—i.e., the *next-state values* of x, y and z—are all determined by the value of the tuple (z, c_1', c_2', c_3'). A closer look reveals that they are also all determined by the value of (x, c_1', c_2', c_3'). As we will argue later, this suggests that it is enough to consider just z or just x as a mode variable for invariant generation purposes. In contrast, it would not be advantageous to consider just y because the next-state value of x is not determined by $(y, c_1', \ldots, c_3')$. In T_2, no tuple consisting of the Boolean variables and just one of the integer variables determines the next-state value of all the other variables. However, a tuple made of b and z and the Boolean variables will do. We formalize this intuition in the following and discuss a mode variable selection heuristics based on it.

Definition 2. *Let $F[z]$ be a formula in $\mathcal{L}$ and let $F^{\mathcal{L}}$ be the relation denoted by F in $\mathcal{L}$. A variable y in z depends (in F) on a tuple $\boldsymbol{x}$ of variables from z, if the projection $\pi_{\boldsymbol{x},y}(F^{\mathcal{L}})$ of $F^{\mathcal{L}}$ over $\boldsymbol{x}, y$, in the sense of relational algebra, is functional; that is, if $\pi_{\boldsymbol{x},y}(F^{\mathcal{L}})$ contains no two distinct tuples of the form $(\boldsymbol{v}, u_1), (\boldsymbol{v}, u_2)$; the variable y strictly depends on $\boldsymbol{x}$ if, additionally, it depends on no proper subtuple of $\boldsymbol{x}$.*

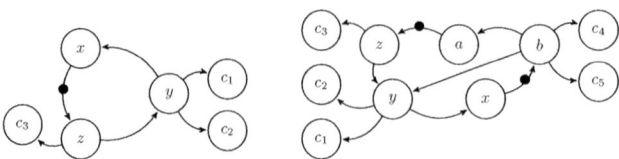

Fig. 4. Dependency graphs for the formulas T_1 and T_2, respectively, from Figure 3

Now, let's consider a formula $T[\boldsymbol{x}, \boldsymbol{x}']$ of $\mathcal{L}$ encoding the transition relation of some system $\mathcal{S}$. Suppose we are given a mapping dep from each variable y' of $\boldsymbol{x}'$ to a tuple of variables (of F) that y' strictly depends on. This mapping induces a directed labelled multigraph (V, E), a *dependency graph for* T, where

$$V = \boldsymbol{x} \qquad E = \{y \longrightarrow z \mid z' \in \mathrm{dep}(y')\} \cup \{y \dashrightarrow z \mid z \in \mathrm{dep}(y')\} \,.$$

Intuitively, there is an edge $\longrightarrow$ in the graph between y and z iff the next-state value of y depends on the next-state value of z, and there is an edge $\dashrightarrow$ between them iff the next-state value of y depends on the current-state value of z. For the transition relation T_1 in Figure 3, a suitable mapping dep would be $\{x' \mapsto (z), y' \mapsto (c'_1, c'_2, x'), z' \mapsto (c'_3, y')\}$. That mapping and its analogous for T_2 induce the multigraphs depicted in Figure 4.

We assume here that, for a given transition relation formula T, it is possible to compute from it a mapping dep, and hence its induced dependency graph. The ease, or in fact the possibility, of doing this automatically depends in general on T's format. In our experiments, where transition relation formulas are generated from system models written in Lustre [6], the process is straightforward because there each variable is given an explicit equational definition, as in the formulas of Figure 3.

Definition 3. *Let* $G = (V, E)$ *be a dependency graph and let* C *be a strongly connected component (SCC) of* G.[4] *The* base *of* C *is the set*

$$\{y \in C \mid x \dashrightarrow y \in E \text{ for some } x, y \in C\} \qquad (3)$$

if this set is non-empty; otherwise it is C *itself. A variable is a* base variable *of* G *if it is in the base of one of* G's *SCCs; it is a* stateful base variable *if it is in a base like (3).*

The SCCs of the left-hand graph of Figure 4 are $\{c_1\}, \{c_2\}, \{c_3\}$, and $\{x, y, z\}$; their respective bases are $\{c_1\}, \{c_2\}, \{c_3\}$, and $\{z\}$. The SCCs of the other graph are $\{c_1\}, \dots, \{c_5\}, \{a, b, c, x, y, z\}$; their respective bases are are $\{c_1\}, \dots, \{c_5\}$, and $\{b, z\}$.

It is not difficult to show that the following holds.

Proposition 1. *Let* $G = (V, E)$ *be a dependency graph for a transition relation* T, *and let* $N = \{x_1, \dots, x_k\}$ *be the union of all the base variables of* G. *Then, every variable in* $V \setminus N$ *depends on* $(x_1, \dots, x_k)$ *in* T.

The proposition above suggests that for invariant generation purposes it is enough to restrict attention to base variables only since they determine the values of all the other variables. Therefore, it is enough to constrain their values only. In fact, one can go

---

[4] With respect to paths built with any of the two edge types of G.

even further and ignore any invariants containing only non-stateful base variables. For instance, invariants over just the base variables $c_1, \ldots, c_3$ and $c_1, \ldots, c_5$ in the graphs of Figure 4. The reason is that the current state values of such variables constraints only current state values of other variables, but no next state values. This means that invariants containing only such variables will be entailed by the transition relation. Such invariants are useless for induction because they do not not strengthen the transition relation.[5] In this work we take a more draconian approach and simply discard *all* non-stateful base variables, to reduce as much as possible the number of predicates $x = v$ fed to our invariant discovery procedure. Of course, we also discard all stateful base variables that do not (or that we cannot determine to have) a finite domain.

The rationale behind this selection heuristics is that each independently defined state machines in the original system model—in particular, submachines of a hierarchical state machine—typically end up generating separate SCCs over mode variables in the dependency graph. Our conjecture is that enough useful invariants about these submachines and their relationships, are captured by considering just the finite-domain stateful base variables of each SCC.

A Variable Selection Procedure. To summarize, to limit the number of variables used for mode invariant generation for a transition relation formula T we use a procedure that (i) computes a dependency graph $G = (V, E)$ for T, (ii) identifies G's strongly connected components, and (iii) collects and returns all and only the finite-domain stateful base variables of these components.

4 Multi-property Incremental Verification

In this section we present a technique to verify simultaneously and incrementally multiple safety properties. Its relevance in this work is that it combines synergistically with the invariant generation techniques described in the previous sections.

Given a transition system encoding (I, T) and a list of properties $P^1, \ldots, P^n$ all to be checked for invariance, there are two possible ways of doing that with k-induction. One is to check each property individually. This is however time consuming and not very effective because a conjunction of formulas is usually easier to prove by induction that its individual conjuncts. Another way then is to check the property $P = P^1 \wedge \cdots \wedge P^n$. But this has its drawbacks as well. To start, if even just one of the individual properties fails to be invariant so does the whole P. However, even when P is indeed invariant, it is often the case that its individual constituents are k-inductive for different values of k. So the k-induction procedure has to wait until the largest of these values is reached before succeeding. In the worst case, one of the individual properties may not be k-inductive for any k, forcing the basic k-induction procedure to diverge.

Our solution to the problems above is to work with all properties at the same time but also keep track in each iteration of the k induction loop of the current status of each property P^i. In the base case, all properties that are falsified for a particular k are removed from consideration before increasing the value of k. In the induction step, all

---

[5] Note, however, that they might nevertheless be useful to *speed up* queries to an $\mathcal{L}$-solver, the way auxiliary (deductive) lemmas generally do.

```
proc base_proc ≡
    k := 0; props := {P¹,...,Pⁿ};  for P ∈ props  P.level := ∞
    while (props ≠ ∅)
        model := SAT(T₀ ∧ ··· ∧ Tₖ ∧ ¬ ⋀_{P∈props}(P₀ ∧ ··· ∧ Pₖ))
        if (model = Unsat) then  k := k + 1
        else
            invalid := filter_sat(model, props)
            send(INVALID(invalid), ind_proc)
            props := props \ invalid
            print_out  invalid
        if receive(VALID(possibly_valid, k'), ind_proc) then
            for P ∈ possibly_valid  P.level := k'
            valid := {P ∈ props | P.level ≤ k}
            props := props \ valid
            if props ≠ ∅ then  send(INVAR(valid), ind_proc)
            print_out  valid
    send(STOP, [ind_proc, inv_gen_proc])
```

Fig. 5. Base step process. For each i, T_i abbreviates $I[\boldsymbol{x}_i]$ if $i = 0$ and $T[\boldsymbol{x}_{i-1}, \boldsymbol{x}_i]$ otherwise. P_i abbreviates $P[\boldsymbol{x}_i]$.

properties that are validated for a particular k (see later for details on how we check this) are also removed from the list of properties to be checked but immediately added back as invariants, to aid the verification of the remaining ones.

Our incremental approach builds on the parallel k-induction-based model checking architecture we developed in recent work [11]. That architecture is designed to minimize synchronization delays and facilitate the incorporation of concurrent invariant generators, and has the following basic structure:

$$\text{base\_proc} \parallel \text{ind\_proc} \parallel \text{inv\_gen\_proc}_i$$

The base and the inductive step of k-induction execute concurrently respectively in the base_proc and the ind_proc process, as do one or more independent processes inv_gen_proc$_i$ that incrementally generate auxiliary invariants for the system being verified. These invariants are fed to the k-induction loop as soon as they are produced and used to strengthen the induction hypothesis. The processes communicate with one another by asynchronous messages passing, with non-blocking send and receive operations relying on message queues. The operation receive(pat, $source$) matches the pattern pat with a message from process $source$, if any; it returns true if there is a message and the matching succeeds, and returns false otherwise. Some more details on each process are described below, assuming for simplicity just one invariant generator.

Base Case Process. Figure 5 shows the pseudo-code for base_proc. Its main task is to partition incrementally the initial set of properties, the initial value of $props$, into valid (i.e., invariant) properties and invalid (i.e., non-invariant) ones.

The process checks the entailment in Case (1) of Definition 1 for increasing values of k starting from 0. The function SAT implements the $\mathcal{L}$-solver. It takes a formula F over n states and returns either unsat or a model $model$ of F, i.e., a sequence of n states that satisfies F. The function filter_sat returns the set of properties in $props$ that

proc ind_proc $\equiv$
   $k := 0;\ props := \{P^1, \ldots, P^n\};\ invs := \emptyset$
   while $(props \neq \emptyset)$
      assert$(T_{k+1} \wedge \bigwedge_{Y \in invs} Y_{k+1} \wedge \bigwedge_{P \in props} P_k)$
      if entailed$(\bigwedge_{P \in props} P_{k+1})$ **then** send(VALID$(props, k)$, base_proc); **exit**
      else
         $possibly_valid := $ recheck_validity$(props, k)$
         send(VALID$(possibly\_valid, k)$, base_proc)
         $props := props \setminus possibly\_valid$
      if receive$(msg, \_)$ **then**
         match msg **with**
            STOP $\rightarrow$ **exit**
           | INVALID$(invalid) \rightarrow props := props \setminus invalid$
           | INVAR$(new\_invs) \rightarrow$ **for** $i = 0$ **to** $k+1$ assert$(\bigwedge_{Y \in new\_invs} Y_i)$
                                   $invs := invs \cup new\_invs$
      $k := k + 1$

Fig. 6. Inductive Step Process. For each i, Y_i abbreviates $Y[x_i]$.

are falsified by one of the states in *model*. Those properties are definitely invalid. They are both printed for the user and sent the inductive step process, and then removed from the current set *props* of properties. Note that the counter k left unchanged as long as the solver keeps finding counter-models for some of the current properties.

Before repeating the main loop the process checks its message queue; a message from ind_proc stating that it has successfully proven the inductive step (2) of Definition 1 for a some k' and a subset *possibly_valid* of the input properties. The value k' need not be the same as k since the two processes increase their own induction level independently. As a consequence, the base_proc first annotates each property in *possibly_valid* with k', storing it in the level field of the property. Then it collects in *valid* all properties from *props* whose level is at that point smaller or equal to the current k. Each property P in *valid* has been cooperatively shown by the two processes to be (P.level)-inductive. So it is removed from the list of properties to be proven and sent back to ind_proc to be used as an invariant, provided there are still properties to be proven. The process terminates when *props* becomes empty, sending a termination signal to the other processes as well.

Inductive Step Process. Pseudo-code for this process is provided in Figure 6. There we assume a stateful $\mathcal{L}$-solver that allows one to assert formulas (with the assert procedure) and then check (with the entailed Boolean function) whether the current set of asserted formulas entails a given one.

The process checks the inductive step entailment for increasing values of k. However, it strengthens the induction hypothesis with any invariants at its disposal (in *invs*). If the entailment holds for the current k and set *props* of properties, they are both sent to the base case process, and the inductive process terminates. As discussed earlier, base_proc will confirm their individual invariance, or not, by checking that they have no counter-examples of length up to k. If the entailment fails, the process passes the properties to the auxiliary function recheck_validity which (using a separate copy of the $\mathcal{L}$-solver) computes the largest subset of *props* for which the entailment test succeeds. This set is sent to base_proc as in the previous case, and removed from *props*.

The remaining properties are rechecked for an increased value of k. Before proceeding, however, the process checks its message queue. If it sees a message (from base_proc) with a set of properties found to be invalid, it removes them from *props*. If it sees a message from an invariant generation process, providing a set of auxiliary invariants, or from base_proc, providing a set of properties confirmed to be valid and so usable as invariants, it asserts all those invariants for all steps from 0 to $k + 1$ and then adds them to the current invariant set *invs*. The process terminates if it sees a termination message from base_proc.

Incremental Invariant Generator. This process can be any incremental invariant generator for the given transition system. It is supposed to keep sending any newly discovered invariants to the induction step process until it can generate no more, or it receives a termination message from the base case process. In our current implementation, we have one such process that essentially implements the general template-based invariant discovery procedure seen in Section 2. The process is composed of three main modules: the *Candidate generator*, which constructs the initial set C of candidate invariants from predefined templates, the *Int invariant generator*, which produces from C invariants of the form $s \leq t$ where s and t are integer terms, and the *Bool invariant generator*, which produces invariants of the forms $F \rightarrow G$ as discussed in Section 3.

5   Experimental Results

To evaluate experimentally the techniques presented in the previous sections, we have implemented it as new options in our k-induction-based model checker KIND.[6] KIND can simultaneously check multiple invariant properties of programs written in an idealized version of the specification/programming language Lustre [6].[7] The underlying logic of KIND is a quantifier-free logic that includes both propositional logic and mixed real-integer linear arithmetic. Lustre programs can be readily encoded as transition systems in this logic [5]. KIND uses the SMT solvers CVC3 and Yices, in alternative, as satisfiability solvers for this logic. The version discussed here is based on the incremental parallel architecture discussed in the previous section.

The experiments discussed below were run, using Yices version 1.0.9 as the background solver, on a 12-core 2.10 GHz AMD Opteron machine under Ubuntu 11.10. The experiments used benchmark derived from the following problem.

NASA Docking Approach Example. This is a complex hierarchical problem that describes the approach behavior of the Space Shuttle when docking with the International Space Station [14]. As the shuttle approaches the ISS it goes through several operational modes related to how the shuttle is to orient itself for capture, dock with the ISS, and capture the ISS docking latch, among several other operational modes. The model describing this behavior is quite intricate and consists of a hierarchical and parallel state machine with three levels of hierarchy and multiple parallel state machines, including a

---

[6] Tools and experimental data can be found at http://clc.cs.uiowa.edu/Kind.

[7] The idealization consists in treating Lustre's numerical types as infinite precision types.

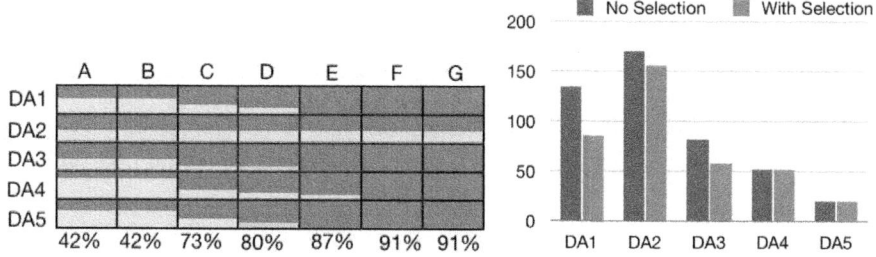

Fig. 7. The left graph illustrates the distribution of solved and unsolved properties for the different benchmarks, DA1 ... DA5, using configurations A though G for KIND. Darker areas indicate the portion of solved properties. The right graph indicates the number of variables considered for mode invariant generation before and after applying the selection procedure from Section 3 to the 5 benchmarks.

total of 64 states. For the purposes of this experiment, we created five reduced versions of the docking approach model in which we replaced one of the complex hierarchical states with a simple state that approximates its behavior. This allows us to examine the behavior of the invariant generation over a range of state machine models with different characteristics (the hierarchical states vary substantially in size). Note that it also causes some of the original properties to be violated.

We ran KIND on the five problems above in different configurations: (A) *single-prop, no invars*; (B) *multi-prop, no invars*; (C) *single-prop, no mode invars*; (D) *multi-prop, no mode invars*; (E) *single-prop, mode invars*; (F) *multi-prop, mode invars*; (G) *multi-prop, selected mode invars*. In the *single-prop* configurations, each property was checked individually; in *multi-prop* configurations the properties were checked together incrementally as discussed in Section 4. In *no invars*, no invariants were generated at all. In *no mode invars*, invariants were generated, but no mode invariants. In *mode invars*, invariants included mode invariants generated for *all* finite domain variables. In the *selected mode invars* configuration, invariants included mode invariants generated only for those variables selected by the procedure discussed in Section 3.

Precision Results. The first graph of Figure 7 summarizes the precision achieved by KIND under the configurations above. In cases A and B, KIND is able to solve 42% of all the properties without relying on auxiliary invariants. The percentage of solved properties goes up to 73%, 80% and 87% in cases C, D and E, respectively, illustrating the effectiveness of invariant generation and of incremental multi-property verification. In particular, general invariants (case C) increase precision by 31 percentage points over configurations A and B. The further addition of mode invariants in the single property case increases precision by 14 more points (from C to E). Going from single to incremental multi-property verification but without mode invariants (from C to D) increases precision by 7 points. Finally, the combination of multi-property verification and mode invariants does noticeably better than each of them alone (91% vs 80% and 87%).

Runtime Results. As we conjectured, reducing the number of variables to generate mode invariants using our variable selection procedure reduces runtimes in general without impacting precision. In particular, in case F, KIND can to solve all the valid properties in a total time of 15s; in case G, such value goes down to 14.3s. As shown by the right-hand side graph of Figure 7, our selection procedure reduces the number of mode variables to consider in problems DA1, DA2 and DA3—although not in DA4 and DA5, perhaps because of their small number there. As a result, the total time for the first three benchmarks goes respectively from 4089ms, 57ms and 6025ms before the selection of mode variables (case F) to 3728ms, 33ms and 5752ms after (case G).

6 Conclusion

We have presented two complementary techniques for the verification of safety properties in synchronous systems with complex modal behavior. A first technique allows the automatic generation of certain inductive invariants for system variables identified heuristically as containing mode information. A second technique allows the simultaneous verification of multiple properties in an incremental fashion. The synergy between these two techniques allowed us to verify safety properties of complex systems like the NASA docking benchmarks.

References

1. Damas, C., Lambeau, B., Dupont, P., van Lamsweerde, A.: Generating annotated behavior models from end-user scenarios. IEEE Transactions on SE 31(12), 1056–1073 (2005)
2. Damas, C., Lambeau, B., Roucoux, F., van Lamsweerde, A.: Analyzing critical process models through behavior model synthesis. In: ICSE 2009, pp. 441–451. IEEE (2009)
3. de Moura, L., Rueß, H., Sorea, M.: Bounded Model Checking and Induction: From Refutation to Verification. In: Hunt Jr., W.A., Somenzi, F. (eds.) CAV 2003. LNCS, vol. 2725, pp. 14–26. Springer, Heidelberg (2003)
4. Gurfinkel, A., Chechik, M., Devereux, B.: Temporal logic query checking: a tool for model exploration. IEEE Transactions on SE 29(10), 898–914 (2003)
5. Hagen, G., Tinelli, C.: Scaling up the formal verification of Lustre programs with SMT-based techniques. In: FMCAD 2008, pp. 1–9. IEEE Press, Piscataway (2008)
6. Halbwachs, N., Caspi, P., Raymond, P., Pilaud, D.: The synchronous data-flow programming language Lustre. Proceedings of the IEEE 79(9), 1305–1320 (1991)
7. Hangal, S., Narayanan, S., Chandra, N., Chakravorty, S.: IODINE: a tool to automatically infer dynamic invariants for hardware designs. In: DAC 2005, pp. 775–778 (June 2005)
8. Jeffords, R., Heitmeyer, C.: Automatic generation of state invariants from requirements specifications. SIGSOFT Softw. Eng. Notes 23, 56–69 (1998)
9. Jeffords, R., Heitmeyer, C.: An algorithm for strengthening state invariants generated from requirements specifications. In: RE 2001, pp. 182–191 (2001)
10. Kahsai, T., Ge, Y., Tinelli, C.: Instantiation-Based Invariant Discovery. In: Bobaru, M., Havelund, K., Holzmann, G.J., Joshi, R. (eds.) NFM 2011. LNCS, vol. 6617, pp. 192–206. Springer, Heidelberg (2011)
11. Kahsai, T., Tinelli, C.: PKIND: a parallel k-induction based model checker. In: PDMC 2011. EPTCS, vol. 72, pp. 55–62 (2011)
12. Khasidashvili, Z., Nadel, A., Palti, A., Hanna, Z.: Simultaneous SAT-Based Model Checking of Safety Properties. In: Ur, S., Bin, E., Wolfsthal, Y. (eds.) HVC 2005. LNCS, vol. 3875, pp. 56–75. Springer, Heidelberg (2006)
13. Manna, Z., Pnueli, A.: Temporal Verification of Reactive Systems: Safety. Springer (1995)
14. Sampson, M., Derevenko, V.: Interface definition document (IDD) for international space station (ISS) visiting vehicles (VVs). Technical report, NASA (2000)
15. Sheeran, M., Singh, S., Stålmarck, G.: Checking Safety Properties Using Induction and a SAT-Solver. In: Johnson, S.D., Hunt Jr., W.A. (eds.) FMCAD 2000. LNCS, vol. 1954, pp. 108–125. Springer, Heidelberg (2000)

A Semantic Analysis of Wireless Network Security Protocols

Damiano Macedonio and Massimo Merro

Dipartimento di Informatica, Università degli Studi di Verona, Italy

Abstract. Gorrieri and Martinelli's *tGNDC* is a general framework for the formal verification of security protocols in a concurrent scenario. We generalise their *tGNDC* schema to verify wireless network security protocols. Our generalisation relies on a simple *timed broadcasting process calculus* whose operational semantics is given in terms of a labelled transition system which is used to derive a standard *simulation theory*. We apply our *tGNDC* schema to perform a security analysis of LiSP, a well-known *key management protocol* for wireless sensor networks.

1 Introduction

Wireless communications are vulnerable to several kinds of threats and risks. An adversary can compromise a device, alter the integrity of the data, eavesdrop on messages, inject fake messages, and waste network resource. Designing security protocols for wireless systems requires a deep understanding of their resource limitations to achieve acceptable performances.

In this paper, we adopt a process calculus approach to formalise and verify wireless network security protocols. We propose a simple *timed broadcasting process calculus*, called aTCWS, for modelling wireless networks. The time model we adopt is known as the *fictitious clock* approach (see e.g. [7]): A global clock is supposed to be updated whenever all nodes agree on this, by globally synchronising on a special timing action σ. Broadcast communications span over a limited area, called *transmission range*. Both broadcast actions and internal actions are assumed to take no time. This is a reasonable assumption whenever the duration of those actions is negligible with respect to the chosen time unit. The operational semantics of our calculus is given in terms of a labelled transition semantics in the SOS style of Plotkin. The calculus enjoys standard time properties, such as: *time determinism, maximal progress*, and *patience* [7]. The labelled transition semantics is used to derive a co-inductive (weak) *simulation theory* which can be easily *mechanised* by relying on well-known interactive theorem provers such as Isabelle/HOL [13] or Coq [2]. Based on our simulation theory, we generalise Gorrieri and Martinelli's *timed Generalized Non-Deducibility on Compositions* (*tGNDC*) schema [5,6], a well-known general framework for the formal verification of timed security properties. The basic idea of *tGNDC* is the following: a protocol M satisfies $tGNDC^{\rho(M)}$ if the presence of an arbitrary *attacker* does not affect the behaviour of M with respect to the abstraction $\rho(M)$. By varying

A. Goodloe and S. Person (Eds.): NFM 2012, LNCS 7226, pp. 403–417, 2012.

$\rho(M)$ it is possible to express different timed security properties for the protocol M. Examples are the *timed integrity* property, which ensures the freshness of authenticated packets, and the *timed agreement* property, when agreement between two parties must be reached within a certain deadline. In this paper, we will focus on the first property. In order to avoid the universal quantification over all possible attackers when proving *tGNDC* properties, we provide a sound proof technique based on the notion of *the most powerful attacker*.

As a main application of our theory, we provide a formal specification of LiSP [14], a well-known key management protocol for *wireless sensor networks* that, through an efficient mechanism of re-keying, provides a good trade-off between resource consumption and network security. We perform our *tGNDC*-based analysis on LiSP showing that old packets can be authenticated as a consequence of a *replay attack*. To our knowledge this attack has never appeared in the literature. Then, we formally prove that similar attacks can be avoided if nonces are added to the original LiSP protocol.

Related Work. A number of process calculi have been proposed for modelling different aspects of wireless systems [8,16,9,4,3,10]. The paper [12] proposes an algebraic approach to perform security analysis of communication protocols for ad hoc networks. The paper [1] proposes a first formalisation of *tGNDC* in our setting and a security analysis of the authentication protocols μTESLA [15] and LEAP+ [17]. μTESLA has also been studied within the calculus tCryptoSPA [5,6], an extension of Milner's CCS where node distribution, local broadcast communication, and message loss are codified in terms of point-to-point transmission and a notion of discrete time. As a consequence, specifications and security analyses of wireless network protocols in tCryptoSPA are much more involved than ours.

2   The Calculus

Table 1 provides the syntax of our *applied Timed Calculus for Wireless Systems*, aTCWS, in a two-level structure: A lower one for *processes* and an upper one for *networks*. We assume a set *Nds* of logical node names, ranged over by m, n. *Var* is the set of *variables*, ranged over by x, y, z. We define *Val* to be the set of values, and *Msg* to be the set of *messages*, i.e., closed values that do not contain variables. Letters $u, u_1 \ldots$ range over *Val*, and $w, w' \ldots$ range over *Msg*.

Both syntax and operational semantics of aTCWS are parametric with respect to a given *decidable* inference system, i.e. a set of rules to model operations on messages by using constructors. For instance, the rules

$$\text{(pair)}\ \frac{w_1 \quad w_2}{\text{pair}(w_1, w_2)} \qquad \text{(fst)}\ \frac{\text{pair}(w_1, w_2)}{w_1} \qquad \text{(snd)}\ \frac{\text{pair}(w_1, w_2)}{w_2}$$

allow us to deal with pairs of values. We write $w_1 \ldots w_k \vdash_r w_0$ to denote an application of rule r to the closed values $w_1 \ldots w_k$ to infer w_0. Given an inference system, the *deduction function* $\mathcal{D} : 2^{Msg} \to 2^{Msg}$ associates a (finite) set ϕ of

Table 1. Syntax of aTCWS

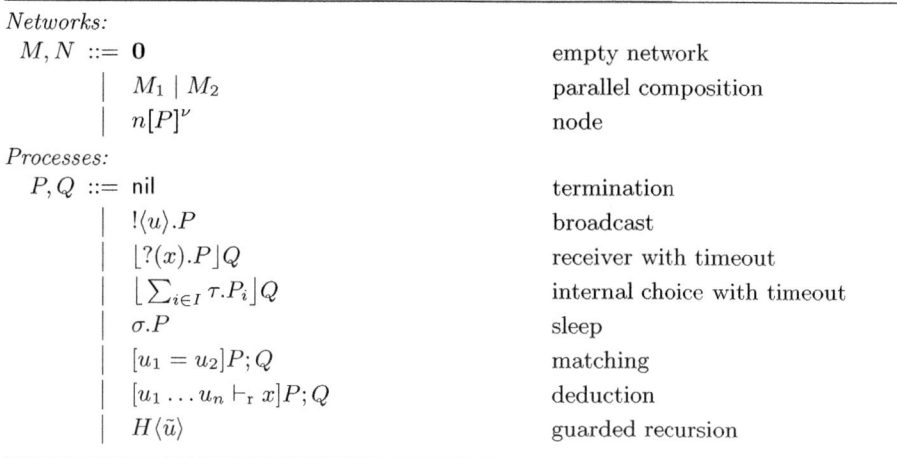

Networks:		
$M, N ::=$	**0**	empty network
	$\mid \quad M_1 \mid M_2$	parallel composition
	$\mid \quad n[P]^{\nu}$	node
Processes:		
$P, Q ::=$	nil	termination
	$\mid \quad !\langle u \rangle.P$	broadcast
	$\mid \quad \lfloor ?(x).P \rfloor Q$	receiver with timeout
	$\mid \quad \lfloor \sum_{i \in I} \tau.P_i \rfloor Q$	internal choice with timeout
	$\mid \quad \sigma.P$	sleep
	$\mid \quad [u_1 = u_2]P; Q$	matching
	$\mid \quad [u_1 \ldots u_n \vdash_{\mathrm{r}} x]P; Q$	deduction
	$\mid \quad H\langle \tilde{u} \rangle$	guarded recursion

messages to the set $\mathcal{D}(\phi)$ of messages that can be deduced from ϕ, by applying instances of the rules of the inference system.

Networks are collections of nodes running in parallel and using a unique common channel to communicate with each other. All nodes have the same transmission range (this is a quite common assumption in models for ad hoc networks [11]). The communication paradigm is *local broadcast*: only nodes located in the range of the transmitter may receive data. We write $n[P]^{\nu}$ for a node named n (the device network address) executing the sequential process P. The tag ν contains the neighbours of n ($\nu \subseteq Nds \setminus \{n\}$). Our wireless networks have a fixed topology as node mobility is not relevant to our analysis.

Processes are sequential and live within the nodes. In the processes $!\langle w \rangle.P$, $\lfloor ?(x).P \rfloor Q$, $\lfloor \sum_{i \in I} \tau.P_i \rfloor Q$ and $\sigma.Q$, the occurrences of P, P_i and Q are said to be *guarded*; the occurrences of Q are also said to be *time-guarded*. In the processes $\lfloor ?(x).P \rfloor Q$ and $[w_1 \ldots w_n \vdash_{\mathrm{r}} x]P$ the variable x is said to be *bound* in P. A variable which is not bound is said to be *free*. We adopt the standard notion of α-conversion on bound variables and we identify processes up to α-conversion. We assume there are no free variables in our networks. The absence of free variables will be maintained as networks evolve. We write $\{^w/_x\}P$ for the substitution of the variable x with the message w in P. We write $H\langle w_1, \ldots, w_k \rangle$ to denote a recursive process H defined via an equation $H(x_1, \ldots, x_k) = P$, where (i) the tuple $x_1, \ldots, x_k$ contains all the variables that appear free in P, and (ii) P contains only guarded occurrences of the process identifiers, such as H itself. H is said to be *time-guarded* if P contains only time-guarded occurrences of the process identifiers.

We report some notational *conventions*. We write $\prod_{i \in I} M_i$ to mean the parallel composition of all M_i, for $i \in I$. We write $\sigma^k.P$ for the process $\sigma.\ldots.\sigma.P$, where prefix σ occurs k times. The process $[w_1 = w_2]P$ is an abbreviation for $[w_1 = w_2]P; \mathsf{nil}$. Similarly, $[w_1 \ldots w_n \vdash_{\mathrm{r}} x]P$ denotes $[w_1 \ldots w_n \vdash_{\mathrm{r}} x]P; \mathsf{nil}$.

Table 2. LTS - Transmissions, internal actions and time passing

$$(\text{Snd}) \ \frac{-}{m[!\langle w\rangle.P]^\nu \xrightarrow{m!w\triangleright\nu} m[P]^\nu} \qquad\qquad (\text{Rcv}) \ \frac{m\in\nu}{n[\lfloor?(x).P\rfloor Q]^\nu \xrightarrow{m?w} n[\{^w/_x\}P]^\nu}$$

$$(\text{RcvEnb}) \ \frac{m\notin \mathsf{nds}\,(M)}{M \xrightarrow{m?w} M} \qquad\qquad (\text{RcvPar}) \ \frac{M \xrightarrow{m?w} M' \quad N \xrightarrow{m?w} N'}{M \mid N \xrightarrow{m?w} M' \mid N'}$$

$$(\text{Bcast}) \ \frac{M \xrightarrow{m!w\triangleright\nu} M' \quad N \xrightarrow{m?w} N' \quad \mu := \nu\backslash\mathsf{nds}\,(N)}{M \mid N \xrightarrow{m!w\triangleright\mu} M' \mid N'}$$

$$(\text{Tau}) \ \frac{h\in I}{m[\lfloor\sum_{i\in I}\tau.P_i\rfloor Q]^\nu \xrightarrow{\tau} m[P_h]^\nu} \qquad\qquad (\text{TauPar}) \ \frac{M \xrightarrow{\tau} M'}{M \mid N \xrightarrow{\tau} M' \mid N}$$

$$(\sigma\text{-nil}) \ \frac{-}{n[\mathsf{nil}]^\nu \xrightarrow{\sigma} n[\mathsf{nil}]^\nu} \qquad\qquad (\text{Sleep}) \ \frac{-}{n[\sigma.P]^\nu \xrightarrow{\sigma} n[P]^\nu}$$

$$(\sigma\text{-Rcv}) \ \frac{-}{n[\lfloor?(x).P\rfloor Q]^\nu \xrightarrow{\sigma} n[Q]^\nu} \qquad\qquad (\sigma\text{-Sum}) \ \frac{-}{m[\lfloor\sum_{i\in I}\tau.P_i\rfloor Q]^\nu \xrightarrow{\sigma} m[Q]^\nu}$$

$$(\sigma\text{-Par}) \ \frac{M \xrightarrow{\sigma} M' \quad N \xrightarrow{\sigma} N'}{M \mid N \xrightarrow{\sigma} M' \mid N'} \qquad\qquad (\sigma\text{-0}) \ \frac{-}{0 \xrightarrow{\sigma} 0}$$

In the sequel, we will make use of a standard notion of structural congruence to abstract over processes that differ for minor syntactic differences.

Definition 1. Structural congruence *over networks, written* $\equiv$, *is defined as the smallest equivalence relation, preserved by parallel composition, which is a commutative monoid with respect to parallel composition and internal choice, and for which* $n[H\langle\tilde{w}\rangle]^\nu \equiv n[\{^{\tilde{w}}/_{\tilde{x}}\}P]^\nu$, *if* $H(\tilde{x}) = P$.

Here, we provide some definitions that will be useful in the remainder of the paper. Given a network M, $\mathsf{nds}\,(M)$ returns the node names of M. More formally, $\mathsf{nds}\,(0) = \emptyset$, $\mathsf{nds}\,(n[P]^\nu) = \{n\}$ and $\mathsf{nds}\,(M_1 \mid M_2) = \mathsf{nds}\,(M_1) \cup \mathsf{nds}\,(M_2)$. For $m \in \mathsf{nds}\,(M)$, the function $\mathsf{ngh}(m, M)$ returns the set of the neighbours of m in M. Thus, if $M \equiv m[P]^\nu \mid N$ then $\mathsf{ngh}(m, M) = \nu$. We write $\mathsf{Env}\,(M)$ to mean all the nodes of the environment reachable by the network M. Formally, $\mathsf{Env}\,(M) = \cup_{m\in\mathsf{nds}(M)}\mathsf{ngh}(m, M) \setminus \mathsf{nds}\,(M)$.

The syntax provided in Table 1 allows us to derive networks which are somehow ill-formed. The following definition identifies well-formed networks.

Definition 2 (Well-formedness). M *is said to be* well-formed *if (i)* $M \equiv N \mid m_1[P_1]^{\nu_1} \mid m_2[P_2]^{\nu_2}$ *implies* $m_1 \neq m_2$; *(ii)* $M \equiv N \mid m_1[P_1]^{\nu_1} \mid m_2[P_2]^{\nu_2}$, *with* $m_1 \in \nu_2$, *implies* $m_2 \in \nu_1$; *(iii) for all* $m, n \in \mathsf{nds}\,(M)$ *there are* $m_1, \ldots, m_k \in \mathsf{nds}\,(M)$, *such that* $m=m_1$, $n=m_k$, $m_i \in \mathsf{ngh}(m_{i+1}, M)$, *for* $1 \leq i \leq k-1$.

In Table 2, we provide a labelled transition system (LTS) for aTCWS, in the SOS style of Plotkin. Intuitively, the computation proceeds in lock-steps: between every global synchronisation all nodes proceeds asynchronously by performing actions with no duration, which represent either broadcast or input or internal

Table 3. LTS - Matching, recursion and deduction

$$(\text{Then})\ \dfrac{n[P]^\nu \xrightarrow{\lambda} n[P']^\nu}{n[[w=w]P;Q]^\nu \xrightarrow{\lambda} n[P']^\nu} \qquad (\text{Else})\ \dfrac{n[Q]^\nu \xrightarrow{\lambda} n[Q']^\nu \quad w_1 \neq w_2}{n[[w_1=w_2]P;Q]^\nu \xrightarrow{\lambda} n[Q']^\nu}$$

$$(\text{Rec})\ \dfrac{n[\{\tilde{w}/\tilde{x}\}P]^\nu \xrightarrow{\lambda} n[P']^\nu \quad H(\tilde{x}) \overset{\text{def}}{=} P}{n[H\langle\tilde{w}\rangle]^\nu \xrightarrow{\lambda} n[P']^\nu}$$

$$(\text{DT})\ \dfrac{n[\{w/x\}P]^\nu \xrightarrow{\lambda} n[R]^\nu \quad w_1\ldots w_n \vdash_r w}{n[[w_1\ldots w_n \vdash_r x]P;Q]^\nu \xrightarrow{\lambda} n[R]^\nu} \quad (\text{DF})\ \dfrac{n[Q]^\nu \xrightarrow{\lambda} n[R]^\nu \quad \nexists w.\ w_1\ldots w_n \vdash_r w}{n[[w_1\ldots w_n \vdash_r x]P;Q]^\nu \xrightarrow{\lambda} n[R]^\nu}$$

actions. Communication proceeds even if there are no listeners: Transmission is a *non-blocking* action. Moreover, communication is *lossy* as some receivers within the range of the transmitter might not receive the message. This may be due to several reasons such as signal interferences or the presence of obstacles.

The metavariable λ ranges over the set of labels $\{\tau, \sigma, m!w \triangleright \nu, m?w\}$ denoting internal action, time passing, broadcasting and reception. Let us comment on the transition rules of Table 2. In rule (Snd) a sender m dispatches a message w to its neighbours ν, and then continues as P. In rule (Rcv) a receiver n gets a message w coming from a neighbour m, and then evolves into process P, where all the occurrences of the variable x are replaced with w. If no message is received in the current time slot a timeout fires and the node n will continue with process Q, according to the rule (σ-Rcv). The rule (RcvPar) models the composition of two networks receiving the same message from the same transmitter. Rule (RcvEnb) says that every node can synchronise with an external transmitter m. Notice that a node $n[\lfloor ?(x).P \rfloor Q]^\nu$ might execute rule (RcvEnb) instead of rule (Rcv); in this manner we model message loss. Rule (Bcast) models the propagation of messages on the broadcast channel. Rules (Tau) and (TauPar) model internal computations. Rule (σ-Rcv) models timeout on receivers, and similarly rule (σ-Sum) describes timeout on internal activities. Rule (σ-Par) models time synchronisation. Rules (Bcast) and (TauPar) have their symmetric counterparts. Table 3 reports the standard rules for matching, recursion and deduction.

Below, we report a number of basic properties of our LTS.

Proposition 1. *Let M, M_1 and M_2 be well-formed networks.*

1. *$m \notin \mathsf{nds}(M)$ if and only if $M \xrightarrow{m?w} N$, for some network N.*
2. *$M_1 \mid M_2 \xrightarrow{m?w} N$ if and only if there are N_1 and N_2 such that $M_1 \xrightarrow{m?w} N_1$, $M_2 \xrightarrow{m?w} N_2$ with $N = N_1 \mid N_2$.*
3. *If $M \xrightarrow{m!w \triangleright \mu} M'$ then $M \equiv m[!\langle w\rangle.P]^\nu \mid N$, for some m, ν, P and N such that $m[!\langle w\rangle.P]^\nu \xrightarrow{m!w \triangleright \nu} m[P]^\nu$, $N \xrightarrow{m?w} N'$, $M' \equiv m[P]^\nu \mid N'$ and $\mu = \nu \setminus \mathsf{nds}(N)$.*
4. *If $M \xrightarrow{\tau} M'$ then $M \equiv m[\lfloor \sum_{i \in I} \tau.P_i \rfloor Q]^\nu \mid N$, for some m, ν, P_i, Q and N such that $m[\lfloor \sum_{i \in I} \tau.P_i \rfloor Q]^\nu \xrightarrow{\tau} m[P_h]^\nu$, for some $h \in I$, and $M' \equiv m[P_h]^\nu \mid N$.*

5. $M_1 \mid M_2 \xrightarrow{\sigma} N$ if and only if there are N_1 and N_2 such that $M_1 \xrightarrow{\sigma} N_1$, $M_2 \xrightarrow{\sigma} N_2$ and $N = N_1 \mid N_2$.

Proposition 2. *Let M be well-formed. If $M \xrightarrow{\lambda} M'$ then M' is well-formed.*

Based on the LTS of Section 2, we define a standard notion of *timed labelled similarity* for aTCWS. We distinguish between transmissions which may be observed and those which may not be observed by the environment. We extend the set of rules of Table 2 with the following two rules:

$$\text{(Shh)} \quad \frac{M \xrightarrow{m!w \triangleright \emptyset} M'}{M \xrightarrow{\tau} M'} \qquad\qquad \text{(Obs)} \quad \frac{M \xrightarrow{m!w \triangleright \nu} M' \quad \mu \subseteq \nu \quad \mu \neq \emptyset}{M \xrightarrow{!w \triangleright \mu} M'}$$

Rule (Shh) models transmissions that cannot be observed because none of the potential receivers is in the environment. Rule (Obs) models transmissions that can be received (and hence observed) by those nodes of the environment contained in ν. Notice that the name of the transmitter is removed from the label. This is motivated by the fact that nodes may refuse to reveal their identities, e.g. for security reasons or limited sensory capabilities in perceiving these identities.

In the sequel, the metavariable α will range over the following actions: τ, σ, $!w \triangleright \nu$ and $m?w$. We adopt the standard notation for weak transitions: the relation $\Rightarrow$ denotes the reflexive and transitive closure of $\xrightarrow{\tau}$; the relation $\xRightarrow{\alpha}$ denotes $\Rightarrow \xrightarrow{\alpha} \Rightarrow$; the relation $\xRightarrow{\hat{\alpha}}$ denotes $\Rightarrow$ if $\alpha = \tau$ and $\xRightarrow{\alpha}$ otherwise.

Definition 3 (Similarity). *A relation $\mathcal{R}$ over well-formed networks is a simulation if $M \mathcal{R} N$ and $M \xrightarrow{\alpha} M'$ imply there is N' such that $N \xRightarrow{\hat{\alpha}} N'$ and $M' \mathcal{R} N'$. We write $M \precsim N$, if there is a simulation $\mathcal{R}$ such that $M \mathcal{R} N$.*

Our notion of of similarity between networks is a pre-congruence, as it is preserved by parallel composition.

Theorem 1. *Let M and N be two well-formed networks such that $M \precsim N$. Then $M \mid O \precsim N \mid O$ for all O such that $M \mid O$ and $N \mid O$ are well-formed.*

3   A *tGNDC* Schema for Wireless Networks

Gorrieri and Martinelli [5] have proposed a general schema for the definition of timed security properties, called *timed Generalized Non-Deducibility on Compositions* (*tGNDC*). Basically, a system M is $tGNDC^{\rho(M)}$ if for any attacker A

$$M \mid A \precsim \rho(M)$$

i.e. the composed system $M \mid A$ satisfies the abstraction $\rho(M)$.

A wireless protocol involves a set of nodes which may be potentially under attack, depending on the proximity to the attacker. This means that, in general, the *attacker* of a protocol M is a distinct network A of possibly colluding nodes. For the sake of compositionality, we assume that each node of the protocol is attacked by exactly one node of A.

Definition 4. *We say that $\mathcal{A}$ is a set of attacking nodes for the network M if and only if $|\mathcal{A}| = \mathsf{nds}\,(M)$ and $\mathcal{A} \cap (\mathsf{nds}\,(M) \cup \mathsf{Env}\,(M)) = \emptyset$.*

During the execution of the protocol an attacker may increase its initial knowledge by grasping messages sent by the parties, according to Dolev-Yao constrains. The knowledge of a network is expressed by the set of messages that the network can manipulate. Thus, we write $\mathsf{msg}(M)$ (resp. $\mathsf{msg}(P)$) to denote the set of the messages appearing in the network M (resp. in the process P). To ensure that attackers cannot prevent the passage of time, in the following definition we denote Prc_{wt} the set of processes in which summations are finite-indexed and recursive definitions are time-guarded.

Definition 5 (Attacker). *Let M be a network, with $\mathsf{nds}\,(M) = \{m_1, ..., m_k\}$. Let $\mathcal{A} = \{a_1, \ldots, a_k\}$ be a set of attacking nodes for M. We define the set of attackers of M with initial knowledge $\phi_0 \subseteq Msg$ as:*

$$\mathbb{A}^{\phi_0}_{\mathcal{A}/M} \stackrel{\mathrm{def}}{=} \Big\{ \prod_{i=1}^{k} a_i [Q_i]^{\mu_i} : Q_i \in Prc_{\mathsf{wt}},\ \mathsf{msg}(Q_i) \subseteq \mathcal{D}(\phi_0),\ \mu_i = (\mathcal{A} \setminus a_i) \cup m_i \Big\}.$$

Sometimes, for verification reasons, we will be interested in observing part of the protocol M under examination. For this purpose, we assume that the environment contains a fresh node $obs \notin \mathsf{nds}\,(M) \cup \mathsf{Env}\,(M) \cup \mathcal{A}$, that we call the 'observer', unknown to the attacker. For convenience, the observer *cannot* transmit: it can only receive messages.

Definition 6. *Let $M = \prod_{i=1}^{k} m_i [P_i]^{\nu_i}$. Given a set $\mathcal{A} = \{a_1, \ldots, a_k\}$ of attacking nodes for M and fixed a set $\mathcal{O} \subseteq \mathsf{nds}\,(M)$ of nodes to be observed, we define:*

$$M^{\mathcal{A}}_{\mathcal{O}} \stackrel{\mathrm{def}}{=} \prod_{i=1}^{k} m_i [P_i]^{\nu'_i} \quad where \quad \nu'_i \stackrel{\mathrm{def}}{=} \begin{cases} (\nu_i \cap \mathsf{nds}\,(M)) \cup a_i \cup obs & if\ m_i \in \mathcal{O} \\ (\nu_i \cap \mathsf{nds}\,(M)) \cup a_i & otherwise. \end{cases}$$

This definition expresses that (i) every node m_i of the protocols has a dedicated attacker located at a_i, (ii) network and attacker are considered in *isolation*, without any external interference, (iii) only *obs* can observe the behaviour of nodes in $\mathcal{O}$, (iv) node *obs* does not interfere with the protocol as it cannot transmit, (v) the behaviour of the nodes in $\mathsf{nds}\,(M) \setminus \mathcal{O}$ is not observable.

We can now formalise the *tGNDC* family of properties as follows.

Definition 7 (*tGNDC* for wireless networks). *Given a network M, an initial knowledge ϕ_0, a set $\mathcal{O} \subseteq \mathsf{nds}\,(M)$ of nodes under observation and an abstraction $\rho(M)$, representing a security property for M, we write $M \in tGNDC^{\rho(M)}_{\phi_0, \mathcal{O}}$ if and only if for all sets $\mathcal{A}$ of attacking nodes for M and for every $A \in \mathbb{A}^{\phi_0}_{\mathcal{A}/M}$ it holds that $M^{\mathcal{A}}_{\mathcal{O}} \mid A \precsim \rho(M)$.*

It should be noticed that when showing that a system M is $tGNDC^{\rho(M)}_{\phi_0, \mathcal{O}}$, the universal quantification on attackers required by the definition makes the proof quite involved. Thus, we look for a sufficient condition which does not make use of the universal quantification. For this purpose, we rely on a timed notion of term stability [5]. Intuitively, a network M is said to be *time-dependent stable* if the attacker cannot increase its knowledge in a indefinite way when M runs in the space of a time slot. Thus, we can predict how the knowledge of the attacker evolves at each time slot. First, we need a formalisation of computation. For $\Lambda = \alpha_1 \ldots \alpha_n$, we write $\stackrel{\Lambda}{\Longrightarrow}$ to denote $\Rightarrow \stackrel{\alpha_1}{\longrightarrow} \Rightarrow \ldots \Rightarrow \stackrel{\alpha_n}{\longrightarrow} \Rightarrow$. In order to count how many time slots embraces an execution trace Λ, we define $\#^\sigma(\Lambda)$ to be the number of occurrences of σ-actions in Λ.

Definition 8 (Time-dependent stability). *A network M is said to be time-dependent stable with respect to a sequence of knowledge $\{\phi_j\}_{j \geq 0}$ if whenever $M_{\mathsf{nds}(M)}^{\mathcal{A}} \,\big|\, A \stackrel{\Lambda}{\Longrightarrow} M' \,\big|\, A'$, where $\mathcal{A}$ is a set of attacking nodes for M, $\#^\sigma(\Lambda) = j$, $A \in \mathbb{A}_{\mathcal{A}/M}^{\phi_0}$ and $\mathsf{nds}(M') = \mathsf{nds}(M)$, then $\mathsf{msg}(A') \subseteq \mathcal{D}(\phi_j)$.*

The set of messages ϕ_j expresses the knowledge of the attacker at the end of the j-th time slot. Time-dependent stability is the crucial notion that allows us to introduce the notion of most general attacker. Intuitively, given a sequence of knowledge $\{\phi_j\}_{j \geq 0}$ and a network M, with $\mathcal{P} = \mathsf{nds}(M)$, we pick a set $\mathcal{A} = \{a_1, \ldots, a_k\}$ of attacking nodes for M and we define the top attacker $\mathrm{Top}_{\mathcal{A}/\mathcal{P}}^{\phi_j}$ as the network which at (the beginning of) the j-th time slot is aware of the knowledge (derivable) from ϕ_j.

Definition 9 (Top attacker). *Let M be a network with $\mathcal{P} = \mathsf{nds}(M) = \bigcup_{i=1}^{k} m_i$. Let $\mathcal{A} = \{a_1, \ldots, a_k\}$ be a set of attacking nodes for M, and $\{\phi_j\}_{j \geq 0}$ a sequence of knowledge. We define:*

$$\mathrm{Top}_{\mathcal{A}/\mathcal{P}}^{\phi_j} \stackrel{\mathrm{def}}{=} \prod_{i=1}^{k} a_i [\mathrm{T}_{\phi_j}]^{m_i} \quad \text{where} \quad \mathrm{T}_{\phi_j} \stackrel{\mathrm{def}}{=} \left\lfloor \textstyle\sum_{w \in \mathcal{D}(\phi_j)} \tau.!\langle w \rangle.\mathrm{T}_{\phi_j} \right\rfloor \mathrm{T}_{\phi_{j+1}} \ .$$

Basically, from the j-th time slot onwards, $\mathrm{Top}_{\mathcal{A}/\mathcal{P}}^{\phi_j}$ can *replay* any message in $\mathcal{D}(\phi_j)$ to the network under attack. Moreover, every attacking node a_i can send messages to the corresponding node m_i, but, unlike the attackers of Definition 5, it does not need to communicate with the other nodes in $\mathcal{A}$ as it already owns the full knowledge of the system at time j.

Top attackers represent a compositional criterion to guarantee *tGNDC*.

Theorem 2 (Criterion for *tGNDC*). *Let M be time-dependent stable with respect to a sequence $\{\phi_j\}_{j \geq 0}$, $\mathcal{A}$ be a set of attacking nodes for M and $\mathcal{O} \subseteq \mathsf{nds}(M) = \mathcal{P}$. Then $M_{\mathcal{O}}^{\mathcal{A}} \,\big|\, \mathrm{Top}_{\mathcal{A}/\mathcal{P}}^{\phi_0} \precsim N$ implies $M \in tGNDC_{\phi_0, \mathcal{O}}^{N}$.*

Theorem 3 (Compositionality). *Let $M = \prod_{i=1}^{k} M_i$ be time-dependent stable with respect to a sequence of knowledge $\{\phi_j\}_{j \geq 0}$. Let $\mathcal{A}_1, \ldots, \mathcal{A}_k$ be disjoint sets of attacking nodes for $M_1, \ldots, M_k$, respectively. Let $\mathcal{O}_i \subseteq \mathsf{nds}(M_i) = \mathcal{P}_i$, for $1 \leq i \leq k$. Then, $(M_i)_{\mathcal{O}_i}^{\mathcal{A}_i} \,\big|\, \mathrm{Top}_{\mathcal{A}_i/\mathcal{P}_i}^{\phi_0} \precsim N_i$, for $1 \leq i \leq k$, implies $M \in tGNDC_{\phi_0, \mathcal{O}_1 \cup \ldots \cup \mathcal{O}_k}^{N_1 | \ldots | N_k}$.*

4   A Security Analysis of LiSP

LiSP [14] is a well-known key management protocol for *wireless sensor networks*. A LiSP network consists of a *Key Server* (KS) and a set of *sensor nodes*

$m_1, \ldots, m_k$. The protocol assumes a *one way function* F, pre-loaded in every node of the system, and employs two different key families: (i) a set of *temporal keys* $k_0, \ldots, k_n$, computed by KS by means of F, and used by all nodes to encrypt/decrypt data packets; (ii) a set of *master keys* $k_{\text{KS}:m_j}$, one for each node m_j, for unicast communications between m_j and BS. The transmission time is split into *time intervals*, each of them is Δ_{refresh} time units long. Thus, each temporal key is tied to a time interval and renewed every Δ_{refresh} time units. At a time interval i, the temporal key k_i is shared by all sensor nodes and it is used for data encryption. Key renewal relies on *loose node time synchronisation* among nodes. Each node stores a subset of temporal keys in a *buffer* of a fixed size, say s with $s << n$.

The LiSP protocol consists of the following phases.

Initial Setup. At the beginning, KS randomly chooses a key k_n and computes a sequence of temporal keys $k_0, \ldots, k_n$, by using the function F, as $k_i := F(k_{i+1})$. Then, KS waits for reconfiguration requests from nodes. More precisely, when KS receives a reconfiguration request from a node m_j, at time interval i, it unicasts the packet InitKey:

$$\text{KS} \rightarrow m_j \; : \; \text{enc}(k_{\text{KS}:m_j}, \, (s \mid k_{s+i} \mid \Delta_{\text{refresh}})) \mid \text{hash}(s \mid k_{s+i} \mid \Delta_{\text{refresh}}) \; .$$

The operator $\text{enc}(k, p)$ represents the encryption of p by using the key of k, while $\text{hash}(p)$ generates a message digest for p by means of a cryptographic hash function used to check the integrity of the packet p. When m_j receives the InitKey packet, it computes the sequence of keys $k_{s+i-1}, k_{s+i-2}, \ldots, k_i$ by several applications of the function F to k_{s+i}. Then, it activates k_i for data encryption and it stores the remaining keys in its local buffer; finally it sets up a *ReKeyingTimer* to expires after $\Delta_{\text{refresh}}/2$ time units (this value applies only for the first rekeying).

Re-Keying. At each time interval i, with $i \leq n$, KS employs the active encryption key k_i to encode the key k_{s+i}. The resulting packet is broadcast as an UpdateKey packet:

$$\text{KS} \rightarrow * \; : \; \text{enc}(k_i, \, k_{s+i}) \; .$$

When a node receives an UpdateKey packet, it tries to authenticate the key received in the packet; if the node succeeds in the authentication then it recovers all keys that have been possibly lost and updates its key buffer. When the time interval i elapses, every node discards k_i, activates the key k_{i+1} for data encryption, and sets up the *ReKeyingTimer* to expire after Δ_{refresh} time units for future key switching (after the first time, switching happens every Δ_{refresh} time units).

Authentication and Recovery of Lost Keys. The one-way function F is used to authenticate and recover lost keys. If l is the number of stored keys in a buffer of size s, with $l \leq s$, then $s - l$ represents the number of keys which have been lost by the node. When a sensor node receives an UpdateKey packet carrying a new key k, it calculates $F^{s-l}(k)$ by applying $s - l$ times the function F. If the result matches with the last received temporal key, then the node stores k in its buffer and recovers all lost keys.

Reconfiguration. When a node m_j joins the network or misses more than s temporal keys, then its buffer is empty. Thus, it sends a RequestKey packet in order to request the current configuration:

$$m_j \to \text{KS} \; : \; \text{RequestKey} \mid m_j \; .$$

Upon reception, node KS performs authentication of m_j and, if successful, it sends the current configuration via an InitKey packet.

Encoding. In Table 4, we provide a specification in aTCWS of the entire LiSP protocol. We introduce some slight simplifications with respect to the original protocol. We assume that (i) the temporal keys $k_0, \ldots, k_n$ have already been computed by KS, (ii) both the buffer size s and the refresh interval Δ_{refresh} are known by each node. Thus, the InitKey packet can be simplified as follows:

$$\text{KS} \to m_j \; : \; \text{enc}(k_{\text{KS}:m_j}, \; k_{s+i}) \mid \text{hash}(k_{s+i}) \; .$$

Moreover, we assume that every σ-action models the passage of $\Delta_{\text{refresh}}/2$ time units. Therefore, every two σ-actions the key server broadcasts the new temporal key encrypted with the key tied to that specific interval. Finally, we do not model data encryption.

When giving our encoding in aTCWS we will require some new deduction rules to model an hash function and encryption/decryption of messages:

$$(\text{hash}) \; \frac{w}{\text{hash}(w)} \qquad (\text{enc}) \; \frac{w_1 \quad w_2}{\text{enc}(w_1, w_2)} \qquad (\text{dec}) \; \frac{w_1 \quad w_2}{\text{dec}(w_1, w_2)} \; .$$

The protocol executed by the key server is expressed by the following two threads: a key distributor D_i and a listener L_i waiting for reconfiguration requests from the sensor nodes, with i being the current time interval. Every Δ_{refresh} time units (that is, every two σ-actions) D_i broadcasts the new temporal key k_{s+i} encrypted with the key k_i of the current time interval i. The process L_i replies to reconfiguration requests by sending an initialisation packet.

At the beginning of the protocol, a sensor node runs the process Z, which broadcasts a request packet to KS, waits for a reconfiguration packet q, and then checks authenticity by verifying the hash code. If the verification is successful then the node starts the broadcasting new keys phase. This phase is formalised by the process $R(k_{\text{C}}, k_{\text{L}}, l)$, where k_{C} is the current temporal key, k_{L} is the last authenticated temporal key, and the integer l counts the number of keys that are actually stored in the buffer.

To simplify the presentation, we formalise the key server as a pair of nodes: a key disposer KD, which executes D_i, and a listener KL, which executes L_i. Thus, the LiSP protocol, in its initial configuration, can be represented as:

$$\text{LiSP} \; \overset{\text{def}}{=} \; \prod_{j \in J} m_j[\sigma.Z]^{\nu_{m_j}} \mid \text{KS}[\sigma.D_0]^{\nu_{\text{KS}}} \mid \text{KL}[\sigma.L_0]^{\nu_{\text{KL}}}$$

where for each node m_j, with $j \in J$, $m_j \in \nu_{\text{KD}} \cap \nu_{\text{KL}}$ and $\{\text{KD}, \text{KL}\} \subseteq \nu_{m_j}$.

Table 4. The LiSP protocol

Key Server:

$$D_0 \overset{\text{def}}{=} \sigma.D_1 \qquad \text{synchronise and move to } D_1$$

$$D_i \overset{\text{def}}{=} [k_i \ k_{s+i} \vdash_{\text{enc}} t_i] \qquad \text{for } i \geq 1, \text{ encrypt } k_{s+i} \text{ with } k_i$$
$$[\text{UpdateKey } t_i \vdash_{\text{pair}} u_i] \qquad \text{build the UpdateKey packet } u_i$$
$$!\langle u_i \rangle.\sigma.\sigma.D_{i+1} \qquad \text{broadcast } r_i, \text{ and move to } D_{i+1}$$

$$L_i \overset{\text{def}}{=} \lfloor ?(r).I_{i+1} \rfloor \sigma.L_{i+1} \qquad \text{wait for request packets}$$

$$I_i \overset{\text{def}}{=} [r \vdash_{\text{fst}} r_1] I_i^1; \sigma.\sigma.L_i \qquad \text{extract first component}$$

$$I_i^1 \overset{\text{def}}{=} [r_1 = \text{RequestKey}] I_i^2; \sigma.\sigma.L_i \qquad \text{check if } r_1 \text{ is a RequestKey}$$

$$I_i^2 \overset{\text{def}}{=} [r \vdash_{\text{snd}} m] \qquad \text{extract node name}$$
$$[k_{\text{KS}:m} \ k_{s+i} \vdash_{\text{enc}} w_i] \qquad \text{encrypt } k_{s+i} \text{ with } k_{\text{KS}:m}$$
$$[k_{s+i} \vdash_{\text{hash}} h_i] \qquad \text{calculate hash code for } k_{s+i}$$
$$[w_i \ h_i \vdash_{\text{pair}} r_i] \qquad \text{build a pair } r_i$$
$$[\text{InitKey } r_i \vdash_{\text{pair}} q_i] \qquad \text{build a InitKey packet } q_i$$
$$\sigma.!\langle q_i \rangle.\sigma.L_i \qquad \text{broadcast } q_i, \text{ move to } L_i$$

Receiver at node m:

$$Z \overset{\text{def}}{=} [\text{RequestKey } m \vdash_{\text{pair}} r] \qquad \text{send a RequestKey packet}$$
$$!\langle r \rangle.\sigma.\lfloor ?(q).T \rfloor Z \qquad \text{wait for a reconfig. packet}$$

$$T \overset{\text{def}}{=} [q \vdash_{\text{fst}} q'] T^1; \sigma.Z \qquad \text{extract fst component of } q$$

$$T^1 \overset{\text{def}}{=} [q' = \text{InitKey}] T^2; \sigma.Z \qquad \text{check if } q \text{ is a InitKey packet}$$

$$T^2 \overset{\text{def}}{=} [q \vdash_{\text{snd}} q''] \qquad \text{extract snd component of } q$$
$$[q'' \vdash_{\text{fst}} w] T^3; \sigma.Z \qquad \text{extract fst component of } q''$$

$$T^3 \overset{\text{def}}{=} [q'' \vdash_{\text{snd}} h] \qquad \text{extract snd component of } q''$$
$$[k_{\text{KS}:m} \ w \vdash_{\text{dec}} k] T^3; \sigma.Z \qquad \text{extract the key}$$

$$T^4 \overset{\text{def}}{=} [k \vdash_{\text{hash}} h'][h = h'] T^5; \sigma.Z \qquad \text{verify hash codes}$$

$$T^5 \overset{\text{def}}{=} \sigma.\sigma.R\langle F^{s-1}(k), k, s-1 \rangle \qquad \text{synchronise and move to } R$$

$$R(k_{\text{c}}, k_{\text{L}}, l) \overset{\text{def}}{=} \lfloor ?(u).E \rfloor F \qquad \text{wait for incoming packets}$$

$$E \overset{\text{def}}{=} [u \vdash_{\text{fst}} u'] E^1; \sigma.F \qquad \text{extract fst component of } u$$

$$E^1 \overset{\text{def}}{=} [u' = \text{UpdateKey}] E^2; \sigma.F \qquad \text{check UpdateKey packet}$$

$$E^2 \overset{\text{def}}{=} [u \vdash_{\text{snd}} u''] \qquad \text{extract snd component of } u$$
$$[k_{\text{c}} \ u'' \vdash_{\text{dec}} k] E^3; \sigma.F \qquad \text{decrypt } u'' \text{ by using } k_{\text{c}}$$

$$E^3 \overset{\text{def}}{=} [F^{s-l}(k) = k_{\text{L}}] E^4; \sigma.F \qquad \text{authenticate } k$$

$$E^4 \overset{\text{def}}{=} \sigma.\sigma.R\langle F^{s-1}(k), k, s-1 \rangle \qquad \text{synchronise and move to } R$$

$$F \overset{\text{def}}{=} [l = 0] Z; \sigma.R\langle F^{l-1}(k_{\text{L}}), k_{\text{L}}, l-1 \rangle \qquad \text{check if buffer key is empty}$$

Security Analysis. In LiSP, a node should authenticate only keys sent by the key server in the last Δ_{refresh} time units. Otherwise, a node needing a reconfiguration would authenticate an obsolete temporal key and it would not be synchronised with the rest of the network. Here, we show that key authentication in LiSP may take longer than Δ_{refresh} time units, as a consequence of a replay attack.

For our analysis, without loss of generality, it suffices to focus on a part of the protocol composed by the KL node of the key server and a single sensor node m. Moreover, in order to make observable a successful reconfiguration, we replace the process T^4 of Table 4 with the process

$$T^{4'} \stackrel{\text{def}}{=} \sigma.\sigma.[\text{auth } k \vdash_{\text{pair}} a]!\langle a \rangle.R\langle F^{s-1}(k), k, s-1 \rangle \ .$$

Thus, the part of the protocol under examination can be defined as follows:

$$\text{LiSP}' \stackrel{\text{def}}{=} m[\sigma.Z']^{\nu_m} \mid \text{KL}[\sigma.L_0]^{\nu_{\text{KL}}} \ .$$

Our freshness requirement on authenticated keys can be expressed by the following abstraction of the protocol:

$$\rho(\text{LiSP}') \stackrel{\text{def}}{=} m[\sigma.\hat{Z}_0]^{obs} \mid \text{KL}[\sigma.\hat{L}_0]^{obs}$$

where

- $\hat{Z}_i \stackrel{\text{def}}{=} !\langle r \rangle.\sigma.\lfloor \tau.\sigma.\sigma.!\langle \text{auth}_{i+1} \rangle.R(k_{i+1}, k_{s+i}, s-1) \rfloor \hat{Z}_{i+1}$,
  with $r = \text{pair}(\text{RequestKey}, m)$ and $\text{auth}_i = \text{pair}(\text{auth}, k_{s+i})$ as in Table 4;
- $\hat{L}_i \stackrel{\text{def}}{=} \lfloor \tau.\sigma.!\langle q_{i+1} \rangle.\sigma.\hat{L}_{i+1} \rfloor \sigma.\hat{L}_{i+1}$, and q_i defined as in Table 4:
  $q_i = \text{pair}(\text{InitKey } r_i)$ with $r_i = \text{pair}(\text{enc}(k_{\text{KS}:m}, k_{s+i}), \text{hash}(k_{s+i}))$.

It is easy to see that $\rho(\text{LiSP}')$ is a correct abstraction of key authentication within the protocol, as the action auth_i occurs exactly Δ_{refresh} time units (that is two σ-actions) after the disclosure of key k_{s+i} through packet q_i.

Proposition 3. $\rho(\text{LiSP}') \stackrel{\Lambda}{\Longrightarrow} \stackrel{!q_i \triangleright obs}{\longrightarrow} \stackrel{\Omega}{\Longrightarrow} \stackrel{!\text{auth}_i \triangleright obs}{\longrightarrow} implies \ \#^\sigma(\Omega) = 2.$

In order to show that LiSP$'$ satisfies our security analysis, we should prove that

$$\text{LiSP}' \in tGNDC^{\rho(\text{LiSP}')}_{\phi_0, \mathcal{O}}$$

for $\mathcal{O} = \text{nds}\left(\text{LiSP}'\right)$ and initial knowledge $\phi_0 = \emptyset$. However, this is not the case.

Theorem 4 (Replay attack to LiSP). $\text{LiSP}' \notin tGNDC^{\rho(\text{LiSP}')}_{\emptyset, \{\text{KL}, m\}} \ .$

Proof. Let us define the set of attacking nodes $\mathcal{A} = \{a, b\}$ for LiSP$'$. Let us fix the initial knowledge of the attacker $\phi_0 = \emptyset$. We set $\nu_a = \{m, b\}$ and $\nu_b = \{\text{KL}, a\}$, and we assume that $\mathcal{O} = \{\text{KL}, m\}$. We give an intuition of the replay attack in Table 5. Basically, an attacker may prevent the node m to receive the InitKey packet within Δ_{refresh} time units. As a consequence, m may complete the protocol only after $2\Delta_{\text{refresh}}$ time units (that is, four σ-actions), so authenticating an old key. Formally, we define the attacker $A \in \mathbb{A}^{\phi_0}_{\mathcal{A}/\{\text{KL}, m\}}$ as $A = a[\sigma.\sigma.\sigma.X]^{\nu_a} \mid b[\sigma.\sigma.X]^{\nu_b}$ where $X = \lfloor ?(x).\sigma.!\langle x \rangle.\text{nil} \rfloor \text{nil}$. We then consider the system $(\text{LiSP}')^{\mathcal{A}}_{\mathcal{O}} \mid A$ which admits the following execution trace:

$$\sigma.!r \triangleright obs.\sigma.!q_1 \triangleright obs.\sigma.\tau.!r \triangleright obs.\sigma.\tau.\sigma.\sigma.!\text{auth}_1 \triangleright obs$$

containing four σ-actions between the packets q_1 and auth_1. By Proposition 3, this trace cannot be matched by $\rho(\text{LiSP}')$. So, $(\text{LiSP}')^{\mathcal{A}}_{\mathcal{O}} \mid A \not\lesssim \rho(\text{LiSP}')$. $\qquad\square$

Table 5. Replay attack to LiSP

$m \longrightarrow$ KL	:	r	m sends a RequestKey and KL correctly receives the packet
$\stackrel{\sigma}{\longrightarrow}$			the system moves to the next time slot
KL $\longrightarrow m$	:	q_1	KL replies with an InitKey which is lost by m and grasped by b
$\stackrel{\sigma}{\longrightarrow}$			the system moves to the next time slot
$b \rightarrow a$	:	q_1	b sends q_1 to a
$m \rightarrow$ KL	:	r	m sends a new RequestKey which gets lost
$\stackrel{\sigma}{\longrightarrow}$			the system moves to the next time slot
$a \rightarrow m$	:	q_1	a *replays* q_1 to m
$\stackrel{\sigma}{\longrightarrow} \stackrel{\sigma}{\longrightarrow}$			after Δ_{refresh} time units
$m \rightarrow *$	:	auth$_1$	m authenticates q_1 and signals the end of the protocol

4.1 LiSP with Nonces

Replay attacks as those described above appears also in other key management protocols, such as μTESLA [15] and LEAP+ [17]. These protocols have been amended by adding nonces to guarantee freshness. We propose to do the same in LiSP. For this purpose, we extend our inference system with a new deduction rule to model a *pseudo-random function*: The application $\text{prf}(m, w_i)$ returns a pseudo-random value w_{i+1} associated to a node m and the last generated value w_i. In our amended specification of LiSP, we add a nonce to the RequestKey packet. The nonce is then included in the corresponding InitKey packet to guarantee the freshness of the reply. These changes affect only those processes which model the key request at the node side and the reply at the server side. We modify these processes as shown in Table 6. The requesting nodes run the process Z_j, where j is the number associated to the current key request. At each request j, the receiver generates a nonce n_j which will be used to check the freshness of the received key. The process $\bar{L}_i$, running at the key server, now includes the received nonce in the InitKey packet. Notice that, as done before, the process T_j^7 signals a successful reconfiguration. Again, for our analysis, it suffices to analyse the following fragment of the protocol:

$$\text{LiSP}'' \stackrel{\text{def}}{=} m[\sigma.Z_1]^{\nu_m} \mid \text{KL}[\sigma.\bar{L}_0]^{\nu_{\text{KL}}} .$$

According to Definition 8, the system LiSP'' is time-dependent stable with respect to the following sequence of knowledge:

$$\begin{aligned}
\phi_0 &= \emptyset \\
\phi_1 &= \{r_1\} \\
\phi_i &= \phi_{i-1} \cup \{q_j\} & \text{if } j > 0 \text{ and } i = 2j \\
\phi_i &= \phi_{i-1} \cup \{\text{auth}_j, r_{j+1}\} & \text{if } j > 0 \text{ and } i = 2j+1
\end{aligned} \quad (1)$$

where

$$\begin{aligned}
\text{auth}_j &= \text{pair}(\text{auth}, k_{s+j}) \\
r_j &= \text{pair}(\text{RequestKey}, \text{pair}(m, n_j)) \\
q_j &= \text{pair}(\text{InitKey}, \ \text{pair}(\text{enc}(k_{\text{KS}:m}, \text{pair}(k_{s+j}, n_j)), \ \text{hash}(k_{s+j}))) .
\end{aligned}$$

Intuitively, ϕ_i consists of ϕ_{i-1} together with the set of messages an intruder can get by eavesdropping on a run of the protocol during the time slot i.

Table 6. LiSP with nonces

Key Server:

$\bar{L}_i \stackrel{\text{def}}{=} \lfloor ?(r).\bar{I}_{i+1}\rfloor\sigma.\bar{L}_{i+1}$		wait for request packets
$\bar{I}_i \stackrel{\text{def}}{=} [r \vdash_{\text{fst}} r_1]\bar{I}_i^1; \sigma.\sigma.\bar{L}_{i+1}$		extract first component
$\bar{I}_i^1 \stackrel{\text{def}}{=} [r_1 = \text{RequestKey}]\bar{I}_i^2; \sigma.\sigma.\bar{L}_{i+1}$		check if r_1 is a RequestKey
$\bar{I}_i^2 \stackrel{\text{def}}{=} [r \vdash_{\text{snd}} t]$		extract second component
$\quad\quad [t \vdash_{\text{fst}} m]\bar{I}_i^3; \sigma.\sigma.\bar{L}_{i+1}$		extract node name
$\bar{I}_i^3 \stackrel{\text{def}}{=} [t \vdash_{\text{snd}} n]$		extract nonce
$\quad\quad [k_{s+i}\, n \vdash_{\text{pair}} p]$		build a pair
$\quad\quad [k_{\text{KS}:m}\, p \vdash_{\text{enc}} w_i]$		encrypt p with $k_{\text{KS}:m}$
$\quad\quad [k_{s+i} \vdash_{\text{hash}} h_i]$		calculate hash code for k_{s+i}
$\quad\quad [w_i\, h_i \vdash_{\text{pair}} r_i]$		build a pair r_i
$\quad\quad [\text{InitKey}\, r_i \vdash_{\text{pair}} q_i]$		build a InitKey packet q_i
$\quad\quad \sigma.!\langle q_i\rangle.\sigma.\bar{L}_{i+1}$		broadcast q_i, move to $\bar{L}_{i+1}$

Receiver at node m:

$Z_j \stackrel{\text{def}}{=} [m\, n_{j-1} \vdash_{\text{prf}} n_j]$		build a random nonce n_j
$\quad\quad [m\, n_j \vdash_{\text{pair}} t]$		build a pair t with name m and nonce n_j
$\quad\quad [\text{RequestKey}\, t \vdash_{\text{pair}} r]$		send a RequestKey packet
$\quad\quad !\langle r\rangle.\sigma.\lfloor ?(q).T_j\rfloor Z_{j+1}$		wait for a reconfig. packet
$T_j \stackrel{\text{def}}{=} [q \vdash_{\text{fst}} q']T_j^1; \sigma.Z_{j+1}$		extract fst component of q
$T_j^1 \stackrel{\text{def}}{=} [q' = \text{InitKey}]T_j^2; \sigma.Z_{j+1}$		check if q is a InitKey packet
$T_j^2 \stackrel{\text{def}}{=} [q \vdash_{\text{snd}} q'']$		extract snd component of q
$\quad\quad [q'' \vdash_{\text{fst}} w]T_j^3; \sigma.Z_{j+1}$		extract fst component of q''
$T_j^3 \stackrel{\text{def}}{=} [q'' \vdash_{\text{snd}} h]$		extract snd component of q''
$\quad\quad [k_{\text{KS}:m}\, w \vdash_{\text{dec}} p]T_j^4; \sigma.Z_{j+1}$		decrypt w
$T_j^4 \stackrel{\text{def}}{=} [p \vdash_{\text{fst}} k]T_j^5; \sigma.Z_{j+1}$		extract the key
$T_j^5 \stackrel{\text{def}}{=} [p \vdash_{\text{snd}} n][n = n_j]T_j^6; \sigma.Z_{j+1}$		verify nonces
$T_j^6 \stackrel{\text{def}}{=} [k \vdash_{\text{hash}} h'][h = h']T_j^7; \sigma.Z_{j+1}$		verify hash codes
$T_j^7 \stackrel{\text{def}}{=} \sigma.\sigma.[\text{auth}\, k \vdash_{\text{pair}} a]!\langle a\rangle.\text{nil}$		reaching of synchronisation

With the introduction of nonces, the abstraction expressing key authentication within Δ_{refresh} time units becomes the following:

$$\rho(\text{LiSP}'') \stackrel{\text{def}}{=} m[\sigma.\hat{Z}_1']^{obs} \mid \text{KL}[\sigma.\hat{L}_0']^{obs}$$

where
$$\hat{Z}_i' = [m\, n_{i-1} \vdash_{\text{prf}} n_i][m\, n_i \vdash_{\text{pair}} t][\text{RequestKey}\, t \vdash_{\text{pair}} r]!\langle r\rangle.\sigma.\lfloor \tau.\sigma^2.!\langle \text{auth}_{i+1}\rangle\rfloor\hat{Z}_{i+1}'$$
$$\hat{L}_i' = \lfloor \sum_{v\in\mathcal{D}(\phi_{2i+1})} \tau.\sigma.!\langle q_{i+1}^v\rangle.\sigma.\hat{L}_{i+1}'\rfloor\sigma.\hat{L}_{i+1}'$$
with $q_i^v = \text{pair}(\text{InitKey}\ \text{pair}(\text{enc}(k_{\text{KS}:m}, \text{pair}(k_{s+i}, v)), \text{hash}(k_{s+i})))$.

In $\rho(\text{LiSP}'')$ keys are authenticated after Δ_{refresh} time units (two σ-actions).

Proposition 4. $\rho(\text{LiSP}'') \stackrel{\Lambda}{\Longrightarrow} \xrightarrow{!q_i^v \,\triangleright\, obs} \stackrel{\Omega}{\Longrightarrow} \xrightarrow{!\text{auth}_i \,\triangleright\, obs} M$ *implies* $\#^\sigma(\Omega)=2$.

Now, everything is in place to prove the safety of the LiSP protocol with nonces.

Lemma 1. *Given two attacking nodes a and b, for m and* KL *respectively, and fixed the sequence of knowledge* $\{\phi_i\}_{i \geq 0}$ *as in* (1), *then*

1. $\mathrm{KL}[\sigma.\bar{L}_0]^{\{b,obs\}} \mid \mathrm{Top}^{\phi_0}_{b/\mathrm{KL}} \lesssim \mathrm{KL}[\sigma.\hat{L}'_0]^{obs}$
2. $m[\sigma.Z_1]^{\{a,obs\}} \mid \mathrm{Top}^{\phi_0}_{a/m} \lesssim m[\sigma.\hat{Z}'_1]^{obs}.$

Theorem 5 (Safety of LiSP with nonces). $\mathrm{LiSP}'' \in tGNDC^{\rho(\mathrm{LiSP}'')}_{\emptyset,\mathrm{nds}(\mathrm{LiSP}'')}$.

Proof . By an application of Lemma 1 and Theorem 3. $\qquad \square$

References

1. Ballardin, F., Merro, M.: A Calculus for the Analysis of Wireless Network Security Protocols. In: Degano, P., Etalle, S., Guttman, J. (eds.) FAST 2010. LNCS, vol. 6561, pp. 206–222. Springer, Heidelberg (2011)
2. Bertot, Y.: A Short Presentation of Coq. In: Mohamed, O.A., Muñoz, C., Tahar, S. (eds.) TPHOLs 2008. LNCS, vol. 5170, pp. 12–16. Springer, Heidelberg (2008)
3. Ghassemi, F., Fokkink, W., Movaghar, A.: Equational Reasoning on Ad Hoc Networks. In: Arbab, F., Sirjani, M. (eds.) FSEN 2009. LNCS, vol. 5961, pp. 113–128. Springer, Heidelberg (2010)
4. Godskesen, J.C.: A Calculus for Mobile Ad Hoc Networks. In: Murphy, A.L., Ryan, M. (eds.) COORDINATION 2007. LNCS, vol. 4467, pp. 132–150. Springer, Heidelberg (2007)
5. Gorrieri, R., Martinelli, F.: A simple framework for real-time cryptographic protocol analysis with compositional proof rules. Science of Computer Programming 50(1-3), 23–49 (2004)
6. Gorrieri, R., Martinelli, F., Petrocchi, M.: Formal models and analysis of secure multicast in wired and wireless networks. Journal of Automated Reasoning 41(3-4), 325–364 (2008)
7. Hennessy, M., Regan, T.: A Process Algebra for Timed Systems. Information and Computation 117(2), 221–239 (1995)
8. Lanese, I., Sangiorgi, D.: An Operational Semantics for a Calculus for Wireless Systems. Theoretical Computer Science 411, 1928–1948 (2010)
9. Merro, M.: An Observational Theory for Mobile Ad Hoc Networks (full paper). Information and Computation 207(2), 194–208 (2009)
10. Merro, M., Sibilio, E.: A Timed Calculus for Wireless Systems. In: Arbab, F., Sirjani, M. (eds.) FSEN 2009. LNCS, vol. 5961, pp. 228–243. Springer, Heidelberg (2010)
11. Misra, S., Woungag, I.: Guide to Wireless Ad Hoc Networks. Computer Communications and Networks. Springer, London (2009)
12. Nanz, S., Hankin, C.: A Framework for Security Analysis of Mobile Wireless Networks. Theoretical Computer Science 367(1-2), 203–227 (2006)
13. Nipkow, T., Paulson, L.C., Wenzel, M.: Isabelle/HOL - A Proof Assistant for Higher-Order Logic. LNCS, vol. 2283. Springer, Heidelberg (2002)
14. Park, T., Shin, K.G.: LiSP: A lightweight security protocol for wireless sensor networks. ACM Transactions in Embedded Computing Systems 3(3), 634–660 (2004)
15. Perrig, A., Szewczyk, R., Tygar, J.D., Wen, V., Culler, D.: SPINS: Security Protocols for Sensor Networks. Wireless Networks 8(5), 521–534 (2002)
16. Singh, A., Ramakrishnan, C.R., Smolka, S.A.: A process calculus for mobile ad hoc networks. Science of Computer Programming 75(6), 440–469 (2010)
17. Zhu, S., Setia, S., Jajodia, S.: Leap+: Efficient security mechanisms for large-scale distributed sensor networks. ACM Trans. on Sensor Networks 2(4), 500–528 (2006)

Runtime Verification with Predictive Semantics

Xian Zhang[1], Martin Leucker[2], and Wei Dong[1]

[1] School of Computer, National University of Defense Technology, Changsha, China
{zhangxian,wdong}@nudt.edu.cn
[2] Institute for Software Engineering, University Lüeck, Lüeck, Germany
leucker@isp.uni-luebeck.de

Abstract. Runtime verification techniques are used to continuously check whether software execution satisfies or violates a given correctness property. In this paper, we extend our previous work of three-valued semantics for Linear Temporal Logic (LTL) to predictive semantics. Combined with the static analysis to the monitored program, the predictive semantics are capable of predicting monitored property's satisfaction/violation even when the observed execution does not convince it. We instrument the monitored program based on its Program Dependence Graph representation in order to emit "predictive word" at runtime. We also implement a prototype tool to support predictive semantics and use it to find predictive words in real, large-scale project. The result demonstrates that the predictive semantics are generally applicable in these projects.

Keywords: Runtime Verification, Three-Valued Semantics, Predictive Semantics, Program Dependence Graph.

1 Introduction

As software systems become more and more pervasive in everyday life, it is becoming increasingly necessary to guarantee their security and reliability. An approach of continuously monitoring software execution is becoming more and more popular. Runtime verification techniques do not assume the deployed software is correct, but continuous check whether software execution satisfies or violates a given correctness property. Runtime verification techniques are frequently used to prevent damage from happening when software is going to malfunction and are quite effective in practice.

We propose a predictive semantics for runtime verification in this paper. The predictive semantics enable monitors to foresee a property satisfaction or violation before the observed execution convince it. We formally define the predictive semantics used in runtime verification. We also give an algorithm on how to generate a monitor from a LTL formula and describe how to use it to check program's execution with predictive semantics.

The remainder of this paper is organized as follows: some preliminary information are first introduced in section 2; in section 3, we formally define runtime verification with predictive semantics; then in section 4, we give an algorithm to generate a monitor from a LTL formula and describe how to check monitored

A. Goodloe and S. Person (Eds.): NFM 2012, LNCS 7226, pp. 418–432, 2012.

program's execution with predictive semantics; in section 5, we implement a prototype tool to support predictive semantics; in section 6, we use this tool to do some experiments on some real projects; finally, we review related work in section 7 and conclude in section 8.

2   Preliminaries

2.1   Linear Temporal Logic and Büchi Automata

LTL is a widely used formalism for specifying and verifying correctness of computer programs. For the remainder of this section, let AP be a finite set of atomic proposition symbols and $\Sigma = 2^{AP}$ be a finite alphabet. Σ^* stands for all finite traces over Σ and Σ^ω stands for all infinite traces over Σ. Usually, u, v, u' and v' are used to denote elements in Σ^* and w and w' in Σ^ω. For a word $w \in \Sigma^\omega$, the prefix set of w is $\text{pref}(w) = \{u | \exists w' \in \Sigma^\omega : (w = uw')\}$.

The syntax of LTL is defined as follows.

$$\varphi ::= \text{true} \mid p \mid \neg\varphi \mid \varphi \vee \varphi \mid \varphi \, \text{U} \, \varphi \mid \text{X} \, \varphi$$

where $p \in AP$. There are some syntax sugar: $\text{F}\,\varphi = \text{true} \, \text{U}\,\varphi$; $\text{G}\,\varphi = \neg\text{F}\,\neg\varphi$. We define the semantics of a formula φ of LTL with respect to infinite traces. Let $w = a_0 a_1 ... \in \Sigma^\omega$ be a infinite word with $i \in \mathbb{N}$ being a position. We define the semantics of LTL formulae inductively as follows: $w, i \vDash \text{true}$; $w, i \vDash p$ iff $p \in a_i$; $w, i \vDash \neg\varphi$ iff $w, i \nvDash \varphi$; $w, i \vDash \varphi_1 \vee \varphi_2$ iff $w, i \vDash \varphi_1$ or $w, i \vDash \varphi_2$; $w, i \vDash \varphi_1 \, \text{U} \, \varphi_2$ iff $\exists k \geq i$ with $w, k \vDash \varphi_2$ and $\forall i \leq l < k$ with $w, l \vDash \varphi_1$; $w, i \vDash \text{X}\,\varphi$ iff $w, i+1 \vDash \varphi$.

In addition, $w, 0 \vDash \varphi$ can be abbreviated as $w \vDash \varphi$. We also use $L(\varphi) = \{w \in \Sigma^\omega \mid w \vDash \varphi\}$ to denote the set of models of a LTL formula φ. Two LTL formulae φ and ψ are called *equivalent* ($\varphi \equiv \psi$) iff $L(\varphi) = L(\psi)$. The language $L(\varphi)$ can be recognized by a corresponding *Nondeterministic Büchi Automaton*(NBA). *NBA* is defined as a tuple $\mathcal{A} = (\Sigma, Q, Q_0, \delta, F)$, where: Σ is a finite alphabet; Q is a finite non-empty set of states; $Q_0 \subseteq Q$ is a set of initial states; $\delta : Q \times \Sigma \to 2^Q$ is a transition function; $F \subseteq Q$ is a set of accepting states.

In order to state conveniently, we extend the transition function $\delta : Q \times \Sigma \to 2^Q$ to $\delta' : 2^Q \times \Sigma^* \to 2^Q$ by $\delta'(Q', \epsilon) = Q'$ and $\delta'(Q', ua) = \bigcup_{q' \in \delta'(Q', u)} \delta(q', a)$ for $Q' \subseteq Q, u \in \Sigma^*$, and $a \in \Sigma$.

A *run* of an automaton $\mathcal{A}$ on a word $w = a_0 a_1 ... \in \Sigma^\omega$ is a sequence of states and input symbols $\rho = q_0 a_0 q_1 a_1 q_2 ...$, where q_0 is an initial state of $\mathcal{A}$ and $q_{i+1} \in \delta(q_i, a_i)$ for all $i \in N$. For a run ρ, we use $\text{Inf}(\rho)$ denote the states visited infinitely often. A run ρ of a NBA $\mathcal{A}$ is called *accepting* iff $\text{Inf}(\rho) \cap F \neq \emptyset$.

A *Nondeterministic Finite Automata* (NFA) $\mathcal{A} = (\Sigma, Q, Q_0, \delta, F)$ is an automaton where Σ, Q, Q_0, δ and F are defined as for a Büchi automata. A NFA only accept finite words. A run of a NFA on a word $u = a_0 ... a_n \in \Sigma^*$ is a sequence of states and input symbols $\rho = q_0 a_0 q_1 a_1 ... a_n q_{n+1}$ where q_0 is an initial state and $q_{i+1} \in \delta(q_i, a_i)$ for all $i \in N$. The run is called accepting if $q_{n+1} \in F$. A NFA is called deterministic iff for all $q \in Q$, $a \in \Sigma$, $|\delta(q, a)| = 1$ and $|Q_0| = 1$.

2.2   Three-Valued Semantics for LTL in Runtime Verification

Andreas Bauer, Martin Leucker and Christian Schallhart have defined a three-valued semantics for LTL on finite traces in article [1], which is tailored to the use in runtime verification. The intuition is as follows: in theory, we observe an infinite sequence w of some system. Thus, for a given formula φ, either $w \vDash \varphi$ holds or not. In practice, however, we can only observe a finite prefix u of w. The three-valued semantics are based on whether the observed finite prefix will definitely lead to a *satisfactory* or *violation* verdict. The three-valued semantics are defined as follows.

Definition 1 (Three-valued semantics). $u \in \Sigma^*$ is a finite word. The truth value of a LTL formula φ with respect to u, denoted by $[u \vDash \varphi]$, is an element of $\mathbb{B}_3$ ($\mathbb{B}_3 = \{\top, \bot, ?\}$) defined as follows:

$$[u \vDash \varphi] = \begin{cases} \top & if \; \forall \sigma \in \Sigma^\omega : u\sigma \vDash \varphi \\ \bot & if \; \forall \sigma \in \Sigma^\omega : u\sigma \nvDash \varphi \\ ? & otherwise \end{cases}$$

Note that in the above definition and the remainder of this paper, we use the notation $[u \vDash \varphi]$ to give a three-valued semantic to a formula based on a finite word u. Further details about three-valued semantics can be found in [1].

3   Predictive Semantics for LTL in Runtime Verification

In a lot of application areas, it is quite appreciated that the verdict can be predicted when the error prefix has not appeared. The three-valued semantics given in the previous section hold an assumption that: the observed word increase incrementally and letters arrive one at a time. When we judge the semantic value of φ with respect to the observed prefix u, we do not know what the following word is. Runtime verification is a body of verification techniques that check whether program's execution satisfies a formal property. The execution is generated by the program's code. If we can do some kinds of static analysis to program code and draw some summary information, it is possible to *predict* what the following word is. Based on this assumption, we propose a predictive semantics for LTL formulae. Although the predictive semantics are only given to LTL formulae in this paper, it can be easily extended to other property specification languages.

Our predictive semantics for LTL formulae are also based on the finite prefix word. But instead of without knowing what the future word is in three-valued semantics, future words are predictable in predictive semantics. The predictive semantics for LTL formulae are defined as follows.

Definition 2 (Predictive semantics). $u \in \Sigma^*$ is an observed word generated by the monitored program so far, $v \in \Sigma^*$ is a finite predictive word which will be *generated in the following time. The truth value of a LTL formula φ with respect to u and v, denoted by $[u \vDash_p^v \varphi]$, is an element of $\mathbb{B}_3$ defined as follows:*

$$[u \vDash_p^v \varphi] = \begin{cases} \top & \text{if } \forall \sigma \in \Sigma^\omega : uv\sigma \vDash \varphi \\ \bot & \text{if } \forall \sigma \in \Sigma^\omega : uv\sigma \nvDash \varphi \\ ? & \text{otherwise} \end{cases}$$

In runtime verification, the observed word u increases as the monitored program runs. There is a distinction between the observed word (which is *known*) and the word which is going to be generated (which is *unknown*). In the predictive semantics definition, v is the word which is going to be generated by the monitored program, but it is already known *now*. That is to say, the predictive word v can be *predicted* at present. That is why we name our semantics as *predictive semantics*. v is a predictive word.

The predictive semantics demand that the predictive word v should be identified every time the observed word u increases. That means, every time a monitored letter happens in execution, we have to predict the word which is going to be generated by the monitored program. Given the monitored program's complexity and dynamic feature, it is not realistic to predict a future word every time a monitored letter happens. In order to make predictive semantics feasible, we propose a implementation predictive semantics.

Only a fixed set of words are predictable in implementation predictive semantics. In predictive semantics definition, we are able to predict the future word every time a monitored letter happens. While in implementation predictive semantics definition, we only have to predict it when it belongs to a *fixed set*. This restriction releases the burden of predicting a future word every time a monitored letter happens. Because we only predict the future word when it falls into a fixed words set.

The implementation predictive semantics for LTL formulae are defined as follows.

Definition 3 (Implementation predictive semantics). *$u \in \Sigma^*$ is an observed word generated by the monitored program so far, $R \subseteq \Sigma^*$ is a fixed predictive words set found in the monitored program. Let v' denote the predictive word which is going to be generated by the monitored program. The truth value of a LTL formula φ with respect to u and R, denoted by $[u \vDash_i^R \varphi]$, is an element of $\mathbb{B}_3$ defined as follows:*

$$[u \vDash_i^R \varphi] = \begin{cases} \top & \text{if } v' \text{ is proven} \in R : (\forall \sigma \in \Sigma^\omega : (uv'\sigma \vDash \varphi)) \\ & \text{or } (\forall \sigma' \in \Sigma^\omega : (u\sigma' \vDash \varphi)) \\ \bot & \text{if } v' \text{ is proven} \in R : (\forall \sigma \in \Sigma^\omega : (uv'\sigma \nvDash \varphi)) \\ & \text{or } (\forall \sigma' \in \Sigma^\omega : (u\sigma' \nvDash \varphi)) \\ ? & \text{otherwise} \end{cases}$$

In implementation predictive semantics definition, R is a fixed predictive words set found in the monitored program through some kinds of static analysis. Different monitored programs have different predictive words sets. R contains all predictable words for formula φ in the monitored program. The predictive word v' is quite different from the the word v in predictive semantics definition: v is an identified word in predictive semantics definition; while v' is only a place holder in implementation predictive semantics, we can only identify it when it is proven to belong to R. If v' does not belong to R, we can not predict the word which is going to be generated by the monitored program and the implementation predictive semantics degenerate into three-valued semantics. Hence, the implementation predictive semantics definition is a compromise between three-valued semantics and predictive semantics. This restriction sacrifices a kind of prediction (when v' is not proven to belong to R), but it also makes this semantics more applicable.

In implementation predictive semantics definition, not every predictive word's (the word belongs to R) occurrence in program's execution is predictable. If v' can not be *proven* $\in R$ through some kinds of static analysis, the implementation predictive semantics do not possess predictive capability. There are some situations in which v' is actually belongs to R but we can not prove it through static analysis.

4   Monitor Construction for LTL with Predictive Semantics

Predictive semantics definition is useless if we can not put it into practice. In this section, we will describe the process of generating a monitor from a LTL formula and describe how to use this monitor to check program's execution with predictive semantics. The monitor generating process is almost identical to the "monitor construction" section in article [1]. We only sketch it here.

4.1   Generating Monitor from LTL Formulae

Given a formula φ, we will construct a monitor (a Finite State Machine(FSM)) M^φ that reads finite word $u \in \Sigma^*$, and produces $[u \models_p^v \varphi]$ with respect to the predictive word v.

For a Nondeterministic Büchi Automata (*NBA*) A, we denote by $A(q)$ the NBA that coincides with A except for the set of initial states Q_0, which is redefined in $A(q)$ as q. Let us fix $\varphi \in LTL$ for the rest of this section, and let $A^\varphi = (\Sigma, Q^\varphi, Q_0^\varphi, \delta^\varphi, F^\varphi)$ denote the NBA that accepts all models of φ and let $A^{\neg\varphi} = (\Sigma, Q^{\neg\varphi}, Q_0^{\neg\varphi}, \delta^{\neg\varphi}, F^{\neg\varphi})$ denote the NBA, which accepts all words falsifying φ. The corresponding construction is standard and explained, for example in [2].

For the automaton A^φ, we define a function $F^\varphi : Q^\varphi \to \mathbb{B}$(with $\mathbb{B} = \{\top, \bot\}$) where we set $F^\varphi(q) = \top$ iff $L(A^\varphi(q)) \neq \emptyset$. We evaluate a state q to $\top$ iff the language of the automaton starting in state q is not empty. To determine $F^\varphi(q)$, we identify in linear time the strongly connected components in A^φ which can be done using Tarjan's algorithm [3]. Using F^φ, we define the NFA $\hat{A}^\varphi = (\Sigma, Q^\varphi, Q_0^\varphi, \delta^\varphi, \hat{F}^\varphi)$ with $\hat{F}^\varphi = \{q \in Q^\varphi \mid F^\varphi(q) = \top\}$. Analogously, we set $\hat{A}^{\neg\varphi} = (\Sigma, Q^{\neg\varphi}, Q_0^{\neg\varphi}, \delta^{\neg\varphi}, \hat{F}^{\neg\varphi})$ with $\hat{F}^{\neg\varphi} = \{q \in Q^{\neg\varphi} \mid F^{\neg\varphi}(q) = \top\}$.

Lemma 1 (The evaluation of LTL formulae with predictive semantics). [1] *Let $u \in \Sigma^*$ is an observed word generated by the monitored program so far, $v \in \Sigma^*$ is a finite predictive word which will be generated in the following time. The truth value of a LTL formula φ with respect to u and v, denoted by $[u \models_p^v \varphi]$, is an element of $\mathbb{B}_3$ defined as follows:*

$$[u \models_p^v \varphi] = \begin{cases} \top & \text{if } uv \notin L(\hat{A}^{\neg\varphi}) \\ \bot & \text{if } uv \notin L(\hat{A}^{\varphi}) \\ ? & \text{otherwise} \end{cases}$$

The lemma yields the following procedure to evaluate the semantics of φ for a given finite word u and a predictive word v: we evaluate both $uv \notin L(\hat{A}^{\neg\varphi})$ and $uv \notin L(\hat{A}^{\varphi})$ and use the above lemma to determine whether $[u \models_p^v \varphi]$. Similarly, we have the same conclusion to the evaluation of LTL formulae with implementation predictive semantics as follows.

Lemma 2 (The evaluation of LTL formulae with implementation predictive semantics). $u \in \Sigma^*$ *is an observed word generated by the monitored program so far, $R \subseteq \Sigma^*$ is a fixed predictive words set found in the monitored program. Let v' denote the future word which is going to be generated by the monitored program. The truth value of a LTL formula φ with respect to u and R, denoted by $[u \models_i^R \varphi]$, is an element of $\mathbb{B}_3$ defined as follows:*

$$[u \models_i^R \varphi] = \begin{cases} \top & \text{if } (v' \text{ is proven} \in R \wedge uv' \notin L(\hat{A}^{\neg\varphi})) \vee (u \notin L(\hat{A}^{\neg\varphi})) \\ \bot & \text{if } (v' \text{ is proven} \in R \wedge uv' \notin L(\hat{A}^{\varphi})) \vee (u \notin L(\hat{A}^{\varphi})) \\ ? & \text{otherwise} \end{cases}$$

As a final step, we now define a (deterministic) FSM M^φ that outputs for each finite word u and predictive word v their associated predictive semantical evaluation. Let $\tilde{A}^\varphi$ and $\tilde{A}^{\neg\varphi}$ be the deterministic versions of $\hat{A}^\varphi$ and $\hat{A}^{\neg\varphi}$, which can be computed in a standard manner using the power-set construction. Then, we define $\bar{A}^\varphi$ as a product of $\tilde{A}^\varphi$ and $\tilde{A}^{\neg\varphi}$.

Definition 4 (Monitor M^φ for a LTL formula φ with predictive semantics). *Let φ be a LTL formula and $\hat{A}^\varphi$ be a NFA. Let $\tilde{A}^\varphi = (\Sigma, Q^\varphi, \{q_0^\varphi\}, \delta^\varphi, \tilde{F}^\varphi)$ be a deterministic automaton with $L(\tilde{A}^\varphi) = L(\hat{A}^\varphi)$, $\tilde{A}^{\neg\varphi} = (\Sigma, Q^{\neg\varphi}, \{q_0^{\neg\varphi}\}, \delta^{\neg\varphi}, \tilde{F}^{\neg\varphi})$ be a deterministic automaton with $L(\tilde{A}^{\neg\varphi}) = L(\hat{A}^{\neg\varphi})$. The product automaton $\bar{A}^\varphi = \tilde{A}^\varphi \times \tilde{A}^{\neg\varphi}$ is a FSM $(\Sigma, \bar{Q}, \bar{q}_0, \bar{\delta}, \bar{\lambda})$ where*

- $\bar{Q} = Q^\varphi \times Q^{\neg\varphi}$,
- $\bar{q}_0 = (q_0^\varphi, q_0^{\neg\varphi})$,
- $\bar{\delta}((q, q'), a) = (\delta^\varphi(q, a), \delta^{\neg\varphi}(q', a))$ *and*
- $\bar{\lambda} : \bar{Q} \to \mathbb{B}_3$ *is defined by*

$$\bar{\lambda}((q, q'), a) = \begin{cases} \top & \text{if } q' \notin \tilde{F}^{\neg\varphi} \\ \bot & \text{if } q \notin \tilde{F}^\varphi \\ ? & \text{if } q \in \tilde{F}^\varphi \wedge q' \in \tilde{F}^{\neg\varphi} \end{cases}$$

---

[1] The proof is included in section 2.3 in article [1].

The monitor M^φ for a LTL formula φ is the unique FSM obtained by minimizing the product automaton $\bar{A}$.

Just as proved in article [1], the following theorem holds (there are similar theorem holds for implementation predictive semantics). According to this theorem, if we want to check whether finite word u and predictive word v satisfy or violate the property φ, we just need to check whether uv drives the monitor M^φ into conclusive states (states labeled by $\top$ or $\bot$ according to $\bar{\lambda}$).

Lemma 3 (LTL monitor correctness). *Let φ be a LTL formula, let $M^\varphi = (\Sigma, \bar{Q}, \bar{q}_0, \bar{\delta}, \bar{\lambda})$ be the corresponding monitor and let v be a predictive word. Then, for all $u \in \Sigma^*$, the following holds:*

$$[u \vDash_p^v \varphi] = [uv \vDash \varphi] = \bar{\lambda}(\bar{\delta}(\bar{q}_0, uv))$$

4.2   Checking Process

Once we generate the monitor from LTL formula, we can use it to check whether the monitored program's execution satisfies (driving the monitor into $\top$ state) or violates (driving the monitor into $\bot$ state) the formula. The checking process of monitor with predictive semantics is slightly different from the traditional runtime monitors. The traditional runtime monitors receive letters while in predictive semantics, they receive predictive words. When a monitor receives a predictive word, it first checks whether the run of predictive word goes through a conclusive state. If so, the monitor can give a verdict before the observed word driving it into conclusive states. If not, the monitor changes its current state $\bar{q}$ to $\bar{\delta}(\bar{q}, v)$ ($\bar{\delta}$ is the transition function, v is the predictive word). For runtime monitors in implementation predictive semantics, there are two different types of letters arising in program's execution: an ordinary letter and a predictive word. When a monitor receives an ordinary letter, it just changes its current state according to the transition function and judge whether it arrives the conclusive states. When receives a predictive word, it acts the same as the monitor in predictive semantics.

4.3   A Demonstration Example

In Java library, `Vector` is a common class to hold a variable collection of objects. The `Vector` class provides an `iterator` method to get an `Iterator` instance. The `Iterator` instance allows users to enumerate all the elements belonging to `Vector`. There is usage demand that once the `Vector` creates an `Iterator` instance, `Iterator`'s next method should not be called if the underlying `Vector` is modified though its own methods, such as `remove` etc, . If we use `create` to stand for the letter of creating an `Iterator` instance, `update` for modifying the `Vector`, and `next` of enumerating the `Vector`, the usage rule can be depicted as the formula: G(`create` $\rightarrow$ G(`update` $\rightarrow$ $\neg$F(`next`))). Using the algorithm described in the preceding section, we get the monitor in Fig. 1.

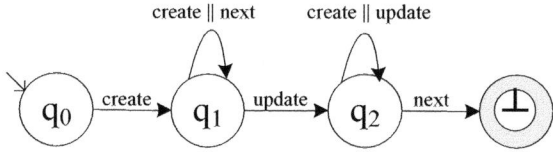

Fig. 1. A monitor generated from G(create → G(update →! F(next)))

If the predictive words set found in the monitored program is `update.next` (where . means concatenation). Then, the monitor can find the violation at state q_1 when it receives the predictive word. While for monitors without predictive semantics, they can only find the violation at the state labeled with $\bot$.

5   Implementation

We have implemented a prototype tool to support the implementation predictive semantics. Our tool integrates our previous work of generating monitors from LTL formulae and extends the work from Hossein Sadat-Mohtasham [4] of finding and instrumenting predictive word in monitored program.

Fig. 2 depicts a typical usage scenario of our tool. Users first specify the properties they want to monitor through LTL formulae. LTL formulae are then converted into monitors through $LTL3^2$ tool. Predictive words sets are found in monitored program according to the monitor's alphabet. Finally, monitors are injected into the monitored program by an AspectJ compiler *abc* [5] and predictive words set are injected by a modified version of *Transcut* [4]. The result program is the desired program with predictive runtime monitoring capability. It is able to detect property's satisfaction or violation before the software execution driving monitor into conclusive states.

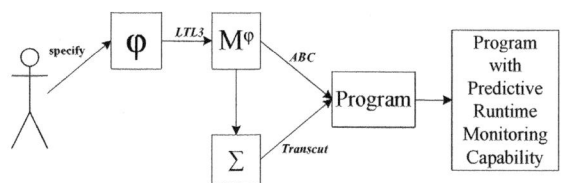

Fig. 2. A typical usage scenario of our prototype tool

5.1   Converting a LTL Formula into a Monitor

We have described the process of generating a monitor from a LTL formula in section 4. The monitor generation process is almost the same as the work

---

in [1] and we have already implemented a *LTL3* tool to realize this process. We also adopt this tool to generate a monitor from a LTL formula. For example, the monitor generated from formula G(create → G(update → ¬ F(next))) by *LTL3* tool is depicted in Fig. 3.

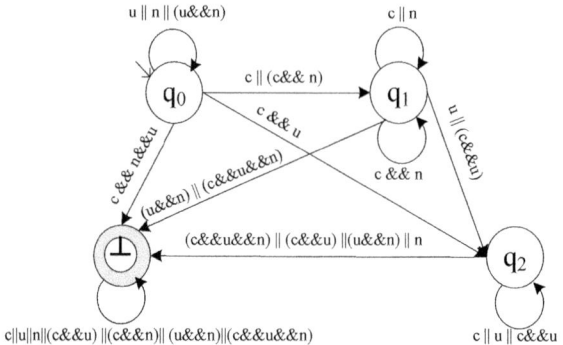

Fig. 3. A monitor generated by *LTL3* tool from G(create → G(update →! F(next))), where: c denotes create, u denotes update and n denotes next

In Fig. 3, edges are labeled with a set of APs. The $\|$ operator means there are two edges between these two states. Each is labeled by one operand. The && operator means the two operands have to be true at same time when the transition happens. For example, the edge labeled with $(c\|(c\&\&n))$ between state q_0 and q_1 denotes that there are actually two edges connecting these two states. One is labeled with c and the other is labeled by $c\&\&n$.

In *LTL3* tool, the alphabet of the monitor generated from φ is 2^{AP}, where AP is the set of automatic propositions in φ. In our implementation, an automatic proposition stands for a certain kind of events in program execution and is specified by a pointcut expression in AspectJ[6]. The problem of judging whether two pointcut expressions would select a common event can be partly solved through pointcut's syntax analysis. We use this analysis to further simplify the monitor. The edges whose label are the intersection of two or more automatic propositions can be simply dropped off if the corresponding pointcut expressions can not select a common event. For instance, the edge labeled by $c\&\&n$ in Fig. 3 denotes events which belong to create and next at the same time. This is actually impossible because the two pointcut expressions for create and next (they are described in Fig. 4) can not select any common events. We also omit the self-loop edges in the initial state q_0 and the conclusive states. The self-loop edge in the initial state can be omitted as we check the program's whole trace (instead of suffix traces). The self-loop edge in the conclusive states can be omitted as there are no edges going out it. Through this kind of simplification, we get the monitor in Fig. 1.

5.2   Instrumenting Monitored Program

Once getting the monitor, we can use its alphabet to find a predictive words set from monitored program and then instrument the monitored program according to the predictive words set. As a letter in monitor's alphabet corresponds to an event in program's execution. We first briefly introduce the map between a letter in alphabet and an event in execution.

The Map Between a Letter in Alphabet and an Event in Execution. A letter in monitor's alphabet corresponds to a certain kind of event in program's execution. We use AspectJ's pointcut to build this connection. A pointcut is a set of well-defined points in the execution of the program and corresponds to a letter in alphabet. A pointcut is defined in terms of an enumeration of method signatures, wildcards and control flow relations [6]. For example, The pointcut expressions for letters (`create`, `update` and `next`) in Fig. 1 are described in Fig. 4. The first line is a pointcut expression denoting the event of calling `iterator` method on *Collection* class or its subClass and corresponding to the letter `create`. The following two `pointcut` expressions have similar meanings and denote `updage` and `next` respectively.

```
pointcut create : call(* java.util.Collection+.iterator()) && target(c) ;
pointcut update : call(* java.util.Collection+.add*(..)) && target(c) ||
        call(* java.util.Collection+.clear()) && target(c) ||
            call(* java.util.Collection+.remove*(..) ) && target(c);
pointcut next :   call(* java.util.Iterator+.next());
```

Fig. 4. A map between letters in monitor's alphabet and events in program's execution

Finding Predictive Words Set on Monitored Program's PDG Representation. While it is easy for AspectJ compiler to find the code place where a letter (event) happens, it is not easy to find the code place where the subsequent happening letter sequence (word) belongs is a predictive word (its length is bigger than 1). We find the predictive words set from monitored program's *control flow graph* (CFG) and *program dependence Graph* (PDG) representation.

A *CFG* is a directional graph in which nodes represent basic blocks. A CFG can be depicted as $< V, E >$, where V denotes the nodes set, E denotes the control flow edge set. A *region* includes the instructions in a program that execute under the same control conditions. Two nodes in a CFG are in the same region if they have the same set of control dependence predecessors [7]. Regions have an important property that makes them very ideal for finding predictive words: if the normal flow of control enters the region (which occurs only through the head node of the region), it will go through all the nodes in the region, and eventually exit through the tail node of the region. This is similar to the property of basic blocks [8] with the difference that regions can consist of non-contiguous pieces of code.

There are two different types of region: *strong region* and *weak region* [7]. We only concern strong region in this paper. Two nodes n_1 and n_2 are in the same *strong region* iff n_1 and n_2 occur the same number of times in any complete control-flow path. Thomas Ball provides an algorithm to find strong regions in $O(V + E)$ time for all CFGs [7]. Predictive words are found in strong regions. Fig. 5 illustrates the concrete finding process. In this figure, the dashed ellipse stands for a strong region. b1,b2,b3 and b4 are four basic blocks or statements. They belongs to the same strong region. r1, r2 and r3 are three potential finding area. Predictive words are found in these potential areas. For example, if b1 and b4 generate letter a and b respectively, and b3, b4 do not generate monitored letters, then we can find the predictive word ab in this strong region.

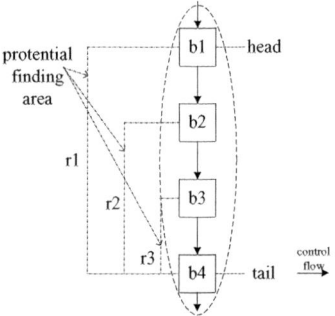

Fig. 5. Finding predictive word in strong region

The predictive word finding process is complicated by the fact that strong region are non-contiguous pieces of code. We not only have to ensure that the code fragments surrounded by the predictive word in the same strong region (such as b2, b3 in Fig. 5) do not generate monitored letters, but also that the code fragments surrounded by the predictive word in CFG also do not generate monitored letters. We ensure the latter condition based on the monitored program's *PDG* representation. *PDG* is an intermediate program representation. It gives a hierarchial relation among regions. The code fragments surrounded by predictive word in CFG are the subregions of the region generating predictive word. If a predictive word is found in a region and the region has subregions, we have to check whether the subregions are surrounded by predictive word and generating monitored letters.

Our implementation for finding predictive words set in monitored program is based on a modified version of the work from Hossein Sadat-Mohtasham [4] and can only find predictive word in each method at present. The major modifications includes matching predictive words in strong region instead of weak region and checking whether the code fragments surrounded by predictive word generates monitored letters. In the future, we are planing to find predictive words across methods.

6   Experiments

6.1   Experiments on Illustration Example

We have described a monitor in Fig. 1 and its alphabet is {create, update, next}. We can use this alphabet to find predictive words in monitored program's PDG representation. For instance, to the code fragment in Fig. 6a and its corresponding PDG representation in Fig. 6b, the found predictive words set is {create.update.next}. That is because the code generating create (statement 2), update (statement 3) and next (statement 4) are in the same strong region and there are no extra code surrounded by them. Fig. 7 describes another code fragment (7a) and its PDG representation (7b), whose predictive words set is also {create.update.next}. The code generating monitored letters (statements 2, 3, 7) are in the same strong region but are not contiguous in CFG. The code (statements 4, 5) surrounded by predictive word do not generate monitored letters. If we instrument the monitored program according to the predictive words set, the monitor can find the violation at statement 2 in Fig. 6a and statement 3 in Fig. 7a.

It is necessary to point out that the code fragments in Fig. 6a, 7a only denote code templates rather than concrete code fragments. Users can scatter any statements in these code fragments as long as they neither contain the monitored letters nor change the monitored program's PDG structure.

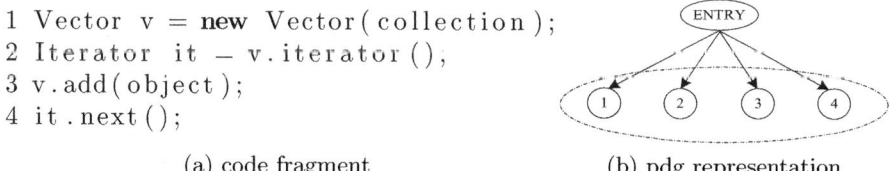

```
1 Vector v = new Vector(collection);
2 Iterator it = v.iterator();
3 v.add(object);
4 it.next();
```

(a) code fragment   (b) pdg representation

Fig. 6. A code fragment and its PDG representation whose predictive words set is create.update.next

6.2   Experiments on Real Projects

We also use the monitor in Fig. 1 to find predictive words set on several big open-source projects. These projects are chosen from Dacapo benchmark [9] and quite representative for real projects.

We summarize the project size (class number, method number) and the ratio of predictable shadows [3] and predictable regions in table 1. The row of predictable shadow's ratio describes the percentage of the shadows for letters in predictive words (compared to all the shadows for monitored letters). The fraction in parentheses list the shadow's actual number. The denominator is the number of monitored letter's shadow found in monitored program, while the

---

[3] A shadow is a code area which may generate monitored letter at runtime.

```
1 Vector  v = new  Vector(col)
2 Iterator  it = v.iterator( );
3 v.add(object);
4 for(each  element  in  col){
5      process  each  element ;
6 }
7 it.next();
```

(a) code fragment   (b) pdg representation

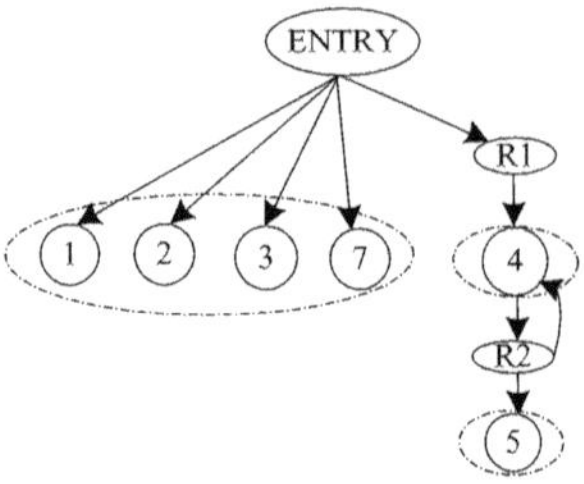

Fig. 7. A code fragment and its PDG representation whose predictive words set is also `create.update.next`

Table 1. The result of finding predictive words on real, large-scale projects

	antlr	eclipse	fop	hsqldb	bloat	lucene
class number	224	344	967	385	263	311
method number	2972	3978	6889	5859	3986	3013
predictable shadow ratio	0% (0/23)	7.92% (53/391)	24.65% (83/288)	28.23% (45/124)	17.06% (608/1495)	25% (61/224)
predictable region ratio	0% (0/23)	3.33% (22/360)	7.86% (24/229)	11.83% (14/93)	7.88% (204/1091)	7.3% (15/178)

numerator is the number of the shadow for letters which are included in predictive words. Generally speaking, the percentage is more bigger, there are more letters belongs to the predictive words and the predictive semantics are more applicable. From the table's fourth row, we can see that if the project contains a large number of shadows, there are a considerable percentage (from 7.92% to 28.23%) of letters (shadows) are predictable. The percentage of predictable letters in *antlr* project is 0%. That is because the antlr project seldom generate monitored letters (there are only 23 places generating monitored letters, and they belong to 23 different regions).

We also list the percentage of the predictable region (the region which contains predictive word) compared to the region which contains monitored letter. The percentage of predictable regions is a bit low (less than 12%). That is because we only find the predictive word in methods at present. In the future, we plans to inline more methods and find predictive word across methods.

7   Related Work

As runtime verification is becoming more and more popular to guarantee software's reliability, many runtime verification frameworks are proposed. The Tracematches [10] framework uses regular expression to specify verification property.

The Java-MOP [11] framework is designed in a way that is neutral among different logical formalisms. Regular Expressions, Context-free Grammars and LTL are all supported in Java-MOP. The Java-MaC [12] framework defines its own property specification languages: Primitive Event Definition Language(PEDL) and Meta Event Definition Language(MEDL). However, the semantics of above frameworks' property specification languages are not predictive. They all match the observed word against the property. What the future word would like is totally ignored.

Let us take *Tracematches* as an example. *Tracematches* is a popular runtime verification framework developed by Oxford and McGill university [10]. Tracematches gives the verdict when the observed letter sequence satisfies/violates the property. Hence, it catches the violation at statement *4* in Fig. 6a and at statement *7* in Fig. 7a respectively. While our implementation catches the violation at statement *2* in these two figures. Obviously, our implementation can catch the problem more earlier.

The semantics used in runtime verification which have a kind of prediction are Feng Chen etc's work [13]. Their work is mainly used in concurrent programs. The prediction in their work means an un-execution trace can be inferred from an execution path based on the program casual model. Hence, it is able to "predict" potential violations of monitored property even when the violations are not encountered in the observed executions. This kind of prediction is not about the events which will happen in the future and is totally different from our work.

8   Conclusions

In this paper, we propose predictive semantics for LTL formulae in runtime verification. We give an algorithm on how to convert a LTL formulae into a monitor and describe how to use it to check monitored program's execution with predictive semantics. We also implement a prototype tool to support predictive semantics. We use it to do some experiments on several real, large-scale projects. The result demonstrates that our predictive semantics is generally applicable in these projects. As far as we know, this is the first predictive semantics definition in runtime verification.

Acknowledgment. The work is supported by National Natural Science Foundation of China under Grant No.60970035, No.90818024, No.91018013, by the Hi-Tech Research and Development Program of China (863 Plan, 2011AA010106), and by the German BMBF Grant CHN 09/003.

References

1. Bauer, A., Leucker, M., Schallhart, C.: Runtime verification for LTL and TLTL. ACM Transactions on Software Engineering and Methodology (TOSEM) (2009) (in press)

2. Vardi, M.Y.: An Automata-Theoretic Approach to Linear Temporal Logic. In: Moller, F., Birtwistle, G. (eds.) Logics for Concurrency. LNCS, vol. 1043, pp. 238–266. Springer, Heidelberg (1996)

3. Tarjan, R.: Depth-First Search and Linear Graph Algorithms. SIAM Journal on Computing 1(2), 146–160 (1972)

4. Sadat-Mohtasham, H.: Transactional Pointcuts for Aspect-Oriented Programming. phd thesis, Department of Computer Science, University of Alberta (2010)

5. Avgustinov, P., Christensen, A.S., Hendren, L., Kuzins, S., Lhoták, J., Lhoták, O., de Moor, O., Sereni, D., Sittampalam, G., Tibble, J.: abc: An Extensible AspectJ Compiler. In: Rashid, A., Aksit, M. (eds.) Transactions on Aspect-Oriented Software Development I. LNCS, vol. 3880, pp. 293–334. Springer, Heidelberg (2006)

6. Kiczales, G., Hilsdale, E., Hugunin, J., Kersten, M., Palm, J., Griswold, W.G.: An Overview of AspectJ. In: Lee, S.H. (ed.) ECOOP 2001. LNCS, vol. 2072, pp. 327–353. Springer, Heidelberg (2001)

7. Ball, T.: What's in a region?: or computing control dependence regions in near-linear time for reducible control flow. ACM Lett. Program. Lang. Syst. 2, 1–16 (1993)

8. Aho, A.V., Sethi, R., Ullman, J.D.: Compilers: principles, techniques, and tools. Addison-Wesley Longman Publishing Co., Inc., Boston (1986)

9. Blackburn, S.M., Garner, R., Hoffmann, C., Khang, A.M., McKinley, K.S., Bentzur, R., Diwan, A., Feinberg, D., Frampton, D., Guyer, S.Z., Hirzel, M., Hosking, A., Jump, M., Lee, H., Moss, J.E.B., Phansalkar, A., Stefanović, D., VanDrunen, T., von Dincklage, D., Wiedermann, B.: The dacapo benchmarks: java benchmarking development and analysis. In: Proceedings of the 21st Annual ACM SIGPLAN Conference on Object-Oriented Programming Systems, Languages, and Applications, OOPSLA 2006, pp. 169–190. ACM, New York (2006)

10. Allan, C., Avgustinov, P., Christensen, A.S., Hendren, L., Kuzins, S., Lhoták, O., de Moor, O., Sereni, D., Sittampalam, G., Tibble, J.: Adding trace matching with free variables to AspectJ. In: OOPSLA 2005: Proceedings of the 20th Annual ACM SIGPLAN Conference on Object Oriented Programming Systems, Languages, and Applications, pp. 345–364. ACM Press, New York (2005)

11. Chen, F., Jin, D., Meredith, P., Roşu, G.: Monitoring oriented programming - a project overview. In: Proceedings of the Fourth International Conference on Intelligent Computing and Information Systems (ICICIS 2009), pp. 72–77. ACM (2009)

12. Kim, M., Viswanathan, M., Kannan, S., Lee, I., Sokolsky, O.: Java-mac: A runtime assurance approach for java programs. Form. Methods Syst. Des. 24, 129–155 (2004)

13. Chen, F., Şerbănuţă, T.F., Roşu, G.: jPredictor: a predictive runtime analysis tool for Java. In: ICSE 2008: Proceedings of the 30th International Conference on Software Engineering, pp. 221–230. ACM, New York (2008)

A Case Study in Verification
of Embedded Network Software

Kalyan C. Regula[1], Hampton Smith[1], Heather H. Keown[1],
Jason O. Hallstrom[1], Nigamanth Sridhar[2], and Murali Sitaraman[1]

[1] School of Computing, Clemson University, Clemson, SC 29634
[2] Electrical and Computer Engineering, Cleveland State University, Cleveland, OH 44115

Abstract. Embedded network systems support a variety of application domains, including environmental monitoring, social networking, and healthcare. These large networks of low-powered microcontroller-based nodes present challenges in ensuring correctness of the software that runs on these systems. Most embedded networked systems are programmed in C. Verifying software written in C is difficult. In this paper, we take a different approach: We report on our work using the RESOLVE language to program embedded networked systems. Our compiler leverages the RESOLVE verification system and maintains the correctness guarantees established during verification. The verified code is then translated into property-preserving C code that can run on the target hardware.

1 Introduction

Embedded network systems represent technology advances in computing capabilities and communications with physical processes. These systems are essentially large networks (thousands of nodes) of devices that are individually small and resource-constrained. Embedded networked systems are typically deployed in settings where maintenance efforts following deployment need to be performed remotely; physical access to every single node in a large system is not feasible.

In order to cope with this near-inability to perform software updates post-deployment, a lot of research energy has been invested in creating ways of increasing confidence in the software before it is deployed in an embedded network system setting. Much of this work has centered around simulation [12, 17] and laboratory testbeds [1, 4, 19]. Despite such efforts, producing reliable software for embedded networked systems still proves to be difficult. An alternative to testing is formal behavioral verification: if the software program installed on the devices can be proven correct against a formal specification, maintenance costs will be reduced and system failures can be avoided.

Most embedded network systems are implemented using the C language. Several approaches for verifying the correctness of C programs have been proposed, including model checking [14], automated reasoning [3], and low-level memory models [18]. These approaches largely rely on hand-written annotations interspersed in the C code, which may introduce complexities of their own. Rather than operate directly on C code, we begin with code in a higher-level verification language that can be verified more easily and directly, then compile down to correct-by-construction C code that targets

A. Goodloe and S. Person (Eds.): NFM 2012, LNCS 7226, pp. 433–448, 2012.

embedded systems. We thus achieve software for embedded network systems that can be certified with high confidence. The technique of generating correct-by-construction code is well studied but to our knowledge this is the first attempt to explore the particular challenges associated with small embedded systems.

As our verification language, we use RESOLVE [5], an integrated specification and programming language. RESOLVE affords a number of advantages including clean semantics, an extensible, modular specification language, an integrated minimalist prover, and a web-based IDE. For a more detailed description and some basic examples, see Appendix A. By design, RESOLVE separates specification from implementation and thus supports multiple interchangeable realizations of a given component [2]. This is a decided advantage for embedded systems where a large number of hardware vendors supply components that compose a single embedded network system. We have used RESOLVE to mathematically model a handful of low-level physical components used by our embedded devices, presenting two here. We then present a RESOLVE-to-C translator that permits application logic implemented and proved correct in RESOLVE to be translated into corresponding C code that can execute on a microcontroller.

The rest of the paper is organized as follows. We present an overview of the MoteStack embedded network platform in Section 2 and overview of the verification process in Section 3. We describe the correctness-preserving translation strategy to generate C code from RESOLVE in Section 4. After presenting related research in Section 5, we conclude with a summary of our contributions in Section 6.

2   The MoteStack Platform

The MoteStack platform is representative of most embedded network system platforms. The MoteStack includes a microcontroller that is connected to sensors that can capture signals from the environment; analog sensors are interfaced to the microcontroller using analog-to-digital converters. Data captured from sensors can be transmitted either via a wired serial communication channel such as a UART, or more commonly, via a wireless radio connection. Actuators that provide sensory output (e.g., LEDs) or modify the surrounding environment (e.g., air conditioner) are connected to the microcontroller. In addition to these peripherals, the Motestack has an internal clock that is used for periodic or timed operation. Figures 1 and 2 show two of these interfaces as RESOLVE concepts (the others are omitted for lack of space).

In RESOLVE, a *concept* defines an interface that a component provides via an abstract state model and operations defined in terms of that model. A *realization* of a concept is a component that implements that interface. An *enhancement* is a component that adds additional behavior to a realization (similar to a subclass). Concepts, realizations, and enhancements can all be parameterized. A *facility* is used to instantiate a parameterized component.

The model for the ADC peripheral is a cartesian product of three boolean variables, respectively representing (a) the on/off status of the sensor, (b) whether or not the sensor has been properly initialized, and (c) whether the sensor is in an error state. When the component is initialized (Figure 1, Line 7), the sensor is off, and the hardware peripheral has not yet been initialized. To begin using the ADC, it must first be initialized

```
1 Concept ADC_Template;
2   uses Std_Integer_Fac, Std_Clock_Fac;
3   Var ADC:Cart_Prod
4       Sensor_On, Init, Error : B;
5   end;
6   Facility_Initialization ensures
7       ADC.Sensor_On = false and ADC.Init = false and ADC.Error = false;
8   Operation Sensor_On();
9       ensures ADC.Sensor_On;
10  Operation ADC_Init();
11      ensures (ADC.Sensor_On = false and ADC.Init) or ADC.Error;
12  Operation Read_ADC(evaluates I: Integer): Integer;
13      requires  ADC.Init = true and ADC.Sensor_On and 0 <= I <= 7;
14      ensures Read_ADC > 0 and ADC.Sensor_On = false;
15 end ADC_Template;
```

Fig. 1. RESOLVE Concept for ADC

```
1 Concept Leds_Template;
2   uses Integer_Theory, Boolean_Theory, Std_Boolean_Fac, Std_Integer_Fac;
3   Var L:Cart_Prod
4       L0:B; L1:B; L2:B; L3:B; L4:B;
5   end;
6   Facility_Initialization ensures
7       L.L0 = false and L.L1 = false and L.L2 = false and L.L3 = false and L.L4 = false;
8   Operation LED0_Init();
9   Operation LED0_Set(evaluates b: Boolean);
10      ensures L.L0 = b;
11  Operation LED0_Toggle();
12      ensures L.L0 = not(L.L0);
13  Operation LED0_Status(): Boolean;
14      ensures LED0_Status = L.L0;
15  // Analogous operations for LED1 .. LED4
16 end Leds_Template;
```

Fig. 2. RESOLVE Concept for LEDs

using ADC_Init(). It can then be turned on by calling Sensor_On(). Once the sensor is ready for use, the sensor can be queried for its value by invoking Read_ADC(), which returns the current value of the sensor as a positive integer, then turns the sensor off. Read_ADC() takes one parameter, which is annotated with the evaluates keyword, indicating that the parameter is pass-by-value. Other parameter passing modes exist. For example, alters indicates that the parameter may be changed in an unspecified way over the course of the operation's call. For more complete information, see the discussion in Appendix A on the details of RESOLVE.

The Leds_Template concept describes the LEDs on the MoteStack, which are representative of actuators that the microcontroller can control. The MoteStack has five LEDs and, correspondingly, the model of the Leds_Template component is a cartesian product of five boolean variables, one for each LED. When the component is initialized, all five LEDs are off. The component provides three operations to operate each of the LEDs: a Set() operation to turn the LED on or off, a Toggle() operation to toggle the LED, and a Status() operation to query the current status of the LED.

```
 1 Facility SenseAndBroadcastAverage;
 2   uses Std_Boolean_Fac, Std_Integer_Fac;
 3   Facility ADC is ADC_Template realized by Std_ADC_Realiz;
 4   Facility XBEE is XBEE_Template realized by Std_XBEE_Realiz;
 5   Facility Clock is Clock_Template realized by Std_Clock_Realiz;
 6   Facility LED is Leds_Template realized by Std_Leds_Realiz;
 7   Facility Integer_Queue_Fac is Queue_Template(Integer, 9)
 8     realized by Circular_Array_Realiz;
 9
10   Operation Main();
11   Procedure
12     Var Sample, Total, Average, Count: Integer;
13     Var On, Off: Boolean;
14     Var Data_Samples: Integer_Queue_Fac.Queue;
15
16     ADC.ADC_Init();
17     XBEE.XBEE_Init();
18     On := True();
19     Off := False();
20     While (True())
21       changing Sample, Total, Average;
22       maintaining True;
23     do
24       Clock.Wait_1S();
25       Sample := ADC.Read_ADC(0);
26       If (Sample mod 256 >= 0) then
27         LED.LED1_Set(On);
28       else
29         LED.LED1_Set(Off);
30       end;
31       // >=64 -> LED2; >=128 -> LED3;
32       // >=192 -> LED4, analogously
33
34       if (Integer_Queue_Fac.Rem_Capacity(Data_Samples) > 0) then
35         Integer_Queue_Fac.Enqueue(Sample, Data_Samples);
36       end else
37         Average := Average_Queue(Data_Samples);
38         Integer_Queue_Fac.Clear(Data_Samples);
39         XBEE_Send_Data(Average, 2);
40       end;
41     end;
42   end Main;
43 end SenseAndBroadcastAverage;
```

Fig. 3. RESOLVE Facility for the Sense-and-Broadcast application

2.1   Application Example

Figure 3 presents a canonical embedded network system example written in RESOLVE. This sense-and-broadcast example simply queries its sensor through the ADC interface every second and sends an average of every ten data samples to the rest of the network via radio broadcast using the XBEE interface.

The SenseAndBroadcastAverage facility depends on five other components, four of which are specific to embedded systems: a timer component (Clock_Template), an interface to the sensor(s) (ADC_Template), a radio interface to broadcast sensor readings (XBEE_Template), and an interface to the LED actuators (LED_Template). In addition to these components, the facility also depends on a Queue_Template component and a method for averaging the values in a queue, which we will present momentarily. On startup, the application first initializes the ADC and radio components (lines 16 and 17, respectively). Once that is done, the application executes a continuous non-terminating loop. The changing clause is required to clearly state which variables in the program may be changed within the loop (line 21).

In each iteration, it first waits for a duration of one second (line 24). At the end of this duration, a sample is read from the sensor by way of the ADC interface on channel 0 (line 25). Based on the value returned from the ADC, the application actuates one or more of the LEDs on the sensor node. As samples are collected, they are enqueued into Data_Samples. When the queue is full, the average of the data samples in the queue is computed and sent out via broadcast using the XBEE radio interface (line 39). Importantly, each of these operations is formally and completely specified in a RESOLVE component, so a proof obligation is raised, for example, that before Read_ADC() is called, the associated peripheral has been initialized and is on.

To readers familiar with writing software for embedded systems, some aspects of this application code may raise concerns. In particular, the code in Figure 3 seems to be performing several blocking input/output operations (waiting on the clock, waiting for sensor readings, etc.). Such blocking implementations are typically bad form on an embedded microcontroller. The current version of the compiler does not deal with this limitation, and the generated code does execute these operations as blocking calls. We made this choice largely because we gave preference to preserving correctness over performance. Non-blocking implementations of these operations are possible using wrappers that present the same abstraction as blocking ones, for example, using techniques described in [13].

3   The Verification Process

The first step toward verifying RESOLVE code is to augment it with mathematical assertions reflecting the proof obligations raised by that module's corresponding specification (for example, assertions are generated at the end of procedure implementations to assert that they meet their ensures clauses.) This represents an intermediate verification step wherein all proof obligations and assumptions are made explicit.

Next, this assertive code is passed to the verification condition generator. A *verification condition*, or VC, is a mathematical statement that must be true for code to be correct. VCs are generated in several situations, including the preconditions of called methods, postconditions of methods being proved, and mathematical assertions related to termination. The primary job of the VC generator is to generate those VCs both necessary and sufficient to demonstrate the correctness of the program. Intermediate mathematical variables (which are different from program variables and are only used in the mathematical proof) must be introduced to reflect the changing state of program variables. A more in-depth discussion of the nature and generation of VCs can be found in [8].

Consider the Average_Queue operation. We provide this functionality as a helper procedure that takes a queue from our Integer_Queue_Fac (i.e, a Queue of up to 9 integers) and returns the floor of its average, restoring the queue to its original value. We provide its specification here:

```
1 Operation Average_Queue(restores Q : Integer_Queue_Fac.Queue) : Integer;
2          requires |Q| < max_int / 2 and
3                   Q /= empty_string and
4                   Holds_for_All(Q, Non_Negative);
5          ensures Average_Queue = Fold_String_Right(Sum, 0, Q) / |Q|;
```

Note that the Average_Queue operation has a requires clause that client code must satisfy before it may be called–the length of the Queue, $|Q|$, must be less than half the maximum representable integer for the platform. The reason for these requirements will be discussed later. In addition, the queue must not be empty, for the obvious reason that it is unclear what the average of a set of zero numbers should be. Finally, all the elements of Q must be non-negative, which removes the complexity of dealing with negative modulos. Note that Holds_for_All() and Fold_String_Right() are higher-order definitions imported from String_Theory, which contains theorems for reasoning about such constructions, and Non_Negative() is a predicate imported from Integer_Theory. We then implement this operation as follows:

```
1  Procedure
2      Var Cur_Entry : Integer;
3  Var Included : Queue;
4  Var Running_Average : Integer;
5  Var Running_Remainder : Integer;
6  Var Remainder_Adjust : Integer;
7  Var Weighted_Cur_Entry : Integer;
8  Var Cur_Entry_Remainder : Integer;
9  Var Q_Length : Integer;
10
11 Q_Length := Length(Q);
12
13 While (Length(Q) > 0)
14   changing Cur_Entry, Included, Q, Running_Average, Running_Remainder,
15     Weighted_Cur_Entry, Cur_Entry_Remainder, Remainder_Adjust;
16   maintaining Included o Q = #Q and
17     Running_Average = Fold_String_Right(Sum, 0, Included) / |#Q| and
18     Running_Remainder =
19                     Fold_String_Right(Sum, 0, Included) mod |#Q| and
20     0 <= Running_Average and
21             Running_Average <= max_int / |#Q| * |Included|;
22   decreasing |Q|;
23 do
24   Dequeue(Cur_Entry, Q);
25
26   Weighted_Cur_Entry := Div(Cur_Entry, Q_Length);
27   Cur_Entry_Remainder := Mod(Cur_Entry, Q_Length);
28
29   Running_Average := Running_Average + Weighted_Cur_Entry;
30   Running_Remainder := Running_Remainder + Cur_Entry_Remainder;
31   Remainder_Adjust := Div(Running_Remainder, Q_Length);
32   Running_Remainder := Mod(Running_Remainder, Q_Length);
33   Running_Average := Running_Average + Remainder_Adjust;
34
35   Enqueue(Cur_Entry, Included);
36 end;
37
38 Q :=: Included;
39
40 Average_Queue := Running_Average;
41 end;
```

The queue declared in this method shares a facility, and thus an implementation, with the one from Section 2.1. This doubles our memory consumption, since we create two circular arrays. Because RESOLVE verifies modularly, we could easily swap out for a linked implementation like the one presented in [10] if our design called for different performance characteristics.

Note the maintaining clause, which asserts expressions that must be true at the beginning and end of each loop iteration. The first conjunct asserts that as we transfer elements out of the original queue (#Q) and into Included, the concatenation of Included and the current version of the queue (Q) will maintain the same contents and order, together, as the original queue. The second indicates that Running_Average will always contain the integer average of the numbers that have already been transferred into Included. The third indicates that Running_Remainder will always contain the remainder of the integer average operation. The final two conjuncts establish the bounds on Running_Average. Since it is the average of positive integers, clearly it will, itself, be positive, and since each number can be at most max_int, clearly it can be no greater than if it were the average of a series of max_ints. These are important for establishing that the running average never strays from the integer bounds.

It is the Running_Remainder calculation that necessitates the, perhaps counter-intuitive, limit on the length of the queue in the enhancement. An inspection of the code will assure the reader that Running_Remainder is upper-bounded by two less than twice the length of the given queue. With the queue-length capped at half max_int, this value can never exceed max_int.

The clause introduced by decreasing also bears explanation. This is called the progress metric and expresses some natural-number-valued expression that is guaranteed to strictly decrease with each iteration through the loop. This also raises a proof obligation for demonstrating that this while loop will terminate.

The above code produces 23 VCs. Many of these arise from the maintaining clause or the decreasing clause. Some correspond to the requires clause on called operations such as Dequeue(). One simple and one complicated example of these VCs are reproduced here.

```
1 VC: 6_1:
2 Base Case of the Invariant of While Statement
3
4 Goal:
5 (empty_string  o Q) = Q
6
7 Given:
8 (((((min_int   <= 0) and
9 (0   < max_int)) and
10 ((Last_Char_Num   > 0) and
11 ((9   > 0) and
12 (((|Q|   <= 9) and
13 ((min_int   <= 9) and
14 (9   <= max_int)))))) and
15 (((|Q|   <= (max_int   / 2)) and
16 Q /= empty_string) and
17 Holds_For_Each(Q, Non_Negative)))
```

```
1 VC: 7_1:
2 Ensures Clause of Queue_Average
3
4 Goal:
5 (Fold_String_Right(Sum, 0, Included')   / |Q|) = (Fold_String_Right(Sum, 0, Included')
   / |Included'|)
6
7 Given:
8 ((((((min_int   <= 0) and
```

```
 9 (0   < max_int)) and
10 ((Last_Char_Num  > 0) and
11 ((9  > 0) and
12 (((|Q|   <= 9) and
13 ((min_int   <= 9) and
14 (9   <= max_int)))))) and
15 ((((|Q|   <= (max_int   / 2)) and
16 Q /= empty_string) and
17 Holds_For_Each(Q, Non_Negative))) and
18 (((((Included'   o Q') = Q and
19 Running_Average' = (Fold_String_Right(Sum, 0, Included')   / |Q|)) and
20 Running_Remainder' = (Fold_String_Right(Sum, 0, Included')   mod |Q|)) and
21 (0   <= Running_Average')) and
22 (Running_Average'   <= ((max_int   / |Q|)   * |Included'|)))) and
23 not((|Q'|   > 0)))
```

The final step in the verification process is for the generated VCs to be recorded in a format accepted by one or more automated provers. RESOLVE can target a variety of back-end provers, some of which are discussed in [16], though this example cannot be verified strictly mechanically. We note, however, that the VCs are straightforward to discharge by hand. As a future direction for research, we would like to expand RESOLVE's own integrated prover to address these VCs more easily.

Assuming each VC is proved, the program can be certified correct. Once certified, the original RESOLVE code is translated into code in another language for execution. This code is thus correct by construction, assuming the translator can be trusted to be correct, and as long as the generated code in the target language respects the design principles prescribed by RESOLVE. Our compiler that generates C code to run on the MoteStack platform preserves these principles; formally certifying it is left for future work. We describe the key pieces of the translation (those pieces that are directly relevant to the RESOLVE principles) in the next section.

It is important to realize that even a simple routine such as averaging presents complicated challenges under verification, and this code went through many versions before arriving at its final form. In particular, the presence of integer overflow and inexact division complicated the process of verifying the code. For example, we began with the most obvious procedure for averaging—sum the integers in the queue, then divide by the length. However, in this case, if we are to avoid overflow, we must require, very arbitrarily, that each element in the queue be no greater than $1 / n * max\_int$, where n is the length of queue. Not only is this restriction confusing to the client, it is likely to be difficult to assure in many cases. A better attempt might be to divide as you go, summing the weighted values, but the compounding round-off error means you are no longer actually calculating the value required by $Average\_Queues$'s ensures clause. An incrementally better attempt might be to divide as you go, summing the remainder of each division until all of the values have been included, then finally accounting for the accumulated remainder, but this leads to an even more counterintuitive restriction on the use of the method that the sums of the remainders of the elements in the queue when divided by the length of the queue be less than max_int. Finally, we arrived at the given example, whose restriction—while perhaps still perceived as arbitrary—is easy to satisfy under most reasonable valuations of max_int and min_int. We note that it is possible to further simplify the requires clause of our averaging operation to remove the length restriction, but doing so complicates the logic significantly and thus makes verification even more challenging.

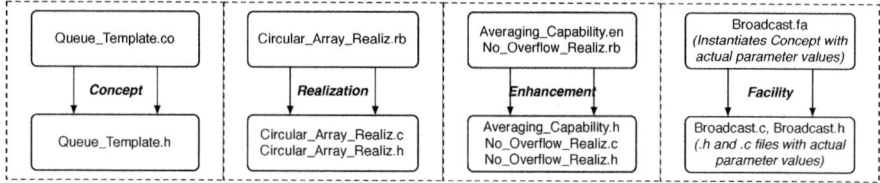

Fig. 4. Overview of Compiler Translation

4   Translation to C

Once all proof obligations have been met, either mechanically or by hand, the compiler translates the RESOLVE source into equivalent C. The process is bottom-up; included modules are translated first in order to correctly resolve dependencies. Figure 4 provides an overview of the translation process for each type of RESOLVE module. A common header file is created to include shared libraries.

4.1   The MoteStack API

The MoteStack API is a standard C library designed to expose the fundamental hardware services provided by the platform. It comprises a collection of body files and corresponding headers, with each such pair realizing a driver for a particular hardware service. More precisely, the API includes drivers for the platform's clock, the onboard analog-to-digital converter, the onboard Zigbee radio, LEDs, and other services commonly used when developing embedded network applications. The constituent functions used to realize each driver provide a high-level service interface, hiding low-level implementation details (e.g., register access, timing-accurate delays). The design was motivated by a desire to simplify the RESOLVE-to-C translation process.

4.2   Data Movement and the Swap Operation

The basic data movement operator in RESOLVE is the *swap* (:=:) operator. This design choice is fundamental to the proof system as it guarantees the absence of runtime aliasing. Providing a constant-time implementation of the operator is straightforward in a heap-based programming model (e.g., as provided by Java and C#). It is more challenging, however, in a heap-free programming model. Indeed, the C code generated by our compiler may appear unusual. The departure from standard C-style programming is attributable to the need to support an efficient swap implementation.

To support a constant-time swap implementation, all program variables must behave as heap-allocated variables. This is in direct conflict with C, which maintains locals on the stack. To overcome this obstacle, we first introduce an additional level of indirection. All translated variables are defined as r_type (resolve type) variables, as shown below – effectively void pointers.

```
1 typedef void* r_type;
2 typedef r_type* r_type_ptr;
```

To ensure object persistence across stack frames, all RESOLVE variables, including locals, are translated to global storage locations; the corresponding r_type pointers reference those locations and serve as the only access mechanism. For instance, for an Integer variable I_VAR declared in RESOLVE, the following C code is generated:

```
1 long int int_i_var = 0 ;
2 r_type i_var = &int_i_var ;
```

This uniform abstraction enables a simple, constant-time implementation of swap that accommodates both scalar and composite types:

```
1 void swap(r_type_ptr var1, r_type_ptr var2) {
2     r_type temp = *var1;
3     *var1 = *var2;
4     *var2 = temp;
5 }
```

Elevating all program variables to global scope may raise red flags in the reader's mind; this is hardly the hallmark of modular software. This code is not, however, intended for programmer use.

4.3 Preserving the Component Model

Consider, for example, a snippet from the Queue_Concept declaration in RESOLVE:

```
1    requires Max_Length > 0;
```

Queue_Concept takes two parameters, corresponding to the type of queue entry and queue capacity, respectively. The requires clause imposes a constraint on the allowable values that may be passed for Max_Length at the point of concept instantiation. The model is expressed using String theory:

```
1    exemplar Q;
2    constraint |Q| <= Max_Length;
3    initialization ensures Q = <>;
```

The exemplar represents a prototypical instance of Queue, used to define state constraints (invariants) and initialization conditions. At the point of declaration, each instance will be empty; it will never hold more than Max_Length entries.

Each realization must define a state representation for the concept and provide operation bodies implemented in terms of that representation. Consider a circular-array-based realization of Queue_Concept that relies on a record containing an array of unknown type (Contents) and two integer variables (Front, Length):

```
1 Realization Circular_Array_Realiz for Queue_Concept;
2 Type Queue = Record
3    Contents: Array 1..Max_Length of Entry;
4    Front, Length: Integer;
5 end;
```

Actual parameters for Entry and Max_Length, defined by Queue_Concept, are supplied at the point of facility instantiation. For example, the following definition instantiates a queue of integers with a maximum length of 5:

```
1 Facility Integer_Queue_Fac is Queue_Concept(Integer, 5)
2    realized by Array_Realiz;
```

The instantiated Queue type is realized in the generated C as follows:

```
1 typedef struct iqf_queue_rep {
2   r_type contents[5-1+1];
3   long int contents_store[5-1+1];
4 } Queue_rep;
```

Note that the Contents array is translated into two arrays. Again, this indirection is introduced to support constant-time swapping over the array entries. One time, at the point of declaration in the generated C code, each of the r_type pointers within contents is initialized to point at the corresponding entry within contents_store. (This process should not be repeated for a given local, say on subsequent calls to an operation, since the element values may have been swapped out of the current scope[1].) The translation strategy mirrors, on an element-by-element basis, the treatment of local variables.

Each operation defined within Queue_Concept is translated to a corresponding C function based on a facility declaration. For example, Enqueue in RESOLVE:

```
1 Operation Enqueue(alters E: Entry; updates Q: Queue);
```

is translated to the following C code based on the Integer_Queue_Fac instantiation:

```
1 void iqf_enqueue(long int* e, iqf_queue_rep* q);
```

Each generated function takes a parameter of type iqf_queue_rep* — the target of the operation. This is, in effect, an explicit version of the implicit this parameter in object-oriented languages.

4.4 Example

We now present the generated source for the sense/broadcast application presented in Figure 3. The original Queue instance (data_samples) is translated to a pointer variable of queue type (a renamed version of r_type) and an associated storage variable (Lines 3–4). Variable initialization is triggered prior to the point of first use (Line 8).

```
1   Var Q1, Q2 : Queue;
2   Q1 :=: Q2;
```

In the case of a swap between two RESOLVE Queue variables, the translator generates a call to the uniform swap implementation discussed previously, passing the addresses of the r_type pointers corresponding to the variables being swapped.

4.5 Evaluation

To evaluate the efficiency of the compiler, we implemented two standard sensing applications. The first, discussed in Section 2, periodically samples onboard sensors and broadcasts the results via radio. The second is similar, but transmits the data to a PC via a serial connection. Table 1 shows the memory usage of the compiled applications.

---

[1] This is a correctable deficiency in our current implementation, but introduces a significant initialization performance penalty.

```
 1// #include "Common.h", "Leds_Template.h", "ADC_Template.h", ...
 2...
 3queue data_samples;
 4iqf_queue_rep data_samples_rep;
 5
 6void broadcast_average_main() {
 7     ...
 8     iqf_queue_initialize(&data_samples_rep);
 9     data_samples = &data_samples_rep;
10     leds_template_led0_init();
11     ...
12     adc_template_adc_init();
13     uart_template_uart_init(9600);
14     xbee_template_xbee_init();
15     ...
16     while(boolean_template_true()) {
17         clock_template_wait_1s();
18         if(integer_template_are_equal(iqf_rem_capacity(&data_samples), 0)){
19             iqf_dequeue(&garbage, &data_samples);
20             leds_template_led0_set(off);
21         }
22         sample = adc_template_read_adc(0);
23         iqf_enqueue(&sample, &data_samples);
24         average = iqf_average(&data_samples);
25         average = integer_template_mod(average, 256);
26         if(integer_template_greater_or_equal(average, 0)){
27             leds_template_led1_set(on);
28         } else{
29             leds_template_led1_set(off);
30         }
31         // >=64 -> LED2; >=128 -> LED3;
32         // >=192 -> LED4, analogously
33         ...
34         uart_template_uart_send_bytes_blocking(&average, 1);
35         clock_template_wait_1s();
36         leds_template_led0_set(on);
37     }
38}
```

Fig. 5. Generated C Code for the SenseAndBroadcast Example

To further improve the compiler's efficiency, we included a translation optimization for scalar variables. Scalars (i.e., Integers, Booleans, and Characters) need not be accessed indirectly to ensure constant-time data movement; value assignment is equally efficient. Hence, scalar variables (and fields) are not translated to r_types; they are translated to standard value types and swapped through data assignment.

5   Related Work

Our experiment is based on work on the RESOLVE language, and in particular, recent work on push-button verification for RESOLVE. As such, RESOLVE provides a unified specification and programming language to build provably correct programs.

The *Slede* project [7] involves automatically verifying sensor network security protocols. The approach is based on formal verification using model-checking techniques applied to nesC programs. *VCC* [15] performs formal verification based on logical inference. It generates verification conditions from annotated C programs, which are proved using an automatic theorem prover. In [11], a dialect of C (*C0*) is compiled and verified

Table 1. Memory Usage of Applications Compiled for the ATmega644 Processor

	No Optimization		With Optimization	
Application	**Program (ROM)**	**Data (RAM)**	**Program (ROM)**	**Data (RAM)**
SenseBroadcastAvg	10860	686	6990	282
SenseBroadcast	9664	432	6348	126
Receiver	9634	432	5804	126

to check the correctness of program implementations in pervasive systems. This work mainly focused on proving logical blocks that involve dynamic memory allocation, address alignments, and function calls using Hoare's partial correctness logic.

In [18], the authors present work on verifying system C code based on its low-level memory model, and improved techniques to prove correctness of code, especially programs with pointer address arithmetic and structure types. The input C source code is annotated with pre- and post-conditions and invariants for program functional blocks. Similar work on verifying C programs with pointers, along with a prototype implementation based on Burstall's model for structures is presented in [6]. This work inserts annotated pre- and post-conditions, global invariants, and loop variants to C programs and uses the *Why* tool for generating verification conditions.

6   Conclusions

In this paper, we have described our work on implementing a compiler for embedded network systems that allows developers to create applications that are provably correct. As the basis for the verification effort, we use the RESOLVE language, which is an integrated specification and programming language that is targeted primarily at provably correct programs. Thus far, research in the RESOLVE language has largely centered on the functional correctness of generic software components. The specification and verification of performance characteristics is an active area of future work. The work presented here uses this foundation as a basis to target a small but useful class of embedded network systems applications. We have demonstrated the use of this verification system and compiler in the context of the MoteStack platform, which uses an Atmel ATmega644 microcontroller.

Acknowledgments. This work was supported in part by the National Science Foundation (CNS-0745846, CNS-0746632, CCF-0811748).

References

[1] Arora, A., et al.: Kansei: A high-fidelity sensing testbed. IEEE Intern. Comp. 10(2), 35–47 (2006)

[2] Bucci, P., et al.: Part III: Implementing components in resolve. SIGSOFT Softw. Eng. Notes 19(4), 40–51 (1994)

[3] Crocker, D., Carlton, J.: Verification of c programs using automated reasoning. In: SEFM 2007, pp. 7–14. IEEE Computer Society, Washington, DC, USA (2007)

[4] Dalton, A., et al.: A testbed for visualizing sensornet behavior. In: IC3N 2008, pp. 1–7. IEEE Computer Society, Washington DC, USA (2008)

[5] Edwards, S., et al.: Part ii: Specifying components in resolve. SIGSOFT Softw. Eng. Notes 19(4), 29–39 (1994)

[6] Filliâtre, J.-C., Marché, C.: Multi-prover Verification of C Programs. In: Davies, J., Schulte, W., Barnett, M. (eds.) ICFEM 2004. LNCS, vol. 3308, pp. 15–29. Springer, Heidelberg (2004)

[7] Hanna, Y., Rajan, H., Zhang, W.: Slede: A domain-specific verification framework for sensor network security protocol implementations. In: WiSec 2008, March 31-April 2, pp. 109–118. ACM, New York (2008)

[8] Harton, H.K.: Modular and Mechanical Verification Condition Generation for Object-Based Software. PhD thesis, Clemson University (2011)

[9] Kulczycki, G.: Direct Reasoning. PhD thesis, Clemson University, Clemson, South Carolina (January 2004)

[10] Kulczycki, G., Smith, H., Harton, H., Sitaraman, M., Ogden, W.F., Hollingsworth, J.E.: The Location Linking Concept: A Basis for Verification of Code Using Pointers. In: Joshi, R., Müller, P., Podelski, A. (eds.) VSTTE 2012. LNCS, vol. 7152, pp. 34–49. Springer, Heidelberg (2012)

[11] Leinenbach, D., Paul, W., Petrova, E.: Towards the formal verification of a c0 compiler: Code generation and implementation correctnes. In: SEFM 2005, pp. 2–12. IEEE Computer Society, Washington, DC, USA (2005)

[12] Levis, P., et al.: Tossim: accurate and scalable simulation of entire tinyos applications. In: SenSys 2003, pp. 126–137. ACM, New York (2003)

[13] McCartney, W.P., Sridhar, N.: Abstractions for safe concurrent programming in networked embedded systems. In: SenSys 2006, New York, USA, pp. 167–180 (2006)

[14] Merz, F., Falke, S., Sinz, C.: LLBMC: Bounded Model Checking of C and C++ Programs Using a Compiler IR. In: Joshi, R., Müller, P., Podelski, A. (eds.) VSTTE 2012. LNCS, vol. 7152, pp. 146–161. Springer, Heidelberg (2012)

[15] Schulte, W., et al.: A glimpse of a verifying C compiler. In: C/C++ Ver. Workshop (July 2007)

[16] Sitaraman, M., et al.: Building a push-button resolve verifier: Progress and challenges. Formal Asp. Comput., 607–626 (2011)

[17] Titzer, B., Lee, D., Palsberg, J.: Avrora: scalable sensor network simulation with precise timing. In: IPSN 2005, p. 67. IEEE Press, Piscataway (2005)

[18] Tuch, H.: Formal verification of C systems code. J. Autom. Reason. 42(2-4), 125–187 (2009)

[19] Werner-Allen, G., Swieskowski, P., Welsh, M.: Motelab: a wireless sensor network testbed. In: IPSN 2005, p. 68. IEEE Press, Piscataway (2005)

A   RESOLVE

RESOLVE is an integrated programming, specification, and proof language intended to explore and balance three goals: verifiability, scalability, and practical usefulness. The design of the system is therefore intended to eliminate those practices that have traditionally led to difficult- or impossible-to-verify code, such as uncontrolled object aliasing, while providing alternatives on the same order of efficiency as the features they replace. For the aliasing example, RESOLVE uses swapping as its basic method of data transfer, rather than deep or reference copying [9]. This discourages aliasing while remaining only a constant factor less efficient. In this way, RESOLVE strives to be a language that is easy to verify while remaining competitive with current industrial languages. This focus on verifiable-but-efficient makes it an excellent choice for embedded applications where both time and memory resources are at a premium.

RESOLVE relies on well-developed, general theories to serve as the basis of the mathematical correspondence of each programming class. As an example, in RESOLVE, each of Stacks, Queues, and Lists may be specified in terms of the String conceptualization and thus make use of the full body of theory development about Strings. These specifications are presented as a Concept in RESOLVE that defines the behavior and interface of that component. For an example of this style of specification, consider this snippet from the specification of Queue:

```
1 Concept Queue_Template(type Entry;   evaluates Max_Length: Integer);
2   uses String_Theory, ...;
3   requires Max_Length > 0;
4   Type Family Queue is modeled by Str(Entry);
5   ...
6   Operation Enqueue(alters E : Entry; updates Q : Queue);
7     requires |Q| < Max_Length;
8     ensures Q = #Q o <#E>;
9   Operation Dequeue(replaces R : Entry; updates Q : Queue);
10    requires |Q| /= 0;
11    ensures #Q = <R> o Q;
12    ...
13 end;
```

In the above example, Queue is parameterized by the type of entries (Entry) it may hold and a maximum capacity (Max_Length). Actual values are provided when the concept is instantiated as a Facility in RESOLVE.

The parameters of each operation use *parameter passing modes* to summarize the effect of the operation on those parameters. A mode of alters indicates that the operation is permitted to modify the parameter in some way. updates indicates that the incoming value may be modified subject to the constraint of the *ensures* clause. replaces indicates that the initial value of the parameter will be ignored, while the final value will be as described in the *ensures* clause.

For each operation, the requires clause expresses the pre-condition that the operation expects the caller to have met, and the ensures clause expresses the post-condition that the operation will terminate in. Note that references to an object in assertions such as requires or ensures statements are references to that object's model. So, the expression $|Q| <$ Max_Length is read as "the length of the String representing Q is less than the mathematical integer Max_Length." The o operator is the concatenation operator on Strings, and the syntax $<X>$ denotes the singleton String containing only X. In ensures clauses, variables preceded with the # symbol are references to the value of the variable at the time the operation was called, while those not preceded by the # symbol are references to the final value of the variable.

A corresponding implementation, called a Realization, must be provided in RESOLVE's imperative programming language. In addition to those operations defined in the class concept, *enhancement operations* may be defined in separate modules and then implemented using the public operations of the classes they extend. In this way, any verified implementation of an enhancement on Queues may be applied to any verified implementation of Queue and be guaranteed to work. As an example, we might find it useful to sort the entries of a Queue according to some ordering, an enhancement for which might look like this:

```
1 Enhancement Sorting_Capability(definition Ordering(X : Entry, Y : Entry) : B);
2   uses String_Theory, Ordering_Theory, ...;
3   requires Is_Transitive(Ordering) and Is_Total(Ordering);
4   Operation Sort(updates Q : Queue);
5     ensures Is_Permutation(Q, #Q) and Is_Conformal_With(Ordering, Q);
6 end;
```

These enhancements inherit the parameterizations of the class specifications to which they are applied and may be further parameterized to serve their unique purposes. Functions like Is_Transitive() and Is_Total() are defined in Ordering_Theory, while functions like Is_Permutation() and Is_Conformal_With() are defined in String_Theory. The corresponding theories contain theorems for manipulating the functions therein.

Checking and Distributing Statistical Model Checking[*]

Peter Bulychev[1], Alexandre David[1], Kim G. Larsen[1], Axel Legay[1,2],
Marius Mikučionis[1], and Danny Bøgsted Poulsen[1]

[1] Computer Science, Aalborg University, Denmark
[2] INRIA/IRISA, Rennes Cedex, France

Abstract. In this paper we propose a general framework for distributed statistical model checking of networks of priced timed automata. The first contribution is a new algorithm to distribute sequential hypothesis testing without introducing bias in the results. The second contribution is an implementation of this algorithm in UPPAAL. The major contribution is an experimental and analytical evaluation of the approach through case studies, including an analysis of the SMC algorithm itself.

1 Introduction

Statistical Model Checking techniques (SMC) [8,12,17], can be seen as a trade-off between testing and formal verification. The core idea of the approach is to conduct some simulations of the system and verify if they satisfy some given property. The results are then used together with statistical algorithms in or der to decide whether the system satisfies the property with some probability. Statistical model checking techniques can also be used to estimate the probability that a system satisfies a given property [8,6]. Of course, in contrast to an exhaustive approach, a simulation-based solution does not guarantee a correct result with 100% confidence. However, it is possible to bound the probability of making an error. SMC is getting widely accepted in various research areas and applied to problems that are beyond the scope of classical formal techniques [1,2,10,11,13,19,20].

Unfortunately, extremely huge sized problems and a demand of extremely high confidence may require generation of a large number of simulation runs, each of which may itself be extremely time consuming. To address this *confidence-explosion* problem, we suggest in this paper to take advantage of PC-clusters and GRID computers. In fact, it is well-known that statistical solutions methods that use samples of independent observations are often trivially parallelizable. As observed in [18], SMC algorithms can be distributed through the help of a master/slave architecture where multiple computers are used to generate the simulations. The idea is as follows: one or more slave processes register their

[*] Work partially supported by VKR Centre of Excellence – MT-LAB, an "Action de Recherche Collaborative" ARC (TP)I and the IDEA4CPS center established on a grant from Danish National Research Foundation.

A. Goodloe and S. Person (Eds.): NFM 2012, LNCS 7226, pp. 449–463, 2012.

ability to generate simulation with a single master process that is used to collect those simulations and perform the statistical test.

However, this process may become complex when considering sequential hypothesis testing (when the number of simulations is not known in advance). The problem is that there might exist a correlation between a time needed to generate a random simulation and the fact that a property is satisfied by this simulation. Thus it is important to guarantee that the technique will not introduce a bias towards the results that are generated by shorter simulations.

In a series of recent works [4], we have extended UPPAAL with SMC algorithms applied to Networks of Priced Timed Automata – hence leading to the first implementation of SMC for real-timed stochastic systems. The objective of this paper is to go one step further and propose the first complete study of distributed SMC, in general, and in the framework of UPPAAL in particular. Our contributions are:

1. A *distributed implementation* of the estimation algorithm proposed in [8]. Building on classical Monte Carlo techniques [7], an estimation algorithm precomputes the number of simulations needed to estimate the probability to satisfy a property with a given confidence. Such an algorithm which is trivially parallelizable amounts to equally distribute the number of simulations to perform between the slave computers.

2. A *new distributed algorithm* for sequential hypothesis testing where simulations are computed on the fly until a threshold is passed and a decision is taken. Here, it is important to avoid introducing bias in the results, which may be potentially complex and eventually decrease the benefit of using several processors. To counter this, [18] proposed a round-Robin solution where the runs are counted in rounds. We generalise the solution in [18] by introducing *batches* and *buffers*. The batch is used to reduce communication by sending an aggregate result of predefined size (instead of individual results). The buffer is used to improve concurrency since the nodes are more loosely synchronized.

3. A *thorough evaluation* of our implementation through new applications of SMC algorithms. In particular, we apply the distributed SMC engine to an analysis of an instance of the LMAC protocol of unprecedented size. Additionally, a thorough evaluation of the distributed SMC framework itself is made aiming at identifying optimal settings of the parameters for the framework. The evaluation is carried out both experimentally (using the implementation) as well as analytically (using SMC) based on a model of the distributed SMC algorithm itself, and with high consistency in identifications made by the two approaches.

2   Statistical Model-Checking in Uppaal

This section introduces the formalisms used in UPPAAL for modeling systems and and specifying properties. Then, we briefly survey existing Statistical Model Checking (SMC) algorithms. Finally, a novel application of SMC is presented.

2.1   Networks of Priced Timed Automata

The new SMC engine of UPPAAL [3] supports the analysis of Priced Timed Automata (PTAs) that are timed automata whose clocks can evolve with different rates, and with no restrictions in guards and invariants. Additionally, we support other features of the UPPAAL model checker's input language such as integer variables, data structures and user-defined functions. We also assume PTAs are input-enabled, deterministic (with a probability measure defined on the sets of successors), and non-zeno. PTAs communicate via broadcast channels and shared variables to generate Networks of Price Timed Automata (NPTA).

Fig. 1 provides an NPTA with three components A, B, and T as specified using the UPPAAL GUI. One can easily see that the composite system $(A|B|T)$ has the transition sequence:

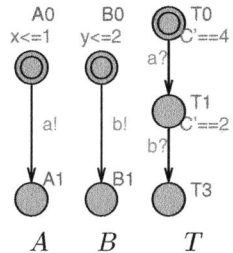

$$((A_0, B_o, T_0), [x = 0, y = 0, C = 0]) \xrightarrow{1} \xrightarrow{a!}$$
$$((A_1, B_0, T_1), [x = 1, y = 1, C = 4]) \xrightarrow{1} \xrightarrow{b!}$$
$$((A_1, B_1, T_2), [x = 2, y = 2, C = 6]),$$

demonstrating that the final location T_3 of T is reachable. In fact, location T_3 is reachable within cost 0 to 6 and within total time 0 and

Fig. 1. An NPTA, $(A|B|T)$

2 in $(A|B|T)$ depending on when (and in which order) A and B choose to perform the output actions $a!$ and $b!$. Assuming that the choice of these time-delays is governed by probability distributions, a measure on sets of runs of NPTAs is induced, according to which quantitative properties such as *"the probability of T_3 being reached within a total cost-bound of 4.3"* become well-defined.

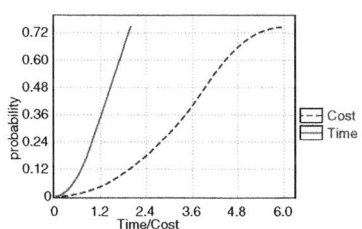

Fig. 2. Cumulative probabilities for `time` and `Cost`-bounded reachability of T_3

In our early works [4], the stochastic semantic of PTA components associates probability distributions on both the delays one can spend in a given state as well as on the transition between states. In UPPAAL uniform distributions are applied for bounded delays and exponential distributions for the case where a component can remain indefinitely in a state. In a network of PTAs the components repeatedly race against each other, i.e. they

independently and stochastically decide on their own how much to delay before outputting, with the "winner" being the component that chooses the minimum delay. For instance, in the NPTA of Fig. 1, A wins the initial race over B with probability 0.75.

Properties. For specifying properties of NPTAs, we use cost-constraint temporal properties over runs of the form $\psi = \Diamond_{C \leq c} \varphi$. Here C is an observer clock (that is never reset and should grow to infinity on any infinite run), $c \in \mathbb{R}_{\geq 0}$

and φ is a state-predicate. We say that a run π satisfies $\psi = \Diamond_{C \leq c} \varphi$ if there exists a state (ℓ, v) in π satisfying φ and with $v(C) \leq c$. For an NPTA M we define $\mathbb{P}_M(\psi)$ to be the probability that a random run of M satisfies ψ.

Reconsider the example of Fig. 1, we can evaluate the the probabilities Pr[time<=2]($\Diamond$ T.T3) and Pr[C<=6]($\Diamond$ T.T3) in UPPAAL, obtaining as expected 0.75 for the composition $(A|B|T)$ for both of these probabilities. In fact Fig. 2 gives the time- and cost-bounded reachability probabilities for $(A|B|T)$ for a range of bounds.

2.2   Statistical Model Checking Algorithms

We briefly recall statistical algorithms allowing to answer the following two types of questions : (1) *Qualitative :* Is the probability that a random run of a given NPTA $\mathcal{A}$ satisfies a property $\Diamond_{C \leq c} \varphi$ greater than a certain threshold θ? and (2) *Quantitative :* What is the probability that a random run of $\mathcal{A}$ satisfies $\Diamond_{C \leq c} \varphi$? For both question a run of the system is encoded as a Bernoulli random variable that is true if the run satisfies the property and false otherwise.

Qualitative Question. This reduces to test the hypothesis H: $p = \mathbb{P}_{\mathcal{A}}(\Diamond_{C \leq c} \varphi) \geq \theta$ against $K : p < \theta$. To bound the probability of making errors, we use strength parameters α and β and we test the hypothesis $H_0 : p \geq p_0$ and $H_1 : p \leq p_1$ with $p_0 = \theta + \delta_0$ and $p_1 = \theta - \delta_1$ (δ_0 and δ_1 are parameters of the algorithm). The interval $p_0 - p_1$ defines an indifference region, and p_0 and p_1 are used as thresholds in the algorithm. The parameter α is the probability of accepting H_0 when H_1 holds and the parameter β is the probability of accepting H_1 when H_0 holds. The above test can be solved by using Wald's sequential hypothesis testing [16]. This test, which is presented in Algorithm 1, computes a proportion r among those runs that satisfy the property. With probability 1, the value of the proportion will eventually cross $\log(\beta/(1 - \alpha)$ or $\log((1 - \beta)/\alpha)$ and one of the two hypothesis will be selected.

Algorithm 1. Hypothesis testing

---

    function hypothesis(S:model , ψ: property)

1 r:=0

2 **while** *true* **do**

3   Observe the random variable x corresponding to $\Diamond_{C \leq c} \varphi$ for a run.

4      $r := r + x * \log(p_1/p_0) + (1 - x) * \log((1 - p_1)/(1 - p_0))$

5   **if** $r \leq \log(\beta/(1 - \alpha))$ **then** accept H_0

6   **if** $r \geq \log((1 - \beta)/\alpha)$ **then** accept H_1

    end

Quantitative Question. This reduces to a Monte Carlo approach that computes the number N of runs needed in order to produce an approximation interval $[p - \epsilon, p + \epsilon]$ for $p = Pr(\psi)$ with a confidence $1 - \alpha$. The values of ϵ and α are chosen by the user and N relies on the Chernoff-Hoeffding bound.

2.3   Analysing SMC in Uppaal

In this section we will use the SMC engine of UPPAAL to our first non-trivial task, namely to analyse itself! More precisely, by suitably modeling the sequentual testing algorithm as well as a sample model M, we will be able to use the SMC engine of UPPAAL to analyse the performance of SMC on M. Later, in Section 4, this will allow us to evaluate various naive (and even faulty) proposals for distributing SMC.

The sample model M given in Fig. 3a[1] makes an initial probabilistic choice between the two branches, each having a looping transition taken repeatedly with a delay choosen uniformly from $]0, 2]$. Performing sequential testing of the hypothesis H_0: $Pr[<=100]$ ($\Diamond$ OK)$\geq$ 0.5 some 10 times with $\alpha = 0.05$ as level of significance and with an indifference region of ± 0.01, we consistently (and correctly) dismiss the hypothesis with an average of 408.6 runs and with standard deviation 127.5.

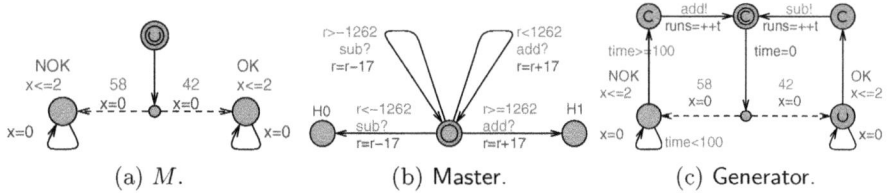

(a) M.   (b) Master.   (c) Generator.

Fig. 3. Sample model M (a) satisfying $Pr[<=100]$ ($\Diamond$ OK)$= 0.42$ and modeling SMC of M (b,c) with respect to H_0: $Pr[<=100]$ ($\Diamond$ OK)$\geq$ 0.5 with 0.05 as level of significance and $[0.49, 0.51]$ as indifference region

Now, aiming at obtaining a better understanding of sequential testing[2] we may simply model the sequential testing algorithm of M directly and analyse its (expected) performance using UPPAAL SMC. The resulting model is given in Fig. 3 and consists of an extension of the sample model M into the component Generator that will repeatedly generate random runs of M (of time-duration 100) and report the outcome to a Master using the channels add (when 100 time-units has elepased without OK having been observed) and sub (used as soon as it is observed that the OK branch has been taken, note the absence of the time>=100 guard on the right side of the Generator model). The Master has the obligation of adjusting appropriately the ratio-variable r according to Alg. 1, and conclude on H_0 or H_1 as soon as the value of r exceeds the given threshold. Given the indifference region $[0.49, 0.51]$ and level of significance 0.05, we find that the approximate values to be used[3] in Alg. 1 are: $- \log(p_1/p_0) =$

---

[1] M is a timed variant of the model proposed in [17] and used to demonstrate bias in a naive distributed approach to SMC.

[2] The performance of sequential testing has been subject to significant studies and is well-understood [15]. The aim here is to demonstrate that our UPPAAL SMC engine is a useful tool for obtaining such an insight.

[3] Those values are obtained by observing Wald's ratio on several application of the SMC algorithm to the same problem, and then take the average of the observations.

$\log(1 - p_0/1 - p_1) = 0.01715$ and $\log((1 - \beta)/\alpha) = -\log(\beta/(1 - \alpha)) = 1.2787$ ($\approx 1.262 + 0.017$). In the model of Fig. 3 we are using scaled integer constants for these values. Now, looking at the estimation of $\Pr[\#<=20000]$ ($\Diamond$ Master.H_1) in Fig. 4, we find – as expected – that the probability of accepting H_1 (H_0) tends to 1 (0) as the number of steps increases. We also see that the average number of runs is estimated to 481.4. The "mismatch" with the experimentally found average 408.6 is due to early termination when the threshold for H_0 is exceeded.

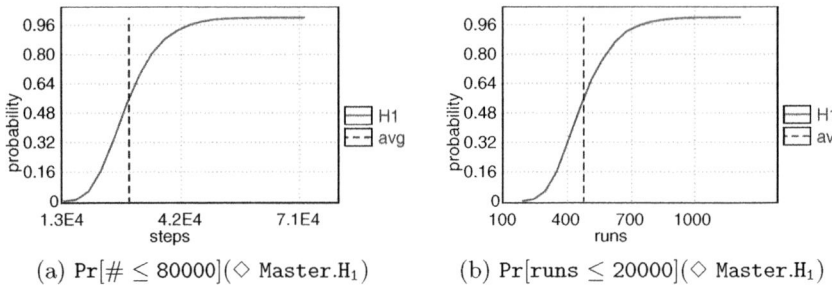

(a) $\Pr[\# \leq 80000](\Diamond$ Master.$H_1)$ (b) $\Pr[\text{runs} \leq 20000](\Diamond$ Master.$H_1)$

Fig. 4. Cumulative probability plots over number of steps and runs

3   Distributed Statistical Model-Checking in Uppaal

SMC suffers from the fact that the high confidence required by an answer may demand a large number of simulation runs, each of which may itself be time consuming. As an example, the first hypothesis test shown later in this section can generate between 14,000 and 2,390,000 runs if the parameters α, β, δ range between 0.01 and 0.001. A possible way to leverage this problem is to use several computers working in parallel using a master/slaves architecture: one or more slave processes register their ability to generate simulation with a single master process that is used to collect those simulations and perform the statistical test.

When working with an estimation algorithm, this collection is trivially performed as the number of simulations to perform is known in advance and can be equally distributed between the slaves. When working with sequential algorithms, the situation gets more complicated. Indeed, we need to avoid introducing bias when collecting the results produced by the slave computers. This means that results should not be collected arbitrarily as illustrated by considering the model of Section 2.3 with several instances of the Generator template.

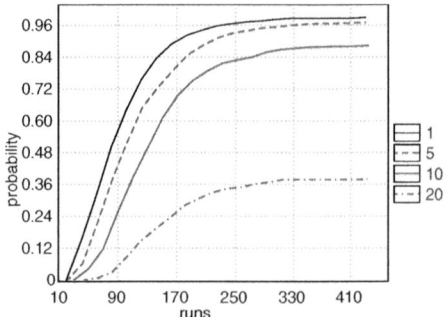

Fig. 5. Probability distributions obtained with 1 (top), 5, 10, and 20 (bottom) generator nodes

Checking the property `Pr[runs<=20000]`($\Diamond$ `Master.H1`) Fig. 5 shows that different distributions can be obtained with different numbers of generator nodes, hence revealing a bias in the results. In fact the probability of accepting H_1 tends (incorrectly) to 0 when the number of `Generator` components increases.

A solution, which was proposed in [17], consists in observing that Wald's ratio r is updated as a function of the Bernouilli random variable x as $r+ = x*r_{acc}+(1-x)*r_{rej}$ with r_{acc} and r_{rej} being constants depending on the tested hypothesis. To reduce blocking and still update r, the non-biased algorithm updates two safe approximations for r (r_1 and r_2). If x is unknown then it updates with $r_1+ = r_{rej}$ and $r_2+ = r_{acc}$, and then testing if $r_1 \leq I$ to accept H_0 or if $r_2 \geq S$ to accept H_1[4]. When all outcomes of a round are known then we can reset $r_1 := r_2 := r$. This allows us to accept H_0 even if some accepting outcomes are missing or conversely to accept H_1 if some rejecting outcomes are missing.

We generalize [17] by aggregating the outcomes x by *batches* (of size B) and also by implementing a *buffer* (of size K) of incoming results. The batch is used to reduce communication by sending B aggregate results. The buffer is used to improve concurrency since the nodes are more loosely synchronized and they can be K runs ahead of the slowest node. Fig. 6 illustrates our algorithm at the master node that receives asynchronous messages from all other nodes in a buffer. A message is an aggregate result containing the outcome of B runs. The master may take a decision as soon as $r_1 \leq I$ or $r_2 \geq S$. When all outcomes at the bottom line of the buffer are known we reset $r_1 := r_2 := r$ with the exact updated value of r with those

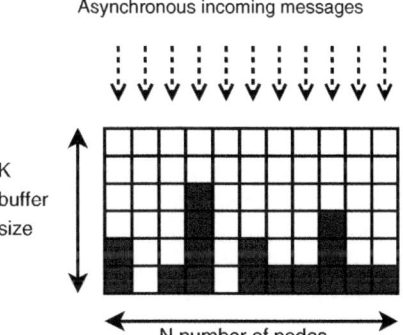

Fig. 6. Buffer of results at the master node

outcomes, and free the bottom line of the buffer. In practice, our algorithm is calibrated to count the runs up to a certain depth in the buffer. Indeed, the outcomes are weighted by B so few missing aggregated outcomes can prevent the algorithm from deciding. We have implemented this algorithm with asynchronous communications (using OpenMPI). There can be at most K pending messages due to the size of the buffer. If a slave tries to send more messages, then the communication will block waiting for a "slot" to be free. The experiment performed in the remainder of the paper has been carried out on varying numbers of nodes on a cluster with dual Xeons 5650 (hexa-cores at 2.66GHz) interconnected with infiniband.

We first make two types of experiments to exhibit the performance characteristics of our algorithm. The experiments are carried out using the train-gate example available as a demo of UPPAAL. This model comprises a number of

---

[4] $I = \log(\beta/(1-\alpha))$ and $S = \log((1-\beta)/\alpha)$ as stated in Alg. 1.

trains crossing a bridge with only one track. A gate controller stops and restart the trains to ensure mutual exclusion on the bridge and absence of starvation for the trains. Our first experiment concerns 6 trains and the property of being in a state where train 5 is crossing while all the other trains are stopped.

```
Pr[<=100](<> Train(5).Cross and
         (forall (i : id_t) i != 5 imply Train(i).Stop)) >= 0.46188
```

The runs are relatively short with few components so they will be cheap to compute and we expect the throughput of messages to be high. In addition, the hypothesis we are testing is not deterministic, which means that the outcomes and computation times of the runs will vary. The property is checked with high confidence (99.999%) and small indifference regions (+/- 0.00001) to have a precise and reliable result – and to stress our distributed algorithm.

Our second experiment considers a "large" instance with 20 trains, where we check if the model satisfies mutual exclusion on the bridge, expressed by the property

```
Pr[<=1000]([] forall (i : id_t) forall (j : id_t)
           Train(i).Cross and Train(j).Cross imply i == j) >= 0.9999
```

Here, the runs are random but bounded by the same large bound and since the inner property []forall(i : id_t)forall(j : id_t)... holds by model-checking, all the runs will all reach their bounds. In addition, we have 20 trains and the runs are long (1000 time units) so they are relatively expensive to generate. This means that all the runs are implicitly synchronized and small deviations are amortized by the long runs. The throughput of messages will be low, which means a low overhead compared to the actual useful work of generating the runs.

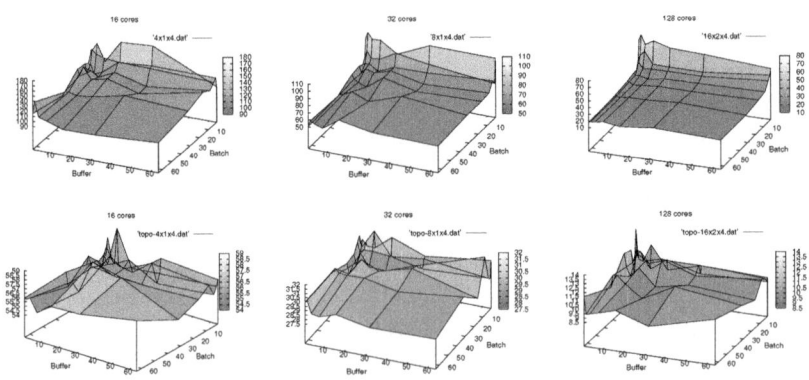

Fig. 7. Verification times on 16, 32, and 128 cores in function of B and K for the "small" model (first row) and the "large" model (second row)

Figure 7 shows our results for different number of cores. The solution in [18] corresponds to the particular case with K and B are equal to one, exhibiting in all the experiments the worst verification time, and with performance deteriorating

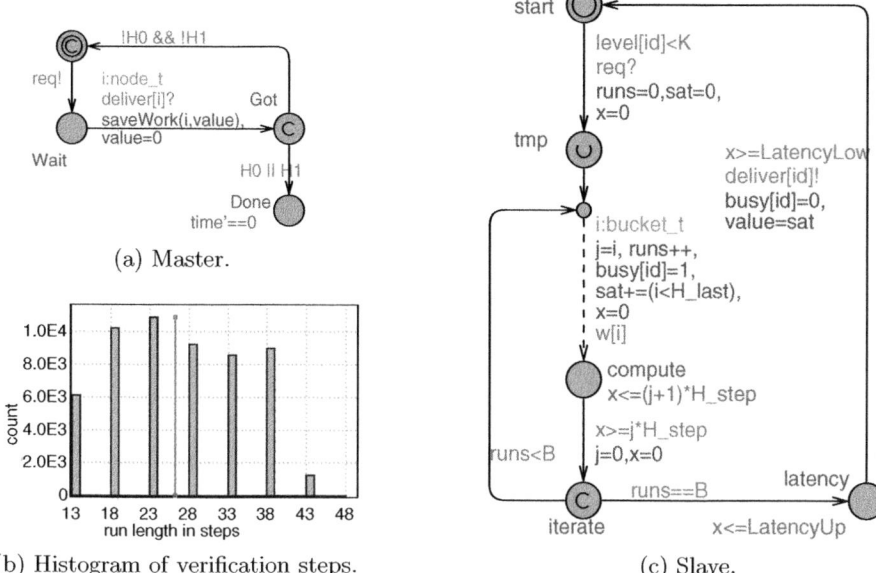

(a) Master.

(b) Histogram of verification steps.

(c) Slave.

Fig. 8. Timed automata model of a statistical model checking process

with increasing number of cores (i.e. for 128 cores performance loss is a factor of 4). Though the impact of the buffer size is less, the experiments indicate that a buffer of size 2-4 will suffice. The results also demonstrate linear scalability of our distributed implementation: for $B = 32$ and $K - 2$ the verification times for 16, 32 and 128 cores are 108, 56 and 19 seconds (respectively).

4   Analyzing Distributed SMC in Uppaal

In this section we model the implemented distributed algorithm of sequential hypothesis testing and we check it using the SMC engine of UPPAAL. The goal is to estimate the verification time and processor utilization, check for bias in the distributed algorithm, and explore the parameters of our distributed SMC algorithm in an analytical manner.

Modeling. We model the master and slave processes described in Section 3 as shown in Fig. 8. The master sends a broadcast request **req!** to verify batches of runs (of size B). We use a standard modeling pattern to synchronize on the corresponding **req?** as soon as possible. The master gathers the results with its **saveWork** function and loops again if neither H_0 nor H_1 is accepted. Listing 1.1 shows this **saveWork** function that implements the distributed hypothesis testing algorithm of Section 3. UPPAAL uses floating point numbers that are not available in the modeling language. Instead we encode fixed point arithmetics with integers and we use precomputed tables for logarithm values. Once the master accepts H_0 or H_1, it moves to the location **Done** and stops the clock **time**.

Listing 1.1. Master code.

```
1  // buffer  portion  for  early  termination:
2  const int   P  = (K<=4)?K : ((K<=8)?5 : ((K<=16)?8 : ((K<=32)?10 : 12)));
3  bool H0 = false,  H1 = false;  // for  hypothesis  H0 and  H1
4  int  batch[N][K];  // buffer  of batches (K batches for  N nodes)
5  long  satisfied   =0,  unsatisfied  =0;  // information about filled    lines
6  long  sat=0, unsat=0, unknown=N*P*B;  // early results in unfilled   lines
7  long  logRatio  = 0, ratioLow  = 0, ratioUp  = 0;  // scaled  by p.scale
8  void  saveWork(const node_t node, const int value) {
9      if  (level [node]<=P) {  // entered the early results   portion
10         sat  += value; unsat += B−value; unknown −= B;
11     }
12     batch[node][level [node]]  = value; level [node]++;  // store
13     if  (level [node]==1) {  // entered at the lowest level
14         bool filled    = forall (i : node_t) level [i]>0;
15         if  (filled  ) {  // line  at the lowest  level  has been filled
16             int  L;
17             for  (i : node_t) {  // shift  all  queues one by one
18                 satisfied   += batch[i][0];  // count as firm  results
19                 unsatisfied   += B−batch[i][0];
20                 sat  −= batch[i][0];   // discount from early  results
21                 unsat −= B−batch[i][0];   unknown += B;
22                 level [i] −−;  // remove from buffer
23                 for  (L=0; L<level[i];  ++L) {
24                     batch[i][ L]  = batch[i][ L+1];  // shift
25                     if  (L==P) {  // entered the early results   portion
26                         sat  += batch[i][L+1]; unsat  += B−batch[i][L+1];
27                     }
28                 }
29                 batch[i][ level [i]]=0;  // cleanup
30             }
31             logRatio  = p.valAcc*satisfied   + unsatisfied *p.valRef;
32             if  (logRatio  <= p.logInf) H0 = true;
33             if  (logRatio  >= p.logSup) H1 = true;
34         }
35     }
36     ratioLow  = p.valAcc*(satisfied      +sat+unknown) +
37                 p.valRef*(unsatisfied   +unsat);
38     ratioUp  = p.valAcc*(satisfied      +sat) +
39                 p.valRef*(unsatisfied    +unsat+unknown);
40     if  (ratioUp   <= p.logInf) H0 = true;
41     if  (ratioLow  >= p.logSup) H1 = true;
42 }
```

Slave processes proceed to compute their batches if their communication buffers are not full (level[id] < K) or wait for the condition to hold. The compute location models the computation time of a run, chosen according to the distribution shown in Fig. 8b. This is encoded using probabilistic edges with weights matching the distribution. The distribution comes from a real verification of the property in Section 3:

```
Pr[<=100](<> Train(5).Cross and
          (forall (i : id_t) i != 5 imply Train(i).Stop)) >= 0.46188
```

The last weighted edge (case i=H) is reserved for the runs that did not satisfy the property.

Verification. In the hypothesis we test, the actual probability is very close to 0.46188. Since the real probability falls in the indifference region of our test, we would expect that a non-biased implementation would accept H_0 or H_1 equally often. Estimating the probability of confirming the hypothesis H_0 with the query Pr[<=10000000](<> master.H0) gives the probability 0.503 ± 0.005 with 99.9% confidence, confirming that our algorithm is not biased as well as the validity of our model.

Similarly, we obtain the distribution of the verification time by the query Pr[<=10000000](<> master.Done) for a model with number of nodes $N = 128$, batch size $B = 64$, and buffer size $K = 4$. The result is 9557.6 time units in average and the distribution histogram is depicted in Fig. 9a. To estimate the processor usage time, we add another process with a single location with the invariant usage'==sum(i:node_t)busy[i]. Here, usage is a clock that grows with a rate equal to the number of busy nodes.

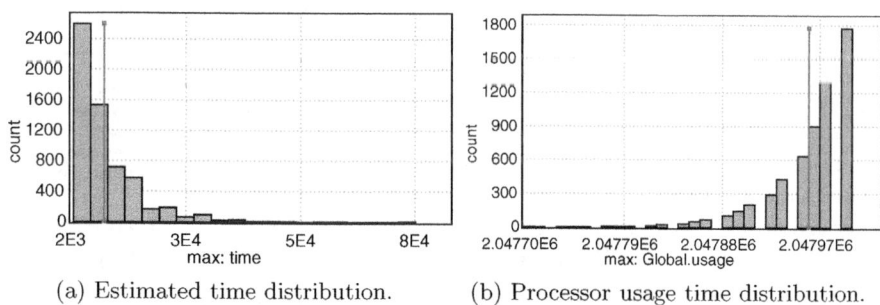

(a) Estimated time distribution.   (b) Processor usage time distribution.

Fig. 9. Time estimation from 6000 runs of DSMC model

The question is now to find a good settings for the parameters of our algorithm (B and K). We perform a parameter sweep to estimate the verification time for values of B and K taking values in $1, 2, 4, 8, 16, 32, 64$ for three topologies with the number of processing nodes $N = 16, 32$, or 128. The results are depicted in Fig. 10, where it is visible that the extremely small batch size requires more time. Large batch sizes can also be detrimental in a large cluster setting (Fig. 10c where too many runs are requested in bulk than actually needed to establish the result). Buffer size of one has a huge penalty of blocking with small blocks, but it is barely noticeable otherwise. This confirms the experimental findings of Section 3.

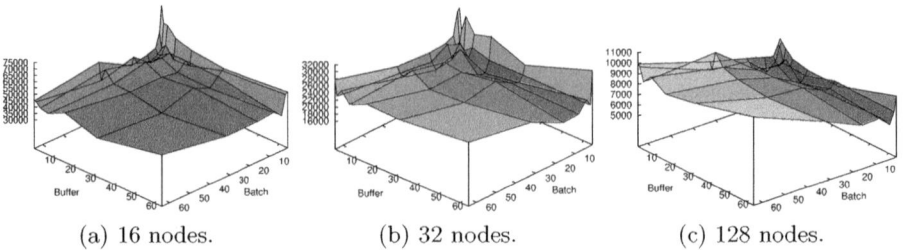

(a) 16 nodes.   (b) 32 nodes.   (c) 128 nodes.

Fig. 10. Estimated verification times in model time units

5   Lightweight Media Access Control

LMAC is a Lightweight Media Access protocol (studied in [4,5]) used for scheduling communication in wireless sensor networks where the topology is determined by physical location and radio connectivity of the individual nodes. One of the goals of the LMAC protocol is to minimize the number of collisions in the network and to reconfigure the network to avoid further collisions.

The original model has been developed in [5] where topologies of 4-5 nodes are studied exhaustively using classical UPPAAL and a number of topologies are identified as problematic, containing perpetual collisions. In this paper we provide new insight as to the likelihood of perpetual collisions in different topologies. This insight could not be delivered by the use of classical UPPAAL and the experiment conducted is of unprecedented size.

In LMAC communication media access time is discretized into time frames and each time frame is divided into time slots. The goal of the protocol is to allocate the time slots to each node efficiently. The challenge is that there is no central node distributing and assigning slots and nodes cannot themselves listen while transmitting, hence neighbours are responsible for detecting and informing each other about collisions.

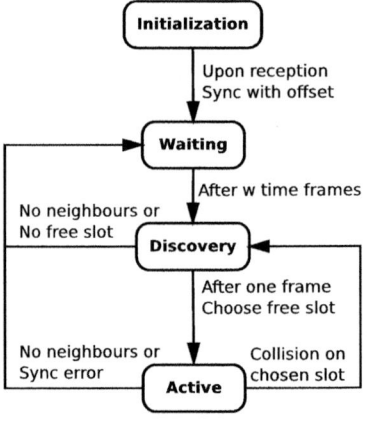

Fig. 11. LMAC protocol phases

After waiting phase, the node moves to a discovery phase and listens for an entire time frame and notes which time slots are used by its neighbours. The collision counting expression `collisions=++cc;` is added on the edge from `rec_one0` to `done0` in Fig. 12b. After one time frame of discovery phase, the node chooses seemingly unused time slot and moves to an active phase. The node falls back to waiting phase if there are no neighbours (no signal received) or all slots are occupied. During active and discovery phases the node listens and notes any collisions (several receptions during the same slot). During active phase the node transmits information about collisions it has detected during its time slot and listens for collisions and information about collisions during other time slots. From the active phase the node may fall back to discovery phase if it is notified about the collisions on its time slot and falls back to the waiting phase if it detects that neighbours are gone.

Figure 11 shows the four phases of the protocol. Initially all nodes except the gateway are listening and waiting for a radio signal from its neighbourhood during the initialization phase. The communication is triggered by a dedicated gateway node. Upon reception of signal, the node notes the relative time offset of the signal and moves to waiting phase, during which it chooses to wait for a random amount of time frames. The random delay is modeled using probabilistic branching (see Fig. 12a) with geometrical weights (`weight` array).

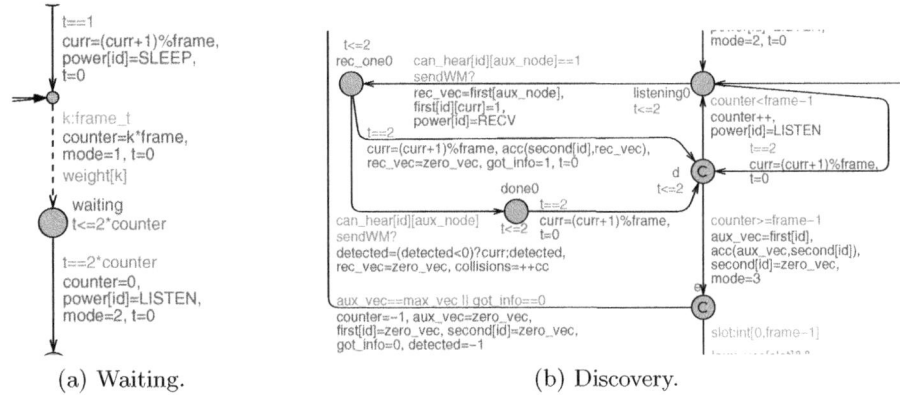

(a) Waiting.   (b) Discovery.

Fig. 12. LMAC phases in the model

Starting from the model[5] of [5], we removed the verification optimizations constraining the parallelism, annotated it with power consumption and collision counting (as cost variables). The model contains twice as many slots as nodes, whereas one slot per node is enough to schedule flawless communication in any topology if nodes were aware of each others choices.

First we examine the distribution of the first collisions over time. The first row of Fig. 13 is a result of a query `Pr[<=1000](◇ collision>0)` and it shows that most collisions happen early in time and in a ring topology some collisions may be discovered later (possibly when the first signal propagation meets at the opposite of the ring). In the second row of Fig. 13 the distribution of possible number of collisions is examined using a query `Pr[collisions<=100](◇ time>=1000)`: in a chain and a ring topologies the collisions are unlikely to occur (> 90% probability of 0 collisions), but in a star it is almost guaranteed to occur (only 8% probability of 0 collisions). The third row of Fig. 13 shows the probability distribution of collision counts after twice as long period of time (using query `Pr[collisions<=100](◇ time>=2000)`). Notice that the shape of distributions has not changed, but the small bumps have shifted to the right at exactly twice the number of collisions and almost identical probability density, which implies that those particular collisions are accumulating proportionally to the progress

---
[5] Thanks to Ansgar Fehnker and Angelika Mader.

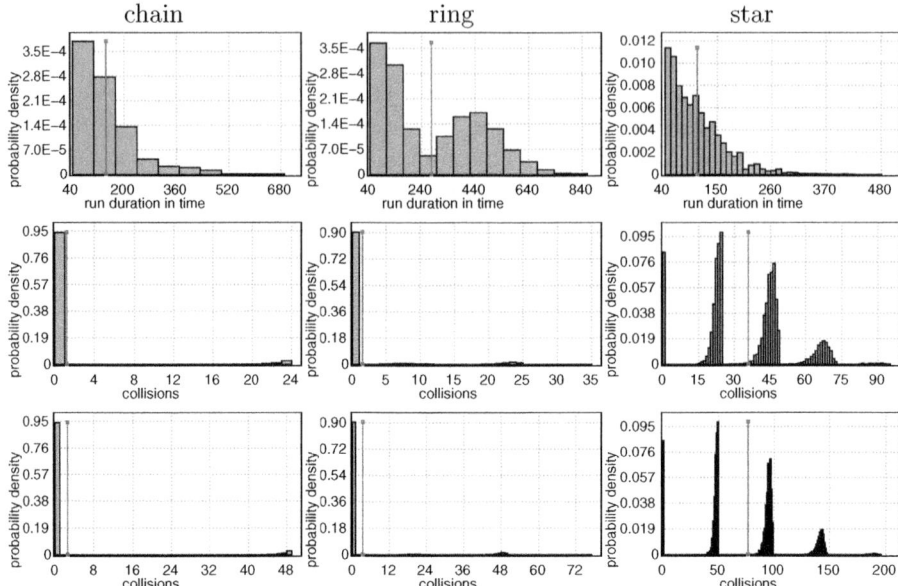

Fig. 13. Collision statistics in three different topologies, in rows: probability of a collision over time, probabilities of a number of collisions up to 1000 and up to 2000 time units

of time, and in other words it means that collisions are reoccurring perpetually without recovery. We checked these three properties on a 128 cores cluster with high precision (with $\alpha = \beta = 0.0001$ and $\varepsilon = 0.0005$) in about 30 minutes, which generated around 19 million runs.

We have demonstrated how UPPAAL SMC can be used to identify problematic topologies and distributed implementation can provide a high degree of accuracy in spotting the reoccurring collisions.

6   Conclusion and Future Work

In this paper we have developed, implemented, applied and evaluated a general and scalable framework for distributed statistical model checking. We have thoroughly investigated the distribution of sequential algorithms where bias can be introduced when collecting the samples produced by slave computers. In particular, we have identified best choices of batch and buffer sizes both experimentally and analytically, with agreement in the findings of the two approaches. In the future, we plan to implement and distribute other SMC algorithms, principaly the Bayesian algorithms introduced in [20,9].

Finally, it is worth mentioning that we have tried to use other distributed SMC model checkers such as Ymer [18] or PVesta [14]. Aside from the fact that the Gui of those two tools is quite restricted, we observed that Ymer does not work anymore and that PVesta only distributes those algorithms where the number of simulations is precomputed in advance.

References

1. Bogdoll, J., Ferrer Fioriti, L.M., Hartmanns, A., Hermanns, H.: Partial Order Methods for Statistical Model Checking and Simulation. In: Bruni, R., Dingel, J. (eds.) FMOODS/ FORTE 2011. LNCS, vol. 6722, pp. 59–74. Springer, Heidelberg (2011)
2. Clarke, E.M., Faeder, J.R., Langmead, C.J., Harris, L.A., Jha, S.K., Legay, A.: Statistical Model Checking in *BioLab*: Applications to the Automated Analysis of T-Cell Receptor Signaling Pathway. In: Heiner, M., Uhrmacher, A.M. (eds.) CMSB 2008. LNCS (LNBI), vol. 5307, pp. 231–250. Springer, Heidelberg (2008)
3. David, A., Larsen, K.G., Legay, A., Mikučionis, M., Wang, Z.: Time for Statistical Model Checking of Real-Time Systems. In: Gopalakrishnan, G., Qadeer, S. (eds.) CAV 2011. LNCS, vol. 6806, pp. 349–355. Springer, Heidelberg (2011)
4. David, A., Larsen, K.G., Legay, A., Mikučionis, M., Poulsen, D.B., van Vliet, J., Wang, Z.: Statistical Model Checking for Networks of Priced Timed Automata. In: Fahrenberg, U., Tripakis, S. (eds.) FORMATS 2011. LNCS, vol. 6919, pp. 80–96. Springer, Heidelberg (2011)
5. Fehnker, A., van Hoesel, L., Mader, A.: Modelling and Verification of the LMAC Protocol for Wireless Sensor Networks. In: Davies, J., Gibbons, J. (eds.) IFM 2007. LNCS, vol. 4591, pp. 253–272. Springer, Heidelberg (2007)
6. Grosu, R., Smolka, S.A.: Monte Carlo Model Checking. In: Halbwachs, N., Zuck, L.D. (eds.) TACAS 2005. LNCS, vol. 3440, pp. 271–286. Springer, Heidelberg (2005)
7. Hammersley, J.M., Handscomb, D.C.: Monte Carlo Methods. Methuen (1975)
8. Hérault, T., Lassaigne, R., Magniette, F., Peyronnet, S.: Approximate Probabilistic Model Checking. In: Steffen, B., Levi, G. (eds.) VMCAI 2004. LNCS, vol. 2937, pp. 73–84. Springer, Heidelberg (2004)
9. Jha, S.K., Clarke, E.M., Langmead, C.J., Legay, A., Platzer, A., Zuliani, P.: A Bayesian Approach to Model Checking Biological Systems. In: Degano, P., Gorrieri, R. (eds.) CMSB 2009. LNCS, vol. 5688, pp. 218–234. Springer, Heidelberg (2009)
10. Legay, A., Delahaye, B.: Statistical model checking: An overview. CoRR, abs/1005.1327 (2010)
11. El Rabih, D., Pekergin, N.: Statistical Model Checking Using Perfect Simulation. In: Liu, Z., Ravn, A.P. (eds.) ATVA 2009. LNCS, vol. 5799, pp. 120–134. Springer, Heidelberg (2009)
12. Sen, K., Viswanathan, M., Agha, G.: Statistical Model Checking of Black-Box Probabilistic Systems. In: Alur, R., Peled, D.A. (eds.) CAV 2004. LNCS, vol. 3114, pp. 202–215. Springer, Heidelberg (2004)
13. Sen, K., Viswanathan, M., Agha, G.: On Statistical Model Checking of Stochastic Systems. In: Etessami, K., Rajamani, S.K. (eds.) CAV 2005. LNCS, vol. 3576, pp. 266–280. Springer, Heidelberg (2005)
14. Sen, K., Viswanathan, M., Agha, G.A.: Vesta: A statistical model-checker and analyzer for probabilistic systems. In: QEST, pp. 251–252. IEEE Computer Society (2005)
15. Wald, A.: Sequential tests of statistical hypotheses. Annals of Mathematical Statistics 16(2), 117–186 (1945)
16. Wald, R.: Sequential Analysis. Dove Publisher (2004)
17. Younes, H.L.S.: Verification and Planning for Stochastic Processes with Asynchronous Events. PhD thesis, Carnegie Mellon (2005)
18. Younes, H.L.S.: Ymer: A Statistical Model Checker. In: Etessami, K., Rajamani, S.K. (eds.) CAV 2005. LNCS, vol. 3576, pp. 429–433. Springer, Heidelberg (2005)
19. Younes, H.L.S., Kwiatkowska, M.Z., Norman, G., Parker, D.: Numerical vs. statistical probabilistic model checking. STTT 8(3), 216–228 (2006)
20. Zuliani, P., Platzer, A., Clarke, E.M.: Bayesian statistical model checking with application to simulink/stateflow verification. In: HSCC, pp. 243–252. ACM (2010)

Author Index

GPSR Compliance

The European Union's (EU) General Product Safety Regulation (GPSR) is a set of rules that requires consumer products to be safe and our obligations to ensure this.

If you have any concerns about our products, you can contact us on ProductSafety@springernature.com

In case Publisher is established outside the EU, the EU authorized representative is:

Springer Nature Customer Service Center GmbH
Europaplatz 3
69115 Heidelberg, Germany

Batch number: 09478804

Printed by Printforce, the Netherlands